HyperTransport™
System Architecture

PC System Architecture Series

MindShare, Inc.

Please see our web site (http://www.awprofessional.com/series/mindshare) for more information on these titles.

80486 System Architecture: Third Edition
0-201-40994-1

AGP System Architecture: Second Edition
0-201-70069-7

CardBus System Architecture
0-201-40997-6

FireWire® System Architecture: Second Edition
0-201-48535-4

InfiniBand System Architecture
0-321-11765-4

ISA System Architecture: Third Edition
0-201-40996-8

PCI System Architecture: Fourth Edition
0-201-30974-2

PCI-X System Architecture
0-201-72682-3

PCMCIA System Architecture: Second Edition
0-201-40991-7

Pentium® Pro and Pentium® II System Architecture: Second Edition
0-201-30973-4

Pentium® Processor System Architecture: Second Edition
0-201-40992-5

Plug and Play System Architecture
0-201-41013-3

Protected Mode Software Architecture
0-201-55447-X

Universal Serial Bus System Architecture: Second Edition
0-201-30975-0

HyperTransport™ System Architecture

MINDSHARE, INC.
Don Anderson
Jay Trodden

⋏⋎ Addison-Wesley

Boston • San Francisco • New York • Toronto • Montreal
London • Munich • Paris • Madrid
Capetown • Sydney • Tokyo • Singapore • Mexico City

Library of Congress Cataloging-in-Publication Data

Trodden, Jay.
  HyperTransport system architecture / Mindshare, Inc., Jay Trodden & Don Anderson.
      p. cm.
    Includes index.
    ISBN 0-321-16845-3 (alk. paper)
    1. Data transmission systems. 2. Integrated circuits. 3. Computer architecture. I.
  Anderson, Don, 1953– II. Mindshare, Inc. III. Title.

TK5105.T745 A53 2003
004.6—dc21   2002043927

Set in 10 point Palatino by MindShare, Inc.

ISBN: 0-321-16845-3
1 2 3 4 5 6 7 8 9 10—MA—0706050403
First printing, February 2003

For our wives, Victoria and Susan

Contents

The MindShare Architecture Series 1
Cautionary Note 2
Intended Audience 2
Prerequisite Knowledge 3
Topics and Organization 3
Documentation Conventions 4
    HyperTransport™ 4
    Hexadecimal Notation 4
    Binary Notation 4
    Decimal Notation 4
    Byte Terminology and Notation 5
    Bits Versus Bytes Notation 5
    Bit Fields and Groups of Signals... 5
    Active Signal States... 5
Visit Our Web Site 6
We Want Your Feedback.. 6

Part One: Overview of HyperTransport

Chapter 1: Introduction to HyperTransport

Background: I/O Subsystem Bottlenecks 9
    Server Or Desktop Computer: Three Subsystems 10
    CPU Speed Makes Other Subsystems Appear Slow 10
        Multiple CPUs Aggravate The Problem.. 10
    DRAM Memory Keeps Up Fairly Well... 10
    I/O Bandwidth Has Not Kept Pace 11
        This Slows Down The Processor 11
        It Also Hurts Fast Peripherals... 11
    Reducing I/O Bottlenecks 11
        The Shared Bus Approach 11
            A Shared Bus Runs At Limited Clock Speeds.. 13
            A Shared Bus May Be Host To Many Device Types 13
            Backward Compatibility Prevents Upgrading Performance 13
            Special Problems If The Shared Bus Is PCI 13
            A Note About PCI-X 14
        The Point-to-Point Interconnect Approach.. 14
            A Note About Connectors 14
What HT Brings... 15
    Key Features Of HyperTransport Protocol 16
    The Cost Factor... 16
    Networking Support.. 17

Contents

Chapter 2: HT Architectural Overview

Transfer Types Supported 21
    Address-Based Semantics 22
    Data Transfer Type and Transaction Flow 23
        Programmed I/O Transfers 24
        DMA Transfers 25
        Peer-to-Peer Transfers 26
HT Signals 27
    Link Packet Transfer Signals 28
    Link Support Signals 29
Scalable Performance 30
    Data Widths 31
    Clock Speeds 32
Extending the Topology 33
Packetized Transfers 35
    Control Packets 36
        Information packet (4 bytes) 36
        Request packet (4 or 8 bytes) 36
        Response packet (4 bytes) 36
    Data Packets 37
HyperTransport Protocol Concepts 37
    Channels and Streams 37
        Virtual Channels 37
        I/O Streams 39
    Transactions (Requests, Responses, and Data) 39
        Transaction Requests 40
        Transaction Responses 40
    Transaction Types 41
        Sized Read Transactions 41
        Sized Write Transactions 42
            Non-Posted Sized Writes 42
            Posted Sized Writes 43
        Flush 44
        Fence 45
        Atomic 46
        Broadcast 46
Managing the Links 47
    Flow Control 47
    Initialization and Reset 49
    Configuration 50
    Error Detection and Handling 50

Contents

Part Two: HyperTransport Core Topics

Chapter 3: Signal Groups

Introduction..53
The Signal Groups 54
The High Speed Signals (One Set In Each Direction) 55
   The CAD Signal Group 55
   Control Signal (CTL)..55
   Clock Signal(s) (CLK) 55
Scaling Hazards: Burden Is On The Transmitter 56
The Low Speed Signals...56
   Power OK (PWROK) And Reset (RESET#) 56
   LDTSTOP# 57
   LDTREQ# 57
Where Are The Interrupt, Error, And Wait State Signals?................................57
   Interrupt Signaling...58
   Error Signaling 58
   Wait State Signaling..58
No Arbitration Signals Either..58

Chapter 4: Packet Protocol

The Packet-Based Protocol 60
   8 Bit Interfaces 60
   Interfaces Narrower Than 8 Bits 62
   Interfaces Wider Than 8 Bits 63
      16 Bit Interfaces 63
      32 Bit Interfaces 64
The Two Packet Types: Control And Data...65
   Control Packet Purpose..65
      Information packets..65
      Request packets...65
      Response packets 66
   Data Packets...66
The Need To Interleave Control And Data Packets 67
   The CTL Signal Indicates Packet Type 67
Packet Format: Control Packets 69
   Control Packets: Information 71
      NOP Packet...71
      Sync/Error Packet 74
   Control Packets: Requests...75

Contents

Sized Read And Sized Write Requests 75
  Generic RdSized And WrSized Request Packet Format 76
RdSized And WrSized Requests: Transaction Limits 78
  RdSized And WrSized (Dword) Transactions 79
  RdSized (Byte) Transactions 79
  WrSized (Byte) Transactions 79
RdSized And WrSized Requests: Other Notes 80
  Coherency 80
  WrSized Requests And The Posted Bit 80
  Errors During RdSized Transactions 81
  PassPW and Response May Pass Posted Requests bits 81
  Compatibility Bit 82
Broadcast Message Requests 82
Flush Requests 84
Flush Requests: Transaction Limits 86
Fence Requests 87
Fence Requests: Transaction Limits 88
Atomic Read-Modify-Write Requests 89
Two Problems In Shared Memory Schemes 89
  Atomic RMW Variants 90
  Compare And Swap 90
  Fetch And Add 90
Atomic RMW Requests: Transaction Limits 93
Control Packets: Responses 94
  Read Responses 94
  Target Done Responses 97

Chapter 5: Flow Control

The Problem 99
How PCI Handles Flow Control 101
  PCI Target Flow Control Problems 101
    PCI Target Not Ready To Start 101
    PCI Target Starts Data Transfer, But Can't Continue 101
    PCI Target Starts, Can Continue, But Needs More Time 102
  PCI Initiator Flow Control Problems 102
    PCI Initiator Starts, But Can't Continue 102
    PCI Initiator Starts, Can Continue, But Needs Wait-States 102
All PCI Flow Control Problems Hurt Performance 103

Contents

HyperTransport Flow Control: Overview 103
   Packets Never Start Unless Completion Assured 103
   Transfer Length Is Always Known 103
   Split Transactions Used When Response Is Required 104
   Flow Control Pins Are Eliminated 104
   Flow Control Buffers Mean No Bus Wait States 104
   Flow Control Buffers For Each Virtual Channel 105
Flow Control, A System View 105
Flow Control Buffer Arrangement 106
   Details Associated With Figure 5-3 107
      Flow Control Buffer Pairs (Item 1) 108
         Posted Request Buffer (Command). 108
         Posted Request Buffer (Data). 108
         Non-Posted Request Buffer (Command). 108
         Non-Posted Request Buffer (Data). 108
         Response Buffer (Command). 108
         Response Buffer (Data). 108
      Receiver Flow Control Counters (Item 2) 108
      Transmitter Flow Control Counters (Item 3) 109
      NOP Packet Update Information (Item 4) 109
      Control Logic (Item 5) 109
      Transmit And Receive FIFO (Item 6) 110
Example: Initialization And Use Of The Counters 110
   Basic Steps In Counter Initialization And Use 110
      Initializing The Flow Control Counter 111
      Device 1 Sends Two Posted Request Packets 112
      New Entries Available: Update Flow Control Information 114
A Few Implementation Notes 115
   Information Packets Not Flow-Controlled 115
   Transmitter Must Be Able To Track 15 Buffer Entries 115
   Sometimes Two Counters Must Be Checked 116
   NOP Packets Cannot Be Completely Blocked 116
   The Isochronous Flow Control Option 116
      How About NOP Updates For Isochronous Buffers? 116
      Isochronous Traffic/Non-Isochronous Flow Control 117
      Isochronous Traffic Disabled At Initialization 117

Contents

Chapter 6: I/O Ordering

The Purpose Of Ordering Rules 120
  Maintain Data Coherency 120
  Avoid Deadlocks 120
  Support Legacy buses 120
  Maximize Performance 120
Introduction: Three Types Of Traffic Flow 121
The Ordering Rules 122
  General I/O Ordering Limits 122
    Ordering Covers Targets At Same Hierarchy Level 122
    Read And Non-Posted Write Completion At Target 123
      What If A Device Requires Response Ordering? 124
    Support For The Producer-Consumer Ordering Model 124
    Producer-Consumer Model Simpler If Flag/Data In Same Place 124
  Upstream Ordering Rules 125
    Reordering Packets In Different Transaction Streams 125
    No Reordering Packets In A Strongly Ordered Sequence 126
    Packets With PassPW Bit Clear Are Restricted In Passing 127
    Packets With PassPW Bit Set May Or May Not Pass 128
    Non-Posted Requests May Pass Each Other 129
    Posted Requests And Responses Must Be Able To Pass 130
    Posted Request Must Be Able To Pass A Response 131
    Non-Posted Requests Or Response May Pass A Response 132
  Host Ordering Requirements 132
    Host Ordering Requirements: General Features 133
    Two Ordering Points Are Defined 134
      Globally Ordered (GO) 134
      Globally Visible (GV) 134
    Ordering Rule Summary 134
    Host Responses To Non-Posted Requests 135
    An Example (Refer to Table 6-1 and Figure 6-13) 136
    Downstream I/O Ordering 136
    Double-Hosted Chain Ordering 137

Chapter 7: Transaction Examples

Packets As Transaction Building Blocks 140
Transaction Examples: Introduction 144
  Packet Format And Optional Features 144
  General Sequence Of Events 144

Example 1: NOP Information Packet.. **145**
  Example 1: NOP Packet Setup 146
    Command[5:0] Field (Byte 0, Bit 5:0) 146
    DisCon Bit Field (Byte 0, Bit 6).. 146
    PostCMD[1:0] Field (Byte 1, Bits 1:0) 146
    PostData[1:0] Field (Byte 1, Bits 3:2)....................................... 146
    Response[1:0] Field (Byte 1, Bits 5:4)....................................... 146
    ResponseData[1:0] Field (Byte 1, Bits 3:2) 147
    Non-PostCMD[1:0] Field (Byte 2, Bits 1:0) 147
    Non-PostData[1:0] Field (Byte 2, Bits 3:2) 147
    Isoc Bit Field (Byte 2, Bit 5) 147
    Diag Bit Field (Byte 2, Bit 6)... 147
    Bits Not Mentioned 147
  Example 1: NOP Sequence Of Events On The Link........................ 148
Generic Request And Response Packet Formats **148**
Example 2: Non-Posted WrSized (Dword) Transaction **150**
  Example 2: WrSized (Dword) Request Packet Setup...................... 151
    Command[5:0] Field (Byte 0, Bit 5:0) 151
    SeqID[3:0] Field (Byte 0, Bit 7:6) and (Byte 1, Bit 6:5) 151
    UnitID[4:0] Field (Byte 1, Bits 4:0) 151
    PassPW Bit Field (Byte 1, Bit 7)... 151
    SrcTag[4:0] Field (Byte 2, Bits 4:0) 151
    Compat Bit Field (Byte 2, Bit 5) 152
    Mask/Count[3:0] Field (Byte 2, Bits 7:6) & (Byte 3, Bits 1:0) 152
    Start Address Field (Bytes 4-7, Bit 7:0) & (Byte 3, Bit 7:2) 152
  Example 2: WrSized (Dword) Transaction Data 152
  Example 2: WrSized (Dword) Request Sequence Of Events 153
  Example 2: The WrSized (Dword) Response Packet 154
    Command[5:0] Field (Byte 0, Bit 5:0) 154
    Isoc Bit Field (Byte 0, Bit 7)... 154
    UnitID[4:0] Field (Byte 1, Bits 4:0) 154
    Bridge Bit Field (Byte 1, Bit 6) 155
    PassPW Bit Field (Byte 1, Bit 7)... 155
    SrcTag[4:0] Field (Byte 2, Bits 4:0) 155
    Error Bit Field (Byte 2, Bit 5)... 155
    NXA Bit Field (Byte 3, Bit 5) 155
  Example 2: WrSized (Dword) Response, Sequence Of Events............ 156
Example 3: Posted Byte Write Request **156**
  Example 3: WrSized (Byte) Request Packet Setup 156
    Command[5:0] Field (Byte 0, Bit 5:0) 156
    SeqID[3:0] Field (Byte 0, Bit 7:6) and (Byte 1, Bit 6:5) 158
    UnitID[4:0] Field (Byte 1, Bits 4:0) 158

Contents

PassPW Bit Field (Byte 1, Bit 7)..158
SrcTag[4:0] Field (Byte 2, Bits 4:0) 158
Compat Bit Field (Byte 2, Bit 5)..158
Mask/Count[3:0] Field (Byte 2, Bits 7:6 and Byte 3, Bits 1:0) 158
Start Address Field (Bytes 4-7, Bit 7:0) and Byte 3, Bit 7:2).............................159
Example 3: Sized (Byte) Write Data Packet And Mask 159
Example 3: WrSized (Byte) Request, Sequence Of Events 160
A Couple Of Notes About WrSized (Byte)..160
Example 4: Dword Read Request..**161**
Example 4: RdSized (Dword) Request Packet Setup 162
Command[5:0] Field (Byte 0, Bit 5:0) 162
SeqID[3:0] Field (Byte 0, Bit 7:6) and (Byte 1, Bit 6:5) 162
UnitID[4:0] Field (Byte 1, Bits 4:0) 162
PassPW Bit Field (Byte 1, Bit 7)..162
SrcTag[4:0] Field (Byte 2, Bits 4:0) 162
Compat Bit Field (Byte 2, Bit 5)..162
Mask/Count[3:0] Field (Byte 2, Bits 7:6) and (Byte 3, Bits 1:0) 163
Start Address Field (Bytes 4-7, Bit 7:0) and (Byte 3, Bit 7:2) 163
Example 4: RdSized (Dword) Request, Sequence Of Events...................................163
Example 4: Dword Read Response Packet Setup 164
Command[5:0] Field (Byte 0, Bit 5:0) 164
Isoc Bit Field (Byte 0, Bit 7)..164
UnitID[4:0] Field (Byte 1, Bits 4:0) 164
Bridge Bit Field (Byte 1, Bit 6) 164
PassPW Bit Field (Byte 1, Bit 7)..164
SrcTag[4:0] Field (Byte 2, Bits 4:0) 165
Error Bit Field (Byte 2, Bit 5)..165
Count[3:0] Field (Byte 2, Bits 7:6) and (Byte 3, Bits 1:0)...................................165
NXA Bit Field (Byte 3, Bit 5)...165
Example 4: Sized (Dword) Read Data Packet...165
Example 4: RdSized (Dword) Response, Sequence Of Events 166
Example 5: Byte Read Request **167**
Example 5: RdSized (Byte) Request Packet Setup..168
Command[5:0] Field (Byte 0, Bit 5:0) 168
SeqID[3:0] Field (Byte 0, Bit 7:6) and (Byte 1, Bit 6:5) 168
UnitID[4:0] Field (Byte 1, Bits 4:0) 168
PassPW Bit Field (Byte 1, Bit 7)..168
SrcTag[4:0] Field (Byte 2, Bits 4:0) 168
Compat Bit Field (Byte 2, Bit 5)..168
Mask/Count[3:0] Field (Byte 2, Bits 7:6) and (Byte 3, Bits 1:0) 169
Start Address Field (Bytes 4-7, Bit 7:0) and Byte 3, Bit 7:2).............................169
Example 5: RdSized (Byte) Request, Sequence Of Events 169

Contents

Example 5: Byte Read Response Packet Setup 170
   Command[5:0] Field (Byte 0, Bit 5:0) 170
   Isoc Bit Field (Byte 0, Bit 7) 170
   UnitID[4:0] Field (Byte 1, Bits 4:0) 170
   Bridge Bit Field (Byte 1, Bit 6) 170
   PassPW Bit Field (Byte 1, Bit 7) 171
   SrcTag[4:0] Field (Byte 2, Bits 4:0) 171
   Error Bit Field (Byte 2, Bit 5) 171
   Count[3:0] Field (Byte 2, Bits 7:6) and (Byte 3, Bits 1:0) 171
   NXA Bit Field (Byte 3, Bit 5) 171
Example 5: Sized (Byte) Read Data Packet 172
Example 5: RdSized (Byte) Response, Sequence Of Events 172
   SrcTag[4:0] Field (Byte 2, Bits 4:0) 173
Example 6: Flush Request **173**
Example 6: Flush Request Packet Setup 175
   Command[5:0] Field (Byte 0, Bit 5:0) 175
   SeqID[3:0] Field (Byte 0, Bit 7:6) and (Byte 1, Bit 6:5) 175
   UnitID[4:0] Field (Byte 1, Bits 4:0) 175
   PassPW Bit Field (Byte 1, Bit 7) 175
Example 6: Flush Request, Sequence Of Events 176
Example 6: Flush Response Packet Setup 176
   Command[5:0] Field (Byte 0, Bit 5:0) 176
   UnitID[4:0] Field (Byte 1, Bits 4:0) 176
   Bridge Bit Field (Byte 1, Bit 6) 177
   PassPW Bit Field (Byte 1, Bit 7) 177
   SrcTag[4:0] Field (Byte 2, Bits 4:0) 177
   Error Bit Field (Byte 2, Bit 5) 177
   NXA Bit Field (Byte 3, Bit 5) 177
Example 6: Flush Response, Sequence Of Events 178
A Few Notes About Flush Operations 178
Example 7: Fence Request **179**
Example 7: Fence Request Packet Setup 180
   Command[5:0] Field (Byte 0, Bit 5:0) 180
   SeqID[3:0] Field (Byte 0, Bit 7:6) and (Byte 1, Bit 6:5) 180
   UnitID[4:0] Field (Byte 1, Bits 4:0) 180
   PassPW Bit Field (Byte 1, Bit 7) 180
Example 7: Fence Request, Sequence Of Events 180
A Few Notes About Fence Operations 181

Contents

Example 8: Atomic Read-Modify-Write..**182**
  Example 8: Atomic RMW Request Packet Setup.......................................183
    Command[5:0] Field (Byte 0, Bit 5:0) 183
    SeqID[3:0] Field (Byte 0, Bit 7:6) and (Byte 1, Bit 6:5) 183
    UnitID[4:0] Field (Byte 1, Bits 4:0) 183
    PassPW Bit Field (Byte 1, Bit 7)...183
    SrcTag[4:0] Field (Byte 2, Bits 4:0) 183
    Compat Bit Field (Byte 2, Bit 5)...183
    Mask/Count[3:0] Field (Byte 2, Bits 7:6) and (Byte 3, Bits 1:0) 184
    Start Address Field (Bytes 4-7, Bit 7:0) and (Byte 3, Bit 7:2) 184
  Example 8: Atomic RMW Request Data Packet 184
  Example 8: Atomic RMW Request, Sequence Of Events 184
  Example 8: Atomic RMW Response Packet Setup 185
    Command[5:0] Field (Byte 0, Bit 5:0) 185
    Isoc Bit Field (Byte 0, Bit 7) 185
    UnitID[4:0] Field (Byte 1, Bits 4:0) 185
    Bridge Bit Field (Byte 1, Bit 6) 186
    PassPW Bit Field (Byte 1, Bit 7)...186
    SrcTag[4:0] Field (Byte 2, Bits 4:0) 186
    Error Bit Field (Byte 2, Bit 5)..186
    Count[3:0] Field (Byte 2, Bits 7:6) and (Byte 3, Bits 1:0)................186
    NXA Bit Field (Byte 3, Bit 5) 186
  Example 8: Atomic RMW Response Data Packet.....................................187
  Example 8: Atomic RMW Response, Sequence Of Events.......................187
  Some Notes About Atomic RMW Operations...188
Example 9: WrSized Request Crosses A Bridge **189**
  Example 9: Request Packet On Bus 1 190
    Command[5:0] Field (Byte 0, Bit 5:0) 190
    SeqID[3:0] Field (Byte 0, Bit 7:6) and (Byte 1, Bit 6:5) 190
    UnitID[4:0] Field (Byte 1, Bits 4:0) 190
    PassPW Bit Field (Byte 1, Bit 7)...190
    SrcTag[4:0] Field (Byte 2, Bits 4:0) 190
    Compat Bit Field (Byte 2, Bit 5)..190
    Mask/Count[3:0] Field (Byte 2, Bits 7:6) & (Byte 3, Bits 1:0) 191
    Start Address Field (Bytes 4-7, Bit 7:0) & (Byte 3, Bit 7:2) 191
  Example 9: Sized (Dword) Write Data Packet: Bus 1 191
  Example 9: Request/Data Sequence Of Events On Bus 1 191
  Example 9: Bridge Reissues Request Packet: Bus 0.................................192
    Command[5:0] Field (Byte 0, Bit 5:0) 192
    SeqID[3:0] Field (Byte 0, Bit 7:6) and (Byte 1, Bit 6:5) 192
    UnitID[4:0] Field (Byte 1, Bits 4:0) 192
    PassPW Bit Field (Byte 1, Bit 7)...192

Contents

SrcTag[4:0] Field (Byte 2, Bits 4:0) 192
Compat Bit Field (Byte 2, Bit 5) 193
Mask/Count[3:0] Field (Byte 2, Bits 7:6) & (Byte 3, Bits 1:0) 193
Start Address Field (Bytes 4-7, Bit 7:0) & (Byte 3, Bit 7:2) 193
Example 9: Sized (Dword) Write Data Packet: Bus 0 193
Example 9: Request/Data Sequence Of Events: Bus 0 193
Example 9: Response Packet On Bus 0 194
Command[5:0] Field (Byte 0, Bit 5:0) 194
Isoc Bit Field (Byte 0, Bit 7) 194
UnitID[4:0] Field (Byte 1, Bits 4:0) 194
Bridge Bit Field (Byte 1, Bit 6) 194
PassPW Bit Field (Byte 1, Bit 7) 194
SrcTag[4:0] Field (Byte 2, Bits 4:0) 194
Error Bit Field (Byte 2, Bit 5) 195
NXA Bit Field (Byte 3, Bit 5) 195
Example 9: Response, Sequence Of Events On Bus 0 195
Example 9: Response Packet On Bus 1 195
Command[5:0] Field (Byte 0, Bit 5:0) 195
Isoc Bit Field (Byte 0, Bit 7) 196
UnitID[4:0] Field (Byte 1, Bits 4:0) 196
Bridge Bit Field (Byte 1, Bit 6) 196
PassPW Bit Field (Byte 1, Bit 7) 196
SrcTag[4:0] Field (Byte 2, Bits 4:0) 196
Error Bit Field (Byte 2, Bit 5) 196
NXA Bit Field (Byte 3, Bit 5) 196
Example 9: Response, Sequence Of Events On Bus 1 196

Chapter 8: HT Interrupts

Introduction 199
Discovering a Device's Interrupt Requirements 200
The Interrupt Message Address Range 201
Interrupt Requests 204
Interrupt Request Packet 204
Interrupt Request Data Packet 206
The End of Interrupt (EOI) Message 207
EOI Packet Format 208
Interrupt Discovery and Configuration Capability Block 209
Interrupt Capability Block Format 209
Last Interrupt Supported 211
Interrupt Definition Registers 211

Contents

Chapter 9: System Management

System Management Transactions 216
Sources of SM Request 216
System Management Address Range 217
The SMC & Upstream Request Packets 218
Upstream Request Packet Format 218
System Management Commands — Upstream 219
The Host Bridge & Downstream Request Packets 220
Downstream Request Packet Format 220
System Management Commands 221
HT Link Disconnect/Reconnect Sequence 223
Reference Information: LDTSTOP# Procedures 223
Example SM Sequence: Link Initialization Disconnect 225
Background 225
Setup and Assumptions 225
The Link Initialization Disconnect Sequence 227

Chapter 10: Error Detection And Handling

Introduction 230
Types Of Errors 230
Reporting Methods 230
The Role Of PCI Configuration Space 230
Most Types Of Error Checking Are Optional 231
System Handling Of HyperTransport Errors Varies 231
The Error Types 231
CRC Errors 231
CRC On 8, 16, or 32 bit Interfaces 232
CRC Generation/Checking: 8/16/32 bit links 232
CRC Generation/Checking: 2/4 bit links 233
4 Bit CAD Width. 233
2 Bit CAD Width. 233
Logging CRC Errors 233
Programming The CRC Error Reporting Policy 234
CRC Interrupts 234
CRC Sync Flood 235
CRC Test Mode 237
Protocol Errors 237
CTL Signal Four-Byte Boundary Violation 237
CTL Deassertion Violation 238
CTL/Data Interleaving Violation 238

Contents

Bad Command Code In Control Packet 238
CTL Deassertion Timeout Violation 238
CTL Deasserted During CRC Transmission 238
Logging Protocol Errors..238
Programming The Protocol Error Reporting Policy 239
Receive Buffer Overflow Errors..241
Logging Receive Buffer Overflow Errors...241
Programming The Buffer Overflow Error Reporting Policy 242
End-Of-Chain Errors 243
How A Device Knows It Is At The End Of A Chain 243
Logging End-Of-Chain Errors 245
Programming The EOC Error Reporting Policy 246
Chain Down Errors...247
Response Errors...248
Response Error Logging And Reporting Policy 249
Error Reporting **250**
Error Responses (Non-Posted Requests Only) 250
Error Response Returned By The Target..251
Error Response Returned By An End-Of-Chain Device 251
Fatal And Non-Fatal Interrupts 252
Sync Flood: When All Else Fails 253
Device Initiating The Sync Flood 253
Devices Detecting Sync Flood..253
Sync Flooding And HyperTransport Bridges...254
Miscellaneous Notes..254
Flooding Continues Until Reset 254
CRC Not Checked During Sync Flood 254
Sync Flood Example 254
Sequence of events: (Figure 10-15 on page 255) 254

Chapter 11: Routing Packets

Packet Routing: Shared Bus vs. Point-Point Topology.................................**258**
Shared Bus Routing 258
HyperTransport Point-Point Routing 259
Review Of Packet Types And Formats **259**
Control Packets...259
Information Packets: No Routing Required...259
Request Packet Routing Information 260
Six Request Types 261
Response Packet Routing Information 262
Data Packet Routing Depends On Control Packets...................................263
Directed vs. Broadcast Requests **263**

Contents

Accepting Packets **264**
Rules For Acceptance 264
A Note About The Subtractive Decoder.. 264
Forwarding Packets... **265**
Rules For Forwarding... 265
Other Notes On Forwarding 265
Forwarding Into The End Of Chain 265
Forwarding If Initialization Is Not Complete.. 266
Rejecting Packets.. **266**
Rules For Rejection 266
Host Bridge Behavior **267**
Directed Request With UnitID = 0... 267
Accepted.. 267
Rejected 267
Response UnitID And Bridge Fields 268
Broadcast Request.. 268
Always Accepted 268
Directed Request With Non-Zero UnitID 268
Accepted Requests.. 268
Internal Target... 268
Peer-to-Peer Target... 269
Compatibility Chain Requests 269
Rejected Requests... 270
Responses Received By The Host Bridge 270
Response With Bridge Bit = 1... 270
Response With Bridge Bit = 0... 270
HyperTransport Bridges: Additional Routing Rules................................... **271**
Tunnel Fairness And Forward Progress **271**
Fairness Is Critical In A Point-Point Topology 271
HyperTransPort Imposes A Fairness Algorithm 272
The Basic Policy.. 272
The Algorithm 272
First, Calculate The Insertion Rate 272
Insertion Rate Calculation Example 273

Chapter 12: Reset & Initialization

General... **276**
Cold Reset.. **276**
Sources of Cold Reset 277
Resetting the Primary HT Bus 277
Resetting Secondary Side of HT-to-HT Bridge... 278

Signalling and Detecting Cold Reset..280
Effects of Cold Reset 281
Link Initialization **282**
Low-Level Link Width Initialization 282
Determining Low-Level Link Width 283
Example 1: 4-bit device connected to 8-bit Device 284
Example 2: 8-Bit Device Connected to 4/8-Bit Device..............................285
Example 3: 32-bit Upstream and 16-bit Downstream 286
Negotiated Link Width Stored in Link Config Registers....................................288
Low-Level Clock Initialization...289
The Default Clock Frequency..291
Control and CAD Sequence after Reset is Removed 291
Clock Synchronization (CTL=0 & CAD=0)..292
Duration of CTL & CAD Driven Low 293
Packet Framing and Initializing the CRC Window 294
Tuning the Link Width (Firmware Initialization)..295
Tuning Example 1: 4-bit device connected to 8-bit Device............................297
Tuning Example 2: 8-bit device connected to 4/8-bit Device 297
Tuning Example 3: 32-bit Upstream and 16-bit Downstream 298
Tuning the Clock Frequency 299
Warm Reset...**302**
Warm Reset Generated by Software 303
LDTSTOP# Disconnect Sequence **304**

Chapter 13: Device Configuration

HyperTransport Uses PCI Configuration...**306**
What PCI Configuration Accomplishes **306**
HyperTransport System Limits **307**
256 Buses In A System..307
32 UnitIDs Per Bus 307
One To Eight Functions Per Device..307
256 Bytes Of Configuration Space 308
Configuration Accesses: Reaching All Devices **308**
Review: How PCI Handles Configuration Accesses **309**
Two Configuration Cycle Types..309
Type 1 Cycle Until Target Bus Is Reached 309
Target Bus Bridge: Convert To Type 0; Assert IDSEL......................................310
An Example: A PCI Configuration Space Access..310
Events In PCI Configuration Space Example (see Figure 13-1) 311

Contents

How HyperTransport Handles Configuration Accesses 311
  Configuration Cycles Are Memory Mapped 311
  How The 32MB Configuration Area Is Used 312
    Upper 16 Address Bits Indicate Type 0 And Type 1 Cycle 313
    HyperTransport Type 1 Configuration Cycle (See Figure 13-3) 314
    HyperTransport Type 0 Configuration Cycle (See Figure 13-3) 314
  No IDSEL Signal Needed In HyperTransport 314
  Example: HT Configuration Space Access 315
    Events In HT Configuration Example (see Figure 13-4) 316
  Initializing Bus Numbers And Unit IDs 316
    Case 1: A Single Chain With One Host Bridge 316
    Case 2: A HyperTransport Bridge Is Discovered 318
      A Note About Bus Numbering In HyperTransport 318
    Case 3: Initializing A Double Hosted Chain 320
      Only One Master Host Bridge 320
      Master Bridge Initialization/Configuration Sequence 320
HyperTransport Configuration Space Format 321
  Two Header Formats Are Used 321
  The Type 0 Header Format 322
  PCI Advanced Capability Registers 323
    Many Advanced Capabilities Are Defined 324
    Discovering The Advanced Capability Blocks 324
  HyperTransport Configuration Type 0 Header Fields 325
    Header Command Register 325
    Header Status Register 326
  Other Fields In The Header 329
    Cache Line Size Register. (Offset 0Ch) 329
    Latency Timer Register. (Offset 0Dh) 329
    Base Address Registers. (Offset 10h-24h) 329
      I/O BAR 329
      Memory BAR 329
    CardBus CIS Pointer. (Offset 28h) 330
    Capabilities Pointer. (Offset 34h) 330
    Interrupt Line Register. (Offset 3Ch) 330
    Interrupt Pin Register. (Offset 3Dh) 330
    Min_Gnt and Max_Latency Registers. (Offsets 3Eh and 3Fh) 330
  HyperTransport Uses Advanced Capability Blocks 330
    HyperTransport Block Types Currently Defined 331
    Block Formats Vary With Capability And Device Type 332
  The Slave/Primary Interface Block 332

Description Of Slave/Primary Interface Fields 333
   Capability ID Register. (Offset 00h) 334
   Capabilities Pointer Register. (Offset 01h) 334
   Slave Command Register. (Offset 02h-03h) 335
   Link Control Registers. (Offset 04h and 08h)..336
   Link Configuration Registers. (Offset 06h and 0Ah) 339
   Revision ID Register. (Offset 0Ch) 343
   Link Frequency Registers. (Offset 0Dh and 11h) 344
   Link Error Registers. (Offset 0Dh and 11h)...345
   Link Frequency Capability Registers. (Offset 0Eh and 12h)..............346
   Feature Capability Register. (Offset 10h) 347
   Enumeration Scratch Pad Register. (Offset 14h) 349
   Error Handling Register. (Offset 16h)...350
   Memory Base Upper Register. (Offset 18h) 352
   Memory Limit Upper Register. (Offset 19h)..352
The Host/Secondary Interface Block 354
Description Of Host/Secondary Interface Fields..355
   Capability ID Register. (Offset 00h) 355
   Capabilities Pointer Register. (Offset 01h) 355
   Host Command Register. (Offset 02h-03h) 356
   Link Control Register. (Offset 04h) 358
   Link Configuration Register. (Offset 06h) 358
   Revision ID. (Offset 08h)..358
   Link Frequency And Link Error Registers. (Offset 09h) 359
   Link Frequency Capability. (Offset 0Ah) 359
   Feature Capability Register. (Offset 0Ch) 359
   Enumeration Scratchpad Register. (Offset 10h) 360
   Error Handling Register. (Offset 12h)..360
   Memory Base/Limits Upper Registers. (Offset 14h, 15h)...................361
Revision ID Capability Block 361

Chapter 14: Electrical

Background and Introduction 364
Power Requirements 366
   Power Supply Voltage...366
   Differential Pair Power Consumption 366
Differential Signaling Characteristics 367
   Differential DC Characteristics 367
      Differential DC Impedance 368
      Differential Output Voltage - DC 369
      Differential Input Voltage - DC 371

Contents

Differential AC Characteristics 372
   Differential AC Impedance 372
   Differential Output Voltage - AC 374
   Differential Input Voltage - AC 376
   Input Rising and Falling Edge Rates 377
Single-Ended Signaling Characteristics **378**
Differential Timing Characteristics **379**
   Differential Signal Skew 379
   Source Synchronous Clock Skew 381
      Source Synchronous Clock Skew at the Transmitter 381
      Source Synchronous Clock Skew at the Receiver 383
   Setup and Hold Timing 384
Testing **385**

Chapter 15: Clocking

Introduction **388**
Clock Initialization **388**
Synchronous Clock Mode **389**
   A Conceptual Example 389
   Sources of Transmit and Receive Clock Variance 390
      Invariant Sources 391
         Cross-byte skew in multi-byte link implementations. 391
         Sampling Error 392
      Variant Sources 392
         Reference Clock Distribution Skew 392
         PLL Variation in Transmitter and Receiver 392
         Transmitter and Link Transfer Variation 393
         Receiver Transfer Variation 393
         Dynamic Cross Byte-Lane Variation 393
      An Example Timing Budget 393
   Clock Variance, FIFO Size, and the Read Pointer 394
      Minimum FIFO Size 395
      Write-to-Read and Read-to-Write Separation 395
         Scenario 1: Tx Out Clock and Rx Clock are in Sync 396
         Scenario 2: Tx Clock Out Lags Rx Clock 397
         Scenario 3: Rx Clock Lags Tx Clock Out 398
   Buffering Width and Speed Differences 399
         CAD/CTL synchronization time: 399

Pseudo-Synchronous Clock Mode 399
  Why Use Pseudo-Synchronous Clock Mode? 400
  Implementation Issues 401
    Methods and Procedures 401
    FIFO Management 401
    Is Support for Pseudo-Sync Mode Required? 402
Asynchronous Clock Mode 402
  Transmit Clock Slower Than Receive Clock 402
  Transmit Clock Faster Than Receive Clock 402

Part Three: HyperTransport Optional Topics

Chapter 16: HyperTransport Bridges

HyperTransport Bridges Uses PCI Configuration 408
Basic Jobs Of A HyperTransport Bridge 408
How Does The Bridge Manage It All? 409
  Same Slave/Primary And Host/Secondary Blocks 409
  HyperTransport Bridge Header Fields 409
    Bridge Header Command Register 410
    Bridge Header Status Register 413
    Secondary Status Register 415
    Memory And Prefetchable Base And Limit Registers 416
    Memory And Prefetchable Memory Base/Limit Notes 419
      Bus-To-Bus Forwarding Rules 419
      64 Bit Addressing And The 40 Bit HyperTransport Space 419
      To Disable Memory or Prefetchable Memory Decoding 419
      The Optional Address Remapping Registers 419
    I/O Base And Limit Registers 420
    Base/Limit Notes 421
      Bus-To-Bus Forwarding Rules 421
      To Disable I/O Decoding 422
    Bridge Control Register 422
  Other Fields In The Header 424
    Primary Latency Timer Register 424
    Base Address Registers 424
      I/O BAR 424
      Memory BAR 424
    Capabilities Pointer 425
    Interrupt Line Register 425
    Interrupt Pin Register 425
    Cache Line Size Register 425

Contents

Chapter 17: Double-Hosted Chains

Introduction..428
  Reasons For Implementing A Double-Hosted Chain 428
  PCI Configuration Plays Key Role In Chain Setup 429
    Slave Command CSR 429
    Host Command CSR 430
Two Types Of Double-Hosted Chains 431
  Sharing Double-Hosted Chain..431
    If Possible, Assign All Devices To Master Host Bridge 432
    If Slave Must Access Devices, It Uses Peer-to-Peer Transfers..................432
  Non-Sharing Double-Hosted Chain...432
    Software May Break The Chain 432
  Additional Notes About Double-Hosted Chains 433
    Initialization In A Double-Hosted Chain 433
    Type 0 Configuration Cycles In A Double-Hosted Chain 434

Chapter 18: HT Power Management

Background 436
Reporting Power Management Events to the Host Bridge 436
Reporting Host Power Management Events to SMC.......................................437
  Processor VID/FID 437
Reporting Power Management Events to HT Devices 437
Signaling Wakeup..439
X86 Power Management Support..441
  Stop Clock Signal 441
    HT Method of STPCLK# Signaling 442

Chapter 19: Networking Extensions Overview

An Important Note..444
Server And Desktop Topologies Are Host-Centric..444
  Upstream And Downstream Traffic 446
  Storage Semantics In Servers And Desktops 446
    Targets Are Assigned An Address Range In Memory Map 446
    Each Byte Transferred Has A Unique Target Address 446
    The Requester Manages Target Addresses 446
    Storage Semantics Work Fine In Servers And Desktops 447
  1.04 Protocol Optimized For Host-Centric Systems 447
Some Systems Are Not Host-Centric 448

The Need For Networking Extensions 448
  Communications Processing Is Often Less Vertical 448
  Communications Processing Example 449
Summary Of Anticipated Networking Extension Features 450
  Network Extensions Adds Message Semantics 450
  16 New Posted Write Virtual Channels 451
  Direct Peer-to-Peer Transfers Added 451
  Link-Level Error Detection And Handling 451
  64 Bit Addressing Option 452
  Increased Number Of Host Transactions 452
  End-To-End Flow Control 452
  Switch Devices Formally Defined 453

Part Four: HyperTransport Legacy Support

Chapter 20: I/O Compatibility

Introduction 457
PCI Bus Issues 458
  PCI Ordering Requirements 458
    Avoiding Deadlocks 460
  Subtractive Decode 461
    Subtractive Decode: The PCI Method 461
    Subtractive Decode: The Simple HT Method 461
    Subtractive Decode: HT Systems Requiring Extra Support 462
      The Problem 462
      The Solution 462
    Subtractive Decode: Behind PCI Bridge 463
    Subtractive Decode: Legacy System Considerations 464
    Subtractive Decode: Without Software Initialization 464
  HT-to-PCI Address Remapping 465
  Transaction Translation 465
  PCI Burst Transactions 466
PCI-X Bus Issues 467
  PCI-X Ordering Requirements 467
  Transaction Translation 468
AGP Bus Issues 470
  AGP Configuration Space Requirements 470
  AGP Ordering Requirements 471
    PCI-Based Ordering 472
    Low Priority Ordering 472
    High Priority Ordering 472
  Transaction Translation 473

Contents

ISA/LPC Buses 473
Deadlocks 473
Deadlock Scenario 1 474
Deadlock Scenario 2 475

Chapter 21: Address Remapping

Introduction 477
The Address Remapping Capability Block 478
I/O Address ReMapping 481
X86 Processor and PCI I/O Remapping Example 481
PowerPC and PCI I/O Remapping Example 483
DownStream HT to Expansion Bus Memory Mapping 485
Downstream Memory Access Without Remapping 485
Memory Accesses with Remapping 486
SBNPCtl and SBPreCtl 488
DMA Mapping 489
Number of DMA Mappings 489
DMA Secondary Base N and DMA Secondary Limit N 489
DMA Primary Base N 489
DMA Control Field 490

Chapter 22: X86 CPU Compatibility

Background 491
Legacy Signals 492
Legacy Special Cycles 493
System Management Messages 493
X86 Interrupt Support 494
APIC Interrupt Support 495
Legacy Method of Handling APIC Interrupts 495
HT Method of Handling APIC Interrupts 498
HT I/O Device Delivery of X86 Interrupts 498
EOI (End of Interrupt) Message 502
Legacy Interrupts (8259 Interrupt Controllers) 503
The Legacy Method of Handling 8259 Interrupts 503
HT Method of Handling 8259 Interrupts 505
Signaling INTR 506
Signaling Interrupt Acknowledge 507
8259 EOI Command 507
Legacy NMI Signaling 508

The A20 Mask **508**
The Legacy Method of Signaling A20M# 509
The HT Method of Signaling A20M# 510
System Management Mode (SMI# & SMIACT#) **511**
SMM Applications 511
Legacy SMM Signals 512
The HT Method of Signaling SMI# & SMIACT# 513
Numeric Error Handling (FERR# and IGNNE#) **515**
Numeric Error Handling (The Original Method) 516
DOS-Compatible Error Handling 517
HT Method of Signaling FERR# and IGNNE# 519
HT Method of FERR# Signaling 520
HT Method of IGNNE# Signaling 520
X86 Instructions and Special Cycles **521**

Appendix

Glossary of Terms **525**
Index **541**

Figures

1-1	Typical PCI North-South Bridge System	12
1-2	Sample HT-based System	15
2-1	Example HyperTransport System	21
2-2	HT Address Map	23
2-3	Transaction Flow During Programmed I/O Operation	24
2-4	Transaction Flow During DMA Operation	25
2-5	Peer-to-Peer Transaction Flow	26
2-6	Primary HT Signal Groups	27
2-7	Link Signals Used to Transfer Packets	28
2-8	Link Support Signals	29
2-9	Scalable Link Width and Speeds	30
2-10	Link Widths Supported	31
2-11	Basic HT Device Types	33
2-12	HyperTransport Topology Supporting All Three Major Device Types	34
2-13	Distinguishing Control from Data Packets	35
2-14	HT Virtual Channels	38
2-15	Example Protocol — Receiving Data from Target	41
2-16	Example Protocol — Non-Posted Sized Write	43
2-17	Example Protocol — Posted Sized Write	43
2-18	Example Protocol — Flush Transaction	44
2-19	Example Protocol — Fence Transaction	45
2-20	Example Protocol — Atomic Operation	46
2-21	Example of Packet Flow During Broadcast Transaction	47
3-1	HyperTransport Signal Groups	54
4-1	Four Byte Packet On An 8-Bit Interface	61
4-2	Four Byte Packet On A 2-Bit Interface	62
4-3	Four Byte Packet On A 16-Bit Interface	63
4-4	Four Byte Packet On A 32-Bit Interface	64
4-5	CAD Bus Control/Data Packet Management Using the CTL Signal	68
4-6	Control Packets: NOP Information	71
4-7	Control Packets: Sync Information	74
4-8	Control Packets: Generic Sized Read/Sized Write Requests	76
4-9	Control Packets: Broadcast Message Request	82
4-10	Control Packets: Flush Request	85
4-11	Control Packets: Fence Request	87
4-12	Control Packets: Atomic Read-Modify-Write Request	91
4-13	Control Packets: Read Response	95
4-14	Control Packets: Target Done Response	97
5-1	PCI Interface Handshake Signals	100
5-2	Flow Control Is On A Link-By-Link Basis	106
5-3	Flow Control Buffers And Counters	107
5-4	Flow Control Counter Initialization	111
5-5	Device 1 Sends Two Packets	113
5-6	Device 2 Updates Flow Control Information	114
6-1	PIO, DMA, And Peer-to-Peer Traffic	121
6-2	Targets At Different Levels In Hierarchy And In Different Chains	122

Figures

6-3   Non-Posted Requests And Responses At Target 123
6-4   Upstream Reordering: Packets From Different Transaction Streams 125
6-5   A Strongly Ordered Sequence Must Be Preserved 126
6-6   Packets With PassPW Clear Can't Pass Posted Requests 127
6-7   Packets With PassPW Set May Or May Not Pass Other Posted Requests 128
6-8   Non-Posted Requests May Pass Each Other 129
6-9   Posted Request Or Response Must Be Able To Pass Non-Posted Requests 130
6-10   Posted Request Must Be Able To Pass An Earlier Response...131
6-11   Non-Posted Request/Response May Pass Earlier Responses...132
6-12   Host Bridge Extends Ordering To Host System..133
6-13   Ordering Example: Read Followed By Posted Write To Cacheable Memory 136
6-14   Double-Hosted Chain Ordering 138
7-1   Example 1: NOP Information Packet With Buffer Updates 145
7-2   Generic RdSized And WrSized Request Packet Format 149
7-3   Generic Read/Target Done Response Packet Format...149
7-4   DMA Non-Posted Write Targeting Main Memory...150
7-5   Posted WrSized (Byte) Write Targeting A Downstream Device 157
7-6   DMA Dword Read Targeting Main Memory 161
7-7   DMA Byte Read Targeting Main Memory..167
7-8   A Flush Request Issued By UnitID 2...174
7-9   A Fence Request Issued By UnitID 3 179
7-10   Atomic Read-Modify-Write Targeting Main Memory...182
7-11   Sized (Dword) Write Transaction Must Cross A Bridge...189
8-1   Interrupt Capability Block Indicates that the Device Supports Interrupts 201
8-2   Interrupt Request and EOI Message Reserved Address Range...202
8-3   Interrupt Request Packet Address Field..203
8-4   Format of Interrupt Request Packet 205
8-5   Format of the Interrupt Request Data Packet 207
8-6   EOI Packet Format 209
8-7   Format of *Interrupt Discovery and Configuration Capability Block* 210
9-1   SM Request Sources...217
9-2   Format of SM Request Packet Issued by the System Management Controller................218
9-3   Format of SM Request Packet Issued by the Host Bridge..221
9-4   Theoretical System with LDTSTOP# Support 226
10-1   8/16/32 Bit Interfaces: CRC Inserted Into CAD Stream Every 512 Bit Times.................232
10-2   Link Control CSR: CRC Error Logging Bits 234
10-3   Error Handling CSR: CRC Error Interrupt Enables..235
10-4   Link Control Register: CRC Sync Flood Enable bit...236
10-5   Link Error Register: Protocol Error Logging Bits...239
10-6   Error Handling CSR: Protocol Error Reporting Enables..240
10-7   Link Error Register: Receive Buffer Overflow Error Logging Bits 241
10-8   Error Handling CSR: Receive Buffer Overflow Error Reporting Enables 242
10-9   End-Of-Chain Device Determination 244
10-10 Link Error Register: End-Of-Chain Error Logging Bits..245
10-11 Error Handling CSR: End-Of-Chain Error Reporting Enables..246
10-12 Error Handling Register: Chain Fail Bit 247

10-13	Error Handling CSR: Response Error Logging And Reporting Policy Bits	249
10-14	Response Packet And Error Bits	250
10-15	Sync Flood Example	255
11-1	Routing: Shared Bus vs. HyperTransport Point-Point	258
11-2	Generic WrSized Or RdSized Request Packet: Key Routing Fields	260
11-3	Generic Read/Target Done Response Packet: Key Routing Fields	262
11-4	Tunnel Inserting Packets Into Upstream Traffic	271
12-1	Example of Reset Distribution in an HT System	277
12-2	Example HT-to-HT Bridge Forwarding Cold Reset	279
12-3	Bridge Control Register Can Force Cold Reset	280
12-4	Cold Reset Signalling	280
12-5	RESET# and PWROK Sequence and Timing Requirements	281
12-6	Low-Level Link Width, Example 1	285
12-7	Low-Level Link Width, Example 2	286
12-8	Low-Level Link Width, Example 3	287
12-9	Link Configuration Register	288
12-10	Link Interface and Clocking	290
12-11	Clock Synchronization and FIFO Load and Unload Pointer Setup	293
12-12	Duration of CTL & CAD Deassertion	294
12-13	Framing and CRC Window Sequence	295
12-14	Link Width Update	296
12-15	Maximum Link Values, Example 1	297
12-16	Maximum Link Values, Example 2	298
12-17	Maximum Link Values, Example 3	299
12-18	Tunnel Example — Link Frequency Capability Registers	300
12-19	Link Frequency Register Location within the HT Capability Register Set	301
12-20	HyperTransport Host Command CSR	304
13-1	PCI Type 1 and Type 0 Configuration Cycles	310
13-2	Configuration Space In The HyperTransport Address Map	312
13-3	Configuration Type 0 And Type 1 Request Packet Format	313
13-4	HyperTransport Type 1 And Type 0 Configuration Cycles	315
13-5	Bus Numbering In A Mixed Topology	319
13-6	Bus Numbering In Double-Hosted Chains	320
13-7	PCI Type 0 Configuration Space Header	322
13-8	PCI Configuration Space With Advanced Capability Register Block(s)	323
13-9	HyperTransport Technology Header Command Register Usage	325
13-10	HyperTransport Technology Header Status Register Usage	327
13-11	HyperTransport Advanced Capability Block Types	331
13-12	Slave/Primary Interface For Tunnel And Cave Devices	333
13-13	Slave/Primary Interface Block Format	334
13-14	HyperTransport Slave Command CSR	335
13-15	Slave Interface Block Link Control Registers 0,1	337
13-16	Slave Interface Link Configuration Registers 0,1	340
13-17	Slave Interface Revision ID Register	343
13-18	Slave Interface Link Frequency Registers 0,1	344
13-19	Slave Interface Link Error Registers 0,1	345

Figures

13-20	Slave Interface Link Frequency Capability Registers 0,1	346
13-21	Slave Interface Feature Capability Register	348
13-22	Slave Interface Enumeration Scratch Pad Register	349
13-23	Slave Interface Error Handling Register	350
13-24	Slave Interface Memory Base/Limit Upper Registers	353
13-25	Host/Secondary Interface For Bridge Devices	354
13-26	Host/Secondary Command Register Format.	355
13-27	HyperTransport Host Command CSR	356
13-28	Host Interface Feature Capability Register	359
13-29	Revision ID Capability Block	361
14-1	Link Signals	364
14-2	HT Link Differential Driver and Receiver	367
14-3	DC Impedance Values	368
14-4	DC Output Voltage Measurements	370
14-5	Differential DC Input Voltage Parameters	371
14-6	AC Impedance Values	373
14-7	Differential AC Output Parameters	374
14-8	Test Setup for AC Output Voltage Measurements	375
14-9	Input Voltage Parameters	377
14-10	Output Skew Measurement	380
14-11	TCADV Minimum and Maximum Measurements	382
14-12	Setup and Hold Time for CAD and Control	384
15-1	Simple Synchronous Clocking Interface	390
15-2	Synchronous Clock Example, Single Direction	394
15-3	FIFO Operation When Tx Clock Out and Rx Clock are in Sync	396
15-4	Effects of Tx Out Clock Lagging Rx Clock	397
15-5	Effects of Rx Clock Lagging Tx Clock Out	398
15-6	Example Pseudo-Synchronous Mode Implementation	400
16-1	HyperTransport-HyperTransport Bridge Interfaces	410
16-2	HyperTransport Bridge Header Command Register	411
16-3	HyperTransport Bridge Header Status Register	413
16-4	HyperTransport Bridge Header Secondary Status Register	415
16-5	HyperTransport Bridge Header Memory And Prefetchable Base/Limit Register	417
16-6	HyperTransport Bridge I/O Base And Limit Register	420
16-7	HyperTransport Bridge Control Register	422
17-1	HyperTransport Double-Hosted Chain Configuration	428
17-2	Slave Command CSR: Key Fields In DHC Configuration	429
17-3	Host Command CSR: Key Fields In DHC Configuration	430
17-4	Sharing Double-Hosted Chain With Master/Slave Host Bridges	431
17-5	Non-Sharing Double-Hosted Chain	433
18-1	LDTSTOP# is an Input to All HT Devices Except the SMC.	438
18-2	LDTREQ# is an Output from All HT Devices and an Input to the SMC.	439
18-3	Example Wakeup Signaled by HT-to-PCI-X Bridge.	440
18-4	SM Request Packet Contents for Delivering STPCLK	442
19-1	Host-Centric HyperTransport System	445
19-2	A HyperTransport-Based Communications Processing System	449

Figures

20-1   Topology Causing Deadlock Scenario for Rows 4 and 5 460
20-2   Subtractive Decode in a Simple HT System...462
20-3   Subtractive Decode Agent Behind PCI Bridge 463
20-4   Subtractive Decode Agent on PCI Bus 0...464
20-5   Legacy Configuration Mapping for Host to HT Bridge
        with AGP and DRAM Controller 470
20-6   Deadlock Scenario 1 Topology..475
21-1   HT Address Space May Exceed that of the Processor and Expansion Bus 479
21-2   Format of the Address Remapping Capability Block..479
21-3   X86 Processor I/O Mapping Example 483
21-4   PowerPC I/O Mapping Example 484
21-5   Example of Direct Memory Mapping between CPU, HT, and PCI...........................486
21-6   Prefetchable Memory Address Remapping Registers...487
21-7   SBNPCtl and SBPreCtl Register Format 488
21-8   DMA Control Field 490
22-1   CPU Signals Routed Between South Bridge and CPU 492
22-2   SM Request Sources...494
22-3   Legacy APIC Implementation...495
22-4   Format of the Interrupt Redirection Register 497
22-5   HT-Based Platform with IO APIC 498
22-6   Example X86 Platform Containing HT I/O Devices and an IO APIC 499
22-7   Format of X86-Based Interrupt Request Message..502
22-8   EOI Request Message Format 503
22-9   Legacy x86 Platform — Single Processor & 8259 Interrupt Controllers 504
22-10   HT-based System with 8259s 505
22-11   Legacy INTR — Interrupt Request and Data Packet Format 506
22-12   Interrupt Acknowledge Request Packet Format..507
22-13   Legacy A20 Mask Signal 509
22-14   HT A20M# Delivery 510
22-15   Format and Contents of the A20M SM Request Packet 511
22-16   SMM Signaling in Legacy Systems..513
22-17   SM Request Content for SMI Message...514
22-18   SM Message Content for SMIACT 515
22-19   Legacy Numeric Coprocessor Error Reporting 517
22-20   DOS Compatible Floating-Point Error Signaling 518
22-21   Example X86 System with Required DOS-Compatible FPU Error Handling.................519
22-22   Format and Content of the SM FERR Message 520

Tables

1 PC Architecture Book Series 1
2-1 Signals Used for Different Link Widths 31
2-2 Maximum Bandwidth Based on Various Speeds and Link Widths 32
4-1 Control Packets And The HyperTransport Command Types 69
4-2 HyperTransport NOP Packet Bit Assignments 72
4-3 HyperTransport Sync Packet Bit Assignments 74
4-4 HyperTransport Sized Read/Write Packet Bit Assignments 76
4-5 HyperTransport Broadcast Message Packet Bit Assignments 83
4-6 HyperTransport Flush Packet Bit Assignments 85
4-7 HyperTransport Fence Packet Bit Assignments 88
4-8 HyperTransport Atomic Read — Modify-Write Packet Bit Assignments 91
4-9 HyperTransport Read Response Packet Bit Assignments 95
4-10 HyperTransport Target Done Response Packet Bit Assignments 97
6-1 Summary Of Host Ordering Rules For Transaction Pairs 134
7-1 Implications Of Sending Information And Request Control Packets 140
8-1 Contents of the Interrupt Definition Registers 212
9-1 Summary of Upstream SysMgtCmd Encodings 219
9-2 Summary of SysMgtCmd Encodings 221
11-1 Definitions Of Request Packet Fields Used In Routing 260
11-2 Request Packet Command Code Summary 261
11-3 Definitions Of Response Packet Fields Used In Routing 262
12-1 Transmitter Value Driven to Indicate Receiver Width 283
12-2 Interpretation of Value Received on the CAD Lines
       to Determine Receiver Width 284
12-3 Encoded Link-Width Values used in the Link Configuration Registers 289
12-4 CTL/CAD Sequence Following Deassertion of RESET# 291
12-5 Encodings for Link Frequency Field of Link Configuration Register 301
12-6 Signal States During Warm Reset 302
13-1 HyperTransport Header Command Register Bit Assignment 326
13-2 HyperTransport Header Status Register Bit Assignment 328
13-3 HyperTransport Advanced Capability Codes 332
13-4 Slave Interface Block Command Register Bit Assignment 335
13-5 Slave Interface Block Link Control Register 0,1 Bit Assignment 337
13-6 Slave Interface Block Link Configuration Register 0,1 Bit Assignment 340
13-7 Slave Interface Link Error Registers 0,1 345
13-8 Slave Interface Link Frequency Capability Registers 0,1 347
13-9 Slave Interface Feature Capability Register 348
13-10 Slave Interface Error Handling Register 350
13-11 Host Interface Block Command Register Bit Assignment 356
13-12 Host Interface Feature Capability Register 360
13-13 Revision ID Capability Block Bit Assignment 361
14-1 Signal Group/Source Synchronous Clock Association 365

14-2	Differential Pair Power Consumption	367
14-3	DC Impedance Specification	369
14-4	Differential DC Output Voltages	370
14-5	Differential DC Input Voltages	372
14-6	AC Impedance Specification	373
14-7	AC Differential AC Output Voltages	376
14-8	Differential DC Input Voltages	377
14-9	Differential Input Edge Rate Parameters and Values	378
14-10	Single-Ended Signaling Characteristics	378
14-11	Maximum Differential Output and Input Skew Values	380
14-12	Source Synchronous CLK Output Skew Values	382
14-13	Source Synchronous CLK Input Skew Values	383
14-14	Setup and Hold Times for CAD and CTL	385
15-1	Timing Variance Budget from Specification for Source of Clock Variation	393
16-1	HyperTransport Bridge Header Command Register Bit Fields	411
16-2	HyperTransport Bridge Header Status Register Bit Fields	414
16-3	HyperTransport Bridge Secondary Status Register Bit Fields	416
16-4	Bridge Memory And Prefetchable Base And Limit Register Bit Fields	418
16-5	Bridge I/O Base And Limit Register Bit Fields	421
16-6	HyperTransport Bridge Control Register Bit Fields	423
17-1	Slave Command CSR: Definitions Of Key Fields In DHC Configuration	429
17-2	Host Interface Block Host Command CSR Bit Assignment	430
20-1	PCI Ordering Rules	459
20-2	PCI to HT Command Conversion	465
20-3	HT to PCI Command Conversion	466
20-4	PCI-X Ordering Rules	468
20-5	PCI-X to HT Command Conversion	468
20-6	HT-to-PCI-X Command Conversion	469
20-7	AGP-to-HT Command Conversion	473
22-1	Summary of Interrupt Request Bit Field Encoding Values	501

Acknowledgments

We wish to thank Jessie Johnson of AMD for his support in the early stages of this project. Also, thanks to Paul Miranda and Renato D'Orfani of AMD, and Brian Holden of PMC-Sierra for providing valuable input and constructive feedback.

The authors wish to thank Ravi Budruk and Dave Dzatko for their major contributions in developing much of the content for MindShare's HyperTransport seminars and for this book.

The MindShare Architecture Series

The MindShare Architecture book series currently includes the books listed in Table 1, "PC Architecture Book Series," on page 1. Rather than duplicating common information in each book, the series uses the building-block approach. Generally-speaking, *ISA System Architecture* is the core book upon which the others build. In a sense, it is a PC-compatibility book. The entire book series is published by Addison-Wesley.

Table 1: PC Architecture Book Series

Category	Title	Edition	ISBN
Processor Architecture	80486 System Architecture	3rd	0-201-40994-1
	Pentium Processor System Architecture	2nd	0-201-40992-5
	Pentium Pro and Pentium II System Architecture	2nd	0-201-30973-4
	PowerPC System Architecture	1st	0-201-40990-9
Bus Architecture	PCI System Architecture	4th	0-201-30974-2
	PCI-X System Architecture	1st	0-201-72682-3
	EISA System Architecture	Out-of-print	0-201-40995-X
	Firewire System Architecture: IEEE 1394a	2nd	0-201-48535-4
	ISA System Architecture	3rd	0-201-40996-8
	Universal Serial Bus System Architecture 2.0	2nd	0-201-46137-4
	HyperTransport System Architecture	1st	0-321-16845-3
Network Architecture	Infiniband Network Architecture	1st	0-321-11765-4

Table 1: PC Architecture Book Series (Continued)

Category	Title	Edition	ISBN
Other Architectures	PCMCIA System Architecture: 16-Bit PC Cards	2nd	0-201-40991-7
	CardBus System Architecture	1st	0-201-40997-6
	Plug and Play System Architecture	1st	0-201-41013-3
	Protected Mode Software Architecture	1st	0-201-55447-X
	AGP System Architecture	1st	0-201-37964-3

Cautionary Note

The reader should keep in mind that MindShare's book series often details rapidly evolving technologies. That being the case, it should be recognized that the book is a "snapshot" of the state of the technology at the time the book was completed. We make every attempt to produce our books on a timely basis, but the next revision of the specification is not introduced in time to make necessary changes.

At the time of this writing, the HyperTransport I/O Link Specification Revision 1.04 is released. The 1.04 revision of the specification does not deal with the networking extensions but much of the work on the HyperTransport 1.05 and 1.1 revisions has been done by the HyperTransport Technical Working Group, and quite a bit of preliminary information on this important addition to the protocol has been released. Some of the more important additions to these specifications are summarized in this book. Please check the MindShare website (www.mindshare.com) for supplemental information, updates, and errata.

Intended Audience

This book is intended for use by hardware and software design and support personnel. The tutorial approach taken may also make it useful to technical personnel not directly involved design, verification, and other support functions.

Prerequisite Knowledge

It is recommended that the reader has a reasonable background in PC architecture, including experience or knowledge of an I/O bus and related protocol. The MindShare publication entitled *ISA System Architecture* focusses on various aspects of PCI architecture and provides the necessary background.

Topics and Organization

Topics covered in this book and the flow of the book are as follows:

Part 1: Overview of HyperTransport
      Chapter 1: Introduction to HyperTransport
      Chapter 2: HT Architectural Overview
Part 2: HyperTransport Core Topics
      Chapter 3: Signal Groups
      Chapter 4: Packet Protocol
      Chapter 5: Flow Control
      Chapter 6: I/O Ordering
      Chapter 7: Transaction Examples
      Chapter 8: HT Interrupts
      Chapter 9: System Management
      Chapter 10: Error Detection and Handling
      Chapter 11: Routing Packets
      Chapter 12: Reset and Initialization
      Chapter 13: Device Configuration
      Chapter 14: Electrical
      Chapter 15: Clocking
Part 3: HyperTransport Optional Topics
      Chapter 16: HyperTransport Bridges
      Chapter 17: Double-Hosted Chains
      Chapter 18: HT Power Management
      Chapter 19: Networking Extensions Overview
Part 4: HyperTransport Legacy Support
      Chapter 20: I/O Compatibility
      Chapter 21: Address Remapping
      Chapter 22: x86 CPU Compatibility
Appendix A: Glossary of Terms
Index

Documentation Conventions

This section defines the typographical convention used throughout this book.

HyperTransport™

HyperTransport™ is a trademark of the HyperTransport Consortium. This book takes the liberty of abbreviating HyperTransport as "HT" to improve readability.

Hexadecimal Notation

All hex numbers are followed by a lower case "h." For example:

89F2BD02h
0111h

Binary Notation

All binary numbers are followed by a lower case "b." For example:

1000 1001 1111 0010b
01b

Decimal Notation

Numbers without any suffix are decimal. When required for clarity, decimal numbers are followed by a lower case "d." Examples:

9
15
512d

Byte Terminology and Notation

This book uses the following terminology regarding quantities of data:

8-bits = 1 byte
16-bits = 2 bytes, Word
32-bits = 4 bytes, Double Word, Dword, or DW
64-bits = 8 bytes, Quad Word, Qword, or QW

Bits Versus Bytes Notation

This book represents bit with lower case "b" and bytes with an upper case "B."
For example:

Megabits/second = Mb/s
Megabytes/second = MB/s

Bit Fields and Groups of Signals

Groups of signals or bits are represented with the high-order bits first followed
by the low-order bits and enclosed by brackets. For example:

[7:0]
Addr[39:0]
CAD[15:0]

Active Signal States

Signals that are active low are followed by #, as in RESET#. Active high signals
have no suffix following the signal, as in PWROK.

Visit Our Web Site

Our web site lists all of our courses and the delivery options available for each course:

Self-paced DVDs and CDs
Live web-delivered classes
Live on-site classes.

All of our books are listed and can be ordered in bound or e-book versions.

www.mindshare.com

We Want Your Feedback

MindShare values your comments and suggestions. Contact us at:

Phone: (719) 487-1417 or within the U.S. (800) 633-1440
Fax: (719) 487-1434
E-mail: don@mindshare.com or jay@mindshare.com

For information on MindShare seminars, DVDs, and books, check our website or contact **nancy@mindshare.com**

Mailing Address:

MindShare, Inc.
4285 Slash Pine Drive
Colorado Springs, CO 80908

Part One

Overview of HyperTransport

Part One discusses the need for and design goals of HT. It also introduces the concepts of HT including the hardware and software elements required for its operation. The chapters included in Part One are:

- Chapter 1: Introduction to HyperTransport
- Chapter 2: HT Architectural Overview

1 *Introduction to HyperTransport*

This Chapter

This chapter discusses some of the motivations leading to the development of HyperTransport. It reviews some of the attributes that limit the ability of older generation I/O buses to keep pace with the increasing demands of new applications and advances in processor and memory technologies. The chapter then summarizes the key features behind the improved performance of HT over earlier buses.

The Next Chapter

The next chapter provides an overview of HT architecture, including the primary elements of HT technology and the relationship between them. The chapter describes the general features, capabilities, and limitations of HT and introduces the terminology and concepts necessary for in-depth discussions of the various HT topics in subsequent chapters.

Background: I/O Subsystem Bottlenecks

New I/O buses are typically developed in response to changing system requirements and to promote lower cost implementations. Current-generation I/O buses such as PCI are rapidly falling behind the capabilities of other system components such as processors and memory. Some of the reasons why the I/O bottlenecks are becoming more apparent are described below.

Server Or Desktop Computer: Three Subsystems

A server or desktop computer system is comprised of three major subsystems:

1. Processor (in servers, there may be more than one)
2. Main DRAM Memory. There are a number of different synchronous DRAM types, including SDRAM, DDR, and Rambus.
3. I/O (Input/Output devices). Generally, all components which are not processors or DRAM are lumped together in this subsystem group. This would include such things as graphics, mass storage, legacy hardware, and the buses required to support them: PCI, PCI-X, AGP, USB, IDE, etc.

CPU Speed Makes Other Subsystems Appear Slow

Because of improvements in CPU internal execution speed, processors are more demanding than ever when they access external resources such as memory and I/O. Each external read or write by the processor represents a huge performance hit compared to internal execution.

Multiple CPUs Aggravate The Problem

In systems with multiple CPUs, such as servers, the problem of accessing external devices becomes worse because of competition for access to system DRAM and the single set of I/O resources.

DRAM Memory Keeps Up Fairly Well

Although it is external to the processor(s), system DRAM memory keeps up fairly well with the increasing demands of CPUs for a couple of reasons. First, the performance penalty for accessing external memory is mitigated by the use of internal processor caches. Modern processors generally implement multiple levels of internal caches that run at the full CPU clock rate and are tuned for high "hit rates". Each fetch from an internal cache eliminates the need for an external bus cycle to memory.

In addition, in cases where an external memory fetch is required, DRAM technology and the use of synchronous bus interfaces to it (e.g. DDR, RAMBUS, etc.) have allowed it to maintain bandwidths comparable with the processor external bus rates.

I/O Bandwidth Has Not Kept Pace

While the processor internal speed has raced forward, and memory access speed has managed to follow along reasonably well with the help of caches, I/O subsystem evolution has not kept up.

This Slows Down The Processor

Although external DRAM accesses by processors can be minimized through the use of internal caches, there is no way to avoid external bus operations when accessing I/O devices. The processor must perform small, inefficient external transactions which then must find their way through the I/O subsystem to the bus hosting the device.

It Also Hurts Fast Peripherals

Similarly, bus master I/O devices using PCI or other subsystem buses to reach main memory are also hindered by the lack of bandwidth. Some modern peripheral devices (e.g. SCSI and IDE hard drives) are capable of running much faster than the busses they live on. This represents another system bottleneck. This is a particular problem in cases where applications are running that emphasize time-critical movement of data through the I/O subsystem over CPU processing.

Reducing I/O Bottlenecks

Two important schemes have been used to connect I/O devices to main memory. The first is the shared bus approach, as used in PCI and PCI-X. The second involves point-to-point component interconnects, and includes some proprietary busses as well as open architectures such as HyperTransport. These are described here, along with the advantages and disadvantages of each.

The Shared Bus Approach

Figure 1-1 on page 12 depicts the common "North-South" bridge PCI implementation. Note that the PCI bus acts as both an "add-in" bus for user peripheral cards and as an interconnect bus to memory for all devices residing on or below it. Even traffic to and from the USB and IDE controllers integrated in the South Bridge must cross the PCI bus to reach main memory.

Figure 1-1: Typical PCI North-South Bridge System

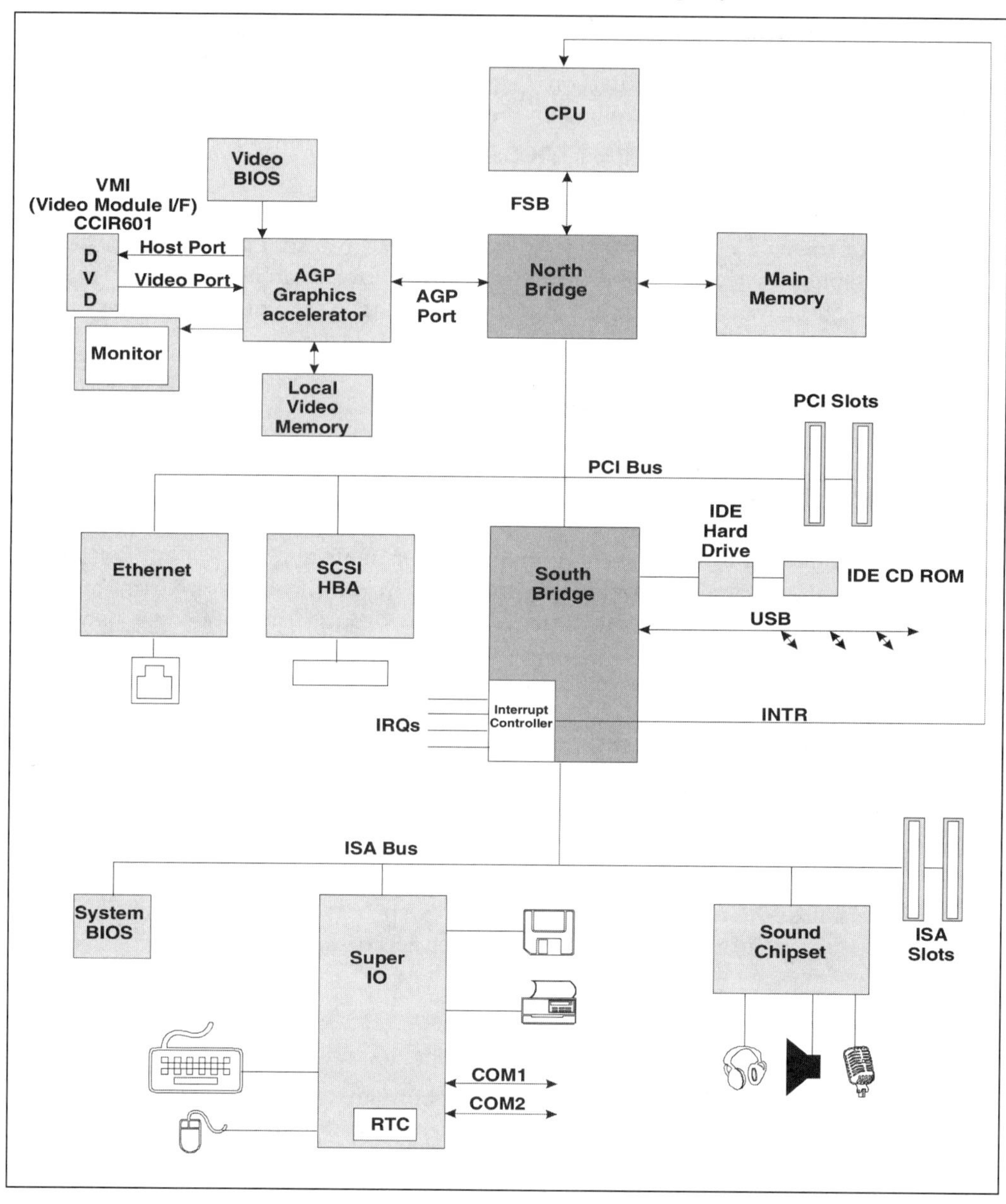

Until recently, the topology shown in Figure 1-1 on page 12 has been very popular in desktop systems for a number of reasons, including:

1. A shared bus reduces the number of traces on the motherboard to a single set.
2. All of the devices located on the PCI bus are only one bridge interface away from the principal target of their transactions — main DRAM memory.
3. A single, very popular protocol (PCI) can be used for all embedded devices, add-in cards, and chipset components attached to the bus.

Unfortunately, some of the things that made this topology so popular also have made it difficult to fix the I/O bandwidth problems which have become more obvious as processors and memory have become faster.

A Shared Bus Runs At Limited Clock Speeds. The fact that multiple devices (including PCB connectors) attach to a shared bus means that trace lengths and electrical complexity will limit the maximum usable clock speed. For example, a generic PCI bus has a maximum clock speed of 33MHz; the PCI Specification permits increasing the clock speed to 66MHz, but the number of devices/connectors on the bus is very limited.

A Shared Bus May Be Host To Many Device Types. The requirements of devices on a shared bus may vary widely in terms of bandwidth needed, tolerance for bus access latency, typical data transfer size, etc. All of this complicates arbitration on the bus when multiple masters wish to initiate transactions.

Backward Compatibility Prevents Upgrading Performance. If a critical shared bus is based on an open architecture, especially one that defines user "add-in" connectors, then another problem in upgrading bus bandwidth is the need to maintain backward compatibility with all of the devices and cards already in existence. If the bus protocol is enhanced and a user installs an "older generation card", then the bus must either revert back to the earlier protocol or lose its compatibility.

Special Problems If The Shared Bus Is PCI. As popular as it has been, PCI presents additional problems that contribute to performance limits:

1. PCI doesn't support split transactions, resulting in inefficient *retries*.
2. Transaction size (there is no limit) isn't known, which makes it difficult to size buffers and causes frequent *disconnects* by targets. Devices are also allowed to insert numerous *wait states* during each data phase.
3. All PCI transactions by I/O devices targeting main memory generally

require a "snoop" cycle by CPUs to assure coherency with internal caches. This impacts both CPU and PCI performance.

4. Its data bus scalability is very limited (32/64 bit data)
5. Because of the PCI electrical specification (low-power, reflected wave signals), each PCI bus is physically limited in the number of ICs and connectors vs. PCI clock speed
6. PCI bus arbitration is vaguely specified. Access latencies can be long and difficult to quantify. If a second PCI bus is added (using a PCI-PCI bridge), arbitration for the secondary bus typically resides in the new bridge. This further complicates PCI arbitration for traffic moving vertically to memory.

A Note About PCI-X. Other than scalability and the number of devices possible on each bus, the PCI-X protocol has resolved many of the problems just described with PCI. For third-party manufacturers of high performance add-in cards and embedded devices, the shared bus PCI-X is a straightforward extension of PCI which yields huge bandwidth improvements (up to about 2GB/s with PCI-X 2.0).

The Point-to-Point Interconnect Approach

An alternative to the shared I/O bus approach of PCI or PCI-X is having point-to-point links connecting devices. This method is being used in a number of new bus implementations, including HyperTransport technology. A common feature of point-to-point connections is much higher bandwidth capability; to achieve this, point-to-point protocols adopt some or all of the following characteristics:

- only two devices per connection.
- low voltage, differential signaling on the high speed data paths
- source-synchronous clocks, sometimes using double data rate (DDR)
- very tight control over PCB trace lengths and routing
- integrated termination and/or compensation circuits embedded in the two devices which maintain signal integrity and account for voltage and temperature effects on timing.
- dual simplex interfaces between the devices rather than one bi-directional bus; this enables duplex operations and eliminates "turn around" cycles.
- sophisticated protocols that eliminate retries, disconnects, wait-states, etc.

A Note About Connectors. While connectors may or may not be defined in a point-to-point link specification, they may be designed into some implementations to connect from board-board or for the attachment of diagnostic equipment. There is no definition of a peripheral add-in card connector for HyperTransport as there is in PCI or PCI-X.

What HT Brings

HyperTransport is a point-to-point, high-performance, "inside-the-box" motherboard interconnect bus. It targets IT, Telecom, and other applications requiring high bandwidth, scalability, and low latency access. Figure 1-2 on page 15 illustrates a single HT bus implementation with a variety of functional devices attached.

Figure 1-2: Sample HT-based System

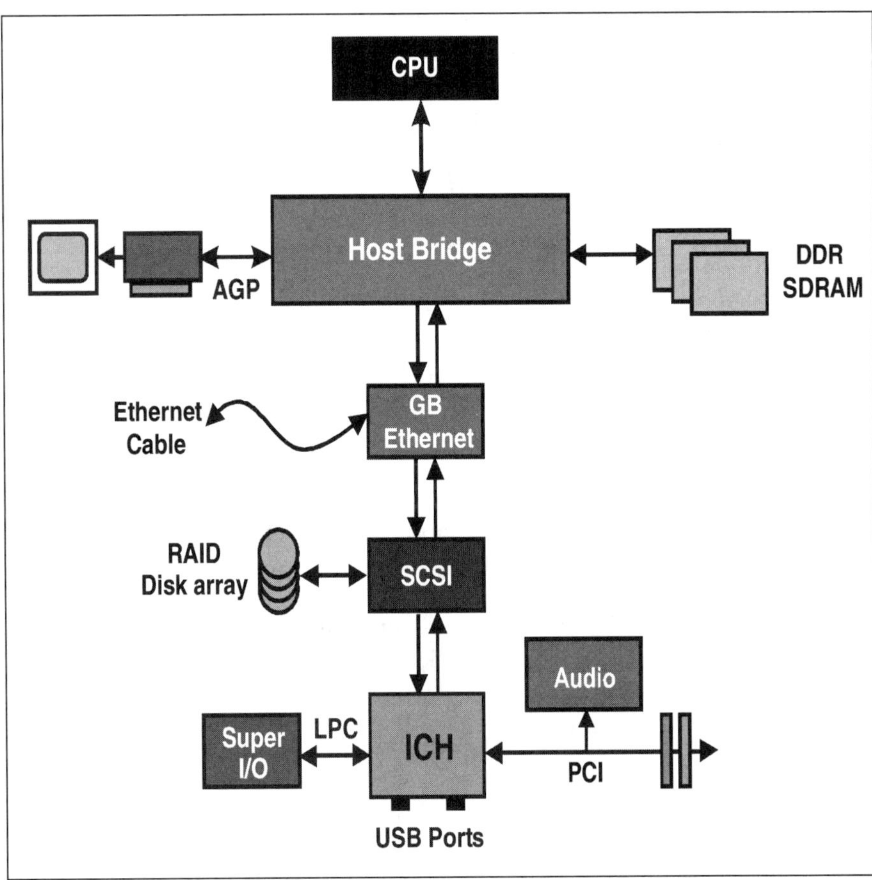

Key Features Of HyperTransport Protocol

The key characteristics of the HT technology include:

- Open architecture, non-proprietary bus
- One or more fast, point-to-point links
- Scaling of individual link width and clock speed to suit cost/performance targets
- Split-transaction protocol eliminates retries, disconnects, and wait-states.
- Standard and optional isochronous traffic support
- PCI compatible; designed for minimal impact on OS and driver software
- CRC error generation and checking
- Programmable error handling strategy for CRC, protocol, and other errors
- Message signalled interrupts
- System Management features
- Support for bridges to legacy busses
- x86 compatibility features
- Device types including tunnels, bridges, and end devices permit construction of a system fabric comprised of independent, customized links.

Formerly known as AMD's Lightning Data Transport (LDT), HyperTransport is backed by a consortium of developers. See ***www.hypertransport.org***.

The Cost Factor

In addition to technology-related issues, there is always pressure on the platform designer to increase performance and other capabilities with each new generation, but to do so at a lower cost than the previous one. One popular method of measuring the success of this effort is to compare the bandwidth of one I/O bus to another, and the number of signals required to achieve it. This bandwidth-per-pin comparison works fairly well because I/O bus bandwidth is a critical factor in determining if system data bottlenecks exist, and a lower pin count translates directly into cost savings due to smaller IC packages, lower power, simplified motherboard routing, etc.

An example:

The bandwidth-per-pin for a generic 32-bit PCI bus during a burst transfer is approximately **3.5 MB/s** (132 MB/s [33MHz x 4 bytes]/38 pins [32 data signals + 5 control lines + 1 clock]). By comparison, a 32 bit HyperTransport interface running at the lowest clock speed of 200MHz yields a per-pin burst bandwidth of approximately **22 MB/s** (1600 MB/s [200Mhz x 2 DDR x 4 bytes]/74 pins [32 CAD signal <u>pairs</u> + 4 clock <u>pairs</u> + 1 CTL <u>pair</u>]).

Networking Support

Finally, at the time of the writing of this book, the HyperTransport I/O Link Specification is at revision 1.04. This specification revision mainly targets I/O subsystem improvements in conventional desktop and server platforms.

A growing number of applications require architectures that integrate well with networking environments. In many of these systems, unlike desktops and servers, processing may be decentralized and features such as message streaming, peer-peer transfers, and assigned isochronous bandwidth become important. In addition, device types such as *switches* help in building topologies suited to communications networking. To accommodate networking applications, work is well underway on the 1.05 and 1.1 revisions of the HyperTransport I/O Link Specification. The 1.05 specification includes the HyperTransport *switch* specification and the 1.1 specification incorporates the *networking extensions* specification. See Chapter 19, entitled "Networking Extensions Overview," on page 443 for a summary of the major features expected to be included in the 1.05 and 1.1 specification revisions.

Visit **www.hypertransport.org** for up-to-date information on all on-going specification revisions.

Also, visit MindShare's website at **www.mindshare.com** for updates to this book relating to this and other HyperTransport topics. Information will be available for free download when the new specification revisions are released and details become publicly available.

2 HT Architectural Overview

The Previous Chapter

To understand why HT was developed, it is helpful to review the previous generation of I/O buses and interconnects. This chapter review the factors that limit the ability of older generation buses to keep pace with the increasing demands of new applications. Finally, this chapter discusses the key factors of the HT technology that provides its improved capability.

This Chapter

This chapter provides an overview of the HT architecture that defines the primary elements of HT technology and the relationship between these elements. This chapter summarizes the features, capabilities, and limitation of HT and provides the background information necessary for in-depth discussions of the various HT topics in later chapters.

The Next Chapter

The next chapter describes the function of each signal in the high- and low-speed HyperTransport signal groups.

General

HyperTransport provides a point-to-point interconnect that can be extended to support a wide range of devices. Figure 2-1 on page 21 illustrates a sample HT system with four internal links. HyperTransport provides a high-speed, high-performance, point-to-point dual simplex link for interconnecting IC components on a PCB. Data is transmitted from one device to another across the link.

The width of the link along with the clock frequency at which data is transferred are scalable:

- Link width ranges from 2 bits to 32-bits
- Clock Frequency ranges from 200MHz to 800MHz (and 1GHz in the future)

This scalability allows for a wide range of link performance and potential applications with bandwidths ranging from 200MB/s to 12.8GB/s.

At the current revision of the spec, 1.04, there is no support for connectors implying that all HyperTransport (HT) devices are soldered onto the motherboard. HyperTransport is technically an "inside-the-box" bus. In reality, connectors have been designed for systems that require board to board connections, and where analyzer interfaces are desired for debug.

Once again referring to Figure 2-1, the HT bus has been extended in the sample system via a series of devices known as tunnels. A tunnel is merely an HT device that performs some function, but in addition it contains a second HT interface that permits the connection of another HT device. In Figure 2-1, the tunnel devices provide connections to other I/O buses:

- Infiniband
- PCI-X
- Ethernet

The end device is termed a cave, which always represents the termination of a chain of devices that all reside on the same HT bus. Cave devices include a function, but no additional HT connection. The series of devices that comprise an HT bus is sometimes simply referred to as an HT chain.

Additional HT buses (i.e. chains) may be implemented in a given system by using a HT-to-HT bridge. In this way, a fabric of HT devices may be implemented. Refer to section entitled, "Extending the Topology" on page 33 for additional detail.

Figure 2-1: Example HyperTransport System

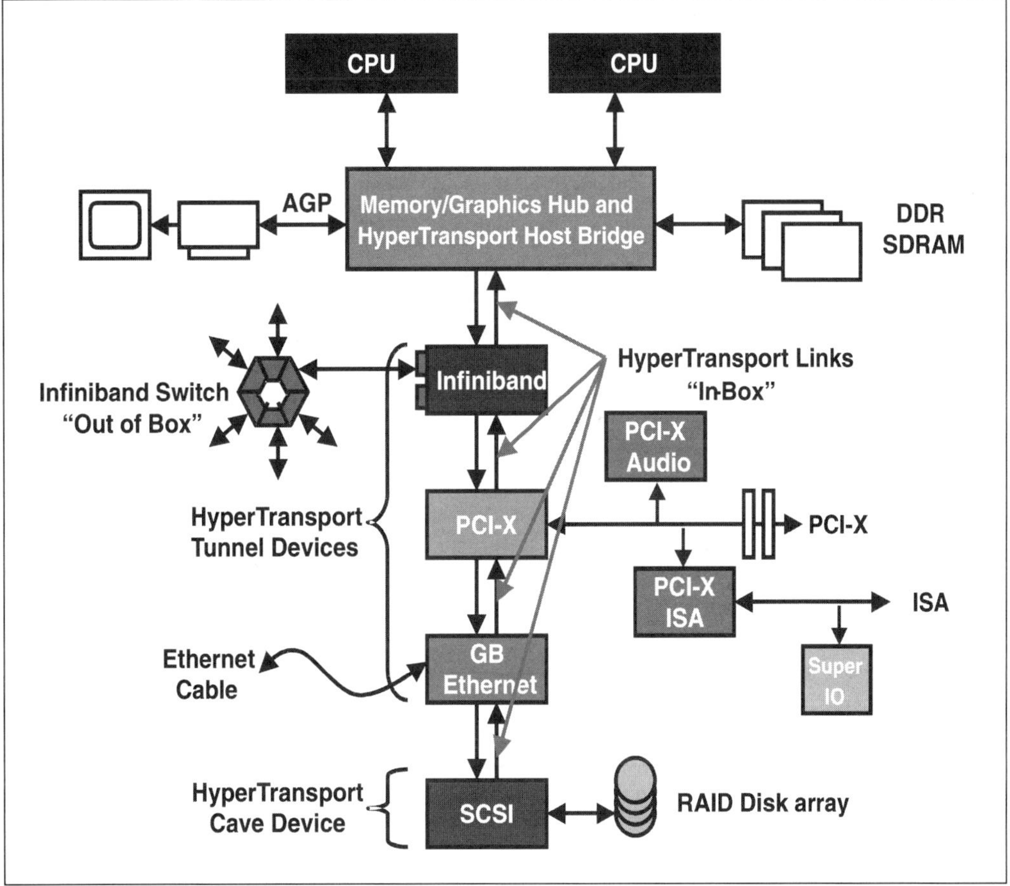

Transfer Types Supported

HT supports two types of addressing semantics:

1. legacy PC, address-based semantics
2. messaging semantics common to networking environments

The first part of this book discusses the address-based semantics common to compatible PC implementations. Message-passing semantics are discussed in Chapter 19, entitled "Networking Extensions Overview," on page 443.

Address-Based Semantics

The HT bus was initially implemented as a PC compatible solution that by definition uses Address-based semantics. This includes a 40-bit, or 1 Terabye (TB) address space. Transactions specify locations within this address space that are to be read from or written to. The address space is divided into blocks that are allocated for particular functions, listed in Figure 2-2 on page 23.

HyperTransport does not contain dedicated I/O address space. Instead, CPU I/O space is mapped to high memory address range (FD_FC00_0000h—FD_FDFF_FFFFh). Each HyperTransport device is configured at initialization time by the boot ROM configuration software to respond to a range of memory address spaces. The devices are assigned addresses via the base address registers contained in the configuration register header. Note that these registers are based on the PCI Configuration registers, and are also mapped to memory space (FD_FE00_0000h—FD_FFFF_FFFFh. Unlike the PCI bus, there is no dedicated configuration address space.

Read and write request command packets contain a 40-bit address Addr[39:2]. Additional memory address ranges are used for interrupt signaling and system management messages. Details regarding the use of each range of address space is discussed in subsequent chapters that cover the related topic. For example, a detailed discussion of the configuration address space can be found in Chapter 13, entitled "Device Configuration," on page 305.

Figure 2-2: HT Address Map

8GB Reserved	**FE_0000_0000h to FF_FFFF_FFFFh**
32MB Configuration	**FD_FE00_0000h to FD_FFFF_FFFFh**
32MB IO	**FD_FC00_0000h to FD_FDFF_FFFFh**
46MB Reserved	**FD_F920_0000h to FD_FBFF_FFFFh**
1MB System Management	**FD_F910_0000h to FD_F91F_FFFFh**
1MB Legacy PIC IACK	**FD_F900_0000h to FD_F90F_FFFFh**
3984MB Interrupt / EOI	**FD_0000_0000h to FD_F8FF_FFFFh**
1012GB DRAM / Memory Mapped IO	**00_0000_0000h to FC_FFFF_FFFFh**

Data Transfer Type and Transaction Flow

The HT architecture supports several methods of data transfer between devices, including:

- Programmed I/O
- DMA
- Peer-to-peer

Each method is illustrated and described below. An overview of packet types and transactions is discussed later in this chapter.

Programmed I/O Transfers

Transfers that originate as a result of executing code on the host CPU are called programmed I/O transfers. For example, a device driver for a given HT device might execute a read transaction to check its device status. Transactions initiated by the CPU are forwarded to the HT bus via the Host HT Bridge as illustrated in Figure 2-3. The example transaction is a write that is posted by the host bridge; thus no response is returned to from the target device. Non-posted operations of course require a response.

Figure 2-3: Transaction Flow During Programmed I/O Operation

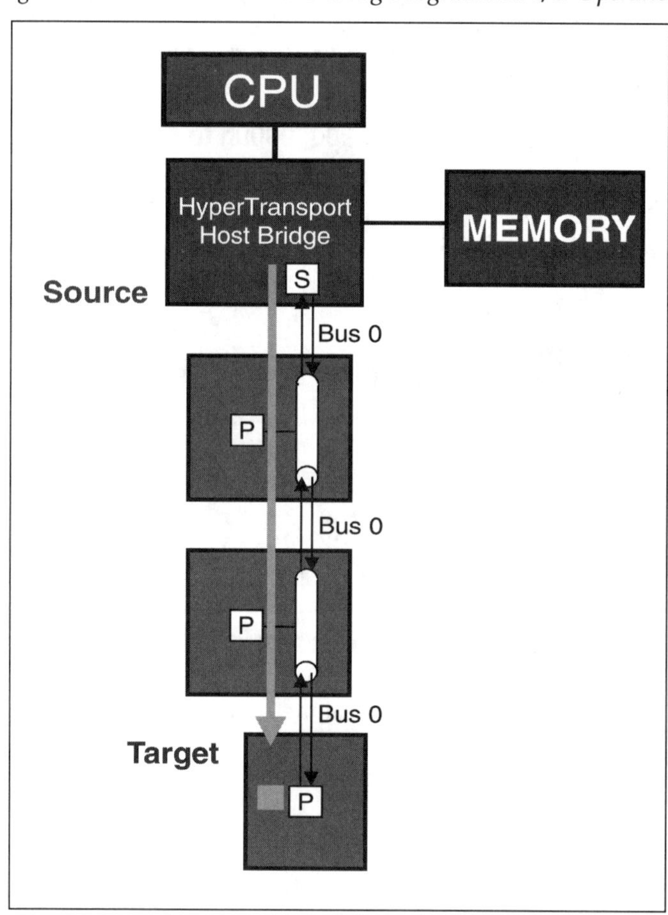

DMA Transfers

HT devices may wish to perform a direct memory access (DMA) by simply initiating a read or write transfer. Figure 2-4 illustrates a master performing a DMA read operation from main DRAM. In this example, a response is required to return data back to the source HT device.

Figure 2-4: Transaction Flow During DMA Operation

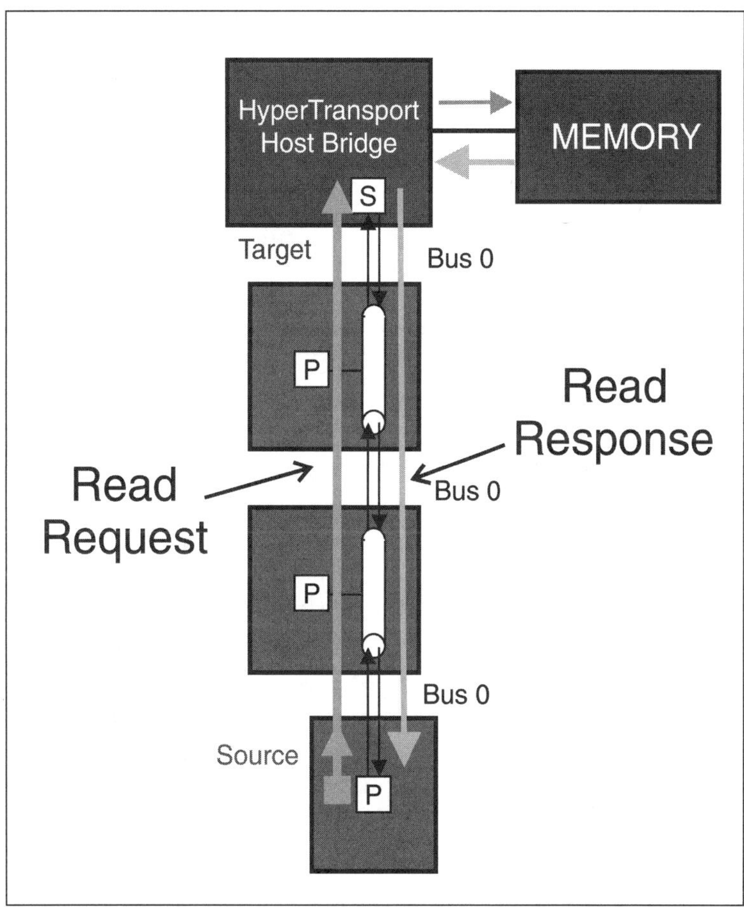

Peer-to-Peer Transfers

Figure 2-5 on page 26 illustrates the initial request to read data from the target device residing on the same bus. Note that even though the target device resides on the same bus, it ignores the request moving in the upstream direction (toward the host processor). When the request reaches the upstream bridge, it is turned around and sent in the downstream direction toward the target device. This time the target device detects the request and returns the requested data in a response packet.

The peer-to-peer transfer does not occur directly between the requesting and responding devices as might be expected. Rather, the upstream bridge is involved in handling both the request and response to ensure that the transaction ordering requirements are managed correctly. This requirement exist to support PCI-compliant ordering. True, or direct, peer-to-peer transfers are supported when PCI ordering is not required as defined by the networking extensions. See Chapter 19, entitled "Networking Extensions Overview," on page 443 for details.

Figure 2-5: Peer-to-Peer Transaction Flow

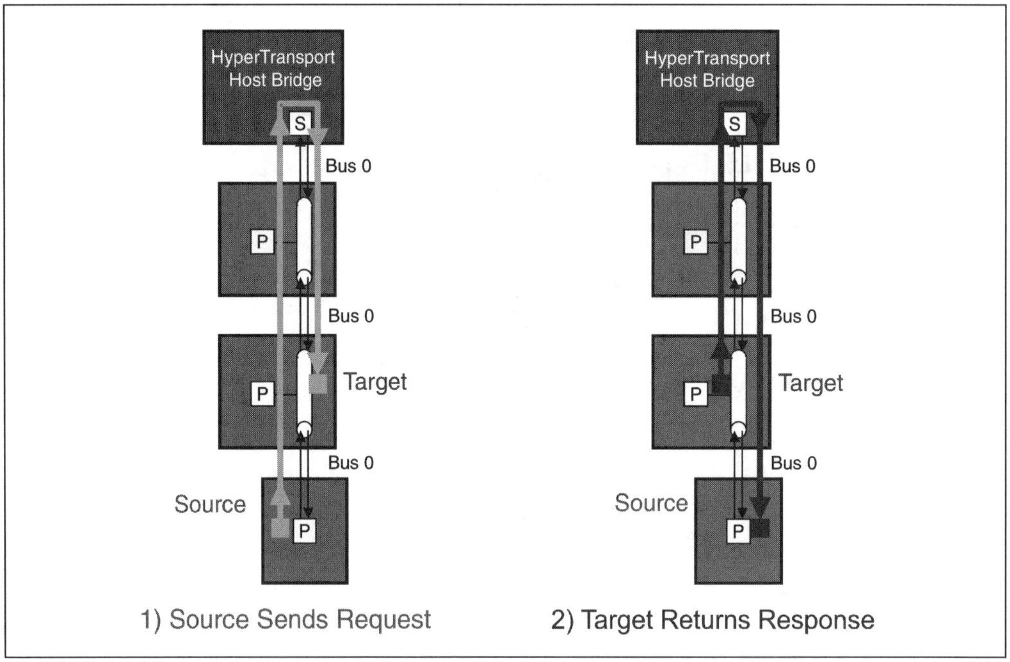

HT Signals

The HT signals can be grouped into two broad categories (See Figure 2-6 on page 27):

- The link signal group — used to transfer packets in both directions (High-Speed Signals).
- The support signal group — that provides required resources such as power and reset, as well as other signals to support optional features such power management (Low-Speed Signals).

Figure 2-6: Primary HT Signal Groups

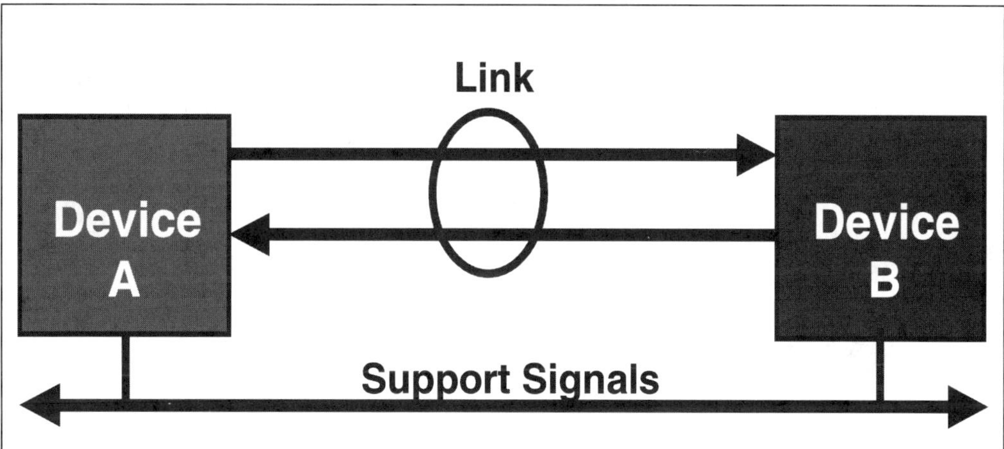

Link Packet Transfer Signals

The high-speed signals used for packet transfer in both directions across an HT link include:

- CAD (command, address, data). Multiplexed signals that carry control packets (request, response, information) and data packets. Note that the width of the CAD bus is scalable from 2-bits to 32-bits. (See "Scalable Performance" on page 30.)
- CLK (clock). Source-synchronous clock for CAD and CTL signals. A separate clock signal is required for each byte lane supported by the link. Thus, the number of CLK signals required is directly proportional to the number of bytes that can be transferred across the link at one time.
- CTL (control). Indicates whether a control packet or data packet is currently being delivered via the CAD signals.

Figure 2-7 illustrates these signals and defines various widths of data bus supported. The variables "n" and "m" define the scaling option implemented. Refer to "Link Initialization" on page 282 for details regarding HT data width and clock speed scaling.

Figure 2-7: Link Signals Used to Transfer Packets

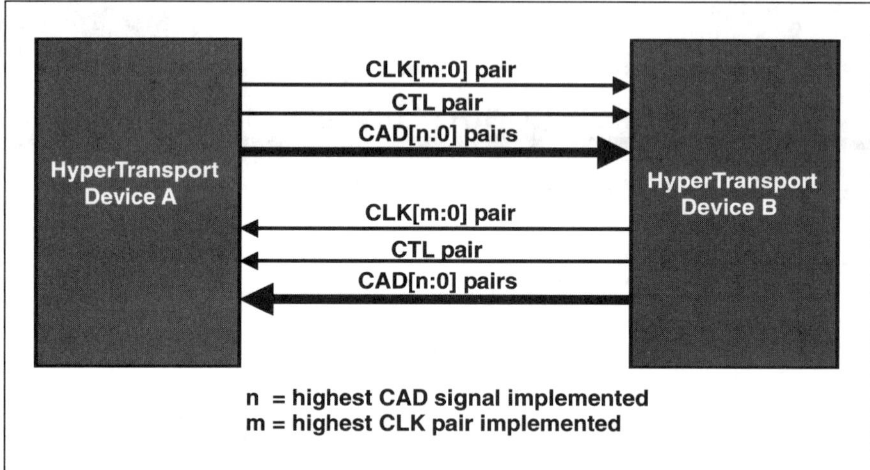

Link Support Signals

The low-speed link support signals consist of power- and initialization-related signals and power management signals. Power- and initialization-related signals include:

- V_{LDT} & Ground — The 1.2 volt supply that powers HT drivers and receivers
- PWROK — Indicates to devices residing in the HT fabric that power and clock are stable.
- RESET# — Used to reset and initialize the HT interface within devices and perhaps their internal logic (device specific).
- Power management signals
  o LDTREQ# — Requests re-enabling links for normal operation.
  o LDTSTOP# — Enables and disables links during system state transitions.

Figure 2-8: Link Support Signals

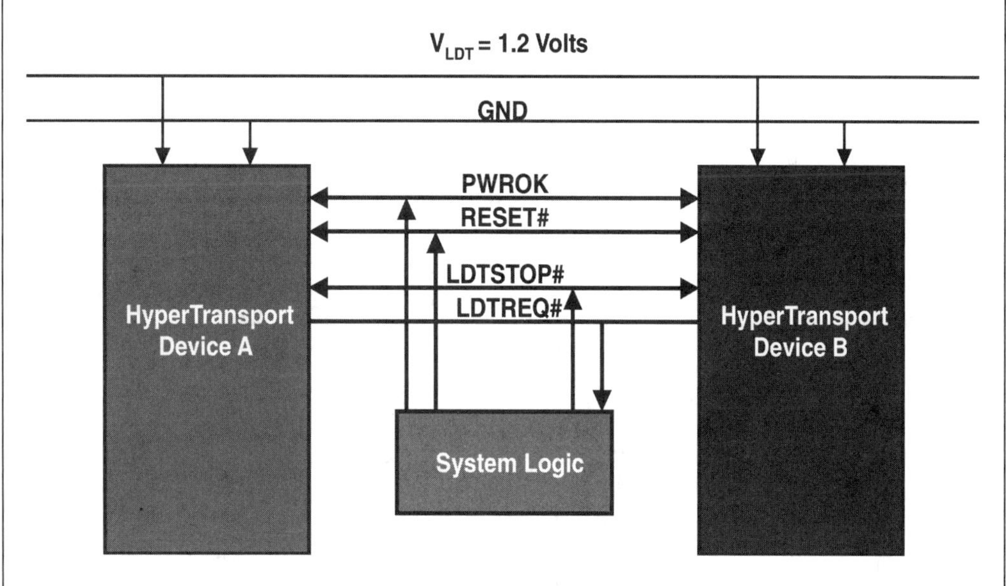

Scalable Performance

The width of the transmit and receive portion of the link (CAD signals) may be different. For example, devices that typically send most of their data to main memory (upstream) and receive limited data from the host can implement a wide path in the high performance direction and narrow path for traffic in the lesser used direction, thereby reducing cost.

The HyperTransport link combines the advantages of both serial and parallel bus architectures. HT provides options for the number of data paths implemented and for the clock rate at which data is transferred (see "Scalable Link Width and Speeds" on page 30); thus, providing scalable link performance ranging from 0.2GB/s to 12.8GB/s. This scalability is helpful to system designers. For example:

- An implementation that needs all the available bandwidth (e.g. system chipsets), can use wide links (up to 32 bits), running at the highest clock frequencies (up to 800MHz now and 1GHz in the future).
- Implementations that don't require high bandwidth but do require low power may use narrow links (as few as 2 bits) and lower frequencies (down to 200MHz).

Figure 2-9: Scalable Link Width and Speeds

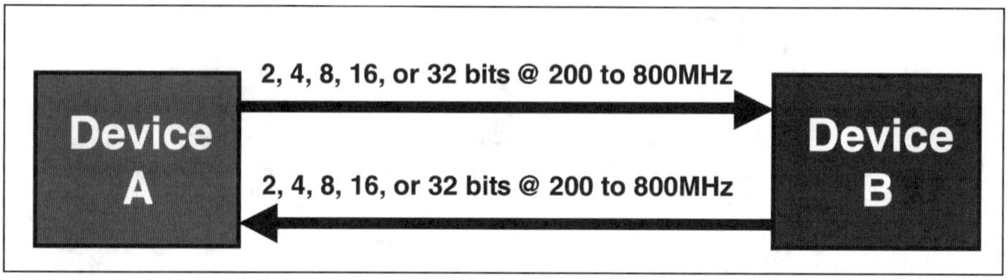

HyperTransport lends itself to scaling well because:

- The high frequency bus translates to fewer pins required to transfer a specific amount of data. The same protocol is used regardless of link width.
- Differential signaling results in a very low current path to ground, thereby reducing the number of power and ground pins required for devices.
- Each additional byte lane added has its own source synchronous clock.
- HT's implementation of ACPI compliant power management and interrupt signaling is message based, reducing pin count. Note that only two additional signals, LDTSTOP# and LDTREQ#, are required for managing power.

Data Widths

HT provides scalable data paths with link widths of 2-, 4-, 8-, 16-, or 32-bits wide in each direction, as pictured in Figure 2-10 on page 31. The link width used immediately following reset is restricted to no wider than 8 bits. Later during software initialization, configuration software determines the maximum link width that can be supported in each direction and configures both devices to use the maximum width supported for each direction. See "Tuning the Link Width (Firmware Initialization)" on page 295 for details.

Figure 2-10: Link Widths Supported

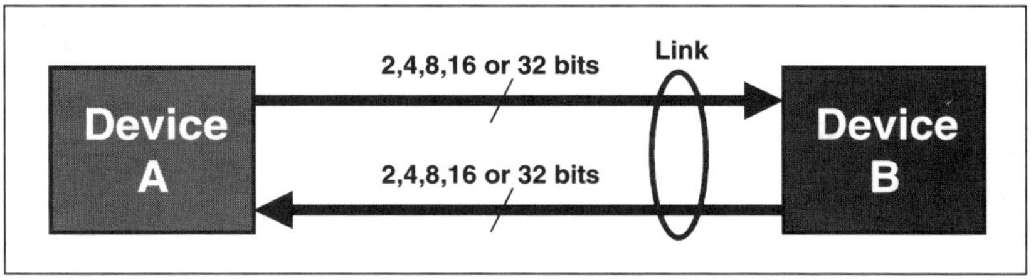

2,4,8,16 or 32 bits

Link

Device A

2,4,8,16 or 32 bits

Device B

Table 2-1: Signals Used for Different Link Widths

Link Widths	2	4	8	16	32
Pin Names	**Number of Pins**				
Data Pins (CAD)	8	16	32	64	128
Clock Pins (CLK)	4	4	4	8	16
Control Pins (CTL)	4	4	4	4	4
LDTSTOP#/LDTREQ#	2	2	2	2	2
RESET#	1	1	1	1	1
PWROK	1	1	1	1	1
V_{HT}	2	2	3	6	10
GND	4	6	10	19	37
Total Pins	26	36	57	105	199

As mentioned earlier, asymmetrical link widths are allowed in HyperTransport. For example, devices that typically send the bulk of their data in one direction and receive limited data in the other direction can save on cost by implementing a wide path in the high bandwidth direction and a narrow path for traffic in the low bandwidth direction. Note that the HyperTransport protocol doesn't change with link width. Packet formats remain the same, although it will obviously require more bit times to shift out a 32 bit value on a 2-bit link vs. a 32-bit link (16 bit times vs. 1 bit time).

Clock Speeds

HyperTransport clock speeds currently supported are 200MHz, 300MHz, 400MHz, 500MHz, 600MHz, and 800MHz. Note that 700MHz is not supported. Both rising edge and falling edges of the clock are used to clock signals. The clocking mechanism is referred to as double data rate (DDR) clocking. DDR clocking translates to an effective clock frequency that is double the actual clock frequency. In addition, because each link is dual simplex, the actual link bandwidth is quadrupled when compared to the clock rate.

Table 2-2 shows the bandwidth numbers based on symmetrical links for selected combinations of clock frequency and link width. For example, consider the bandwidth in GigaBytes/second for a 32-bit link operating at 800MHz:

- 800MHz clock with DDR = effective clock of 1,600MHz/s (1.6GTransfers/s)
- 1.6GTransfers/s x 4 bytes = 6.4GB/s
- 6.4GB/s in both directions = 12.8GB/s.

Table 2-2: Maximum Bandwidth Based on Various Speeds and Link Widths

Link Width (bits)	Bandwidth per Link (in Gbytes/sec)		
	800MHz	400MHz	200MHz
2	0.8	0.4	0.2
4	1.6	0.8	0.4
8	3.2	1.6	0.8
16	6.4	3.2	1.6
32	12.8	6.4	3.2

Extending the Topology

Based on point-to-point links, a HyperTransport chain may be extended into a fabric, using single and multi-link devices together. Devices defined for HT include:

- Single HT link "cave" devices used to implement a peripheral function
- Single or multi-link Bridges; (HT-to-HT, or HT to one or more other protocols such as PCI, PCI-X, AGP or Infiniband)
- Multi-link Tunnel devices used to implement a function and extend a link to a neighboring device downstream, thus creating a chain

These devices are the basic building blocks for the HT fabric and are illustrated in Figure 2-11.

Figure 2-11: Basic HT Device Types

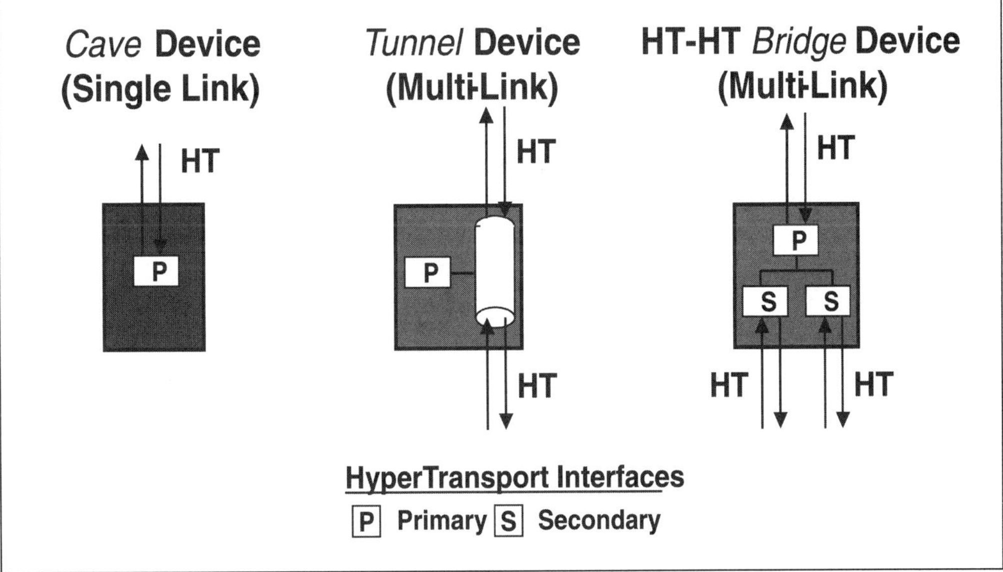

Figure 2-12 exemplifies a HyperTransport topology that includes all three device types previously discussed. The basic difference between an HT-to-HT bridge and a tunnel device is:

- A bridge creates a new link (with its own bus number), and acts as a HyperTransport host bridge for each secondary link.
- A tunnel buffers signals, passes packets, but merely extends an existing link to another device. It is not a host, and the bus number is the same on both sides of the tunnel. It also implements an internal function of its own, which a bridge typically would not.

Figure 2-12: HyperTransport Topology Supporting All Three Major Device Types

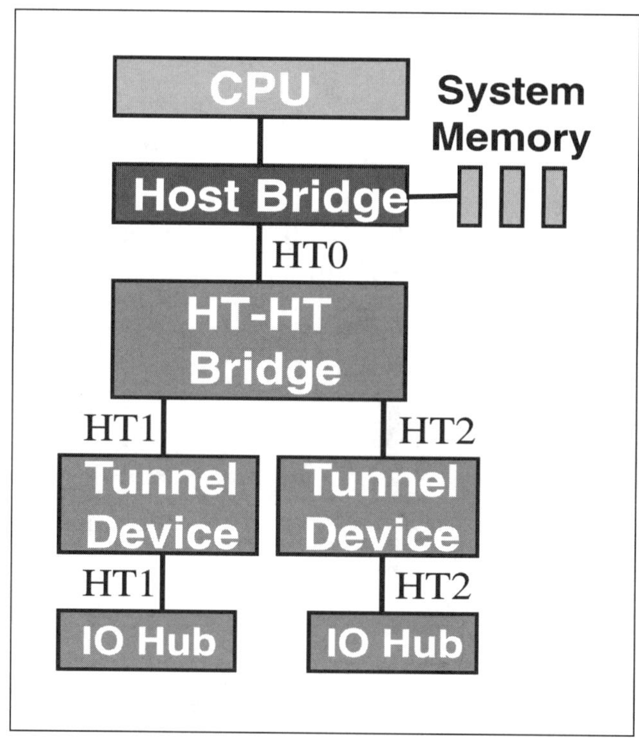

Packetized Transfers

Transactions are constructed out of combinations of various packet types and carry the commands, address, and data associated with each transaction. Packets are organized in multiples of 4-byte blocks. If the link uses data paths that are narrower than 32 bits, successive bit-times are added to complete the packet transfer on an aligned 4-byte boundary. The primary packet types include:

- Control Packets — used to manage various HT features, initiate transactions, and respond to transactions
- Data packets — that carry the payload associated with a control packet (maximum payload is 64 bytes).

As illustrated in Figure 2-13, the control (CTL) signal differentiates control packets from data packets on the bus.

Figure 2-13: Distinguishing Control from Data Packets

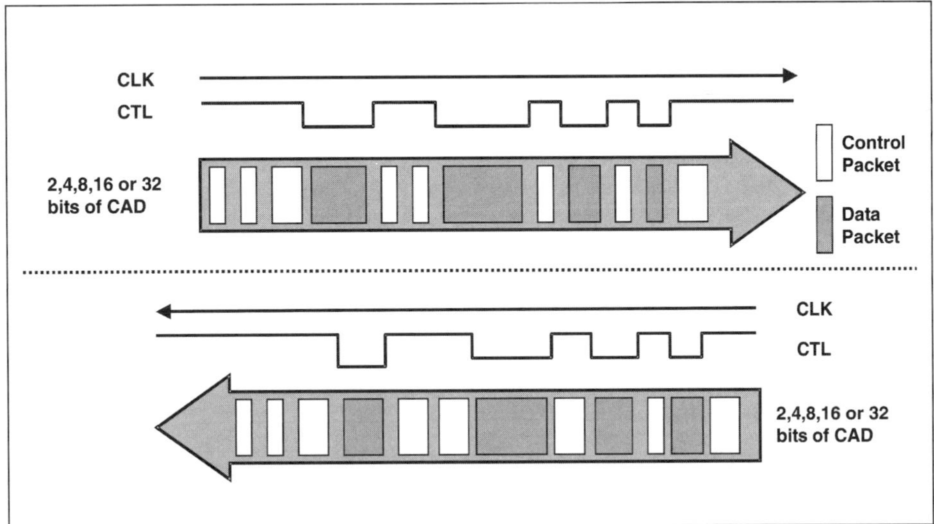

For every group of 8 bits (or less) within the CAD path, there is a CLK signal. These groups of signals are transmitted source synchronously with the associated CLK signal. Source synchronous clocking requires that CLK and its associated group of CAD signals must all be routed with equal length traces in order to minimize skew between the signals.

Control Packets

Control packets manage various HT features, initiate transactions, and respond to transactions as listed below:

- Information packets
- Request packets
- Response packets

Information packet (4 bytes)

Information packets are exchanged between the two devices on a link. They are used by the two devices to synchronize the link, convey a serious error condition using the Sync Flood mechanism, and to update flow control buffer availability dynamically (using tags in NOP packets). The information packets are:

- NOP
- Sync/Error

Request packet (4 or 8 bytes)

Request packets initiate HT transactions and special functions. The request packets include:

- Sized Write (Posted)
- Broadcast Message
- Sized Write (non-posted)
- Sized Read
- Flush
- Fence
- Atomic Read-Modify-Write

Response packet (4 bytes)

Response packets are used in HT split-transactions to reply to a previous request. The response may be a *Read Response* with data, or simply a *Target Done Response* confirming a non-posted write has reached its destination.

Data Packets

Some Request/Response command packets have data associated with them. Data packet structure varies with the command which caused it:

- Sized Dword Read Response or Write data packets are 1-16 dwords (4-64 bytes)
- Sized Byte Read Response data packets are 1 dword (any byte combination valid)
- Sized Byte Write data packets are 0-32 bytes (any byte combination valid)
- Read-Modify-Write

HyperTransport Protocol Concepts

Channels and Streams

In HyperTransport, as in other protocols, ordering rules are needed for read, posted/non-posted write transactions, and responses returning from earlier requests. In a point-point fabric, all of these occur over the same link. In addition, transactions from different devices are also merging over the same links. HyperTransport implements Virtual Channels and I/O Streams to differentiate a device's posted requests, non-posted requests, and responses from each other and from those originating from different sources.

Virtual Channels

HyperTransport defines a set of three required virtual channels that dictate transaction management and ordering:

- Posted Requests — Posted write transactions belong to this channel.
- Non-Posted Requests — Reads, non-posted writes, and flushes belong to this channel.
- Responses — Read responses and target done packets belong to this channel.

An additional set of Posted, Non-Posted and Response virtual channels is required for *isochronous* transactions, if supported. This dedicated set of virtual channels assist in guaranteeing the bandwidth required of isochronous transactions.

When packets are sent over a link, they are sent in one of the virtual channels. Attribute bits in the packets tag them as to which channel they should travel. Each device is responsible for maintaining queues and buffers for managing the virtual channels and enforcing ordering rules.

Each device implements separate command/data buffers for each of the 3 required virtual channels as pictured in Figure 2-14 on page 38. Doing so ensures that transactions moving in one virtual channel do not block transactions moving in another virtual channel. There are I/O ordering rules covering interactions between the three virtual channels of the same I/O stream. Transactions in different I/O streams have no ordering rules (with exception of ordering rules associated with Fence requests). Enforcing ordering rules between transactions in the same I/O stream prevents deadlocks from occurring and guarantees data is transferred correctly. Based on ordering requirements, nodes may not:

- Make accepting a request dependent on the ability of that node to issue an outgoing request.
- Make accepting a request dependent on the receipt of a response due to a request previously issued by that node.
- Make issuing a response dependent on the ability to issue a request.
- Make issuing a response dependent upon receipt of a response due to a previous request.

Figure 2-14: HT Virtual Channels

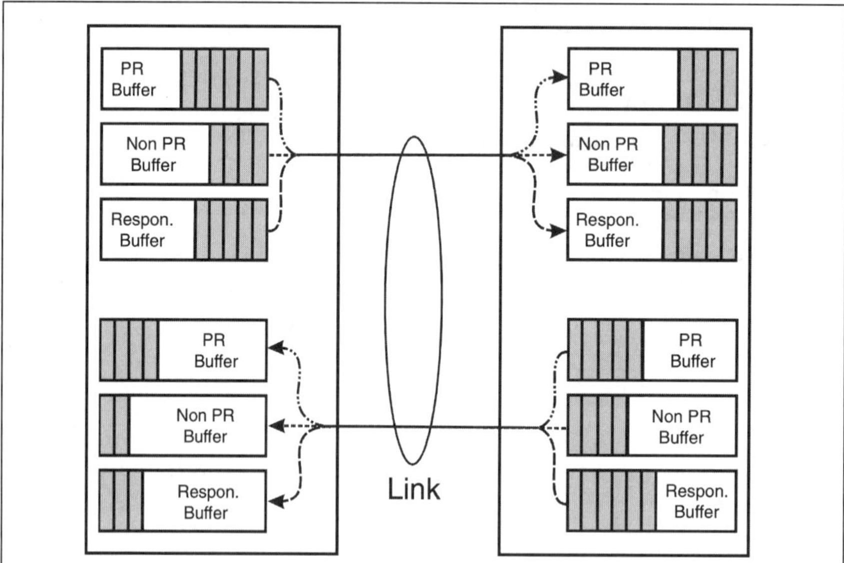

I/O Streams

In addition to virtual channels, HyperTransport also defines I/O streams. An I/O stream consists of the requests, responses, and data associated with a particular UnitID and HyperTransport link. Ordering rules require that I/O streams be treated independently from each other. When a request/response packet is sent, it is tagged with sender attributes (UnitID, Source Tag, and Sequence ID) that are used by other devices to identify the transaction stream in use, and the required ordering within it. Entries within the virtual channel buffers include the transaction stream identifiers (attributes).

Used properly, the independent I/O streams create the effect of separate connections between devices and the host bridge above them — much as a shared bus connection appears.

Transactions (Requests, Responses, and Data)

Transfers initiated by HT devices require one or more transactions to complete. These devices may need to perform a variety of operations that include:

- sending or forwarding data (write)
- requesting that a target return data to it (read)
- performing an *atomic* read/modify/write operation
- wanting additional control over ordering of its posted transactions (using Flush and Fence commands)
- wanting to broadcast a message to all downstream agents (done by bridges only)

The format of these transactions also vary depending on the type of operation (request) specified as listed below:

- Requests that behave like reads and that require a read response and data (i.e., Sized Read, Atomic RMW)
- Requests that behave like writes, and require a target done response to confirm completion (i.e. Non-posted Sized Writes)
- Posted Requests that behave like writes but don't require any target response or data. (i.e. Posted Sized Writes, Broadcast Message, or Fence)

Transaction Requests

Every transaction begins with the transmission of a Request Packet. Note that the actual format of a request packet varies depending on the particular request, but in general each request contains the following information:

- Target address within HyperTransport memory space
- The request type (command)
- Sender's transaction stream ID (UnitID, SeqID)
- The amount of data to be transferred (if any)
- Other attributes: virtual channel to use, etc.

HT defines seven basic request types. The characteristics of each request type is discussed in the following sections.

Transaction Responses

Responses are generated by the target device in cases where data is to be returned from the target device, or when confirmation of transaction completion is required. Specifically, in HyperTransport, a response follows all non-posted requests. A target responds to:

- Return data to satisfy an earlier read or Atomic Read-Modify Write (RMW) request
- Confirm the arrival of non-posted write data
- Confirm the completion of a Flush operation
- Report errors

The information in a response varies both with the Request that causes it, and with the direction the response is traveling in the HyperTransport fabric. However, content of an HT response generally includes:

- Response type (command)
- Response direction (upstream or downstream)
- Transaction stream (UnitID, Source Tag)
- Misc. info: virtual channel to use, error, etc.

Transaction Types

As discussed earlier, HT defines seven basic transaction types. This section introduces the characteristics of each type and defines any sub-types that exist.

Sized Read Transactions

Sized Read transactions permit remote access to a device memory or memory-mapped I/O (MMIO) address space. The operation may be initiated on HT from the host bridge (PIO operation), or an HT device may wish to read data from memory (DMA operation) or from another HT device (peer-to-peer operation). Two types of Sized Read transactions define the different quantities of data to be read.

- Sized (Byte) Read — this request defines an aligned 4 byte block of address space from which 0 to 4 bytes can be read. Any single byte location or any group of bytes within the 4 byte block can be accessed. The typical use of this transaction is for reading MMIO registers.
- Sized (DW) Read — this request identifies an aligned 64 byte block of address space from which 4-64 bytes can be read. Any continuous group of aligned 4 byte groups (DWs) can be accessed.

The protocol associated with Sized Read transactions is illustrated in Figure 2-15 on page 41. These transactions begin with the delivery of a Sized Read Request packet and completes when the target device returns a corresponding response packet followed by data.

Figure 2-15: Example Protocol — Receiving Data from Target

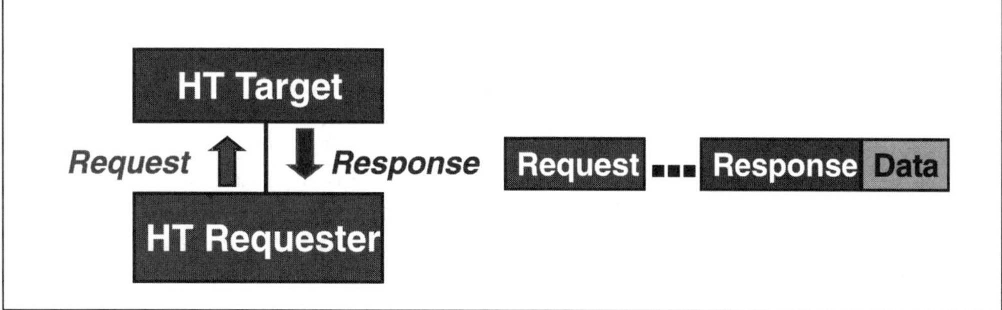

The basic rules for maintaining high performance of HT reads include:

- For reads, the requester won't issue the request until it has buffers available to receive all requested data without wait states.
- The requester won't issue the request until it knows the target has room in its transaction queue to accept it (Flow Control)
- Upon receiving the read request, the target won't issue the read response until it has all requested data and status available to send. Once it starts the response, there will be no wait states until the read response packet and all data (up to 16 dwords) have been sent.
- Upon receiving the response, the requester will check the error bits to make certain the data is valid.
- The target and any bridges in the path de-allocate buffers and queue entries as soon as the response has been sent.

Sized Write Transactions

Sized Write transactions permit the host bridge (PIO operation) to send data to a HyperTransport device, or permits a HyperTransport device to send data to memory (DMA operation) or to another device (Peer-to-peer operation). Two types of Sized Write requests permit different sizes of memory or MMIO space to be accessed.

- Sized (Byte) Write — this request identifies an aligned block of 32 bytes of address space into which data is to be written. The amount of data to be written can be from 0 to 32 bytes. Note that the maximum transfer size of 32 bytes only occurs if the start address is 32 byte aligned. If the start address is not on a 32-byte boundary, the transfer will be less than 32 bytes. Furthermore, no Byte Write transaction crosses a 32 byte address boundary. Any combination of bytes (need not be contiguous) can be written from the start address to the next aligned 32 byte block of address space.
- Sized (DW) Write — this request identifies an aligned block of 64 bytes of address space into which data can be written. The start address must be aligned on 4-byte boundaries, and data to be written is always aligned in 4-byte contiguous groups (DWs). The amount of data written can be from 1 to 16 DW increments.

Non-Posted Sized Writes. The packet protocol associated with Sized Write transactions depends on whether the Sized Write is posted or not. Figure 2-16 on page 43 illustrates the case of a non-posted Sized Write. This diagram illustrates the basic HT split-transaction *request-target done* response sequence.

The basic rules for maintaining high performance in HT writes include:

- The requester won't issue the non-posted write request until it knows the target can accommodate the request and all of the data to be sent. Refer to the section on Flow Control to see how this is managed for writes.
- Upon receiving the write request and data, the target won't issue the *target done* response until it has properly delivered all data. Once it starts the response, there will be no wait states until the four bytes of the *target done* response packet have been sent.
- Upon receiving the response, the requester will check the error bits to make certain delivery is complete.
- The target and any bridges in the path de-allocate request queue entries as soon as the *target done* response has been sent.

Figure 2-16: Example Protocol — Non-Posted Sized Write

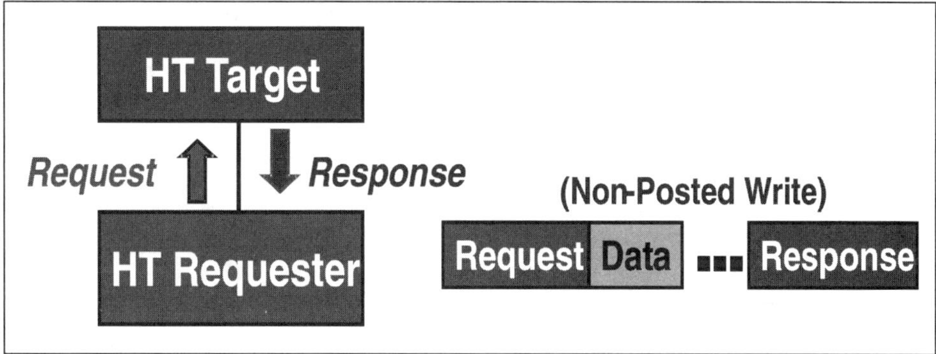

Posted Sized Writes. Figure 2-17 on page 43 depicts a posted Sized Write. In both case the transaction begins with the Sized Write request followed by the data. Non-posted operations include a response packet that is delivered back to the requester as verification that the operation has completed, whereas posted writes end once the data is sent.

Figure 2-17: Example Protocol — Posted Sized Write

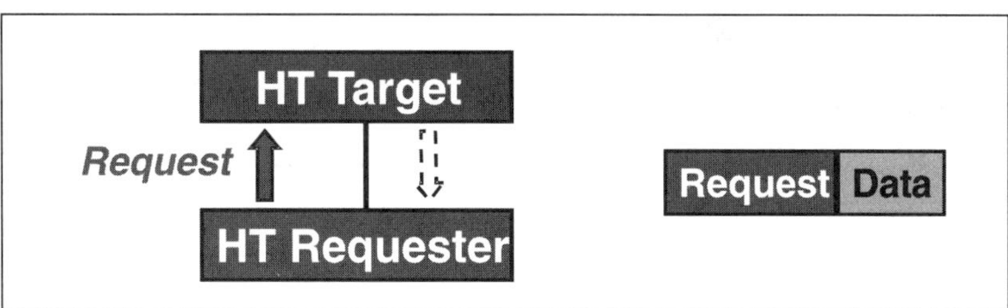

Flush

Flush is useful in cases where a device must be certain that its posted writes are "visible" in host memory before it takes subsequent action. Flush is an upstream, non-posted "dummy" read command that pushes all posted requests ahead of it to memory. Note that only previously posted writes within the same transaction stream as Flush transaction need be flushed to memory. When an intermediate bridge receives a Flush transaction, it generates one or more Sized Write transactions necessary to forward all data in its upstream posted-write buffer toward the host bridge. Ultimately, the host bridge receives the command and flushes the previously-posted writes to memory. Receipt of the read response from the host bridge is confirmation that the flush operation has completed.

The protocol used when performing a Flush transaction is depicted in Figure 2-18. When the Flush request reaches the host bridge it completes previously-posted writes to memory. In this example two previously-posted writes are flushed to memory, after which the Target Done (TgtDone) response is returned to the requester.

Figure 2-18: Example Protocol — Flush Transaction

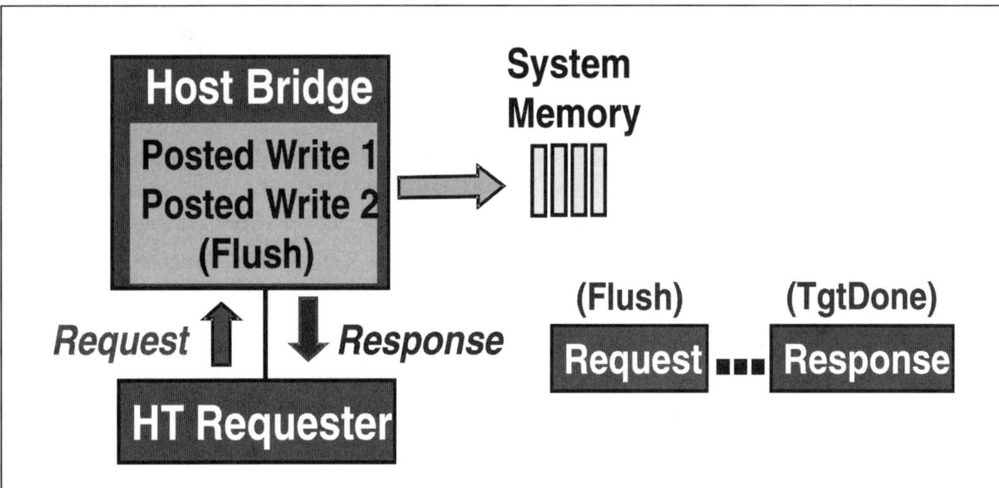

Fence

Fence is designed to provide a barrier between posted writes, which applies across all UnitIDs and therefore across all I/O streams and all virtual channels. Thus, the fence command is global because it applies to all I/O streams. The Fence command goes in the posted request virtual channel and has no response. The behavior of a Fence is as follows:

- The PassPW bit must be clear so that the Fence pushes all requests in the posted channel ahead of it.
- Packets with their PassPW bit **clear** will not pass a Fence regardless of UnitID.
- Packets with their PassPW bit **set** may pass a Fence.
- A nonposted request with PassPW clear will not pass a Fence as it is forwarded through the chain, but it may do so after it reaches a host bridge.

Fence requests are never issued as part of an ordered sequence, so their SeqID will always be 0. Fence requests with PassPW set, or with a nonzero SeqID, are legal, but may have an unpredictable effect. Fence is only issued from a device to a host bridge or from one host bridge to another. Devices are never the target of a fence so they do not need to perform the intended function. If a device at the end of the chain receives a fence, it must decode it properly to maintain proper operation of the flow control buffers. The device should then drop it.

Figure 2-19: Example Protocol — Fence Transaction

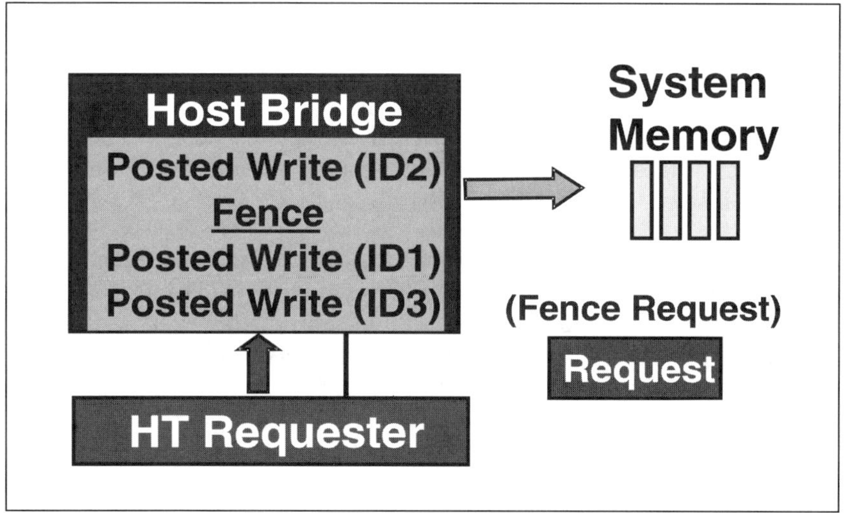

Atomic

Atomic Read-Modify-Write (ARMW) is used so that a memory location may be read, (evaluated and) modified, then conditionally written back — all without the race-condition of another device trying to do it at the same time. HT defines two types of Atomic operation:

- Fetch & Add
- Compare & Swap

The protocol associated with an Atomic Transaction is shown in Figure 2-20 on page 46. The request is followed by a data packet that contains the argument of the atomic operation. The target device performs the request operation and returns the original data read from the target location.

Figure 2-20: Example Protocol — Atomic Operation

Broadcast

Broadcast Message requests are sent downstream by host bridges, and are used to send messages to all devices. They are accepted and forwarded by all agents onto all links.

Figure 2-21 illustrates the operation of a Broadcast transaction. This example shows a broadcast request working its way down the HT fabric. All devices recognize the Broadcast Message request type and the reserved address, accept the message, and pass it along. Examples of Broadcast Messages include Halt, Shutdown, and the End-Of-Interrupt (EOI) message.

Figure 2-21: Example of Packet Flow During Broadcast Transaction

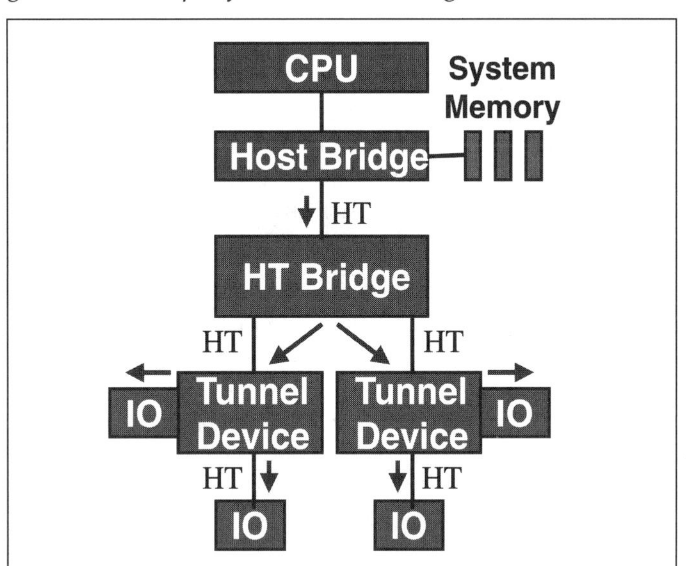

Managing the Links

This section introduces a collection of miscellaneous topics that we have labeled Link Management. They include:

- Flow Control
- Initialization and Reset
- Configuration
- Error Detection and Handling

Each of these topics is discussed in the following sections.

Flow Control

Other than information packets, all packets are transmitted from a transmitter to a buffer in the receiver. The receiver buffer will overflow if the transmitter sends too many packets. Flow control ensures that the transmitter only sends as many packets to the receiver device as buffer space allows.

Information packets are not subject to flow control. They are not transmitted to buffers within a device. Devices are always ready to accept information packets (e.g. NOP packets). Only request packets, response packets and data packets are subject to flow control.

Flow control occurs across each link between the source and the ultimate target device. HyperTransport devices must implement the six types of buffers listed above as part of its receiver state-machine. A designer implements buffers of appropriate size to meet bandwidth/performance requirements. The size of each buffer is conveyed to the transmitter during initialization, and available space is updated dynamically through NOP transmission.

HyperTransport requires transmitters on each link to accept NOP packets from receivers at reset indicating virtual channel buffering capacity, then establish a packet coupon scheme that:

- Guarantees no transmitter will send a packet that the receiver can't accept
- Eliminates the need for inefficient disconnects and retries on the link.
- Requires each receiver to dynamically inform the transmitter (via NOP packets) as buffer space becomes available.

With three virtual channels, there are three pairs of buffers in each receiver to handle request/responses and the data:

- Posted Request Buffer
- Posted Request Data Buffer
- Non-Posted Request Buffer
- Non-Posted Request Data Buffer
- Response Buffer
- Response Data Buffer

Buffer entries are sized according to what will be contained in them.

If A Device Supports the optional Isochronous Channel, it must implement additional flow control buffers to support them. An "ISOC" bit is set in request and response packets indicating routing. If the "ISOC" bit is set, all link devices that support it will use these channels; others will pass Isochronous pacekts along in regular channels.

ISOC traffic is exempt from the fairness algorithm implemented for non-ISOC traffic, resulting in higher performance. Isochronous transactions are serviced by devices before non-isochronous traffic. Theoretically, isochronous traffic may result in starving non-isochronous traffic. Applications must guarantee that iso-chronous bandwidth does not exceed overall available bandwidth.

Initialization and Reset

HyperTransport defines two classes of reset events:

Cold Reset. This occurs on boot and starts when the PWROK and RESET# signals are both seen low. When this happens:

- All devices and links return to default inactive state
- Previously assigned UnitID numbers are "forgotten" and all return to default UnitID of 0.
- All Configuration Space registers return to default state
- All error bits and dynamic status bits are cleared

Warm Reset. This occurs when PWROK is high and RESET is seen low.

- All devices and links return to default inactive state
- Previously assigned UnitID numbers are "forgotten", and all return to default UnitID of 0.
- All Configuration Space registers defined as *persistent* retain previous values. The same is true for Status and error bits defined as persistent.
- All other error bits and dynamic status bits are cleared

Because HyperTransport supports scalable link width and clock speed, a set of default minimum link capabilities are in effect following cold reset.

- Initial link width is conveyed when both devices sample CAD signal inputs from the other at the end of reset. Initial link clock speed is 200MHz.
- Later, Configuration of devices allows optimizing CAD width and clock speeds for each link.
- Refer to the core topic section on Reset and Initialization for details on this process.

It is a motherboard's responsibility to tie upper CAD inputs to 0 if a device receiver is attached to a narrower transmitter CAD interface.

Configuration

At boot time, PCI configuration is used to set-up HyperTransport devices:

- Read in configuration information about device requirements and capabilities.
- Program the device with address range, error handling policy, etc.

Basic configuration of a device is similar to that of PCI devices; however, specific HyperTransport-specific features are handled via the *advanced capability registers*.

Error Detection and Handling

HyperTransport defines required and optional error detection and handling. Key areas of error handling:

- Cycle Redundancy Check (CRC) generation and checking on each link.
- Protocol (violation) errors
- Receive buffer overflow errors
- End-Of-Chain errors
- Chain Down errors
- Response errors

Part Two

HyperTransport Core Topics

Part Two details the required HT mechanisms. The chapters included in Part Two are:

- Chapter 3: Signal Groups
- Chapter 4: Packet Protocol
- Chapter 5: Flow Control
- Chapter 6: IO Ordering
- Chapter 7: Transaction Examples
- Chapter 8: HT Interrupts
- Chapter 9: System Management
- Chapter 10: Error Detection and Handling
- Chapter 11: Routing Packets
- Chapter 12: Reset and Initialization
- Chapter 13: Device Configuration
- Chapter 14: Electrical
- Chapter 15: Clocking

3 *Signal Groups*

The Previous Chapter

The previous chapter provided an overview of the HT architecture that defines the primary elements of HT technology and the relationship between these elements. The chapter summarized the features, capabilities, and limitation of HT and provided the background information necessary for in-depth discussions of the various HT topics in later chapters.

This Chapter

This chapter describes the function of each signal in the high and low speed HyperTransport signal groups. The CAD, CTL, and CLK high speed signals are routed point-to-point as low-voltage differential pairs between two devices (or between a device and a connector in some cases). The RESET#, PWROK, LDTREQ#, and LDTSTOP# low speed signals are single-ended low voltage CMOS and may be bused to multiple devices. In addition, each device requires power supply and ground pins. Because the CAD bus width is scalable, the actual number of CAD and CLK signal pairs varies, as does the number of power and ground pins to the device.

The Next Chapter

The next chapter describes the use of HyperTransport *control* and *data* packets to construct HyperTransport link transactions. Control packet types include Information, Request, and Response variants; data packets contain a payload of 0-64 valid bytes. The transmission, structure, and use of each packet type is presented.

Introduction

Signals on each HyperTransport link fall into two groups: high speed signals associated with the sending and receiving of control and data packets, and miscellaneous low-speed signals required for such things as reset and power management. Whereas the low speed signals are not scalable and employ conventional low voltage CMOS signalling, the high speed signal group is scalable in terms of both bus width and clock rate, and each signal is actually a low-voltage differential signal pair.

While device pin count varies with scaling, signal group functions remain the same; the only real difference in signaling over a 32-bit link vs. a 2-bit link is the number of bit times required to shift information onto the bus.

The Signal Groups

As illustrated in Figure 3-1 on page 54, the high-speed HyperTransport signals on each link consist of an outbound (transmit) set of signals and an inbound (receive) set of signals for each device; these are routed point-to-point. Having two sets of uni-directional signals allows concurrent traffic. In addition, there is one set of low speed signals that may be bused to multiple devices.

Figure 3-1: HyperTransport Signal Groups

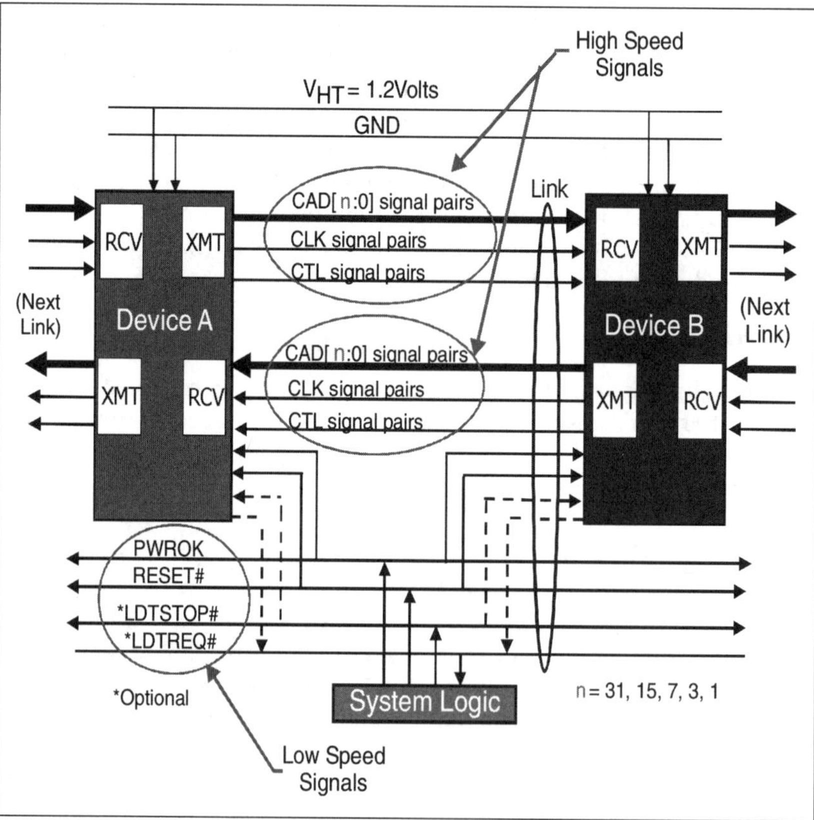

The High Speed Signals (One Set In Each Direction)

Each high-speed signal is actually a differential signal pair. CAD (Command/Address/Data) information consists of the two basic types of HyperTransport packets: control and data. When a link transmitter sends packets on the CAD bus, the receive side of the interface uses the CLK and CTL signals, also supplied by the transmitter, to latch in packet information during each bit time. CTL distinguishes control packets from data packets.

The CAD Signal Group

The CAD bus is always driven by the transmitter side of a link, and is comprised of signal pairs that carry HyperTransport *requests*, *responses*, and *data*. Each CAD bus may consist of between 2 bits (two differential signal pairs) and 32 bits (thirty-two differential signal pairs). The HyperTransport specification permits the CAD bus width to be different (asymmetrical) for the two directions. To enable the corresponding receiver to make a distinction as to the type of information currently being sent over the CAD bus, the transmitter also drives the CTL signal (see the following description).

Control Signal (CTL)

This signal pair is driven by the transmitter to qualify the information being sent concurrently over the CAD signals. If this signal is asserted (high), the transmitter is indicating that it is sending a *control* packet; if deasserted, the transmitter is sending a *data* packet. The receiver uses this information when routing incoming CAD information to appropriate request queues, data buffers, etc. There is one (and only one) CTL signal for each link direction, regardless of the width of the CAD bus.

Clock Signal(s) (CLK)

As a source-synchronous connection, each HyperTransport transmitter sends a differential clock signal along with CAD and CTL signals to the receiver at the other end of the link. There is one CLK signal pair for <u>each byte</u> of CAD width. While the timing on each clock pair is the same, replicating clocks help in routing of CAD signal pairs with respect to their clock signals. The current HyperTransport specification allows clock speeds from 200MHz (default) to 800MHz.

Scaling Hazards: Burden Is On The Transmitter

It is a requirement in HyperTransport that the transmitter side of each link must be aware of the capabilities of its corresponding receiver and avoid the double hazard of a scalable bus: running at a faster clock rate than the receiver can handle <u>or</u> using a wider data path than the receiver supports. Because the link is not a shared bus, the transmitter side of each device is concerned with the capabilities of only one target. Refer to "Link Initialization" on page 282 for a description of how HyperTransport links are initialized and configured to avoid these problems.

The Low Speed Signals

Power OK (PWROK) And Reset (RESET#)

PWROK used with RESET# indicates to HyperTransport devices whether a *Cold or Warm* Reset is in progress. Which system logic component is responsible for managing the PWROK and RESET# signals is beyond the scope of the HyperTransport specification, but timing and use of the signals are defined. The basic use of the signals includes:

- At power up, PWROK is asserted by system logic when it can be guaranteed that system power and clocks related to HyperTransport are within proper limits.
- RESET# is asserted by system logic to indicate that a reset is required. The state of PWROK when RESET# is seen asserted indicates the type of reset to be performed. PWROK and RESET# both asserted is a warm reset; PWROK deasserted and RESET# asserted indicates cold reset.
- After initial system power up, reset, and initialization, a cold or warm reset may also be generated under software control writing configuration registers in the host bridge.

The HyperTransport specification describes the actions to be taken by devices during either type of reset event. Refer to Chapter 12, entitled "Reset & Initialization," on page 275 for a thorough discussion of how PWROK and RESET are used during system power-up and initialization.

LDTSTOP#

(Note: the signal names LDTSTOP# and LDTREQ# were carried forward from the earlier name AMD assigned to HyperTransport technology — Lightning Data Transfer).

LDTSTOP# is an input to HyperTransport devices which is asserted by system logic to enable and disable link activity during power management state transitions. Support for this signal is optional for HyperTransport devices.

A transmitter which detects LDTSTOP# asserted finishes sending any control packet in progress, then commences a disconnect NOP sequence followed by disabling its output drivers (if so enabled in the transmitter's Configuration Space *Tri-State Enable Bit)*. Upon receipt of the disconnect NOP sequence, the target also turns off its input receivers (if similarly enabled in it's Configuration Space *Tri-State Enable Bit)*.

Later, when the transmitter detects LDTSTOP# deasserted, it re-enables its drivers and begins the initialization sequence. A receiver that responds to LDTSTOP# deasserted turns its input receivers on.

LDTREQ#

LDTREQ# is a wire-or'd output from HyperTransport devices that is used to request system logic to re-enable links previously disabled using the LDTSTOP# mechanism. Upon receipt of the LDTREQ# signal from one or more HyperTransport devices, system logic (typically the South Bridge) deasserts LDTSTOP# which triggers the sequence described previously. Specifically, the LDTREQ# signal indicates that a HyperTransport transaction is required somewhere in a system that is currently in the ACPI C3 state; the system is required to transition to the C0 state. Support for this signal is optional for HyperTransport devices.

Where Are The Interrupt, Error, And Wait State Signals?

The HyperTransport specification eliminates a number of control signals that are commonly found on other buses. While devices are not prohibited from implementing signals beyond those defined in the specification, HyperTransport is a generic, simple interface and handles interrupts, errors, and data wait states in the following general way:

Interrupt Signaling

Interrupts are conveyed in HyperTransport as messages sent over the link in the posted request channel. This eliminates the need for dedicated interrupt signal traces. Depending on the architecture, it may also eliminate the need for a separate interrupt controller (e.g. IOAPIC). Refer to Chapter 8, entitled "HT Interrupts," on page 199 for a discussion of HyperTransport interrupt management.

Error Signaling

HyperTransport error handling employs CRC checking of bit traffic across each link interface. In the event of an error, there are several possible handling schemes. All of this is done without any dedicated error signals. Refer to Chapter 10, entitled "Error Detection And Handling," on page 229 for a discussion of HyperTransport error detection and handling.

Wait State Signaling

Wait states during transmission of data are a problem on any bus because they represent wasted time on the part of the devices performing the transfer and for other devices waiting to perform subsequent transfers. In HyperTransport, wait state, disconnect, and retry mechanisms used on other buses are eliminated. This is made possible through a coupon-based *flow control* scheme that guarantees that no transfer will be started by a transmitter which cannot be immediately accepted by the corresponding receiver on the other side of the link. Dynamic flow control information concerning buffer availability is embedded in NOP packets sent by each device — removing the need for dedicated transmitter and receiver *ready* signals. Refer to Chapter 5, entitled "Flow Control," on page 99 for a discussion of HyperTransport flow control.

No Arbitration Signals Either

Unlike a shared bus such as PCI or PCI-X with multiple masters, HyperTransport links are point-to-point connections with only one possible transmitter and one receiver for each direction. Because of this, arbitration signals and related arbitration latency are eliminated; as long as flow control for a link is not violated, a transmitter simply starts a transaction whenever it is required.

4 *Packet Protocol*

The Previous Chapter

The previous chapter described the function of each signal in the high and low speed HyperTransport signal groups. The CAD, CTL, and CLK high speed signals are routed point-to-point as low-voltage differential pairs between two devices (or between a device and a connector in some cases). The RESET#, PWROK, LDTREQ#, and LDTSTOP# low speed signals are single-ended low voltage CMOS and may be bused to multiple devices. In addition, each device requires power supply and ground pins. Because the CAD bus width is scalable, the actual number of CAD and CLK signal pairs varies, as does the number of power and ground pins to the device.

This Chapter

This chapter describes the use of HyperTransport *control* and *data* packets to construct HyperTransport link transactions. Control packet types include Information, Request, and Response variants; data packets contain a payload of 0-64 valid bytes. The transmission, structure, and use of each packet type is presented.

The Next Chapter

The next chapter describes HyperTransport *flow control*, used to throttle the movement of packets across each link interface. On a high-performance connection such as HyperTransport, efficient management of transaction flow is nearly as important as the raw bandwidth made possible by clock speed and data bus width. Topics covered here include background information on bus flow control and the initialization and use of the HyperTransport virtual channel flow control buffer mechanism defined for each transmitter-receiver pair.

The Packet-Based Protocol

HyperTransport employs a packet-based protocol in which all information — address, commands, and data — travel in packets which are multiples of four bytes each. Packets are used in link management (e.g. flow control and error reporting) and as building blocks in constructing more complex transactions such as read and write data transfers.

It should be noted that, while packet descriptions in this chapter are in terms of bytes, the link's bidirectional interface width (2, 4, 8, 16, or 32 bits) ultimately determines the amount of packet information sent during each *bit time* on HyperTransport links. There are two bit times per clock period.

Before looking at packet function and use, the following sections describe the mechanics of packet delivery over 2,4,8,16, and 32 bit scalable link interfaces.

8 Bit Interfaces

For 8-bit interfaces, one byte of packet information may be sent in each bit time. For example, a 4-byte request packet would be sent by the transmitter during four adjacent bit times, least significant byte first as shown in Figure 4-1 on page 61. Total time to complete a four-byte packet is two clock periods.

Figure 4-1: Four Byte Packet On An 8-Bit Interface

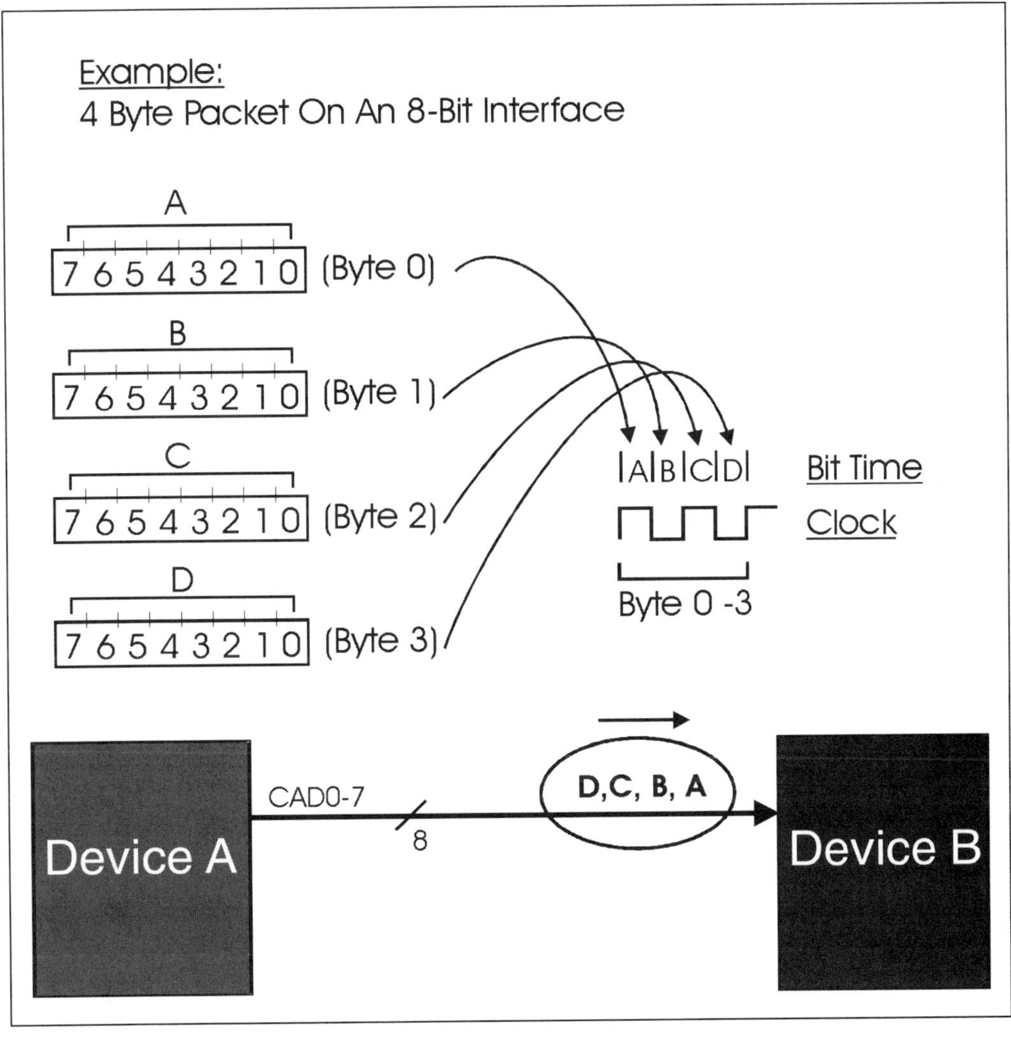

Interfaces Narrower Than 8 Bits

For link interfaces which are narrower than 8 bits, the first byte of packet information is shifted out over multiple bit times, least significant bits first. Referring to Figure 4-2 on page 62, a 2-bit interface would require four bit times to transmit each byte of information. After the first byte is sent, subsequent bytes in the packet are shifted out in the same manner. Total time to complete four byte packet: eight clock periods.

Figure 4-2: Four Byte Packet On A 2-Bit Interface

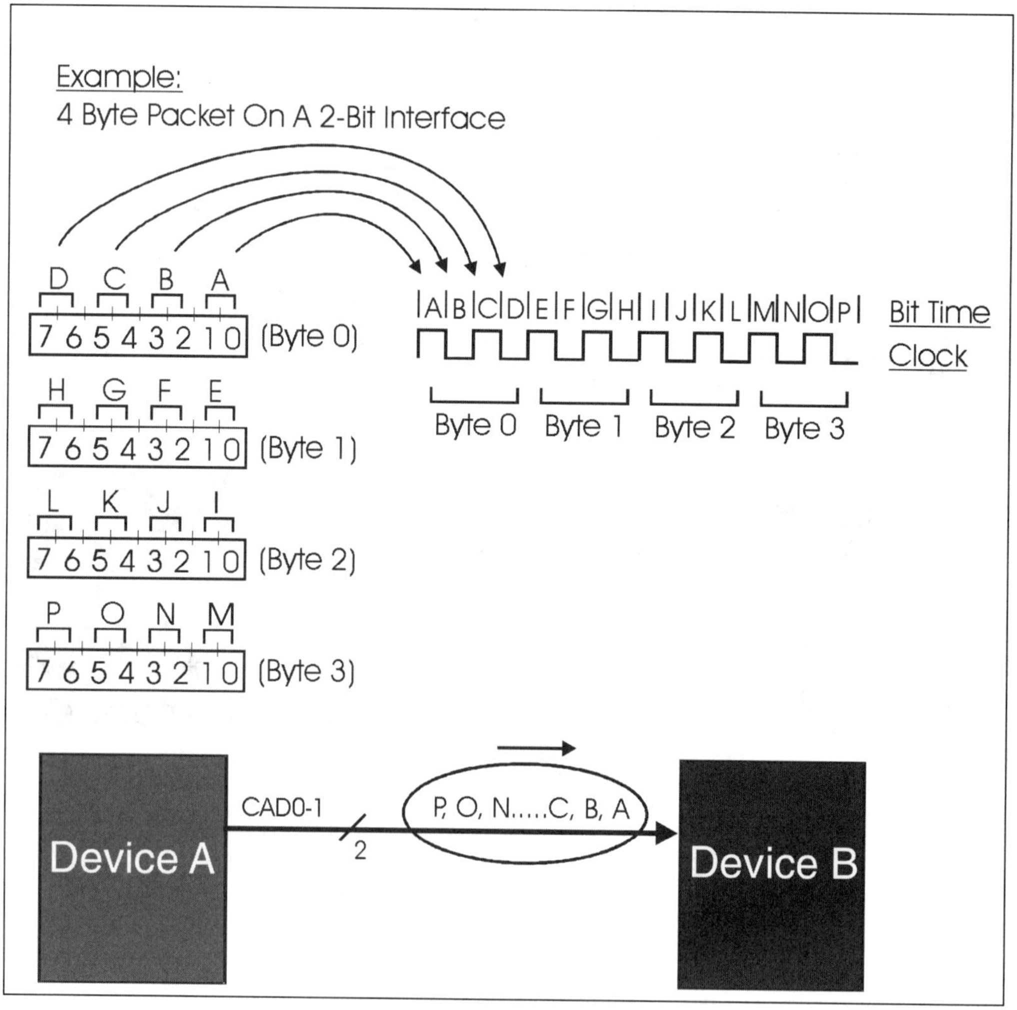

Interfaces Wider Than 8 Bits

For 16 or 32 bit interfaces, packet delivery is accelerated by sending multiple bytes of packet information in parallel with each other.

16 Bit Interfaces

On 16-bit interfaces, two bytes of information may be sent in each bit time. Referring to Figure 4-3 on page 63, note that even numbered bytes travel on the lower portion of the 16 bit interface, odd numbered bytes on the upper portion.

Figure 4-3: Four Byte Packet On A 16-Bit Interface

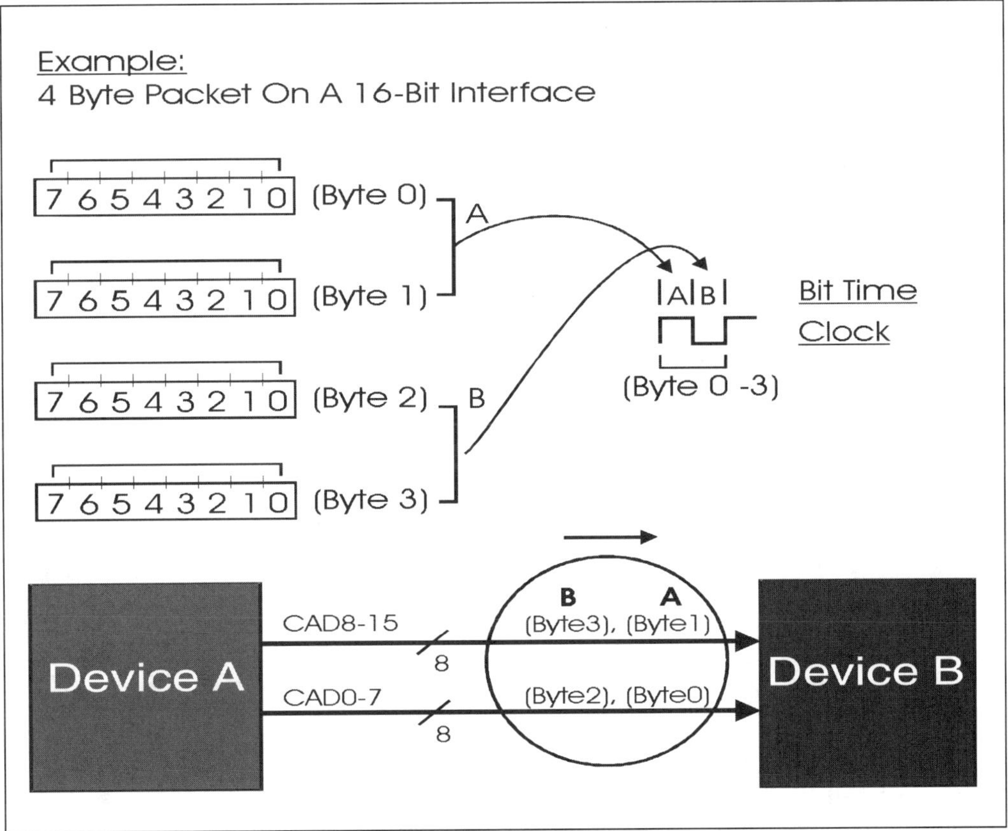

32 Bit Interfaces

Similarly, four bytes of information may be sent in each bit time on a 32-bit interface. This is shown in Figure 4-4 on page 64. Note that Byte 0 travels on the low portion of the interface (CAD0-7), Byte 1 on the second byte lane (CAD8-15), etc.

Figure 4-4: Four Byte Packet On A 32-Bit Interface

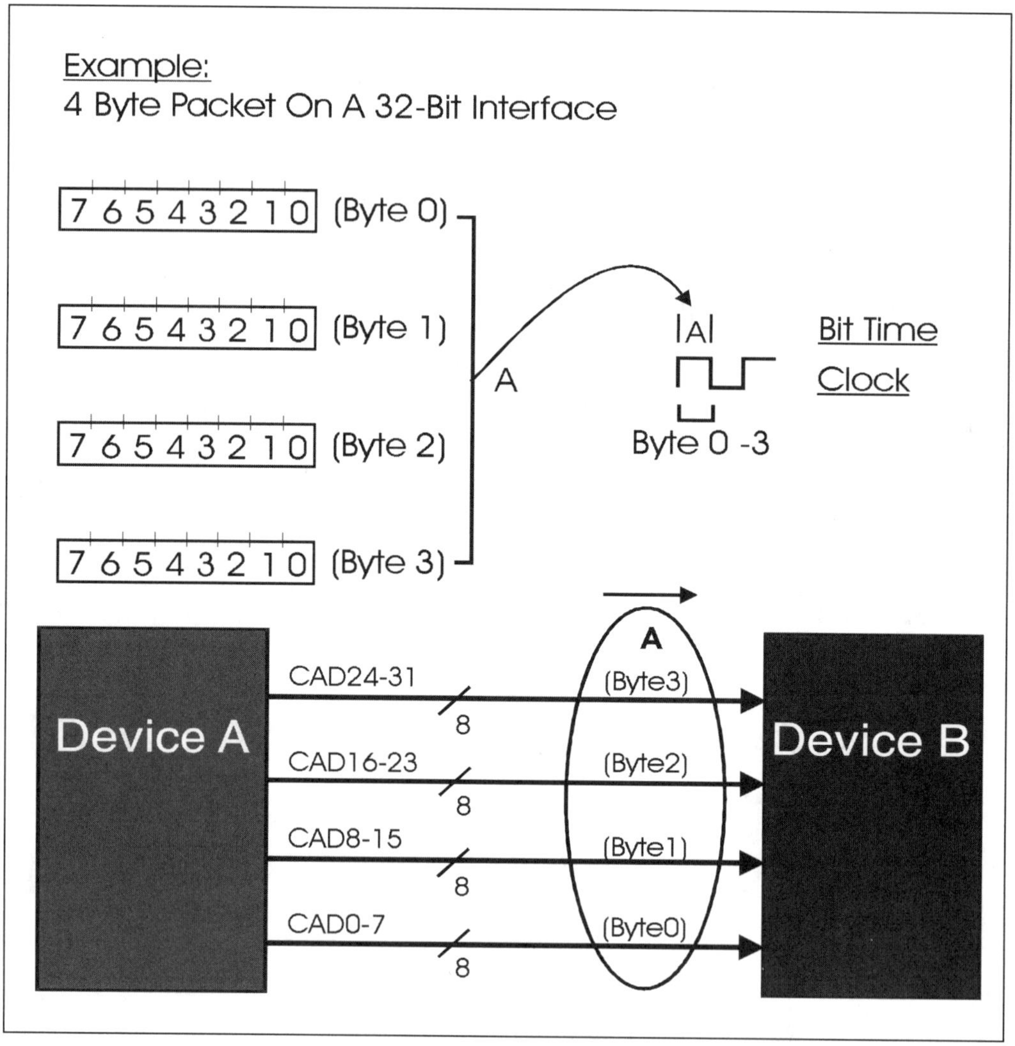

A reminder: Because all HyperTransport packets are multiples of 4 bytes, bits of packet information always divide evenly into the available bus width. There never is a need to "pad" unused bit lanes.

The Two Packet Types: Control And Data

Packets moving across links fall into two groups: control packets and data packets. Control packet types are further divided into three additional classes: Information, Request, and Response.

Control Packet Purpose

The three classes of control packets serve the following purposes on a HyperTransport link:

Information packets

Information packets are always 4 bytes each. They are used for nearest neighbor communication between the transmitter-receiver pairs on each link; communication between these nodes is necessary for dynamic flow control updates and other miscellaneous functions. Information packets are not buffered internally or subject to flow control; when sent by a transmitter they <u>must</u> be accepted by the receiver.

Request packets

Requests are 4 bytes in length if there is no address field, or 8 bytes if the packet does include an address field. They may be either posted or non-posted, and the basic job of a request is to define a pending data or message transaction, or to help bridges manage posted write transactions (through the use of Flush and Fence commands). These packets originate at a source device and are accepted by a target device.

Devices in the path between the source and target forward requests along, subject to HyperTransport rules for ordering.

Response packets

Responses are always 4 bytes each. They are returned by the target after it has serviced a non-posted request. Devices in the path between the response sender and the original requester forward responses along, subject to HyperTransport rules for ordering.

When associated with a non-posted write or flush request, the *target done* response packet acts as a confirmation (returned to the source device) that the operation has completed. In the event of a problem delivering non-posted write data or completing the flush, the response packet will contain an error flag and a bit indicating whether the target done response is being returned by the intended target OR by another device acting on its behalf (e.g. end-of-chain device).

For read transactions, which are always split in HyperTransport, the *read* response packet precedes the returning data and identifies the specific read request being serviced. In the event of an error when fetching the data, the read response will contain an error flag and a bit indicating whether the problem occurred at the intended target or at an end-of-chain device acting on its behalf. If there is an error, all data is driven back as FFh by either the target or the end-of-chain device.

Data Packets

While there is only one type of data packet, consisting of 1-16 Dwords, the payload of valid information within a data packet ranges from 0-64 valid bytes-- depending on the attributes of the request that caused it. The appropriate time to send a data packet also depends on the request/response associated with it:

1. For write requests, the data packet is sent immediately after the request. Because there is no routing information in a data packet, the request is used to deliver the data to the intended target.
2. For read requests, the data packet immediately follows the read response. Key fields in the response are filled in with requester transaction stream information provided in the read request (e.g. UnitID and Source Tag). The response is then used to route the read data packet back to the original requester.
3. Atomic read-modify-write requests are a hybrid. A data packet is sent with the request (as in a write transaction) and another data packet is returned following the read response (as in a read transaction).
4. Finally, some requests don't have data packets at all (e.g. Flush and Fence).

The Need To Interleave Control And Data Packets

An important feature of HyperTransport packet management is that a transmitter may interleave control packets with data packets associated with earlier requests. Interleaving control packets with data helps mitigate "stalls" in sending new control packets on the multiplexed CAD bus when large data transfers are in progress. An example of such as stall is as follows:

1. A transmitter starts a Sized Dword Write of 64 bytes (16 dwords) on a 2-bit HyperTransport link interface.
2. After the write transaction commences, the transmitter realizes it needs to send a read request or NOP information packet over the bus.
3. Without the ability to interleave control packets, the transmitter would have to send the entire data payload first (64 bytes x 4 bit times/byte = 256 bit times). This represents a worst-case latency of 128 clocks to start sending the new control packet.

To avoid such situations, HyperTransport allows a transmitter to insert new control packets into a data payload on four byte boundaries, as long as the control packets do not have any immediate data of their own. For example, read requests and NOP flow control packets are candidates for interleaving; write requests would not be candidates for interleaving because they <u>are</u> accompanied by immediate data.

The CTL Signal Indicates Packet Type

A transmitter uses the CTL signal on a HyperTransport link interface to indicate the presence of control vs. data packets it is sending concurrently on the CAD bus. When CTL is asserted (high), a control packet is in transit on the CAD bus; when CTL is deasserted (low), a data packet is being sent. During idle periods, CTL is asserted and control information NOP packets are sent.

When interleaving control packets and data packets on a link, the transmitter is required to observe the following rules as it asserts and deasserts the CTL signal:

1. CTL is always asserted and deasserted on four byte boundaries.
2. The only time CTL is deasserted is when a data packet associated with an earlier control packet (e.g., request or response packet) is being sent.
3. CTL is asserted during all bit times of a control packet; for control packets which are either 4 or 8 bytes, the packet must be sent in its entirety without

deasserting CTL (there is no interleaving within control packets). This also means that flow control must assure that transmitters never start sending a control packet if the receiver lacks sufficient buffer space to accept all bytes at full speed. Changes in flow control buffer availability are reported by means of NOP packets.

4. CTL is deasserted through all bit times of data packets.

5. Re-assertion of CTL within data packets is permitted on four byte boundaries if the transmitter decides to interleave a new control packet, providing it does not have immediate data of its own. After the control packet is sent, CTL is again deasserted and the current data packet transfer resumes.

6. Only one data packet may be in progress at a time, although it may be paused for the interleaving of control packet(s).

7. Ordering of control packets is not affected by the fact that data packets may be paused to interleave them.

8. The bit time immediately following the end of a data packet is always the start of a control packet, and CTL must be asserted.

Figure 4-5 illustrates packet transmission basics and the use of the CTL signal by a transmitter in accordance with the above rules.

Figure 4-5: CAD Bus Control/Data Packet Management Using the CTL Signal

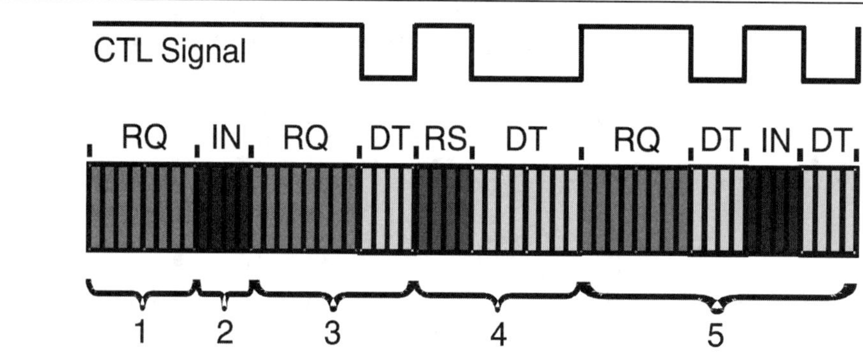

1. 8 byte *Request* Control Packet (no immediate data)
2. 4 byte *Information* Control Packet
3. 8 byte *Request* Control Packet with 4 byte Data Packet
4. 4 byte *Response* Control Packet with 8 byte Data Packet
5. 8 byte *Request* Control Packet with 8 byte Data Packet which is paused on a 4-byte boundary for the insertion of a new 4 byte *Information* Control Packet.

Packet Format: Control Packets

Table 4-1 on page 69 summarizes the HyperTransport *Information, Request,* and *Response* control packet types and the command names associated with them. Some things to note in the table:

- For each packet variant, the virtual channel (VChan) is indicated in the second column: posted, non-posted, or response. Note: information packets do not travel in any of the virtual channels and are not subject to flow control.
- The first byte in each control packet type contains a 6-bit *Command (CMD) Code.* By sending this information at the beginning of a control packet, the receiver is informed immediately of the type of packet being transferred, the number of bytes to expect, and the format of the bit fields contained within. The Command Codes are shown in the left column of Table 4-1.
- In some Command Codes, a number of bits are variables (indicated by ".xxx") which are used to select transaction options: dword vs. byte transfer count, isochronous flag, coherency requirement, etc.; refer to the *Comments* column Table 4-1 for usage of each optional bit.

Table 4-1: Control Packets And The HyperTransport Command Types

CMD Code	V Chan	Command Name	Packet Type	Comments
000000	-----	**NOP**	Info	Used by each receiver to report flow-control information to its transmitter.
111111	-----	**Sync/Error**	Info	Similar to PCI SERR#, indicates need for link reset and re-synchronization.
101xxx	Posted	**Sized Write (Posted)**	Request	**Usage of three least-significant bits:** [2] Dword/Byte (1 = dword; 0 = byte). [1] Isoc request (1 = Isoc; 0 = std.). [0] Coherency (1 = req'd; 0 = not)
001xxx	Non-Posted	**Sized Write (Non-Posted)**	Request	**Usage of three least-significant bits:** [2] Dword/Byte (1 = dword; 0 = byte). [1] Isoc request (1 = Isoc; 0 = std.). [0] Coherency (1 = req'd; 0 = not)

Table 4-1: Control Packets And The HyperTransport Command Types

CMD Code	V Chan	Command Name	Packet Type	Comments
111010	Posted	**Broadcast Message**	Request	Broadcast messages originate at host bridge, and are accepted and propagated downstream by all devices which see them.
01xxxx	Non-Posted	**Sized Read (all reads are non-posted)**	Request	**Usage of four least-significant bits:** [3] Response may pass posted requests (1 = OK; 0 = Do not pass) [2] Dword/Byte (1 = dword; 0 = byte). [1] Isoc(1 = Isochronous; 0 = std.). [0] Coherency (1 = req'd; 0 = not)
000010	Non-Posted	**Flush**	Request	Forces all preceding posted writes <u>in same transaction stream</u> to destination (within host).
111100	Posted	**Fence**	Request	Forces all preceding posted writes to destination (all virtual channels).
111101	Non-Posted	**Atomic RMW**	Request	A non-posted write transaction with a read response. Two variants: *Fetch and Add, Compare and Swap*. Both variants allow reading, modification, and write back of a "locked" memory location semaphore.
110000	Resp	**Read Response**	Response	On read and Atomic RMW transactions, **read response** precedes the data being returned by target. In the event of a failure in completing the read, error bits in the response indicate the nature of the problem.
110011	Resp	**Target Done**	Response	On non-posted write or flush transactions, **target done** response confirms completion. In the event of a failure, error bits in the response indicate the nature of the problem.

Control Packets: Information

There are two types of *Information* control packets, NOP and Sync/Error. These four-byte packets are exchanged between the transmitter-receiver pairs on a single link. Unlike request and response packets, information packets are not flow controlled; when one is sent by a transmitter to its corresponding receiver, it must be accepted.

NOP Packet

The NOP (No Operation) command indicates an idle condition on the link. After the link is initialized, each transmitter issues NOP commands continuously unless another command type is required. In addition to indicating the idle condition, these packets inform the device receiving them about changes in the status of flow control buffers and other miscellaneous information concerning link management and diagnostics. Figure 4-6 on page 71 depicts the various fields of the four-byte NOP packet. Table 4-2 immediately following summarizes the usage of each bit field.

Figure 4-6: Control Packets: NOP Information

Byte	Bits 7	6	5	4	3	2	1	0
0	Reserved	DisCon	Command Type Cmd[5:0] = 000000b					
1	ResponseData[1:0]		Response[1:0]		PostData[1:0]		PostCmd[1:0]	
2	0	Diag	Isoc	Reserved	NonPostData[1:0]		NonPostCmd[1:0]	
3	Reserved							

Table 4-2: HyperTransport NOP Packet Bit Assignments

Byte	Bit	Function
0	5:0	**NOP Command Code**. This is the six bit command code for a NOP information packet. Value = 000000b.
0	6	**DisCon**. When this bit is set to a one, the transmitter is indicating that it is starting a LDTSTOP# disconnect sequence. All six buffer release fields must all be = 0 when this bit is set (see next two bytes in packet format).
0	7	**Reserved**. Tie to low level.
1	1:0	**PostCmd[1:0]**. Number of *posted command* buffer entries released since last NOP. Two bit field is coded as: 00 = 0 posted command buffer entries released since last NOP 01 = 1 posted command buffer entry released since last NOP 10 = 2 posted command buffer entries released since last NOP 11 = 3 posted command buffer entries released since last NOP
1	3:2	**PostData[1:0]**. Number of *posted data* buffer entries released since last NOP. Two bit field is coded as: 00 = 0 posted data buffer entries released since last NOP 01 = 1 posted data buffer entry released since last NOP 10 = 2 posted data buffer entries released since last NOP 11 = 3 posted data buffer entries released since last NOP
1	5:4	**Response[1:0]**. Number of *response command* buffer entries released since last NOP. Two bit field is coded as: 00 = 0 response buffer entries released since last NOP 01 = 1 response buffer entry released since last NOP 10 = 2 response buffer entries released since last NOP 11 = 3 response buffer entries released since last NOP
1	7:6	**ResponseData[1:0]**. Number of *response data* buffer entries released since last NOP. Two bit field is coded as: 00 = 0 response data buffer entries released since last NOP 01 = 1 response data buffer entry released since last NOP 10 = 2 response data buffer entries released since last NOP 11 = 3 response data buffer entries released since last NOP

Table 4-2: HyperTransport NOP Packet Bit Assignments

Byte	Bit	Function
2	1:0	**NonPostCmd[1:0].** Number of *non-posted command* buffer entries released since last NOP. Two bit field is coded as: 00 = 0 non-posted command buffer entries released since last NOP 01 = 1 non-posted command buffer entry released since last NOP 10 = 2 non-posted command buffer entries released since last NOP 11 = 3 non-posted command buffer entries released since last NOP
2	3:2	**NonPostData[1:0].** Number of *non-posted data* buffer entries released since last NOP. Two bit field is coded as: 00 = 0 non-posted data buffer entries released since last NOP 01 = 1 non-posted data buffer entry released since last NOP 10 = 2 non-posted data buffer entries released since last NOP 11 = 3 non-posted data buffer entries released since last NOP
2	4	**Reserved.** Tie to low level.
2	5	**Isoc.** When set, this bit indicates that flow-control information being sent in this NOP applies to the isochronous virtual channels. Isochronous operation is optional; unless it has been enabled on the link, no isochronous flow-control information should be sent. If this bit is = 0, flow-control information being sent in bytes 0,1, and 2 applies to standard posted, non-posted, and response virtual channels.
2	6	**Diag.** (Optional Feature) Software enables CRC testing by writing the *CRC Start Test* bit in the Link Control Register. When Diag bit is first detected set = 1, the CRC diagnostic testing phase commences: The receiver, seeing this NOP bit set, ignores its CAD and CTL signals for 512 bit times. Then the transmitter sends any test pattern on the CAD/CTL lines; CRC is checked by the receiver, and errors are logged. If enabled, sync flood will be also performed on CRC test error. Aside from CRC check, CAD bus data values are ignored during test and not retransmitted.
2	7	**Reserved.** Tie to low level
3	7:0	**Reserved.** Tie to low level

Sync/Error Packet

If a reset or error condition occurs which requires a re-synchronization of HyperTransport devices, a "sync flood" pattern may be issued. All bit fields of a Sync/Error packet are 1's, allowing a device to detect and decode a Sync packet even if it has a corrupt sense of clock rate and link width. Each transmitter that drives the Sync pattern holds it until the link resets and re-synchronizes. Any receiver on an 8-, 16-, or 32-bit link assumes it has detected a Sync event if decodes sync packets <u>or</u> if all 1's are received for 16 bit times on the lowest 8 bits of the link; this time is extended to 32 bit times on a 4-bit link interface and 64 bit times on a 2-bit link interface.

The Sync/Error information packet is illustrated in Figure 4-7 on page 74 using normal decode logic. Table 4-3 on page 74 defines the Sync packet bit fields.

Figure 4-7: Control Packets: Sync Information

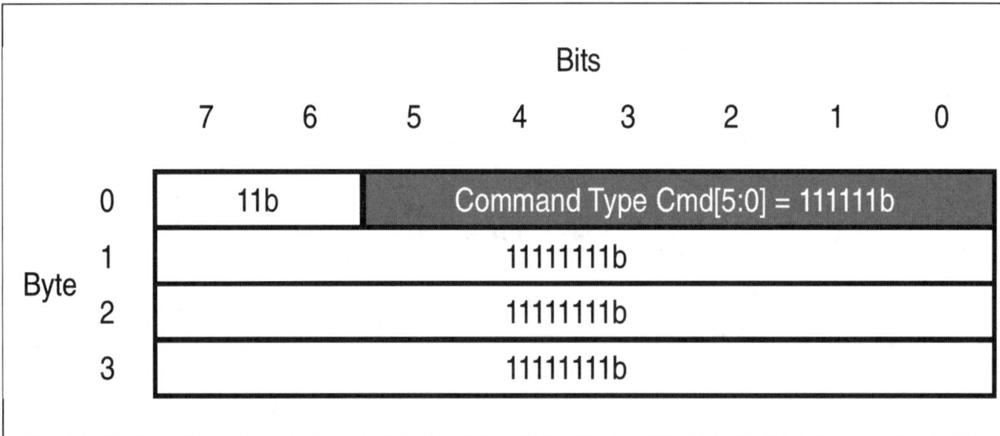

Table 4-3: HyperTransport Sync Packet Bit Assignments

Byte	Bit	Function
0	5:0	**Sync Command Code**. This is the six bit command code for a Sync information packet. Value = 111111b.
0	7:6	**Reserved**. Must be driven to 1's.
3:1	7:0	**Reserved**. Must be driven to 1's.

Control Packets: Requests

As shown previously in Table 4-1 on page 69, there are a number of different request types; each variant has a slightly different way of using the fields within its request packet. In this section, the basic packet format layout used by the principal request types is covered, including Sized Read (always non-posted), Sized Write (posted and non-posted), Broadcast Message (always posted), Flush (always non-posted), Fence (always posted), and Atomic Read-Modify-Write (always non-posted).

Sized Read And Sized Write Requests

The eight-byte sized read and sized write packets (abbreviated *RdSized* and *WrSized* in the Specification) are the mainstream commands used to perform most of the data transfers to both memory or I/O in HyperTransport. Some of the options available with sized read and write requests are:

- Byte or dword read/write data transfers; valid data transferred ranges from 0 bytes to 64 bytes (16 dwords).
- Posted or non-posted virtual channel for writes. Reads are always split transactions traveling in the non-posted virtual channel.
- Isochronous posted or non-posted virtual channels for the request and any subsequent response. Isochronous flow control buffers are required to support this traffic.
- Coherency option bit which indicates whether the transaction requires enforcement of host cache coherency. If the transaction does not target host memory, this feature does not apply.
- Assignment of a non-zero *Sequence ID* attribute to requests forces other devices to maintain strict ordering for all requests from same source. A Sequence ID of 0 indicates that there is no strict ordering required.
- Use of reserved ranges in RdSized and WrSized request packet address fields to support special-case transactions, including configuration cycles, interrupt requests, and End-Of-Interrupt (EOI) messages, etc.

Generic RdSized And WrSized Request Packet Format. Figure 4-8 on page 76 depicts the various fields of the eight-byte Sized Read or Sized Write packet. Table 4-4 on page 76 summarizes the usage of each bit field.

Figure 4-8: Control Packets: Generic Sized Read/Sized Write Requests

Byte	Bits
	7 6 5 4 3 2 1 0
0	SeqID[3:2] | Command Type Cmd[5:0] = x01xxxb or 01xxxxb
1	PassPW | SeqID[1:0] | UnitID[4:0]
2	Mask/Count[1:0] | Compat | SrcTag[4:0]/Reserved
3	Addr[7:2] | Mask/Count[3:2]
4	Addr[15:8]
5	Addr[23:16]
6	Addr[31:24]
7	Addr[39:32]

Table 4-4: HyperTransport Sized Read/Write Packet Bit Assignments

Byte	Bit	Function
0	5:0	**Command Code.** This is the six bit command code for RdSized and WrSized requests. x01xxxb = WrSized Request 001xxxb = RdSized Request Usage of bits marked "x": refer to Table 4-1 on page 69.

Table 4-4: HyperTransport Sized Read/Write Packet Bit Assignments

Byte	Bit	Function
0	7:6	**SeqID[3:2].** (also see Byte 1, bits 5,6). This field tags groups of requests that are part of a strongly ordered sequence. The SeqID value is assigned by the requestor; all transactions within the same transaction stream and virtual channel, and having the same non-zero SeqID value must have their ordering maintained. The SeqID value of 0 is reserved, and indicates a transaction is not part of an ordered sequence.
1	4:0	**UnitID[4:0].** In a request, this field identifies the source of a transaction. UnitID of 0 is used by host bridges; non-zero UnitIDs are for interior devices. Because of this convention, requests with UnitID = 0 are moving downstream (from the bridge), and requests with UnitID > 0 are moving upstream (from an interior device). Physical devices are allowed to consume multiple UnitIDs.
1	6:5	**SeqID[1:0].** (also see Byte 0, bits 6,7). This is the other half of the 4-bit field used to tag groups of requests that are part of a strongly ordered sequence. The SeqID value of 0 is reserved, and indicates a transaction is not part of an ordered sequence.
1	7	**PassPW.** When set, this bit indicates that this packet may pass packets in the posted request virtual channel of the same transaction stream. If the bit is clear, this packet must stay ordered behind them.
2	4:0	**SrcTag[4:0].** This 5-bit field is used as a transaction tag that uniquely identifies all outstanding transactions sourced by the same UnitID. Each UnitID may have up to 32 outstanding transactions at a time. The UnitID and SrcTag values together uniquely identify non-posted requests in a particular transaction stream. The SrcTag field is reserved and not used for posted requests.
2	5	**Compat.** When set, this bit indicates that this request packet should only be claimed by the system subtractive decode device which is responsible for forwarding transactions to legacy devices (e.g. compatibility bridge). Requests with this bit set originate at the host bridge and travel downstream in the part of the topology called the "compatibility chain."

Table 4-4: HyperTransport Sized Read/Write Packet Bit Assignments

Byte	Bit	Function
2	7:6	**Mask/Count[1:0].** (also see Byte 3 bits 0,1). This is the lower half of the 4-bit field that defines dword transfer count or valid bytes in a dword transfer. The meaning of this field depends on whether a byte/dword read or write transfer is being done: **For (Sized) Byte Read transfers:** This field is a 4 bit mask indicating which of the four bytes within the target dword are valid (much like byte enables in PCI). Any mask pattern is valid. **For (Sized) Byte Write transfers**: This (n-1) field indicates the total number of dwords to be transferred, <u>plus</u> the required dword *write mask* that precedes data. Example: If 6 dwords containing bytes of interest are to be transferred, the count field would be ((6 + 1)-1) = 6. **For (Sized) Dword Read or Write transfers**: This field is an n-1 count indicating the total number of dwords to be transferred. Again, a count of 0 = 1 dword; a count of 15d = 16 dwords.
3	1:0	**Mask/Count[3:2].** (also see Byte 2 bits 6,7). This is the upper half of the 4-bit field that defines which bytes are valid during a RdSized or WrSized transfer. The meaning of this field depends on whether a byte or dword transfer is being done. Refer to Byte 2, bits 7:6 above.
3	7:2	**Start Address[7:2]** (also see Bytes 4-7 bits 0-7) This field provides the lowest bits of the dword-aligned, 40 bit HyperTransport target start address. Refer to the HyperTransport address map for a detailed description of the address ranges set aside for memory, I/O, configuration cycles, broadcast messages, interrupts, etc.
7:4	7:0	**StartAddress[39:8]** (also see Byte 3 bits 2-7) This field provides the upper bits of the 40 bit HyperTransport target start address.

RdSized And WrSized Requests: Transaction Limits

Using the various request packet option bits when constructing RdSized and WrSized transactions makes it possible to perform byte and dword read and write transfers in a number of variations. The following section describes some of the key limits associated with RdSized and WrSized requests.

RdSized And WrSized (Dword) Transactions. Sized dword read and write transactions can transfer any number of contiguous dwords within a 64 byte, address-aligned block. The request packet *Mask/Count* field provides the number of dwords to be transferred, beginning at the start address and indexing addresses sequentially upward until the limit defined by the Mask/Count field is reached. All bytes in the range are considered valid. Dword read and write start addresses must be dword aligned. If the start address is 64 byte aligned, the transfer may include the entire 64 byte (16 dword) region; if the start address is not 64 byte aligned, the transfer can only go to the end of the current 64-byte address-aligned block. Dword requests which would cross 64 byte address boundaries must be broken into multiple transactions.

RdSized (Byte) Transactions. Sized byte read transactions can transfer any combination of bytes within one address-aligned dword; requests which would cross an aligned dword address boundary must be broken into multiple transactions. The request packet *Mask/Count* field provides the "byte enable" mask pattern, indicating which bytes are valid. Mask[0] qualifies byte 0, Mask[1] qualifies byte 1, etc. Any mask pattern is legal; mask bits can be ignored by targets reading from "pre-fetchable" locations (all four bytes in the target dword are always returned).

WrSized (Byte) Transactions. Sized byte write transactions can transfer any combination of bytes within a 32-byte address-aligned region. The request packet *Mask/Count* field provides the total number of <u>dwords</u> to be transferred including the required single dword "write mask" pattern. The mask itself is sent just ahead of the data byte payload, and indicates which of the data bytes that follow are valid. Mask bit[0] qualifies byte 0, Mask bit [31] qualifies byte 31, etc. Byte write start address must be dword aligned. If the start address is 32 byte aligned, the write transfer may be as large as the entire 32 byte (8 dword) region; if the start address is not 32 byte aligned, the transfer can only go to the end of the current 32 byte address-aligned block. Basically, start address bits [4:2] identify the first the valid dword of data within the 32-byte region defined by start address bits [39:5]. Byte write requests which would cross 32 byte address boundaries must be broken into multiple transactions. A couple of subtle things about these transfers:

- The entire dword (32 bit) mask is always sent ahead of the data payload, regardless of start address and number of bytes being transferred. Mask bit fields are cleared for all invalid bytes in the 32-byte region ahead of the start address, for all invalid bytes within the transfer range itself, and for all unsent bytes remaining in the 32-byte region beyond the transfer limit implied by the Mask/Count field.
- While it isn't illegal to send invalid dwords at the front and back of a

WrSized (Byte) transfer, it is more efficient to adjust the start address and Mask/Count field to trim off completely invalid dwords in front of the first and after the last dwords containing at least one valid byte in the 32 byte aligned region.

RdSized And WrSized Requests: Other Notes

Coherency. The coherency bit in the Command field of RdSized and WrSized request packets (Byte 0, bit 0) indicates whether host cache coherency is a concern when HyperTransport RdSized and WrSized requests target host memory. Some buses, such as PCI, require coherency enforcement any time a transaction originating in the I/O subsystem targets main memory. This can represent a serious performance hit as processors spend much of their time snooping internal caches for accesses which they may not cache anyway.

HyperTransport uses the coherency bit in the Command field of the request packet to inform the system whether coherency actions are required. If the coherency bit is set:

- All HyperTransport writes targeting host memory result in the CPU updating or invalidating the relevant cache line.
- All HyperTransport reads targeting main memory must result in the latest copy being returned to the requestor. If the CPU has a modified cache line, the system must assure that this is the one returned to the requestor.

If a device has no particular requirement for coherency, it may chose to keep the coherency bit cleared. In this case, the request will complete without any coherency events.

Special Case: Forcing A Coherency Event. A RdSized (byte) targeting host memory with all Mask/Count bits set = 0 (no valid bytes) and coherency bit set = 1 in the request packet Command field causes a host coherency action, using the address provided in the read. One dword of invalid data will be returned.

WrSized Requests And The *Posted* Bit. Sized write request packets may or may not set the *posted* bit (bit 5 of the CMD field). The implications of this bit are as follows:

If set, the bit indicates the write request will travel in the *posted request* virtual channel and that there will not be a response from the target. Each device in the transaction path may de-allocate its buffers as soon as the posted request is transmitted. This also means that the SrcTag field is not used (reserved) because posted writes have no outstanding responses to track. This is in contrast to non-posted requests which require a unique SrcTag field for each request issued.

It the posted bit is not set, the requestor expects a confirmation that the data written has reached the destination — and is willing to suffer the performance penalty and wait for it. Eventually, a Target Done response will be routed back to the original requestor. In HyperTransport, certain address ranges require non-posted writes; this includes configuration and I/O cycles.

Errors During RdSized Transactions. In the event of a read error (SizedRd command), a response and all requested data is returned to the requestor, even though some or all of the data is not valid. Proceeding with a "dummy" read of invalid data is mainly for the benefit of devices in the transaction path that have already allocated flow control buffer space for the returning data. These devices use the return of each byte to simplify de-allocation of buffer space.

***PassPW* and *Response May Pass Posted Requests* bits.** Hyper-Transport supports the strict producer-consumer ordering model found in PCI systems. There are occasions when strict producer/consumer ordering may not be required. In these cases, devices are allowed some flexibility in reordering of posted and non-posted request packets, as well as response packets. Ordering rules, including relaxed ordering, are described in more detail in the chapter entitled Ordering. Relaxing ordering rules is application-specific, and may provide better system performance in some cases.

The source of a transaction indicates whether or non relaxed ordering is permitted through the setting or clearing of two bits in a request:

1. **PassPW bit**. The *PassPW* request packet bit (Byte 1, bit 7) is programmed in the request packet and affects how ordering rules are applied to request as it moves toward the target. If set = 1, relaxed ordering is enabled; if PassPW is clear, relaxed ordering is not allowed.
2. **Response May Pass Posted Requests** bit. For RdSized transactions, there is also a bit in the Command field of the RdSized request packet called *Response May Pass Posted Requests* (Byte 0, bit 3). This bit state will be replicated in the PassPW bit of the returning response and affects how ordering rules are applied to response as it moves back to the original source. The *Response May Pass Posted Requests* bit does not apply to commands other than RdSized. For reads, the bit should be cleared if the strict producer/consumer ordering model is required; otherwise this bit and the PassPW bit should both be set in the request.

Compatibility Bit. In keeping with PCI subtractive decoding, HyperTransport may use the *Compat* bit in RdSized and WrSized request packets (Byte 2, bit 5) to enable them to reach legacy hardware (e.g. boot firmware) behind the system subtractive decoder. When the Compat bit is set, all system devices should pass the request downstream through the "compatibility chain" to the subtractive decoder. Only the subtractive decoder may claim these transactions. The Compat bit is reserved and must not be set for upstream requests or configuration cycles.

Broadcast Message Requests

The eight-byte Broadcast Message request initiates a global message to all enabled HyperTransport devices. They are issued by host bridges, and travel only in the downstream direction. Implementation of Broadcast Message schemes are system-specific, so the use of address and many other fields is left to designers. Basic format is shown in Figure 4-9 on page 82. Table 4-5 on page 83 summarizes the usage of each defined bit field.

Figure 4-9: Control Packets: Broadcast Message Request

Byte	Bits
	7 6 5 4 3 2 1 0
0	SeqID[3:2] / Command Type Cmd[5:0] = 111010b
1	PassPW / SeqID[1:0] / UnitID[4:0]
2	Reserved
3	Addr[7:2] / Reserved
4	Addr[15:8]
5	Addr[23:16]
6	Addr[31:24]
7	Addr[39:32]

Table 4-5: HyperTransport Broadcast Message Packet Bit Assignments

Byte	Bit	Function
0	5:0	**Broadcast Message Request Command Code**. This is the six bit command code for a Broadcast Message request packet. Value = 111010b.
0	7:6	**SeqID[3:2]**. (also see Byte 1, bits 5,6). This field tags groups of requests that are part of a strongly ordered sequence. The SeqID value is assigned by the requestor; all transactions within the same transaction stream and virtual channel, and having the same non-zero SeqID value must have their ordering maintained. The SeqID value of 0 is reserved, and indicates a transaction is not part of an ordered sequence.
1	4:0	**UnitID[4:0]**. Must be 0. In a request, this field identifies the source of a transaction. UnitID of 0 is used by host bridges; non-zero Unit-IDs are for interior devices. Because of this convention, requests with UnitID = 0 (such as Broadcast Message) only move down-stream.
1	6:5	**SeqID[1:0]**. (also see Byte 0, bits 6,7). This is the other half of the 4-bit field used to tag groups of requests that are part of a strongly ordered sequence. The SeqID value of 0 is reserved, and indicates a transaction is not part of an ordered sequence.
1	7	**PassPW**. Reserved because Broadcast Message always travels in posted virtual channel so a response is not required.
2	7:0	These bits are reserved for a Broadcast Message because *SrcTag* isn't needed (posted request), *Mask/Count* isn't needed (no data packet), and the *Compatibility* bit is never set for these messages.
3	1:0	**SeqID[1:0]**. (also see Byte 0, bits 6,7). This is the other half of the 4-bit field used to tag groups of requests that are part of a strongly ordered sequence. The SeqID value is assigned by the requestor; all transactions within the same transaction stream and virtual channel, and having the same non-zero SeqID value must have their ordering maintained. The SeqID value of 0 is reserved, and indicates a transaction is not part of an ordered sequence.
3	1:0	**Reserved**. *Mask/Count* isn't needed for Broadcast Messages (no data packet)

Table 4-5: HyperTransport Broadcast Message Packet Bit Assignments

Byte	Bit	Function
3	7:2	**Start Address[7:2]** (also see Bytes 4-7 bits 0-7) This field provides the lowest bits of the dword-aligned, 40 bit HyperTransport target start address. Broadcast Message usage of this field is system specific.
7:4	7:0	**Start Address[39:8]** (also see Byte 3 bits 2-7) This field provides the upper bits of the 40 bit HyperTransport target start address. Broadcast Message usage of this field is system specific

Flush Requests

One of the hazards of posted write buffers is that there is no certainty about <u>when</u> the data actually arrives at the destination because no response is ever expected (or sent). The four-byte Flush request guarantees that all previous posted writes within the same transaction stream are "globally visible" in host memory. Flush behaves like a dummy read operation in that it is a non-posted request followed by a response (Target Done) which simply indicates that the Flush operation is complete all of the way to the host bridge.

The Flush request format is shown in Figure 4-10 on page 85. Table 4-6 immediately following summarizes the usage of each defined bit field.

Figure 4-10: Control Packets: Flush Request

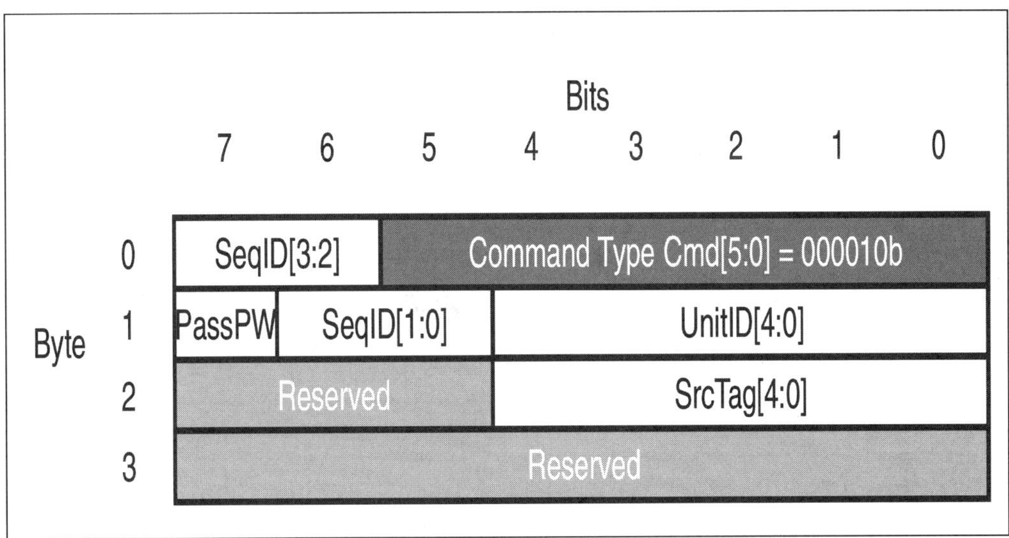

Table 4-6: HyperTransport Flush Packet Bit Assignments

Byte	Bit	Function
0	5:0	**Flush Request Command Code**. This is the six bit command code for a Flush request packet. Value = 000010b.
0	7:6	**SeqID[3:2]**. (also see Byte 1, bits 5,6). **Must be 0**. This is half of the 4-bit field used to tag groups of requests that are part of an ordered sequence within a particular transaction stream and virtual channel. The SeqID value must be 0 for Flush requests because they are never part of an ordered sequence.
1	4:0	**UnitID[4:0]**. This field identifies the source of the Flush request.
1	6:5	**SeqID[1:0]**. (also see Byte 0, bits 6,7). **Must be 0.** This is the other half of the 4-bit field used to tag groups of requests that are part of an ordered sequence within a particular transaction stream and virtual channel. The SeqID value must be 0 for Flush requests because they are never part of an ordered sequence.

Table 4-6: HyperTransport Flush Packet Bit Assignments

Byte	Bit	Function
1	7	**PassPW. Must be 0** in a Flush operation in order for the Flush to accomplish its task of pushing posted writes ahead of it.
2	4:0	**SrcTag[4:0].** This 5-bit field is used as a transaction tag that uniquely identifies all transactions in progress by the same UnitID. Each UnitID may have up to 32 outstanding transactions at a time. The UnitID and SrcTag values together uniquely identify non-posted requests in a particular transaction stream, including Flush.
2	7:5	**Reserved.** *Mask/Count* and *Compat* bits are reserved in Flush request packets because no data is returned with the Target Done response and these requests never target the compatibility bus.
3	7:0	**Reserved.**

Flush Requests: Transaction Limits

The Flush request is a tool used to manage posted writes headed toward host memory. Two important limitations of the Flush request are:

1. If the posted writes target memory other than host memory (e.g. peer-to-peer transfers), then the flush request and response only guarantee that the posted writes have reached the destination host bridge, not the ultimate target. After the host bridge re-issues all peer-to-peer requests downstream towards the intended targets, it sends the target done response back to the original requestor; it is entirely possible the flush response (target done) will reach the original requestor before the request is seen at the target.
2. Flushes have no impact on the isochronous virtual channels. If isochronous flow control is not enabled on a link, then packets which do have the Isoc bit set actually travel in the normal virtual channels and will be affected by Flush requests.

Fence Requests

Another tool in the management of posted write transactions is the HyperTransport Fence command. The main features of the Fence request are:

1. A Fence request provides a barrier between posted writes which applies to all UnitID's (transaction streams). This is different from the Flush which is specific to the posted writes associated with a single transaction stream. When the Fence is decoded by the bridge, it sends any previously posted writes in its buffers toward memory. As always, ordering is maintained for posted writes within individual single transaction streams, but no particular ordering is required for different streams.
2. The Fence request travels in the posted virtual channel, meaning that there is no response expected or sent.

The Fence request format is shown in Figure 4-11 on page 87. Table 4-7 immediately following summarizes the usage of each defined bit field.

Figure 4-11: Control Packets: Fence Request

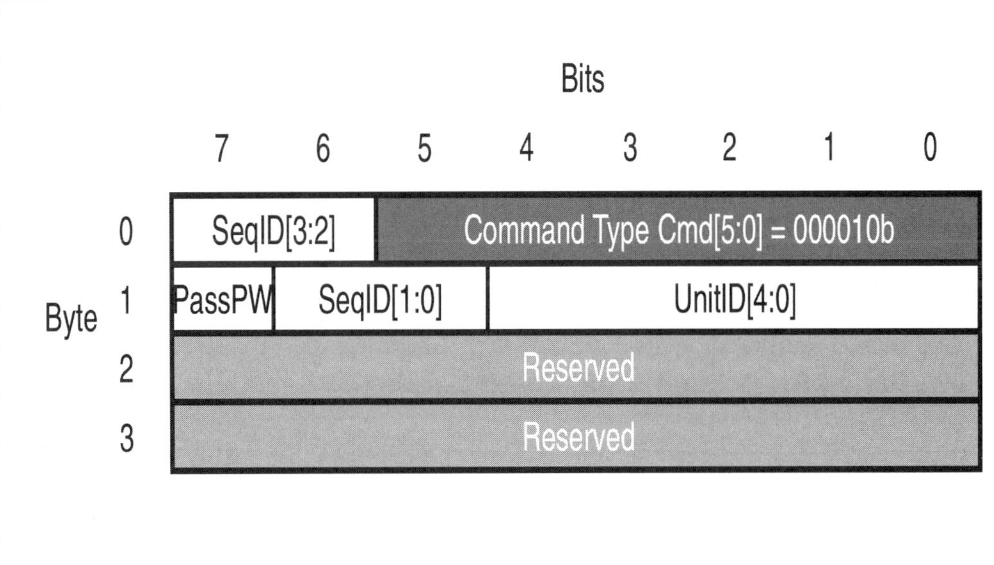

Table 4-7: HyperTransport Fence Packet Bit Assignments

Byte	Bit	Function
0	5:0	**Fence Request Command Code**. This is the six bit command code for a Fence request packet. Value = 000010b.
0	7:6	**SeqID[3:2]**. (also see Byte 1, bits 5,6). **Must be 0**. This is half of the 4-bit field used to tag groups of requests that are part of an ordered sequence within a particular transaction stream and virtual channel. The SeqID value must be 0 for Fence requests because they are never part of an ordered sequence.
1	4:0	**UnitID[4:0]**. This field identifies the source of the Fence request.
1	6:5	**SeqID[1:0]**. (also see Byte 0, bits 6,7). **Must be 0**. This is the other half of the 4-bit field used to tag groups of requests that are part of an ordered sequence within a particular transaction stream and virtual channel. The SeqID value must be 0 for Fence requests because they are never part of an ordered sequence.
1	7	**PassPW. Must be 0** in a Fence operation in order for the Fence to accomplish its task of pushing all previously posted writes ahead of it.
2	7:0	**Reserved**. *SrcTag*, *Mask/Count* and *Compat* bits are reserved in Fence request packets because posted requests don't use SrcTags, no data is associated with the Fence request, and these requests never target the compatibility bus.
3	7:0	**Reserved**.

Fence Requests: Transaction Limits

The Fence request is a tool used to manage posted writes headed toward host memory from all transaction streams. Limitations of the Fence request include:

1. Fence requests are issued from a device to a host bridge, or from one host bridge to another. While a tunnel forwards fence requests it sees, tunnels and single-link cave devices are never the target of a fence request and are never required to perform the fence function internally.
2. Fences have no impact on the isochronous virtual channels. If isochronous flow control is not enabled, then other packets which do have the Isoc bit set actually travel in the normal virtual channels and <u>will be</u> affected by

fence requests.

3. If a fence request is seen by an end-of-chain device, it decodes the transaction and drops it. It may optionally choose to log the event as an end-of-chain error.

Atomic Read-Modify-Write Requests

While sized read and sized write requests can handle most general purpose HyperTransport data transfers, there are times when a combined, or atomic, read/write command is needed.

Two Problems In Shared Memory Schemes

Two problems related to shared memory schemes include:

1. A memory location may be used for storing a "semaphore" to be checked by multiple devices (e.g. CPUs or I/O masters) before using a shared system resource. If the contents of the semaphore location indicate the resource is available, the device which reads it then over-writes the semaphore value to indicate the resource is now busy. If another agent reads the semaphore location and sees it is busy, it must wait until the agent using it clears the semaphore location, thus indicating it is again free. The problem arises when a sharing agent has read the semaphore and found the device is not busy. Before it over-writes the data value to claim the resource, another agent reads the semaphore location and also concludes the device is not busy. Now there is a race condition which can result in both devices attempting to over-write the semaphore and use the resource.

2. The second problem is simpler. If a shared memory location is being used as an accumulator, agents will periodically read the current value, add a constant to it, and write the result back. Again, there is a hazard that the location will be read by one agent and before it can modify it and write it back, another agent may read it with a similar intention. In this case, one of the addends may be lost from the sum.

Most modern bus protocols that support shared memory include a mechanism to avoid the conditions just described. HyperTransport uses the Atomic Read-Modify-Write request for this purpose. The purpose of the Atomic RMW is to force a one-qword (8 byte) memory location to remain "locked" for the duration of the read/modify/write operation required to check and change the targeted location. No other agent is allowed to access the address carried by the Atomic RMW request packet until the entire transaction completes. It is the responsibility of the bridge managing the memory to enforce the locking mechanism.

As a transaction, the Atomic RMW behaves like non-posted write that generates a read response. The read response is accompanied by a single qword of data — the value read from the targeted memory location before any changes are made.

Atomic RMW Variants. The Atomic Read-Modify-Write request has two variants that are designed to address the two cases just described.

Compare And Swap. The *Compare and Swap* variant of the Atomic RMW sends two qwords of data with the request. One qword (the *compare* value) is to be checked against the current value in memory; the other qword (the *input* value) is the data to be written to the memory location if the compare value is equal to the current value. If the compare value is <u>not</u> equal to the current value, the input value is not written to memory. In either case, a read response will be returned accompanied by the original qword read from memory.

Fetch And Add. The *Fetch and Add* variant of Atomic RMW sends a single qword (the *input* value) of data with the request. When the Atomic RMW reaches the bridge to main memory, the bridge unconditionally reads the current value from memory, adds the input value to it, and writes the result back to memory. The memory location remains locked to other transactions while the read-modify-write is in progress. A read response is then returned to the requestor, accompanied by the original qword read from memory.

The Atomic RMW request format is shown in Figure 4-12 on page 91. Table 4-8 on page 91 summarizes the usage of each defined bit field.

Figure 4-12: Control Packets: Atomic Read-Modify-Write Request

<table>
<tr><td rowspan="2"></td><td colspan="8" align="center">Bits</td></tr>
<tr><td>7</td><td>6</td><td>5</td><td>4</td><td>3</td><td>2</td><td>1</td><td>0</td></tr>
<tr><td>0</td><td colspan="2">SeqID[3:2]</td><td colspan="6">Command Type Cmd[5:0] = 111101b</td></tr>
<tr><td>1</td><td>PassPW</td><td colspan="2">SeqID[1:0]</td><td colspan="5">UnitID[4:0]</td></tr>
<tr><td>2</td><td colspan="2">Count[1:0]</td><td>Compat</td><td colspan="5">SrcTag[4:0]</td></tr>
<tr><td>3</td><td colspan="5">Addr[7:3]</td><td>Reserved</td><td colspan="2">Count[3:2]</td></tr>
<tr><td>4</td><td colspan="8">Addr[15:8]</td></tr>
<tr><td>5</td><td colspan="8">Addr[23:16]</td></tr>
<tr><td>6</td><td colspan="8">Addr[31:24]</td></tr>
<tr><td>7</td><td colspan="8">Addr[39:32]</td></tr>
</table>

(Byte labels shown in left margin for rows 0–7.)

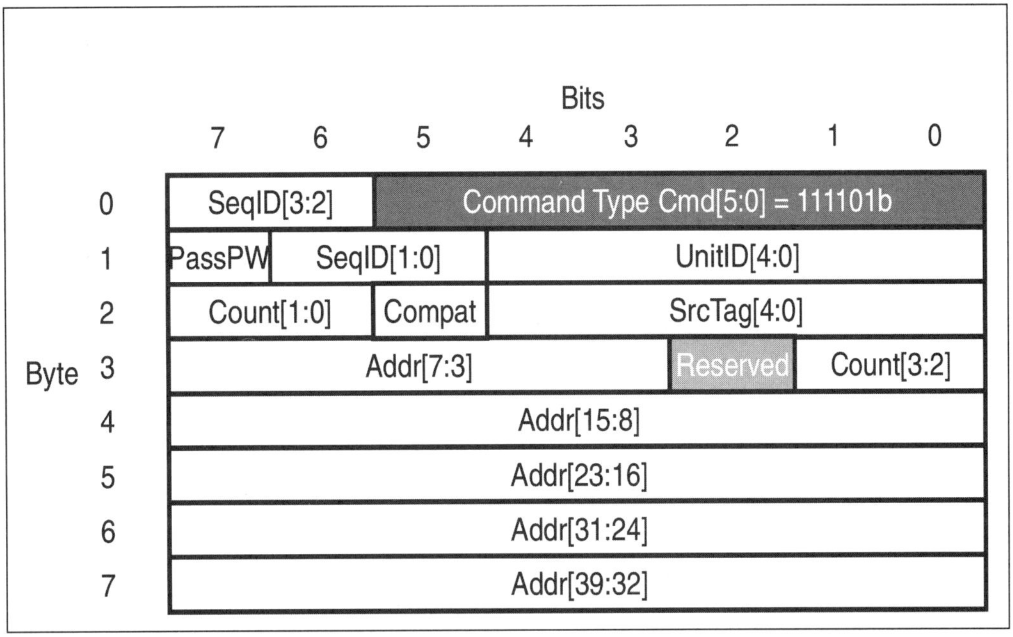

Table 4-8: HyperTransport Atomic Read — Modify-Write Packet Bit Assignments

Byte	Bit	Function
0	5:0	**Atomic RMW Request Command Code.** This is the six bit command code for a Atomic Read-Modify-Write request packet. Value = 111101b.
0	7:6	**SeqID[3:2]**. (also see Byte 1, bits 5,6). This field tags groups of requests that are part of a strongly ordered sequence. The SeqID value is assigned by the requestor; all transactions within the same transaction stream and virtual channel, and having the same non-zero SeqID value must have their ordering maintained. The SeqID value of 0 is reserved, and indicates a transaction is not part of an ordered sequence.

Table 4-8: HyperTransport Atomic Read — Modify-Write Packet Bit Assignments

Byte	Bit	Function
1	4:0	**UnitID[4:0].** This field identifies the source of the Atomic RMW request.
1	6:5	**SeqID[1:0].** (also see Byte 0, bits 6,7). This is the other half of the 4-bit field that tags groups of requests that are part of a strongly ordered sequence. The SeqID value is assigned by the requestor; all transactions within the same transaction stream and virtual channel and having the same SeqID value must have their ordering maintained.
1	7	**PassPW. Must be 0** in an Atomic RMW operation.
2	4:0	**SrcTag[4:0].** This 5-bit field is used a transaction tag that uniquely identifies all transactions in progress by the same UnitID. Each UnitID may have up to 32 outstanding transactions at a time. The UnitID and SrcTag values together uniquely identify non-posted requests in a particular transaction stream, including Flush.
2	5	**Compat. Normally 0.** When set, this bit indicates that this packet should only be claimed by the system subtractive decode device which is responsible for forwarding transactions to legacy devices (e.g. compatibility bridge). Atomic RMW transactions normally target host bridges, so this bit is clear.
2	7:6	**Mask/Count[1:0].** (also see Byte 3 bits 0,1). This is the lower half of the 4-bit field used to define which bytes are valid during a transfer. The value programmed in the count field depends on the variant of Atomic RMW request: **For Fetch And Add RMW:** Count field is set = 1 which indicates 2 dwords (1 qword of data sent with request). **For Compare And Swap RMW:** This field is set = 3 which indicates 4 dwords (2 qwords of data sent with request).

Table 4-8: HyperTransport Atomic Read — Modify-Write Packet Bit Assignments

Byte	Bit	Function
3	1:0	**Mask/Count[3:2].** (also see Byte 2 bits 6,7). This is the upper half of the 4-bit field that defines which bytes are valid during a transfer. The value programmed in the count field depends on the variant of Atomic RMW request: **For Fetch And Add RMW:** Count field is set = 1 which indicates 2 dwords (1 qword of data sent with request). **For Compare And Swap RMW:** This field is set = 3 which indicates 4 dwords (2 qwords of data sent with request).
3	7:3	**Start Address[7:3]** (also see Bytes 4-7 bits 0-7) This field provides the lowest bits of the dword-aligned, 40 bit HyperTransport target start address. For an Atomic RMW, a qword aligned start address must be provided.
7:4	7:0	**Start Address[39:8]** (also see Byte 3 bits 2-7) This field provides the upper bits of the 40 bit HyperTransport target start address. (See previous field).

Atomic RMW Requests: Transaction Limits

The Atomic RMW request locks a qword memory address block while a read-modify-write operation is performed. Limitations of the Atomic RMW request include:

1. The request transfer size, as indicated in the Mask/Count field, is restricted to either one or two qwords. Following the request, a read response returns a single qword of data from memory.
2. These transactions are designed to be generated by I/O devices or bridges, and target system memory. Other than the host bridge, no HyperTransport devices are expected to support atomic operations. If a target detects an unsupported RMW, it may return a one qword read response with the error bit set or perform a non-atomic read-modify-write. The current Hyper-Transport Specification does not require peer-to-peer reflection of Atomic RMW.

Control Packets: Responses

There are two response types used in HyperTransport: Read Response and Target Done. Responses are returned by target devices following a non-posted request, and much of the response packet field information is extracted from the requests that caused them. Because responses are routed back to the original requestor either implicitly or based on UnitID, they don't require a 40 bit address field like requests do. All response packets are four bytes.

Read Responses

The four-byte read response is returned when data requests are made, including RdSized and Atomic RMW requests. All HyperTransport read transactions are non-posted and split; this means that data is never returned immediately as it generally is on buses such as PCI. The advantage of split reads is that the latency involved, in waiting for a target to access its internal memory before returning read data, can be minimized by sending the request, releasing the bus, and waiting for the target to initiate the return of data when it has it.

In HyperTransport, the read response is used by the target to indicate the return of previously requested data. The read response immediately precedes the data, and contains the following general information:

- The response packet type.
- Whether the response should travel in the standard or isochronous virtual channel.
- UnitID which acts as an address for responses.
- A direction bit indicating whether the response is moving upstream or downstream.
- Whether relaxed ordering may be used for this response relative to posted writes moving in the same stream.
- Error bits indicating whether or not the returning data can be considered valid; if it is invalid, error bits indicate whether the error occurred at the target or if the request inadvertently reached an end-of-chain device.

Figure 4-13 on page 95 depicts the various fields of the four-byte read response packet. Table 4-9 on page 95 summarizes the usage of each bit field.

Figure 4-13: Control Packets: Read Response

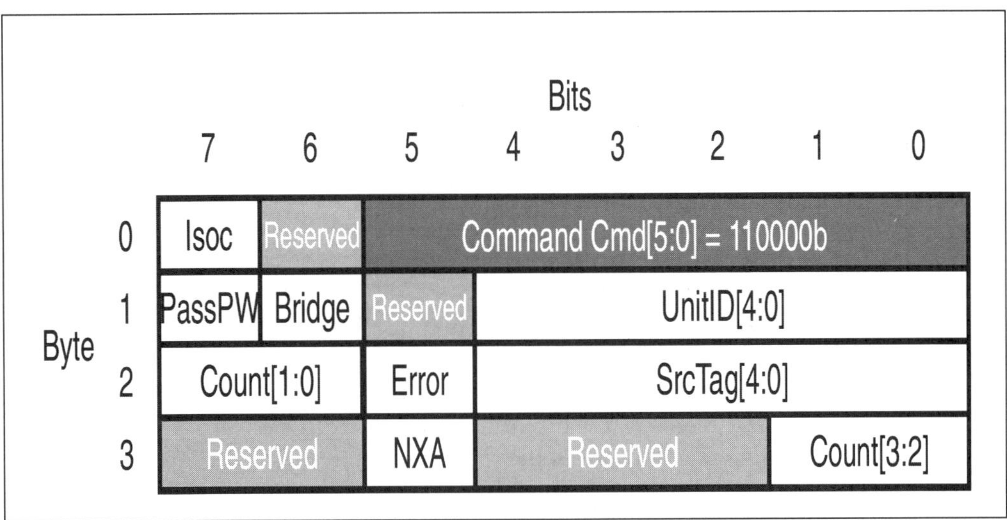

Table 4-9: HyperTransport Read Response Packet Bit Assignments

Byte	Bit	Function
0	5:0	**Command Code**. This is the six bit command code for the Read Response packet. Value: 110000b
0	6	**Reserved**.
0	7	**Isoc**. If set = 1, this response should travel in the isochronous virtual channels for responses and response data. This bit is set in the target response if the Isoc bit was set in the request (Command field) that caused it. *Note: The state of this bit should be preserved even when passing through tunnel devices with isochronous flow control disabled.*
1	4:0	**UnitID[4:0]**. (also see *Bridge* bit below). This field helps route the responses and is programmed in two different ways: **For Upstream Responses (Bridge = 0):** This field contains the UnitID of the node that generated the response (original target) **For Downstream Responses (Bridge = 1):** This field contains the UnitID of the original requestor

Table 4-9: HyperTransport Read Response Packet Bit Assignments

Byte	Bit	Function
1	6	**Bridge**. This bit is set by host bridges to indicate responses which are traveling downstream. Interior devices use Bridge bit and UnitID to claim returning responses. Upstream responses from interior devices have the Bridge bit cleared and carry the UnitID of the responder, meaning that they are routed implicitly to host bridge based only on the fact that the Bridge bit = 0.
1	7	**PassPW**. This bit will be set in the read response if *response may pass posted requests* bit was set in the command field of the read request that caused it. If set, relaxed ordering may be applied.
2	4:0	**SrcTag[4:0]**. This field is copied from the request packet.
2	5	**Error**. When set, this bit indicates that an error occurred during the read transaction. All of the requested data is returned, even if there is an error.
2	7:6	**Count[1:0]**. (also see Byte 3 bits 0,1). This is the lower half of the 4-bit field that indicates the quantity of returning data: **For Dword Read transfers:** This field is a copy of the count field in the request packet **For Byte Read transfers**: Count field is always set = 0 (1 dword) **For Atomic RMW transfers**: Count field is always set = 1 (2 dwords =1 qword).
3	1:0	**Count[3:2]**. (also see Byte 2 bits 6,7). This is the upper half of the 4-bit field that indicates the quantity of returning data: **For Dword Read transfers:** This field is a copy of the count field in the request packet **For Byte read transfers**: Count field is always set = 0 (1 dword) **For Atomic RMW transfers**: Count field is always set = 1 (2 dwords = 1 qword).
3	4:2	**(Reserved.**
3	5	**NXA (Non-Existent Address)** This bit is only valid if Error bit (Byte 2, bit 5) is set. If NXA and Error are both set = 1, error occurred at end-of-chain device due to a non-existent address problem. If NXA = 0 and Error is set = 1, then error occurred at target.
3	7:6	**Reserved.**

Target Done Responses

The four-byte target done response is returned when non-posted WrSized or Flush requests are made. As no data is returned with the target done response, it is routed back to the original requestor as a way to confirm the completion of a write transaction or a Flush operation. The contents of the target done response packet are very similar to the read response packet except that no mask/count information is required because there is no data to transfer.

Figure 4-14 on page 97 depicts the various fields of the four-byte read response packet. Table 4-10 summarizes the usage of each bit field.

Figure 4-14: Control Packets: Target Done Response

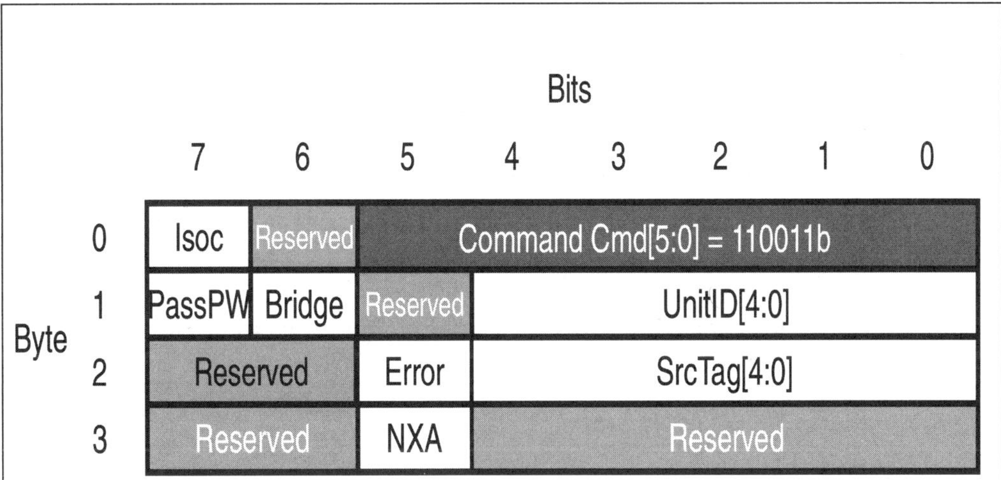

Table 4-10: HyperTransport Target Done Response Packet Bit Assignments

Byte	Bit	Function
0	5:0	**Command Code.** This is the six bit command code for the Target Done Response packet. Value: 110011b
0	6	**Reserved.**
0	7	**Isoc.** If set = 1, this response should travel in the isochronous virtual channels for responses and response data. This bit is set in the target done response if the Isoc bit was set in the request (Command field) that caused it. *Note: The state of this bit should be preserved even when passing through tunnel devices with isochronous flow control disabled.*

Table 4-10: HyperTransport Target Done Response Packet Bit Assignments

Byte	Bit	Function
1	4:0	**UnitID[4:0].** (also see *Bridge* bit below). This field helps route the responses, and is programmed in two different ways: **For Upstream Responses (Bridge = 0):** This field contains the UnitID of the node which generated the response (original target) **For Downstream Responses (Bridge = 1):** This field contains the UnitID of the original requestor
1	6	**Bridge.** This bit is set by host bridges to indicate responses which are traveling downstream. Interior devices use Bridge bit and UnitID to claim returning responses. Upstream responses from interior devices have the Bridge bit cleared and carry the UnitID of the responder, meaning that they are routed implicitly to host bridge based only on the fact that the Bridge bit = 0.
1	7	**PassPW.** This bit is set in the target done response if relaxed ordering of the target done response is permitted. As there is no *response may pass posted requests* bit in write requests, it is device-specific whether this response packet bit is set or not. Generally, it is expected to be set.
2	4:0	**SrcTag[4:0].** This field is copied from the request that caused this target done response.
2	5	**Error.** When set, this bit indicates that an error occurred during the transaction.
2	7:6	**Reserved**
3	4:0	**Reserved**
3	5	**NXA (Non-Existent Address)** This bit is only valid if Error bit (Byte 2, bit 5) is set. If NXA and Error are both set = 1, error occurred at end-of-chain device due to a non-existent address problem. If NXA = 0 and Error is set = 1, then error occurred at target.
3	7:6	**Reserved.**

5 *Flow Control*

The Previous Chapter

The previous chapter described the use of HyperTransport *control* and *data* packets to construct HyperTransport link transactions. Control packet types include Information, Request, and Response variants; data packets contain a payload of 0-64 valid bytes. The transmission, structure, and use of each packet type is presented.

This Chapter

This chapter describes HyperTransport *flow control*, used to throttle the movement of packets across each link interface. On a high-performance connection such as HyperTransport, efficient management of transaction flow is nearly as important as the raw bandwidth made possible by clock speed and data bus width. Topics covered here include background information on bus flow control and the initialization and use of the HyperTransport virtual channel flow control buffer mechanism defined for each transmitter-receiver pair.

The Next Chapter

The next chapter describes the rules governing acceptance, forwarding, and rejection of packets seen by HyperTransport devices. Several factors come into play in routing, including the packet type, the direction it is moving, and the device type which sees it. A related topic also covered in this chapter is the fairness algorithm used by a tunnel device as it inserts its own packets into the traffic it forwards upstream on behalf of devices below it. The HyperTransport specification provides a fairness algorithm and a hardware method for tunnel management packet insertion.

The Problem

On any bus where an agent initiates the exchange of information (commands, data, status, etc.) with a *target*, a number of things can cause a delay (or even end) the normal completion of the intended transfer. The throttling of information delivery on a bus is referred to as *flow control*. PCI is a good example of a bus protocol which has reasonably high burst bandwidth, but is subject to performance hits caused by an unsophisticated flow control mechanism. Before looking at the HyperTransport approach to flow control, some of the general problems in bus flow control are described in the following section in terms of the PCI protocol. Refer to Figure 5-1 on page 100.

Figure 5-1: PCI Interface Handshake Signals

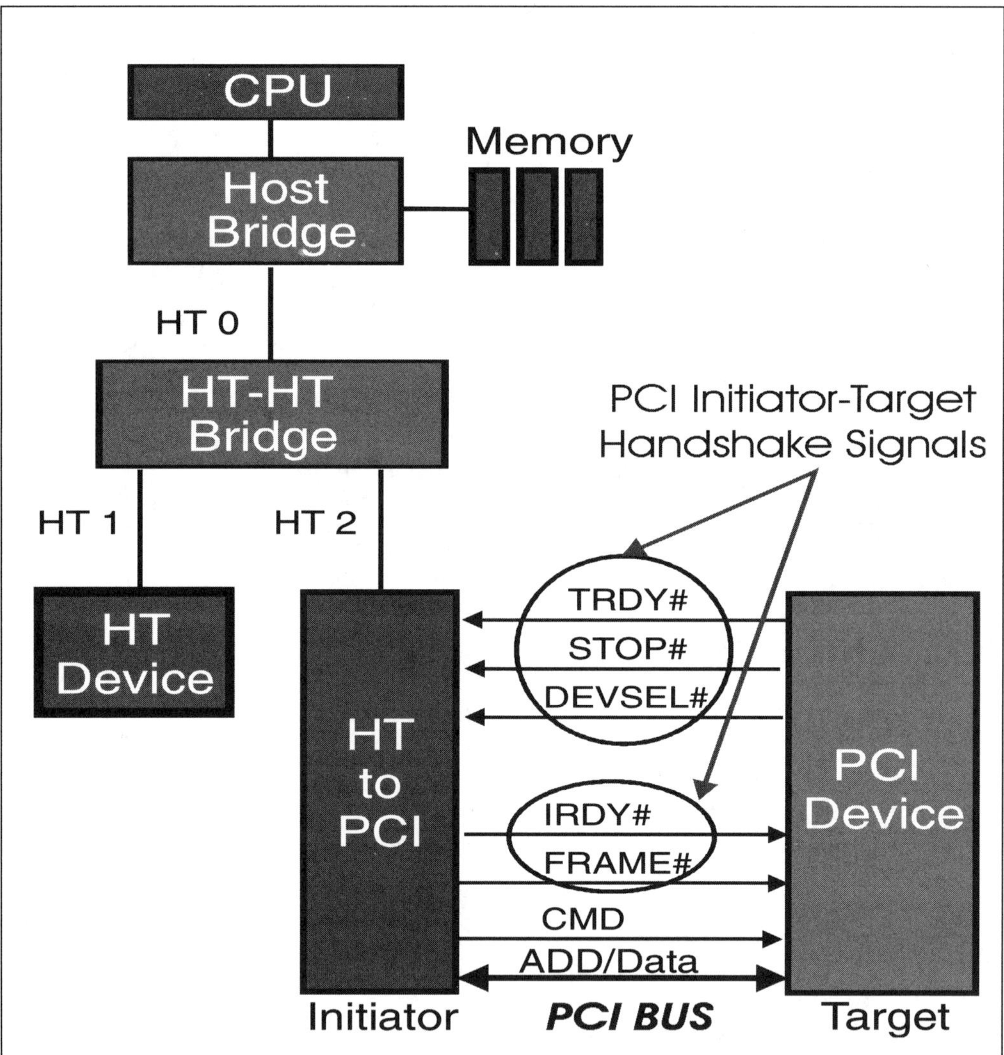

How PCI Handles Flow Control

While the PCI specification permits 64-bit data bus and 66MHz clock options, a generic PCI bus carries only 32 bits (4 bytes) of data and runs at a 33MHz clock speed. This means that the burst bandwidth for this bus is 132MB/s (4 bytes x 33MHz = 132MB/s). In many systems the PCI bus is populated by all sorts of high- and low-performance peripherals such as hard drives, graphics adapters, and serial port adapters. All PCI bus master devices must take turns accessing the shared bus and performing their transfers. The priority of a bus master in accessing the bus and the amount of time it is allowed to retain control of the bus is a function of PCI *arbitration*. In a typical computer system, the PCI arbiter logic resides in the system chipset.

Once a PCI bus master has won arbitration and verifies the bus is idle, it commences its transaction. After decoding the address and command sent by the master, one target claims the cycle by asserting a signal called DEVSEL#. At this point, if both devices are prepared, either *write data* will be sent by the initiator or *read data* will be returned by the target. For cases where either the master or target are not prepared for full-speed transfer of some or all of the data, flow control comes into play. In PCI there are a number of cases that must be dealt with.

PCI Target Flow Control Problems

PCI Target Not Ready To Start. In some cases, a PCI device being targeted for transmission is not prepared to transfer any data at all. This could happen if the target is off-line, does not have buffer space for write data being sent to it, or does not have requested read data available. It may also occur if the transaction must cross a bridge device to a different bus. Many bus protocols, including PCI, place a limit on how long the bus may be stalled before completing a transaction; in cases where a target can't meet the requirement for even the first data, a mechanism is required to indicate the transaction should be abandoned and re-attempted later. PCI calls the target cancellation of a transaction (without transferring any data) a *Retry*; a Retry is indicated when a target asserts the STOP# signal (instead if TRDY#) in the first data phase.

PCI Target Starts Data Transfer, But Can't Continue. Another possibility is that a transaction started properly, some data has transferred, but at some point before completion the target "realizes" it can't continue the transfer within the time allowed by the protocol. The target must indicate to the master that the transaction must be suspended (and resumed later at the point where it left off). PCI calls this target suspension of a transaction (with a partial transfer of data) a *Disconnect.* A Disconnect is signalled when the target asserts the STOP# signal in a data phase after the first one.

PCI Target Starts, Can Continue, But Needs More Time. Sometimes a transaction is underway and the target requires additional time to complete transmission of a particular data item; in this case, it does not need to suspend the transaction altogether, but simply stretch one or more data phases. The generic name for this is *wait-state insertion.* Wait states are a reasonable alternative to Retry and Disconnect if there are not too many of them; when there are excessive wait states, bus performance would be better served by the devices giving up the bus and allowing it to be used by other devices while they prepare for the resumption of the suspended transaction. PCI targets de-assert the TRDY# signal during any data phase to indicate wait states. A target must be prepared to complete each data phase within 8 PCI clocks (maximum of seven wait states), except for the first data phase which it must complete within 16 clocks. If a target cannot meet the "16 and 8 tick" rules for completing a data phase, it must signal Retry or Disconnect instead.

PCI Initiator Flow Control Problems

While many flow control problems are associated with the target of a transaction, there are a couple which may occur on the initiator side. Again, the cases are described in terms of PCI protocol.

PCI Initiator Starts, But Can't Continue. Some bus protocols also allow an initiator to break off a transaction early in the event it can't accept the next read data or source the next write data within the time allowed by the protocol — even with wait states. PCI initiators suspend transactions simply by de-asserting the FRAME# signal early. As a rule, the master will re-arbitrate later for the PCI bus and perform a new transaction which picks up from where it left off previously.

PCI Initiator Starts, Can Continue, But Needs Wait-States. Some bus protocols allow an initiator to insert wait states in a transfer, just as the target may. Other bus protocols (e.g. PCI-X) only allow targets to insert wait states — based on the assumption that a device which starts a transaction should be ready to complete it before requesting the bus. In any case, PCI initiators de-assert the IRDY# signal to indicate wait states. An initiator must be prepared to complete each data phase within 8 clocks (maximum of seven wait states); if it can't meet this rule for any data phase, it must instead suspend the transaction by de-asserting FRAME#.

All PCI Flow Control Problems Hurt Performance

Each of the initiator and target flow control problems just described impact PCI bus performance for both the devices involved in the transfer, and for devices waiting to access the bus. While not every transaction is afflicted with target retries and disconnects, or early de-assertion of FRAME# by initiators, they happen enough to make effective bandwidth considerably less than 132MB/s on the PCI bus. In addition, arbitration and flow control uncertainties make system performance difficult to estimate.

HyperTransport Flow Control: Overview

All of the flow control problems described previously for PCI severely hurt bus performance and would be even less acceptable on a very high-performance connection. The flow control scheme used in HyperTransport applies independently to each transmitter-receiver pair on each link. The basic features include the following.

Packets Never Start Unless Completion Assured

All transfers across HyperTransport links are packet based. No link transmitter ever starts a packet transfer unless it is known the packet can be accepted by the receiver. This is accomplished with the "coupon based" flow control scheme described in this section, and eliminates the need for the Retry and Disconnect mechanisms used in PCI.

Transfer Length Is Always Known

Hypertransport control packets have a fixed size (four or eight bytes) and data packets have a *known* and *maximum* transfer length, unlike PCI data transfers. This makes buffer sizing and flow control much more straightforward as both transmitter and receiver are aware of their actual transfer commitments. It also makes the interleaving of control packets with data packets much simpler.

Split Transactions Used When Response Is Required

HyperTransport performs all read and non-posted write operations as split transactions, eliminating the need for the inefficient Retry mechanism used in PCI. A split transaction breaks a transfer which requires a response (and maybe data) into two parts — the sending of the request packet, followed later by response/data packets returned by the original target. This keeps the link free during the period between request and response, and means that the burden for completing the transaction is on the device best equipped to know when it is possible to do so — the target.

Flow Control Pins Are Eliminated

Because HyperTransport uses a message-based flow control scheme, it eliminates the flow control handshaking pins and signal traces found on other buses. Instead, each pair of devices on a link convey flow control information related to their receivers by sending update NOP packets over their transmitter connections.

Flow Control Buffers Mean No Bus Wait States

All link receiver interfaces are required to implement a set of buffers which are capable of receiving packets at full speed. Once a transmitter has determined that buffer space is available at the receiver, the transfer of the bytes within the packet always proceeds at full bus speed into the receiver buffer. The buffers are sized such that the full packet can always be accepted. Data packets can be as large as 64 bytes (16 dwords) and control packets can be as large as 8 bytes. The one twist to this is the fact that the transmitter has the option of interleaving new control packets into a large data packet on four byte boundaries. Still, this is done at full speed, without any wait states. The transmitter simply asserts the CTL signal to indicate control packets are moving across the CAD bus, and deasserts it to indicate data packets are moving across; the target uses the CTL signal input to determine which buffer the packet should enter.

Flow Control Buffers For Each Virtual Channel

Finally, because there are a minimum of three virtual channels as packets move through HyperTransport, the flow control mechanism maintains separate flow control buffer pairs for the *posted request, non-posted request,* and *response* virtual channels. Each non-posted request has an associated response (and possibly data); that must be tracked internally by the device until the response comes back. Posted requests do not have a response, and may be flushed internally as soon as they are processed. In addition, the separate flow control buffers are important in enforcing the ordering rules that apply to the three virtual channels.

Optionally, devices may also support isochronous transfers; in this case, three additional receiver flow control buffer sets (CMD/Data) would be required to track this traffic.

Flow Control, A System View

While HyperTransport flow control is enforced on a per-link basis, there are system implications if it does not perform properly. As a point-point technology, HyperTransport devices such as tunnels and bridges have responsibilities for generating their own packets upstream as well as forwarding those from devices either above or below them. Refer to Figure 5-2 on page 106. Note that when the peripheral at the bottom of the chain on the right issues a packet upstream towards memory, it is dependent on how well the devices above manage their links.

Figure 5-2: Flow Control Is On A Link-By-Link Basis

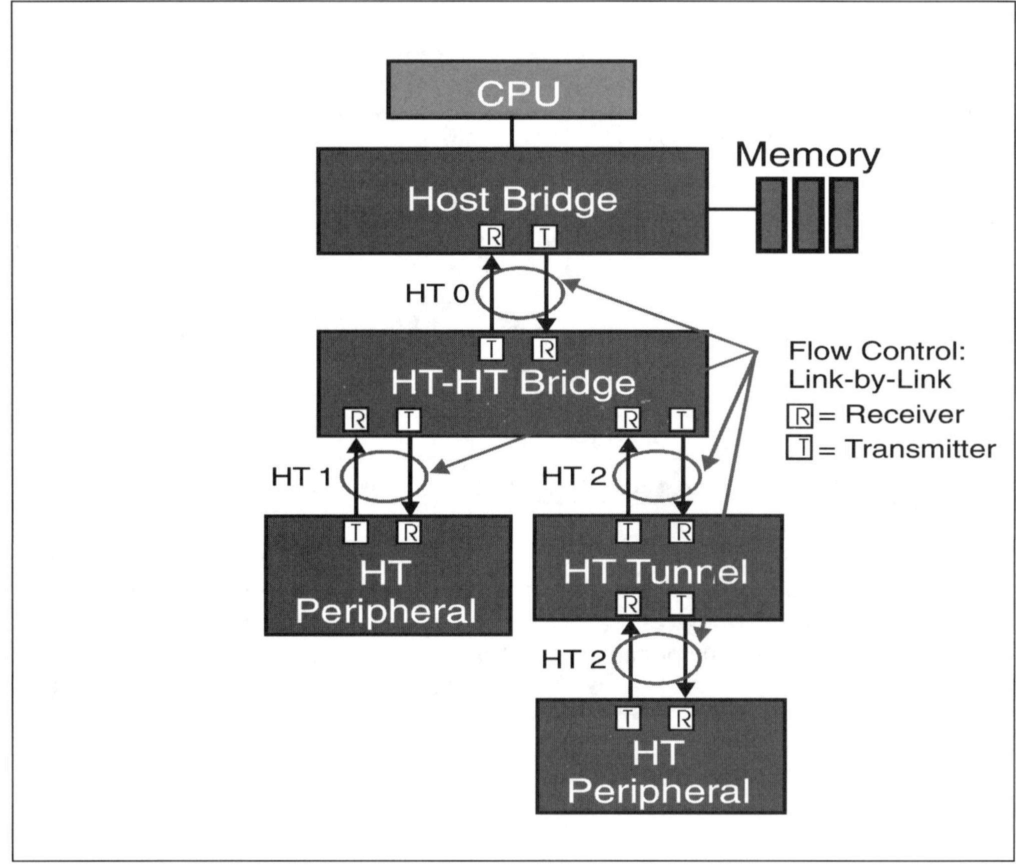

Flow Control Buffer Arrangement

Figure 5-3 on page 107 illustrates the general arrangement of HyperTransport flow control buffers and counters required of each link receiver interface. Note that while the transmit interface maintains flow control counters, the flow control buffers are only on the receive side of each device.

Figure 5-3: Flow Control Buffers And Counters

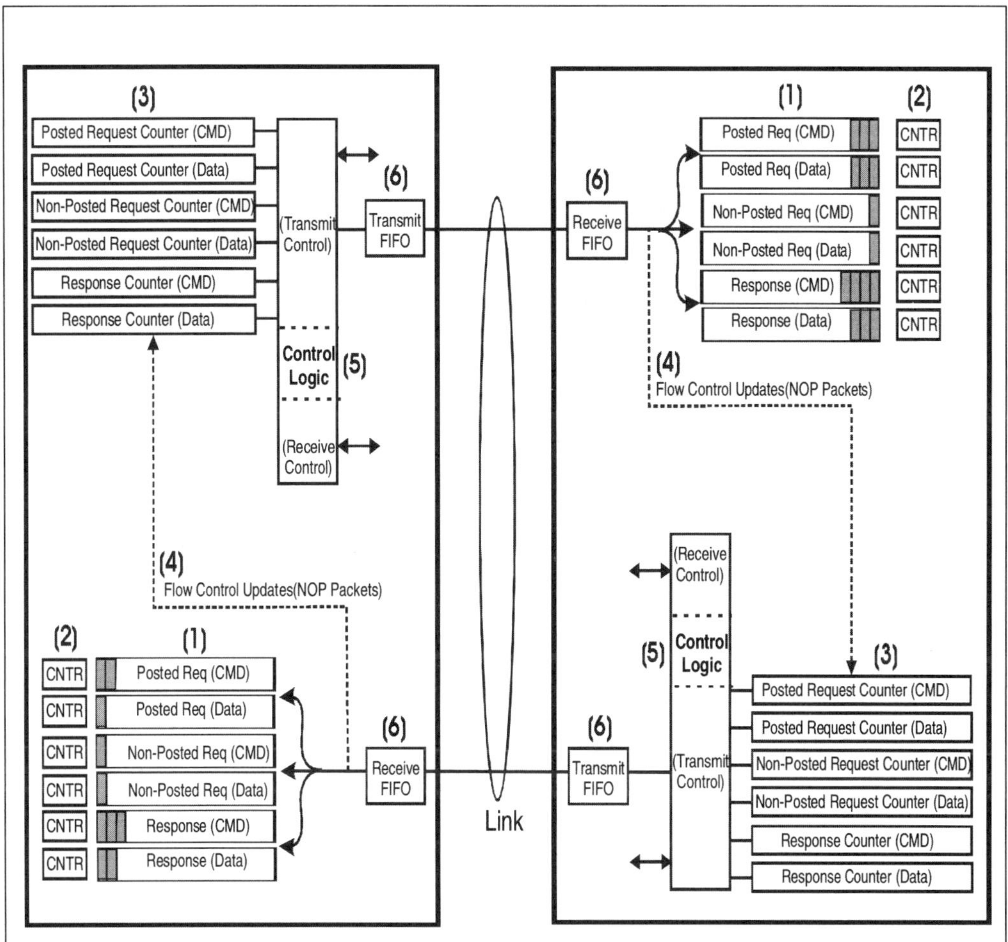

Details Associated With Figure 5-3

The following section describes the architectural features shown in Figure 5-3 on page 107. Note that the drawing is conceptual and is intended to show the major features of flow control on a single link. For multiple-link devices such as tunnels, this logic would be replicated for each interface.

Flow Control Buffer Pairs (Item 1)

Each receiver interface is required to implement six buffers to accept the following packet types being sent by the corresponding transmitter. *The specification requires a <u>minimum depth of one</u> for each buffer, meaning that a receiver is permitted to deal with as few as one packet of each type at a time. It may optionally increase the depth of one or more of the buffers to track multiple packets at a time.*

Posted Request Buffer (Command). This buffer stores incoming posted request packets. Because every request packet is either four or eight bytes in length, each entry in this buffer should be eight bytes deep.

Posted Request Buffer (Data). This buffer is used in conjunction with the previous one and stores data associated with a Posted Request. Because posted request data packets may range in size from 1 dword to 16 dwords (64 bytes), each entry in this buffer should be 64 bytes deep.

Non-Posted Request Buffer (Command). This buffer stores incoming non-posted request packets. Because every request packet is either four or eight bytes in length, each entry in this buffer should be eight bytes deep.

Non-Posted Request Buffer (Data). This buffer is used in conjunction with the previous one and stores data associated with a Non-Posted Request. Because non-posted request data packets may range in size from 1 dword to 16 dwords (64 bytes), each entry in this buffer should be 64 bytes deep.

Response Buffer (Command). This buffer stores returning response packets. Because every response packet is four bytes in length, each entry in this buffer should be four bytes deep.

Response Buffer (Data). This buffer is used in conjunction with the previous one and stores data associated with a returning response. Because responses may precede data packets ranging in size from 1 dword to 16 dwords (64 bytes), each entry in this buffer should be 64 bytes deep.

Receiver Flow Control Counters (Item 2)

The receiver interface uses one counter for each of the flow control buffers to track the availability of new buffer entries. The size of the counter is a function of how many entries were designed into the corresponding flow control buffer. After initialization reports the starting buffer size to the transmnitter, the value in each counter only increments when a new entry becomes available due to a packet being consumed or forwarded; it decrements when NOP packets carrying buffer update information are sent to the transmitter on the other side of the link.

Transmitter Flow Control Counters (Item 3)

It is a transmitter responsibility on each link to check the current state of receiver readiness before sending a packet in any of the three required virtual channels. It does this by maintaining its own set of flow control counters, which track the available entries in the corresponding receiver flow control buffer. For example, if the transmitter wishes to send a read request across the link, it would first consult the Non-Posted Request CMD counter to see the current number of credits. If the counter = 0, the receiver is not prepared to accept any additional packets of this type and the transmitter <u>must wait</u> until the count is updated via the NOP mechanism to a value >0. If the counter value is =1, the receiver will accept one packet of this type, etc. Note that for requests that are accompanied by data (e.g. posted or non-posted writes), the transmitter must consult <u>both</u> its CMD counter and the Data counter for that virtual channel. If either is at 0, it must wait until both counters have been updated to non-zero values.

NOP Packet Update Information (Item 4)

During idle times on the link, each device sends NOP packets to the other. If one or more buffer entries in any of the six receiver flow control buffers have become available, designated fields in the NOP packets are encoded to indicate that fact. Otherwise those fields contain 0, indicating no new buffer entries have become available since the previous NOP transmission. In the next section, use of the NOP packet fields for flow control updates is reviewed. Refer to "NOP Packet" on page 71 for additional discussion of the NOP packet format.

Control Logic (Item 5)

This generic representation of internal control logic is intended to indicate that a number of things related to flow control are under the management of each HyperTransport device. In general:

- Logic associated with the transmit side of a link interface always must consult transmitter flow counters before commencing a packet transfer in any virtual channel. This assures that any packet sent will be accepted.
- Logic monitoring the progress of packet processing in the receiver flow control buffers, must translate new entries that become available into NOP update information to be passed back to the transmitter.
- Logic monitoring the receive side of a link interface must parse incoming

NOPs to determine if the receiver is reporting any changes in buffer availability. If so, then the information is used to update the transmitter's flow control counters to match the available buffer entries on the receiver side.

Transmit And Receive FIFO (Item 6)

The transmit and receive FIFOs are not part of flow control at all, and are shown here as a reminder that all packets moving across the high-speed HyperTransport link pass through an additional layer of buffering to help deal with the effects of clock mismatch within the two devices, skew between multiple clocks sourced by the transmitter on a wide interface, etc. See Chapter 15, entitled "Clocking," on page 387 for a discussion concerning the FIFOs.

Example: Initialization And Use Of The Counters

The following three diagrams and associated descriptions explain the initialization of HyperTransport buffer counts, followed by the actions taken by the transmitter and receiver as two packets are sent across the link. The diagrams have been simplified to show a single flow control buffer and the corresponding receiver and transmitter counters used to track available entries. In this example, assume the following:

- The flow control buffer illustrated is the Posted Request Command (CMD) buffer.
- The designer of the receiver interface has decided to construct this flow control buffer with a depth of five entries. Because this is a buffer for receiving requests, each entry in the buffer will hold up to 8 bytes (this covers the case of either four or eight byte request packets)
- Following initialization, the transmitter wishes to send two Posted Request packets to the receiver.

Basic Steps In Counter Initialization And Use

1. At reset, the transmitter counters in each device are reset = 0. This prevents the initiation of any packet transfers until buffer depth has been established.
2. At reset, the receiver interfaces load each of the RCV counters with a value that indicates how many entries its corresponding flow control buffer supports (shown as N in the diagram). This is necessary because the receiver is allowed to implement buffers of any depth.
3. Each device then transmits its initial receiver buffer depth information to the other device using NOP packets. Each NOP packet can indicate a range

of 0-3 entries. If the receiver buffer being reported is deeper than 3 entries, the device will send additional NOPs which carry the remainder of the count.

4. As each device receives the initial NOP information, it updates its transmitter flow control counters, adding the value indicated in the NOP fields to the appropriate counter total.

5. When a device has a non-zero value in the counter, it can send packets of the appropriate type across the link. Each time it sends packet(s), the device subtracts the number of packets sent from the current transmitter counter value. If the counter decrements to 0, the transmitter must wait for NOP updates before proceeding with any more packet transmission.

Initializing The Flow Control Counter

In Figure 5-4 on page 111, assume that the designer of Device 2 has implemented a Posted Request (CMD) buffer with a depth of 5 entries. After reset, it must convey this initial buffer availability to the transmitter in the other device before any Posted Request packets may be sent.

Figure 5-4: Flow Control Counter Initialization

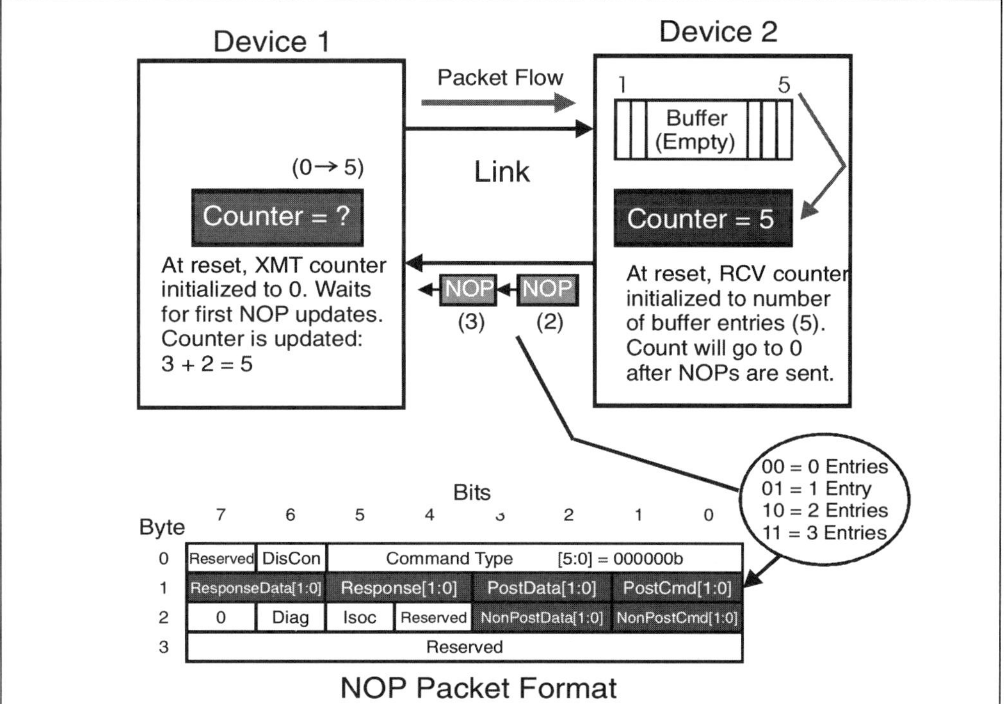

The basic steps in flow control counter initialization are:

1. The transmitter in Device 1 initializes its Posted Request (CMD) counter to 0 at reset (all transmit counters reset = 0). It then waits for the receiver on the other side to update this counter with the starting buffer depth available (this will be the maximum depth the receiver supports).
2. Device 2 loads its receiver Posted Request counter = 5 (its maximum).
3. Device 2 then sends two NOP packets which carry this buffer availability information: the first NOP has a 11b (3) in the Post CMD field (Byte 1, bits 0,1 above), and the second NOP has a 10b (2) in this field. Total = 5.
4. Upon receipt of these two NOPs, the Device 1 has updated its transmit counter, first by three then again by two. It now has 5 "credits" available for sending Posted Request packets — representing five separate Posted Requests which may be initiated.
5. Having sent the NOPs, the Device 2 RCV counter is now at 0, and will remain that way until additional packets are received, processed, and move out of the buffer, thereby creating new entries.

Note that this process will be repeated for each of the six required flow control buffers; it will also be done for the six isochronous flow control buffers if they are supported. In the NOP packet format (see above), six transmit registers can be updated at once using the six fields provided. The *Isoc* bit (Byte 2, bit 5) would be set if the NOP update was to be applied to the isochronous flow control buffer set.

Device 1 Sends Two Posted Request Packets

Figure 5-5 on page 113 shows the Device 1 transmitter sending two Posted Request packets. Also illustrated is the state of the flow control registers after this has been done, but before the receiver has processed the incoming packets.

Figure 5-5: Device 1 Sends Two Packets

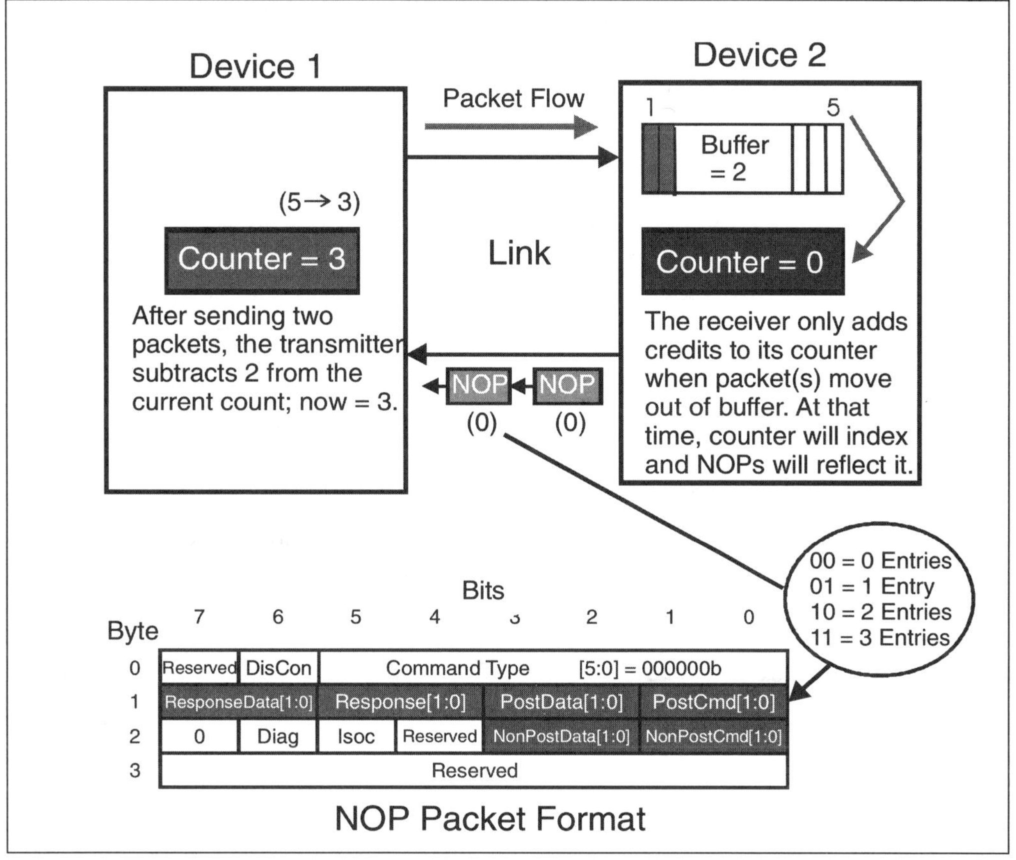

What the diagram shows:

1. After the flow control counters are initialized after reset, the transmitter sends packets for which credits are available in the transmit counter, then subtracts the number of packets sent from the current total. In this case, two credits were subtracted from the starting count of 5. Maintaining its own counter assures the transmitter will never send packets which can't be taken, even if there is a considerable lag in NOP updates from the receiver.
2. The receiver will not update its RCV counter or indicate new entries in its NOP packets until it moves a packet out of the flow control buffer.

New Entries Available: Update Flow Control Information

The last diagram, Figure 5-6 on page 114, shows the updating of flow control information after the receiver has processed the two packets it received previously. These could have been consumed internally, or been forwarded to another link if Device 2 is a bridge or tunnel.

Figure 5-6: Device 2 Updates Flow Control Information

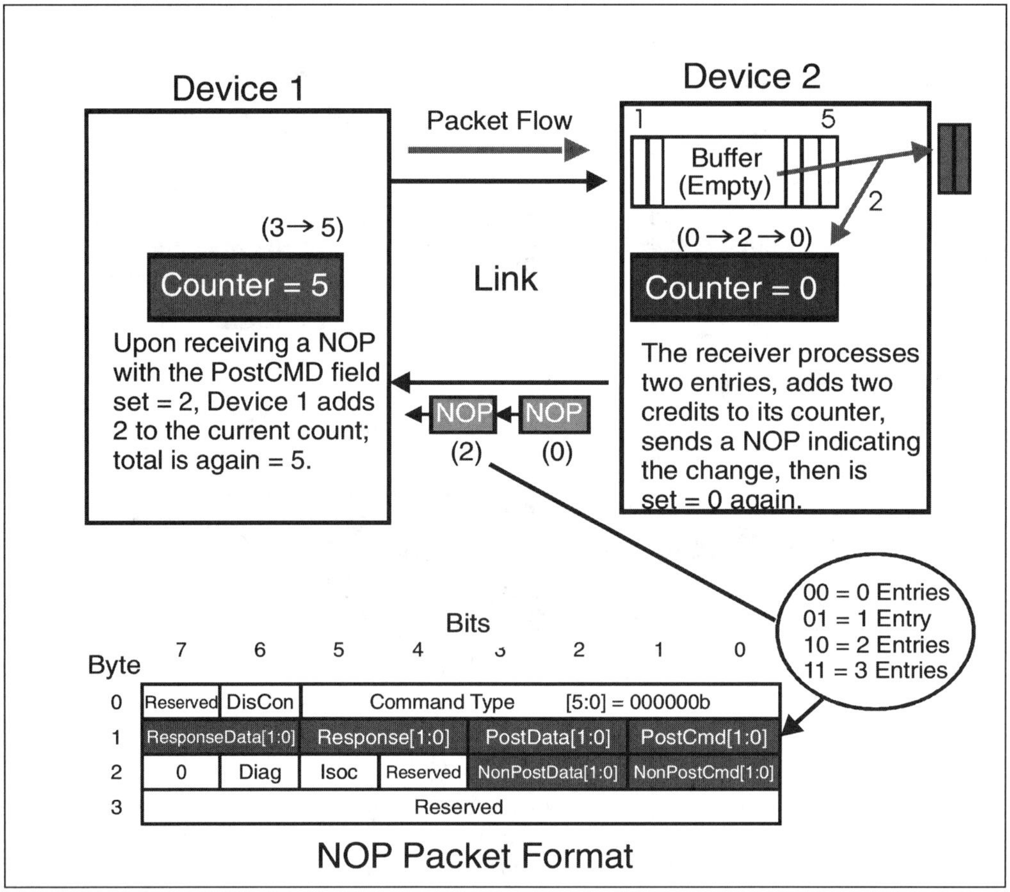

What has happened:

1. When Device 2 moves packets out of its buffer, it indexes its receive counter to reflect the new availability (2), and sends this information to Device 1 with a NOP update packet carrying the value 2 in the PostCMD field. After sending the update NOP, Device 2 clears its receive counter.
2. After sending packets and subtracting credits from its transmit counter, Device 1 will parse incoming NOPs for updates from the receiver enabling it to bump credits in one or more of its counters. In this case, Device 1 bumps its transmit counter by 2 upon receiving the NOP update. It again has 5 credits to use for sending Posted Request command packets.

This dynamic updating of flow control information happens continuously during idle times on the bus. If the receiver has no change in buffer entry availability to report, the NOPs it sends will have all six fields in the NOP packet cleared.

A Few Implementation Notes

Information Packets Not Flow-Controlled

Information packets, including NOP and Sync are not subject to flow control. When sent, they must be accepted, and are used for point-point communication between a transmitter and its corresponding receiver on a given link.

Transmitter Must Be Able To Track 15 Buffer Entries

While a receiver has the option of implementing buffers of any desired depth, including single entry buffers, each transmitter interface must implement its flow control counters such that they can track up to 15 entries in each of the six corresponding receiver flow control buffers (a four bit transmit counter will do this). If the transmitter counter is larger than that of the receiver, only a portion of it will ever be used (because NOP updates are always are based on available receiver flow control buffer entries). In the event that a transmitter implements a counter smaller than that of the receiver, the counter must "saturate" at the maximum value it can handle, and not roll over. The idea is that once the counters are initialized, they will use the maximum count that both devices can accommodate.

Sometimes Two Counters Must Be Checked

A transmitter can't issue a request packet that has data associated with it (e.g. a posted or non-posted write) without assuring the receiver has buffer entries available for both the request and the data. This is necessary because there is no receiver disconnect or retry mechanism once such a transaction starts.

NOP Packets Cannot Be Completely Blocked

The HyperTransport Specification indicates that it is the responsibility of each device to make certain that NOP update packets it sends are not starved (prevented from being sent) because of the sending of other types of traffic. If they are blocked, eventually one or more of the virtual channels may completely stall.

The Isochronous Flow Control Option

In the event that a designer decides to provide Isochronous flow control in addition to the standard three virtual channels, each receiver interface which supports Isochronous will implement six more receiver flow control buffers and counters, and an additional set of transmitter flow control counters as well. The way the receiver determines which flow control buffer (isochronous or standard) a packet should use is determined by a bit in the request packet (*Isoc* bit). If the Isoc bit is asserted in a request, it will also be asserted in the response when it comes back — again identifying the buffer set to use.

How About NOP Updates For Isochronous Buffers?

If a device supports the Isochronous flow control buffers, it will track packet progress through these buffers in the same way as it does for non-isochronous packets it receives. As Isochronous buffer entries become available, the receiver will return NOP update packets to the other device and will set the *Isoc* bit in the NOP packet (byte 2, bit 5) indicating the NOP packet updates should be applied to the isochronous transmit counters.

Isochronous Traffic/Non-Isochronous Flow Control

Receivers which see request packets with the Isoc bit set, but which are not in isochronous flow control mode, do not use the dedicated isochronous flow control buffers to handle them. In this case, the standard six flow control buffers are used and NOP buffer update packets returned to the transmitter all apply to the standard transmitter flow counters. Such devices preserve the Isoc bit in both the request packet and its response as they forward it to the next device; in this way, if there is a device in the path that does support isochronous traffic, it can still be used in that portion of the topology.

Isochronous Traffic Disabled At Initialization

At initialization, all devices are disabled with respect to isochronous traffic. Software can later enable ISOC traffic on a link-by-link basis after support on both sides of the link is determined through configuration space accesses. Once enabled for ISOC traffic, each device which sees isochronous packets and supports them is expected to apply a higher priority to them than for standard virtual channel packets.

6 *I/O Ordering*

The Previous Chapter

The previous chapter described the rules governing acceptance, forwarding, and rejection of packets seen by HyperTransport devices. Several factors come into play in routing, including the packet type, the direction it is moving, and the device type which sees it. A related topic also covered in this chapter is the fairness algorithm used by a tunnel device as it inserts its own packets into the traffic it forwards upstream on behalf of devices below it. The HyperTransport specification provides a fairness algorithm and a hardware method for tunnel management packet insertion.

This Chapter

This chapter describes the ordering rules which apply to packets associated with the three types of HyperTransport I/O traffic: PIO, DMA, and Peer-to-Peer. Depending on whether compatibility with the full producer-consumer ordering model used in PCI is required or relaxed ordering is permissible, attribute bits in request and response packets may be set or cleared. These bits are defined by the requester and are used by devices in the path to the target, and within the target, to enforce proper ordering. HyperTransport applies dedicated sets of ordering rules for upstream I/O traffic, downstream I/O traffic, and the special ordering required of host bridges and in double-hosted chains. Refer to Chapter 20, entitled "I/O Compatibility," on page 457 for a description of the additional ordering requirements when interfacing HyperTransport to other compatible protocols (e.g. PCI, PCI-X, and AGP).

The Next Chapter

In the next chapter, examples are presented which apply the packet principles described in the preceding chapter. The examples also entail more complex system transactions than discussed previously, including reads, posted and non-posted writes, and atomic read-modify-write operations.

The Purpose Of Ordering Rules

Some of the important reasons for enforcing ordering rules on packets moving through HyperTransport include the following:

Maintain Data Coherency

If transactions are in some way dependent on each other, a method is required to assure that they complete in a deterministic way. For example, if Device A performs a write transaction targeting main memory and then follows it with a read request targeting the same location, what data will the read transaction return? HyperTransport ordering seeks to make such events predictable (deterministic) and to match the intent of the programmer. Note that, compared to a shared bus such as PCI, HyperTransport transaction ordering is complicated somewhat by point-to-point connections which result in target devices on the same chain (logical bus) being at different levels of fabric hierarchy.

Avoid Deadlocks

Another reason for ordering rules is to handle cases where the completion of two separate transactions are each dependent on the other completing first. HyperTransport ordering includes a number of rules for deadlock avoidance. Some of the rules are in the specification because of known deadlock hazards associated with other buses to which HyperTransport may interface (e.g. PCI).

Support Legacy buses

One of the principal roles of HyperTransport is to serve as a backbone bus which is bridged to other peripheral buses. HyperTransport explicitly supports PCI, PCI-X, and AGP and the ordering requirements of those buses.

Maximize Performance

Finally, HyperTransport permits devices in the path to the target, and the target itself, some flexibility in reordering packets around each other to enhance performance. When acceptable, relaxed ordering may be enabled by the requester on a per-transaction basis using attribute bits in request and response packets.

Introduction: Three Types Of Traffic Flow

Hypertransport defines three types of traffic: Programmed I/O (PIO), Direct Memory Access (DMA), and Peer-to-Peer. Figure 6-1 on page 121 depicts the three types of traffic.

1. Programmed I/O traffic originates at the host bridge on behalf of the CPU and targets I/O or Memory Mapped I/O in one of the peripherals. These types of transactions often are generated by CPU to set up peripherals for bus master activity, check status, program configuration space, etc.
2. DMA traffic originates at a bus master peripheral and typically targets main memory. This traffic is used so that the CPU may be off-loaded from the burden of moving large amounts of data to and from the I/O subsystem. Generally, the CPU uses a few PIO instructions to program the peripheral device with information about a required DMA transfer (transfer size, target address in memory, read or write, etc.), then performs some other task while the DMA transfer is carried out. When the transfer is complete, the DMA device may generate an interrupt message to inform the CPU.
3. Peer-to-Peer traffic is generated by an interior node and targets another interior node. In HyperTransport, direct peer-to-peer traffic is not allowed. As indicated in Figure 6-1 on page 121, the request is issued upstream and must travel to the host bridge. The host bridge examines the address and determines whether the request should be reflected downstream. If the request is non-posted, the response will similarly travel from the target back up to the host bridge and then be reissued to the original requester.

Figure 6-1: PIO, DMA, And Peer-to-Peer Traffic

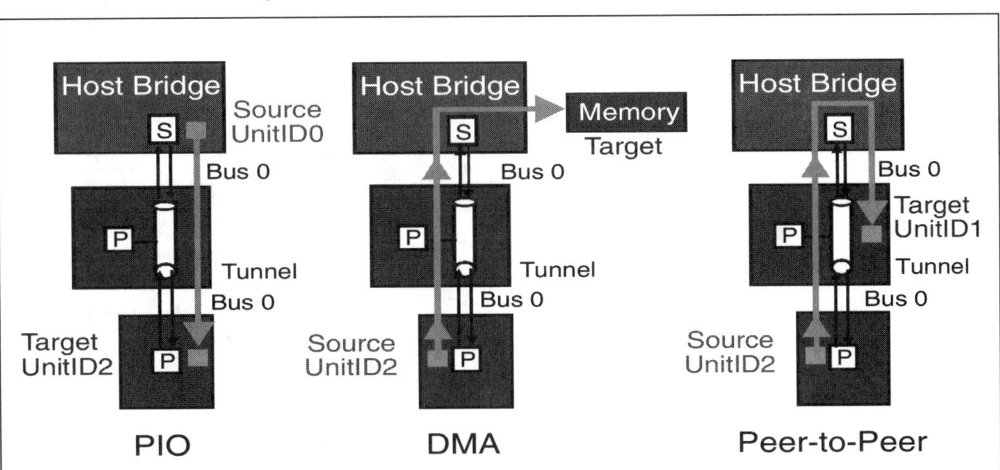

The Ordering Rules

HyperTransport packet ordering rules are divided into groups: general rules, rules for upstream I/O ordering, and rules for downstream ordering. Even the peer-to-peer example in Figure 6-1 on page 121 can be broken into two parts: the request moving to the bridge (covered by upstream ordering rules) and the reflection of the request downstream to the peer-to-peer target (covered by downstream I/O ordering rules). Refer to Chapter 20, entitled "I/O Compatibility," on page 457 for a discussion of ordering when packets move between HyperTransport and another protocol (PCI, PCI-X, or AGP).

General I/O Ordering Limits

Ordering Covers Targets At Same Hierarchy Level

Ordering rules only apply to the order in which operations are detected by targets at the same level in the HyperTransport fabric hierarchy. Referring to Figure 6-2 on page 122, assume that two peer-to-peer writes targeting devices on two different chains have been performed by the end device in chain 0.

Figure 6-2: Targets At Different Levels In Hierarchy And In Different Chains

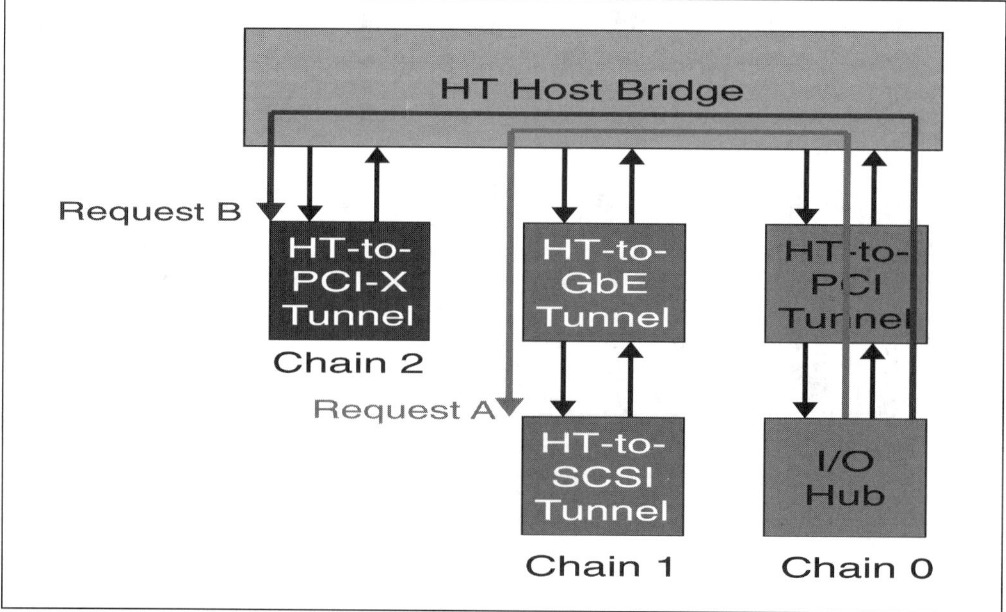

In the illustration Figure 6-2 on page 122, assume that Request A, a write transaction, is sent first. This is immediately followed by Request B, another write request. HyperTransport general ordering rules are then applied:

1. Upstream ordering rules assure that the two writes (Request A and Request B) arrive at the host bridge in the order they were generated.
2. When the host bridge then reflects the two write transactions downstream onto the separate chains (Chain 1 and Chain 2), downstream ordering rules guarantee that they will leave the host bridge in the order they arrived.
3. Once the two writes reach their respective chains, there is no way to guarantee that they will arrive at their respective targets in the order the requester intended because the ultimate targets are at different levels in the hierarchy.
4. The HyperTransport specification indicates that if the requester <u>must be certain</u> of the completion order at the targets, it should either poll the target of Request A for completion before issuing Request B or use a non-posted write for Request A and wait for the response to return before sending Request B.

Read And Non-Posted Write Completion At Target

Non-posted transactions issued by one requester to the same target are required to complete <u>at the target</u> in the order they were issued by the requester. This means that any combination of reads and non-posted writes must complete at the target in the original order they were issued. However, there is no ordering guarantee on the responses which are returned for each.

Figure 6-3: Non-Posted Requests And Responses At Target

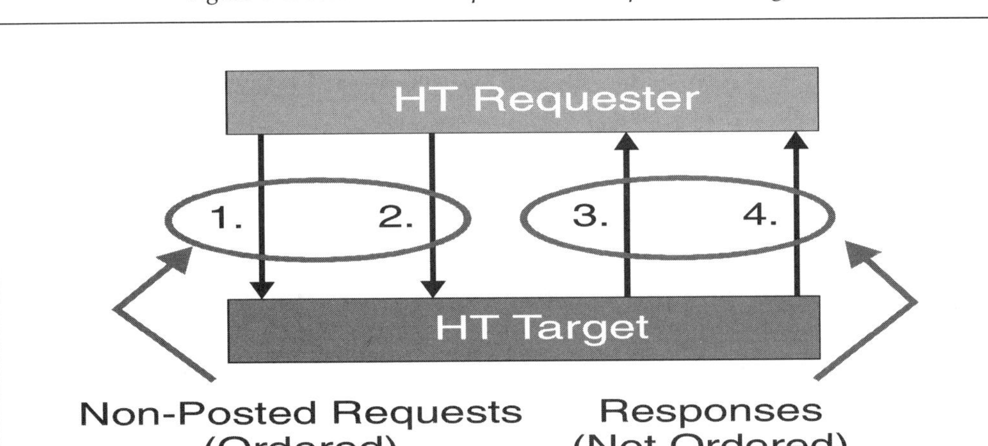

Referring to Figure 6-3 on page 123, ordering rules for target completion of non-posted requests and subsequent responses may be summarized:

1. The requester issues a non-posted write or read request to the target (1)
2. The requester issues another non-posted read or write request (2)

Ordering rules require that the two requests be handled internally by the target in order. When the responses return, they may come back in either order (3) and (4). The results of non-posted transactions must be globally visible (to all system devices) before a response is returned.

What If A Device Requires Response Ordering?　All HyperTransport devices must be able to tolerate out-of-order response delivery or else restrict outstanding non-posted requests to one at a time. This also applies to bridges which sit between HyperTransport and a protocol that requires responses be returned in order. The bridge must not issue more outstanding requests than it has internal buffer space to hold responses it may be required to reorder.

Support For The Producer-Consumer Ordering Model

When the *PassPW* and *Sequence ID* bits are cleared in a request packet, HyperTransport transactions are compatible with the same producer-consumer model PCI employs. Basic features of the model include:

1. A producer device anywhere in the system may send data and modify a flag indicating data availability to a consumer anywhere in the system.
2. The data and flag need not be located in the same device as long as the consumer of the data waits for the response of a flag read before attempting to access the data.
3. In cases where the consumer is allowed to issue two ordered reads <u>without making them part of an ordered sequence</u> (setting SequenceID tag to a non-zero value), the producer-consumer model is only supported if the flag and data are within the same device.
4. Ordering rules guarantee that if the flag is modified after the data becomes available, the flag read will return valid status.

Producer-Consumer Model Simpler If Flag/Data In Same Place

If the flag and data are restricted to being in the same device, the PassPW bit may be set in requests which relaxes the ordering of responses and improves performance. At the same time, the producer-consumer model is maintained.

Upstream Ordering Rules

Posted requests, non-posted requests, and responses travel in independent virtual channels. Each uses a different command, which permits devices to distinguish them from one another. Requests have a *Sequence ID* field. Assigning non-zero sequence ID fields to non-posted requests forces all tunnel and bridge devices in the path to the target to forward these requests in the same order they were received. The target is also required to maintain this order when processing these requests internally. Requests with a Sequence ID of zero are not considered to be part of an ordered sequence. Requests and response packets also carry a *May Pass Posted Writes* (PassPW) bit.

Reordering Packets In Different Transaction Streams

Other than when a Fence command is issued, there is no ordering guarantee for packets originating from different sources. Traffic from each UnitID is considered a separate transaction stream; devices may reorder upstream packets from different streams as necessary. Figure 6-4 on page 125 depicts reordering being done by a tunnel (UnitID1); this device is forwarding packets upstream on behalf of two devices behind it (UnitID2 and UnitID3).

1. UnitID3 issues packet (1) first. It is forwarded by UnitID2 to UnitID1.
2. Next UnitID1 receives a packet (2) from UnitID2.
3. When UnitID 1 forwards the two packets onto its upstream link, it may send packet (2) first. Packet (2) has then been reordered around packet (1).

Figure 6-4: Upstream Reordering: Packets From Different Transaction Streams

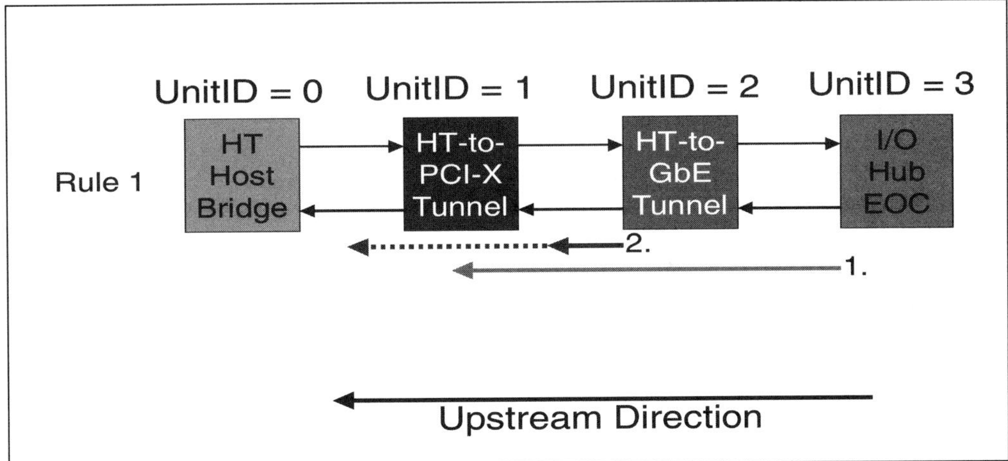

No Reordering Packets In A Strongly Ordered Sequence

If one requester has issued a series of request packets carrying the same non-zero SequenceID, the packets may not be reordered (regardless of the state of the PassPW bit. The sequence only applies to packets within a single transaction stream (UnitID) and VC. Upstream devices still may reorder these packets with respect to those from other streams. Figure 6-5 on page 126 illustrates an ordered sequence issued by an I/O Hub cave device. Key details include:

1. The I/O Hub issues a series of requests (1), (2), (3). All carry the same, non-zero SequenceID in the request.
2. When they are received by the first tunnel device, it checks the sequence ID field and the UnitID (all are identical). When it forwards the three packets to the PCI-X tunnel, it sends them in the same strongly ordered sequence.
3. The HyperTransport-to-PCI-X bridge makes the same determination and forwards packets (1), (2), and (3) through its tunnel interface to the host bridge in the same order.
4. The host bridge is also required to treat the three packets as a strongly ordered sequence internally.
5. If these were non-posted requests, there would be no guarantee of ordering in the responses returned to the I/O hub.

Figure 6-5: A Strongly Ordered Sequence Must Be Preserved

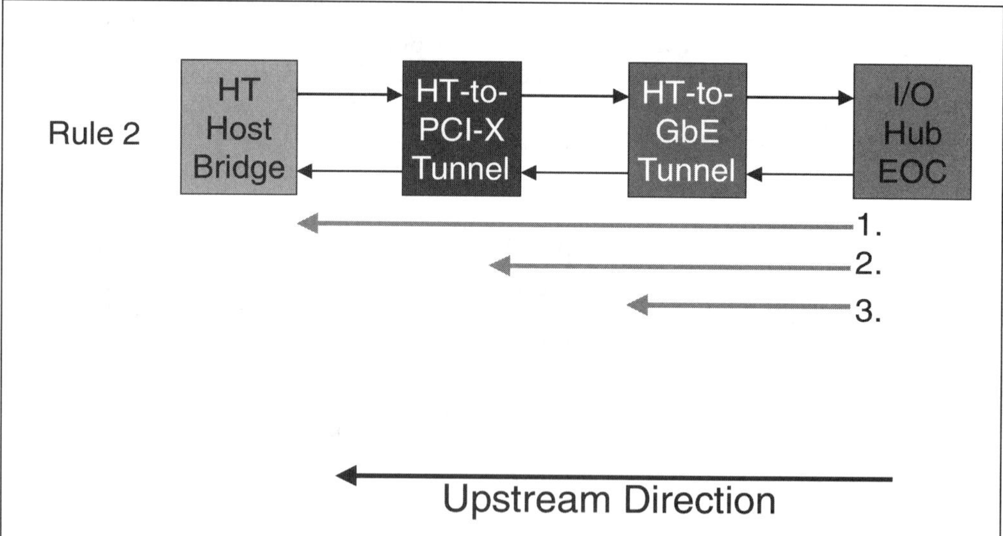

Packets With PassPW Bit Clear Are Restricted In Passing

Packets with the PassPW bit clear must not pass an already posted request in the same stream and not part of the same ordered sequence. This forces packets of all types (posted request, non-posted request, or response) which have not been granted relaxed ordering privileges to remain behind all previously posted requests within the same transaction stream. This guarantees the ultimate target (e.g. host bridge) will see them all in the original order they were issued. Figure 6-6 on page 127 illustrates this case.

1. The I/O Hub first issues a posted request (1)
2. It then issues another packet (2) which has the PassPW bit clear. This could be a posted or non-posted request, or a response packet.
3. When the upstream tunnel devices receive the two packets and determine they are from the same source (UnitID), and that the first is traveling in the posted virtual channel and the second has PassPW disabled, the order will be maintained during forwarding.
4. The host bridge is guaranteed to see the two packets in the original order they were issued, (1) then (2).

Figure 6-6: Packets With PassPW Clear Can't Pass Posted Requests

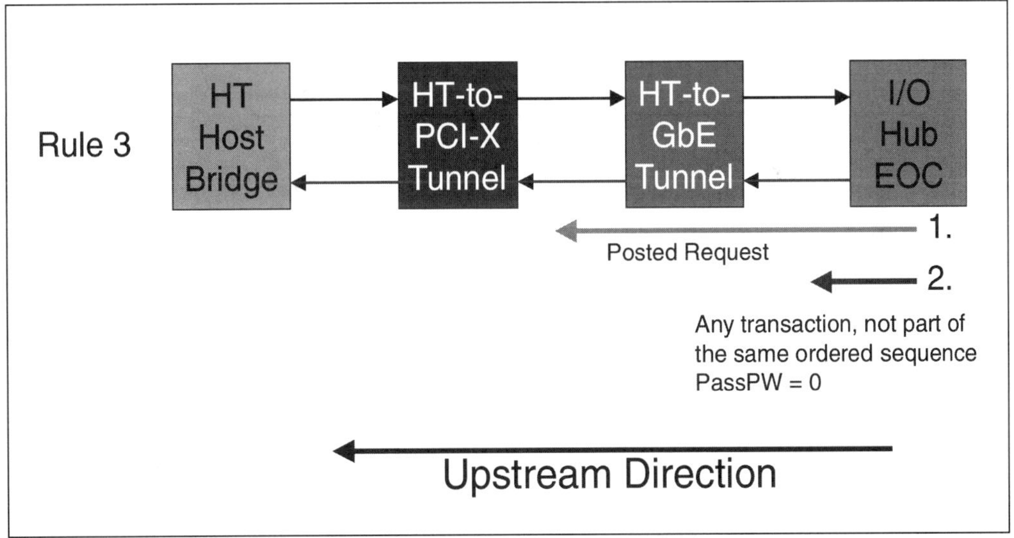

Packets With PassPW Bit Set May Or May Not Pass

Packets with the PassPW bit set may or may not pass an already posted request in the same stream and not part of the same ordered sequence. It is up to the forwarding devices to determine whether there is a benefit to reordering the two. Figure 6-7 on page 128 illustrates this case.

1. The I/O Hub first issues a posted request (1)
2. It then issues another packet (2) which has the PassPW bit set = 1. This could be a posted or non-posted request, or a response packet.
3. When the upstream tunnel devices receive the two packets and determine they are from the same source (UnitID), and that the first is traveling in the posted virtual channel and the second has PassPW set, they may or may not reorder the two packets as they are sent upstream.
4. In this case, it is indeterminate which packet will arrive at the host bridge first. If it matters which arrives first, then the requester would have issued the packets as a strongly ordered sequence (both packets would carry the same, non-zero Sequence ID).

Figure 6-7: Packets With PassPW Set May Or May Not Pass Other Posted Requests

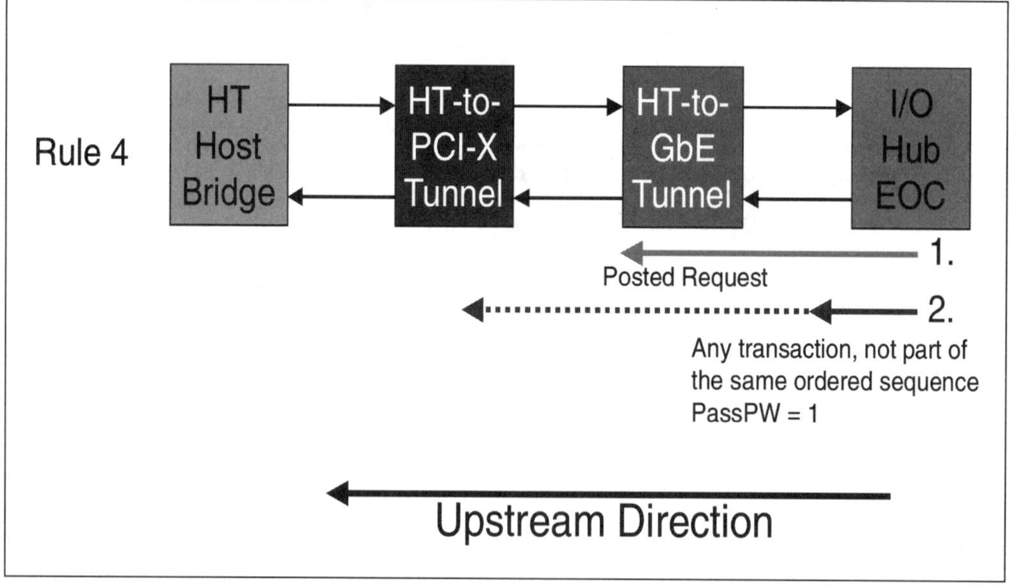

Non-Posted Requests May Pass Each Other

For non-posted requests which are not part of an ordered sequence (Sequence ID = 0), ordering rules allow them to pass other non-posted requests in the same transaction stream. Again, it is up to the forwarding devices to determine whether there is a benefit to reordering the non-posted requests. Figure 6-8 on page 129 illustrates this case.

1. The I/O Hub first issues a non-posted request packet (1)
2. It then issues another non-posted request packet (2).
3. When the upstream tunnel devices receive the two packets and determine they are from the same source (UnitID) and that they are non-posted requests which are not part of an ordered sequence, they may or may not reorder the two packets as they are sent upstream.
4. Again, it is indeterminate which packet will arrive at the host bridge first. If it matters which arrives first, then the requester would have issued the packets as a strongly ordered sequence (both packets would carry the same, non-zero Sequence ID).

Figure 6-8: Non-Posted Requests May Pass Each Other

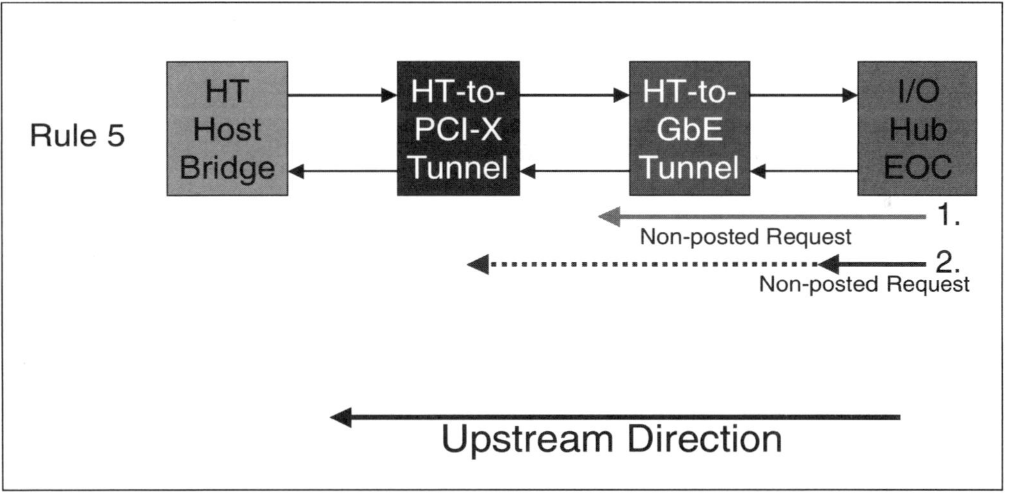

Posted Requests And Responses Must Be Able To Pass

Posted requests and responses must be able to pass previous non-posted requests that are not part of the same ordered sequence. This is part of the HyperTransport deadlock-avoidance strategy. *Note: The HyperTransport specification provides several additional recommendations for designers related to deadlock-avoidance.*

Figure 6-9 on page 130 illustrates the case of posted requests and responses passing previous non-posted requests in the same virtual channel.

1. The I/O Hub first issues a non-posted request packet (1)
2. It then issues another posted request or response packet (2).
3. When the upstream tunnel devices receive the two packets and determine they are from the same source (UnitID), they may or may not reorder the posted request (or response) around the non-posted request as they are sent upstream.
4. Again, a strongly ordered sequence could have been used if determinacy was required.

Figure 6-9: Posted Request Or Response Must Be Able To Pass Non-Posted Requests

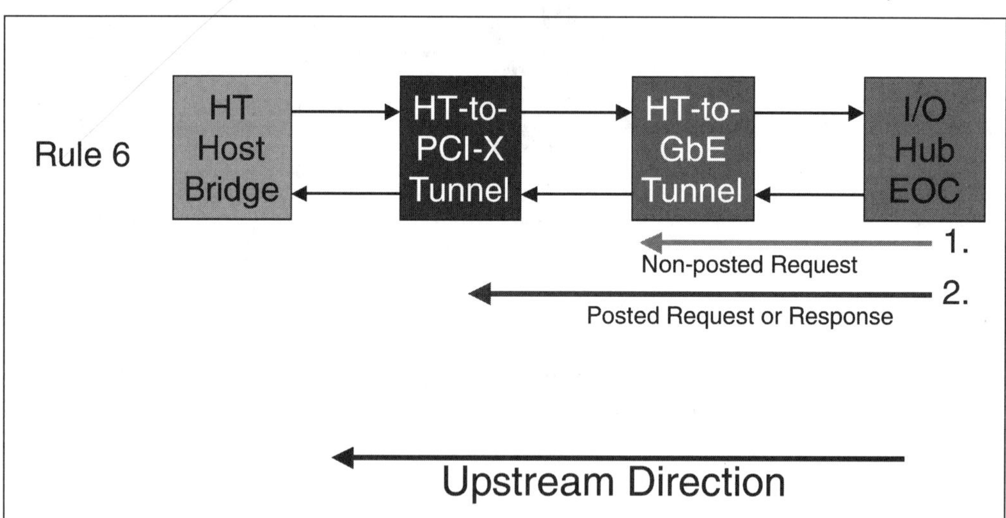

Posted Request Must Be Able To Pass A Response

Posted requests must be able to pass an earlier response which is not part of the same ordered sequence. This is another component of the HyperTransport deadlock-avoidance strategy. *Note: The HyperTransport specification provides several additional recommendations for designers related to deadlock-avoidance.*

Figure 6-10 on page 131 illustrates the case of a posted requests passing an earlier response packet in the same transaction stream

1. The I/O Hub first issues a response packet (1).
2. It then issues a posted request packet (2).
3. When the upstream tunnel devices receive they must be able to reorder the posted request around the earlier response as they are sent upstream.
4. Again, a strongly ordered sequence could have been used if determinacy was required.

Figure 6-10: Posted Request Must Be Able To Pass An Earlier Response

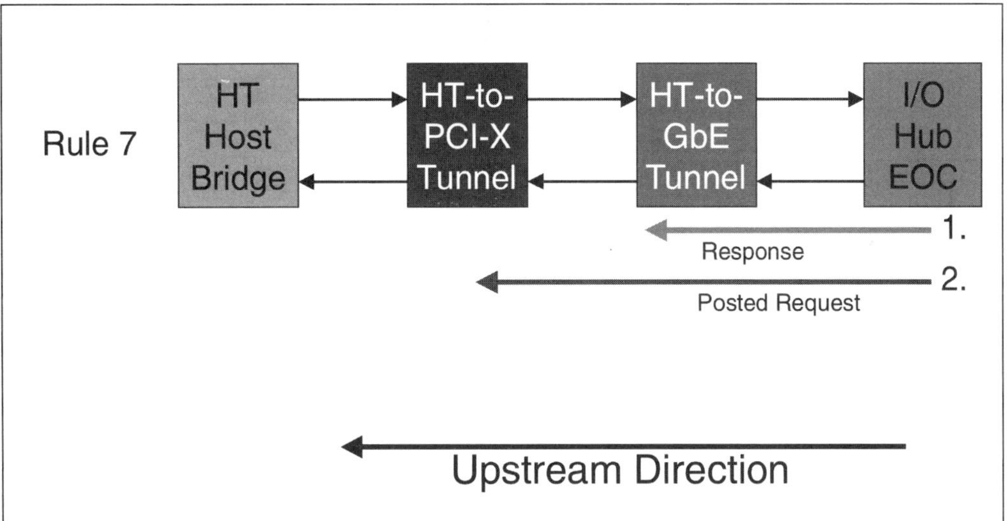

Non-Posted Requests Or Response May Pass A Response

Non-Posted requests or responses may or may not pass an earlier response which is not part of the same ordered sequence. Figure 6-11 on page 132 illustrates the case of a non-posted request or response passing an earlier response packet in the same transaction stream

1. The I/O Hub first issues a response packet (1).
2. It then issues a non-posted request or response packet (2).
3. When the upstream tunnel devices receive the two packets and determine they are from the same source (UnitID), they may or may not reorder the non-posted request or later response around the earlier response as they are sent upstream.
4. Again, a strongly ordered sequence could have been used if determinacy was required.

Figure 6-11: Non-Posted Request/Response May Pass Earlier Responses

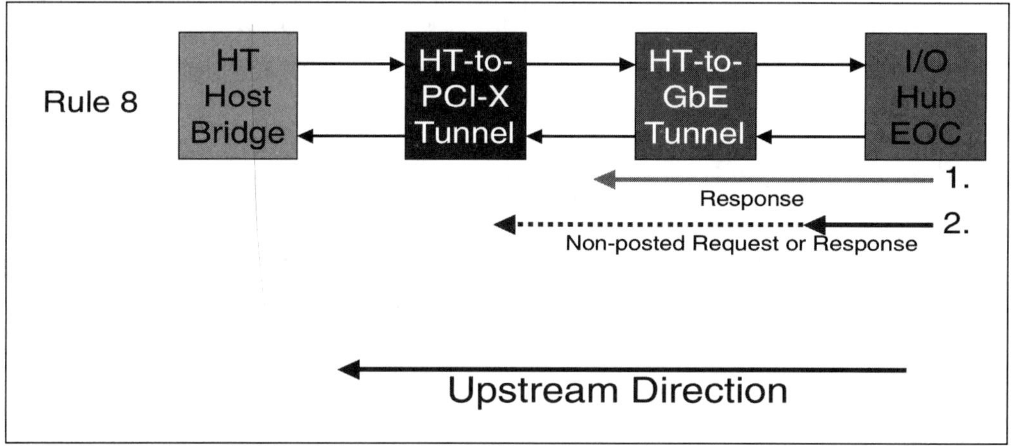

Host Ordering Requirements

A system that hosts HyperTransport must assure that the ordering of virtual channel traffic in the HyperTransport topology is extended to the host system. Because the host bridge interfaces directly to the processor(s), main memory, and the HyperTransport fabric, it plays a central role in enforcing the proper ordering interaction between the host system and HyperTransport. Figure 6-12 on page 133 illustrates the central role played by the host bridge in host system and HyperTransport ordering.

Note that some aspects of host system ordering are system-specific. For example, CPU host bus protocol and cache management vary with the processor. Still, host ordering rules make it possible for any host system to reliably interact with the HyperTransport fabric.

Figure 6-12: Host Bridge Extends Ordering To Host System

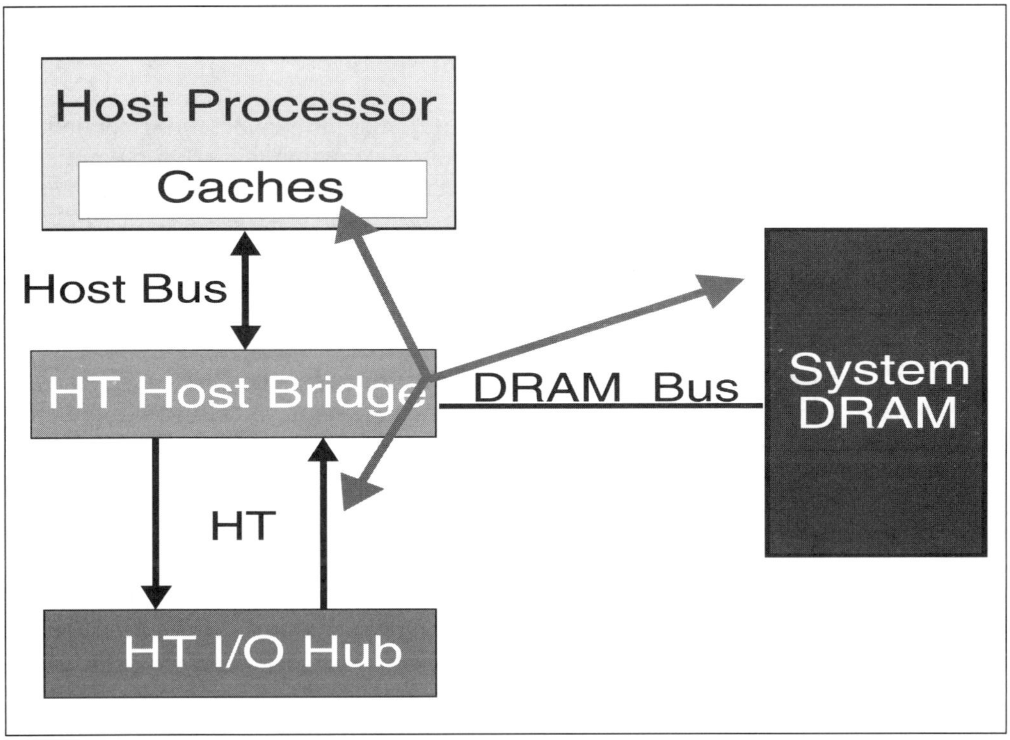

Host Ordering Requirements: General Features

The HyperTransport specification breaks down the ordering rules governing transaction completion in the host system into a set of rules for ordered pairs of transactions. Depending on the request types and where the target locations are, the second request may be received but might have to wait to <u>take effect</u> in the host fabric until the first request reaches a specific point in completion called its *ordering point*. "Taking effect", in this case, means that a read request actually fetches data, a write request actually exposes new data, peer-to-peer requests are actually queued for reissue downstream, etc.

How read and write accesses originating in HyperTransport are handled depends on the type of space they target in the host system.

1. Cacheable address ranges have strongest ordering
2. Non-cacheable memory, I/O, and MMIO have weaker ordering
3. Interrupt and System Management Address ranges have special ordering

Two Ordering Points Are Defined

There are two ordering points (degrees of transaction completion) defined for the first transaction in an ordered pair; this information is used in determining whether the second request of the ordered pair may take effect or must wait. The ordering points are called *Globally Ordered* (GO) and *Globally Visible* (GV).

Globally Ordered (GO). HyperTransport defines the globally ordered point for the first request as the point where it is guaranteed to be observed in the correct order (with respect to the second transaction) from any "observer". While the two transactions are guaranteed to complete in the proper order, they may not have actually done so yet. This means agents such as caches may not have been updated at this ordering point.

Globally Visible (GV). HyperTransport defines the globally visible ordering point for the first request as the point where it is assured to be "visible" to all observers (CPUs, I/O devices, etc.). It also means that all side effects of the first request (cache transitions, etc.) have completed.

Note: If there are no "sideband" agents (caches, etc.), GO and GV are equivalent.

Ordering Rule Summary

Table 6-1 on page 134 summarizes the host ordering rules for various combinations of transaction ordered pairs.

Table 6-1: Summary Of Host Ordering Rules For Transaction Pairs

First Command	Second Command	Second Command Waits For First To Be:
Cacheable Write	Cacheable Write	GV
Cacheable Write	Cacheable Read	GO

Table 6-1: Summary Of Host Ordering Rules For Transaction Pairs

First Command	Second Command	Second Command Waits For First To Be:
Cacheable Read	Cacheable Read or Write	GO
Non-Cacheable	Non-Cacheable	GO
Cacheable Write	Non-Cacheable	GV
Cacheable Read	Non-Cacheable	GO
Non-Cacheable	Cacheable	GO
Cacheable Write	Flush/Interrupt SysMgmt Response	GV
Cacheable Read	Flush/Interrupt SysMgmt Response	No Wait Requirement
Non-Cacheable	Flush/Interrupt SysMgmt Response	GO
Flush/Response	Any	No Wait Requirements
Int/SysMgmt	Fence or Response	GV
Int/SysMgmt	Any Except Fence/ Response	No Wait Requirements
Posted Cacheable	Fence	GV
Posted Non-Cacheable	Fence	GO
Any Non-Posted	Fence	No Wait Requirements
Fence	Any	GV

Host Responses To Non-Posted Requests

Although the second request in an ordered pair may be allowed to take effect in the host system before the first request is globally visible, the host is not allowed to return the response for a non-posted HyperTransport request until all previous ordered requests are complete and the side effects (e.g. cache state transitions) of the current request are similarly globally visible.

An Example (Refer to Table 6-1 and Figure 6-13)

Assume that an ordered pair of requests have been received by the host bridge. The first is a read request targeting a cacheable area of memory; the second is a non-posted write targeting the same location in cacheable memory space. Both requests are part of the same strongly ordered sequence (same stream, same VC, SeqID >0). According to Table 6-1 (see line number 3), the second request (cacheable write) must wait until the first request is *globally ordered* (it is guaranteed to complete first).

1.   The I/O hub issues the read request targeting a cacheable area of memory.
2.   It immediately issues a second request, a non-posted write to the same location. These two requests are part of a strongly ordered sequence.
3.   The host bridge causes a snoop cycle of processor caches before allowing the read of memory (the processor may have a modified cache line)
4.   With cache coherency taken care of, the read of memory completes. The response (not shown) may be returned to the requester because previous ordered requests are complete and side effects of this read are handled.
5.   The cacheable write request is submitted for CPU cache look-up. The cache line will be invalidated in the event of a hit in the cache.
6.   Coherency assured, the cacheable write is allowed to complete to memory.

Figure 6-13: Ordering Example: Read Followed By Posted Write To Cacheable Memory

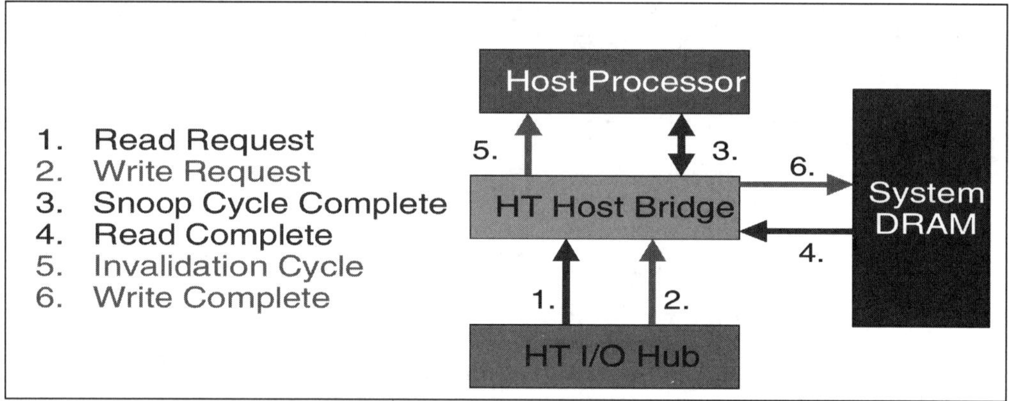

Downstream I/O Ordering

Downstream ordering rules in HyperTransport are much the same as the upstream rules previously described, with a few exceptions:

1.   While the same virtual channels are used (posted request, non-posted

request, and response), downstream I/O streams are determined by the target of the transaction instead of the source.

2. Although *UnitID* uniquely identifies upstream transaction stream requests, it can't be used for this purpose in downstream requests because the UnitID field is always that of the host bridge (UnitID 0). All downstream request traffic is assumed to be part of the same transaction stream (the host bridge's).

3. The *bridge* bit is used to help nodes distinguish downstream from upstream response traffic. It also helps devices interpret the UnitID field in responses. Upstream responses carry the UnitID of the sender (the original target), while downstream responses carry the UnitID of the original requester. Interior nodes are only allowed to claim response packets which carry their UnitID and are moving downstream (bridge bit set = 1).

4. A host bridge (this includes the secondary interface of HyperTransport-HyperTransport bridges) which performs a peer-to-peer reflection must preserve strongly ordered sequences (non-zero Sequence ID) when it reissues them downstream. It is allowed to change the Sequence ID tag, but the same tag will be applied to all requests in the sequence.

Double-Hosted Chain Ordering

Upstream traffic and downstream traffic in HyperTransport have no ordering interaction because they are in different transaction streams. A special case arises in sharing double-hosted chains when one of the host bridges must send traffic to the other host bridge. Refer to Figure 6-14 on page 138.

1. Host bridge A sends a posted write targeting host bridge B.
2. At nearly the same time, host bridge B performs a read from host bridge A.
3. The read response/data will be travelling in the same direction as the posted write (towards host bridge B).
4. Although the posted write request is traveling downstream and the read response is traveling upstream (from the perspective of Device B), the producer-consumer ordering model requires that both must be treated as being in the same transaction stream (response will push posted write request if PassPW is clear).
5. Devices in the path can perform ordering tests on upstream responses based only on UnitID (both 0 in the case of two bridges communicating with each other), and by disregarding the direction of the requests.
6. In the event a host has its *Act as Slave* bit set = 1, then it won't use UnitID 0; for its requests and responses; in this case, conventional ordering based on UnitID will work.

Figure 6-14: Double-Hosted Chain Ordering

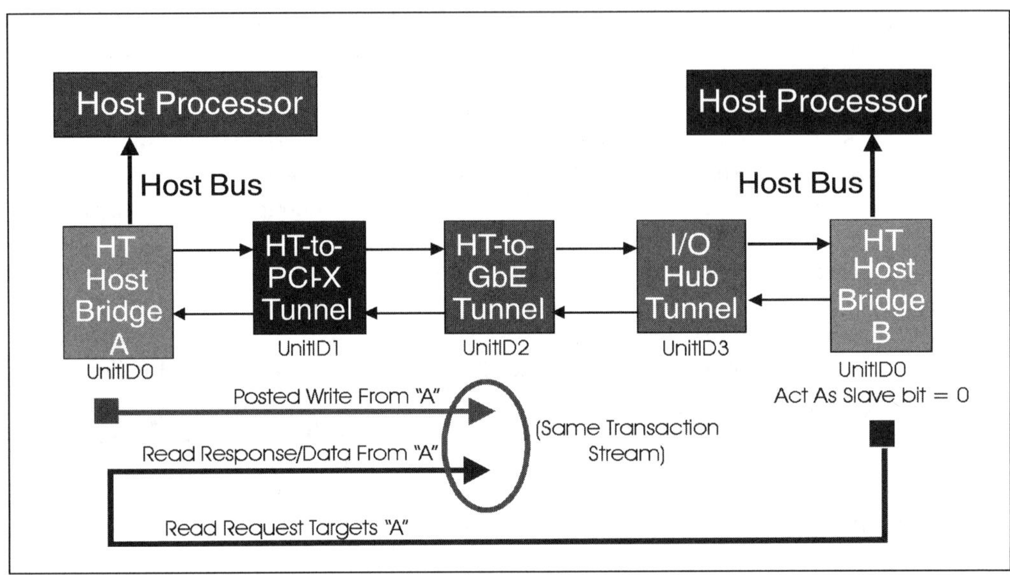

7 *Transaction Examples*

The Previous Chapter

The previous chapter described the ordering rules which apply to packets associated with the three types of HyperTransport I/O traffic: PIO, DMA, and Peer-to-Peer. Depending on whether compatibility with the full producer-consumer ordering model used in PCI is required or relaxed ordering is permissible, attribute bits in request and response packets may be set or cleared. These bits are defined by the requester and are used by devices in the path to the target, and within the target, to enforce proper ordering. HyperTransport applies dedicated sets of ordering rules for upstream I/O traffic, downstream I/O traffic, and the special ordering required of host bridges and in double-hosted chains. Refer to Chapter 20, entitled "I/O Compatibility," on page 457 for a description of the additional ordering requirements when interfacing HyperTransport to other compatible protocols (e.g. PCI, PCI-X, and AGP).

This Chapter

In this chapter, examples are presented which apply the packet principles in the preceding chapters and includes more complex system transactions, not previously discussed. The examples include reads, posted and non-posted writes, and atomic read-modify-write.

The Next Chapter

HT uses an interrupt signaling scheme very similar to PCI's Message Signaled Interrupts. The next chapter defines how HT delivers interrupts to the Host Bridge via posted memory writes. This chapter also defines an End of Interrupt message and details the mechanism that HT uses for configuring and setting up interrupt transactions (which is different from the PCI-defined mechanisms).

Packets As Transaction Building Blocks

HyperTransport *control* packet types — information, request, and response are used in various combinations to accomplish transactions. In many transactions, *data* packets are also used with the control packets to carry a data payload ranging from 0-64 valid bytes. Transactions start when the transmit interface of a device sends an information or request control packet. Any bridges or tunnels in the path between a requester and the ultimate target have responsibilities for forwarding any request, response, and data packets associated with the transfer in the proper direction. *Note:* This chapter highlights key packet fields used in the construction of HyperTransport transactions; refer to Chapter 4, entitled "Packet Protocol," on page 59 for a more complete description of HyperTransport packets and the bit fields associated with them.

Table 7-1 on page 140 summarizes the interaction between a request agent and the ultimate target of a HyperTransport transaction following the sending of various types of information and request control packets.

Table 7-1: Implications Of Sending Information And Request Control Packets

Packet Type	Command Name	Comments
Information	**NOP**	Used by each node transmitter to indicate the idle condition, report receiver flow control updates, and send other miscellaneous information to its corresponding receiver. These packets are not forwarded by the receiver and no response or data is associated with them.
Information	**Sync/Error**	Sent by each node transmitter during link synchronization <u>or</u> by a device enabled to report errors using the Sync flood mechanism to indicate the need for link reset and re-synchronization. During a Sync flood, each recipient re-issues the Sync packets onto all outgoing links on the chain until reset is detected. There are no response or data packets associated with a Sync packet.

Table 7-1: Implications Of Sending Information And Request Control Packets

Packet Type	Command Name	Comments
Request	**Sized Write (Posted)** **dword or byte transfers OK**	A posted sized write is used to initiate a write transfer of dwords or bytes of data to a target. For dword writes, the 1-16 dword data packet immediately follows the write request. For byte writes, a single dword "byte mask" precedes a data packet of 1-8 dwords (containing up to 32 valid bytes). No response is ever returned to a posted write and devices in the target path may deallocate buffers as soon as the request and data are forwarded.
Request	**Sized Write (Non-Posted)** **dword or byte transfers OK**	A non-posted sized write is also used to initiate a write transfer of dwords or bytes of data to a target. For dword writes, the 1-16 dword data packet immediately follows the write request. For byte writes, a single dword "byte mask" precedes a data packet of 1-8 dwords (containing up to 32 valid bytes). The *Target Done* response will be sent when the write completes (either by the target or by an EOC device). Bridges in the target path must track outstanding non-posted write requests until the target done response is returned.
Request	**Broadcast Message**	Broadcast messages originate at the host bridge, and are accepted and propagated downstream on all links by each device which sees them. As they are posted requests, there is no response and devices in the target path may deallocate buffers as soon as the broadcast message request is forwarded.

Table 7-1: Implications Of Sending Information And Request Control Packets

Packet Type	Command Name	Comments
Request	**Sized Read** **dword or byte transfers OK**	A sized read is used to initiate a read transfer of dwords or bytes from a target. For byte reads, a single dword of data is returned immediately after the read response. For dword reads, 1-16 dwords are returned immediately after the read response. Bridges in the target path must track all outstanding read requests until the read response and data are returned.
Request	**Flush**	Issued by a requester to force its preceding posted writes in the same transaction stream to the Host Bridge. This is a non-posted request and the Host Bridge returns a *Target Done Response* when the flush of all previous posted writes for this source is completed to memory (or to the destination chain in a peer-to-peer transaction). Bridges in the target path must track outstanding Flush requests until the *Target Done Response* is returned. There is no data packet associated with the Flush.
Request	**Fence**	Issued to force the host bridge to place a barrier between previous and subsequent posted writes in all transaction streams. The Host Bridge will push previous writes for all streams to memory before allowing <u>any</u> subsequent posted writes with(*PassPW* clear) to be processed. Unlike Flush, this command is posted. There will be no response; devices in the target path may deallocate buffers as soon as Fence request is forwarded. There is no data packet associated with the Fence.

Table 7-1: Implications Of Sending Information And Request Control Packets

Packet Type	Command Name	Comments
Request	**Atomic RMW**	Issued by a requester seeking to perform a read-modify-write of a memory location in a single transaction. This hybrid request causes a transaction made up of a *non-posted write* operation followed by a *read response with data*. There are two variants of Atomic RMW (Fetch & Add and Compare & Swap.) Because Atomic RMW requests are always non-posted, bridges in the path must track outstanding Atomic RMW requests until the response/data are returned.
Response	**Read Response**	Issued by a target when it is ready to either return previously requested read data (see Sized Read and Atomic RMW requests) or an error indication that the request did not complete properly. The read response immediately precedes read data being returned by the target. If an error occurred, the requested amount of data is returned anyway; response error bits will indicate that it is not valid and whether the response was sourced by the original target or an end-of-chain device.
Response	**Target Done Response**	Issued by a target to confirm the completion of an earlier non-posted write or Flush request. There is no data packet associated with a Target Done response. If an error occurred in completing the original request, the target done response error bits indicate the failure and whether it occurred at the original target or the request was inadvertently sent to an end-of-chain device.

Transaction Examples: Introduction

The following section describes some transactions which might be seen in HyperTransport. Not all cases are covered, but the ones which are presented are intended to illustrate two important aspects of each transaction:

Packet Format And Optional Features

Most of the control packet variants have a number of fields used to enable optional features: isochronous vs. standard virtual channels, byte vs. dword data transfers, posted vs. non-posted channels, etc. In each of the examples, the key options are described; in some cases, certain packet fields don't apply to a particular request or response at all. Refer to Chapter 4, entitled "Packet Protocol," on page 59 for low level, bit-by-bit packet field descriptions.

General Sequence Of Events

Once the packet fields are determined, the next topic covered in the transaction examples is a general summary of events which must occur in the HyperTransport topology to perform the transaction. The description starts at the agent initiating an information or request control packet and follows it (and possibly a data packet) to the target; if there is a response required, this packet (and any data associated with it) is similarly followed from the target back to the original requester. Packet movement through HyperTransport involves aspects of topics described in more detail in other chapters, including flow control, packet ordering and routing, error handling, etc.

For each of the examples, assume the link interface CAD bus is eight bits in each direction. This simplifies the discussion of packet bytes moving across the connection — each byte of packet content is sent in one bit time. The only difference, when link width is greater or less than 8 bits, is how the packet bytes are shifted out and received. For example, on a 2-bit interface each byte of packet content is shifted onto the link over 4 bit times. For an interface 32 bits wide, four bytes of packet content are sent in parallel with each other in each bit time.

Example 1: NOP Information Packet

Problem: Device 2 in Figure 7-1 on page 145 is using a NOP packet to inform Device 1 about the availability of two new *Posted Request (CMD)* entries and one new *Response (Data)* entry in its receiver flow control buffers.

Figure 7-1: Example 1: NOP Information Packet With Buffer Updates

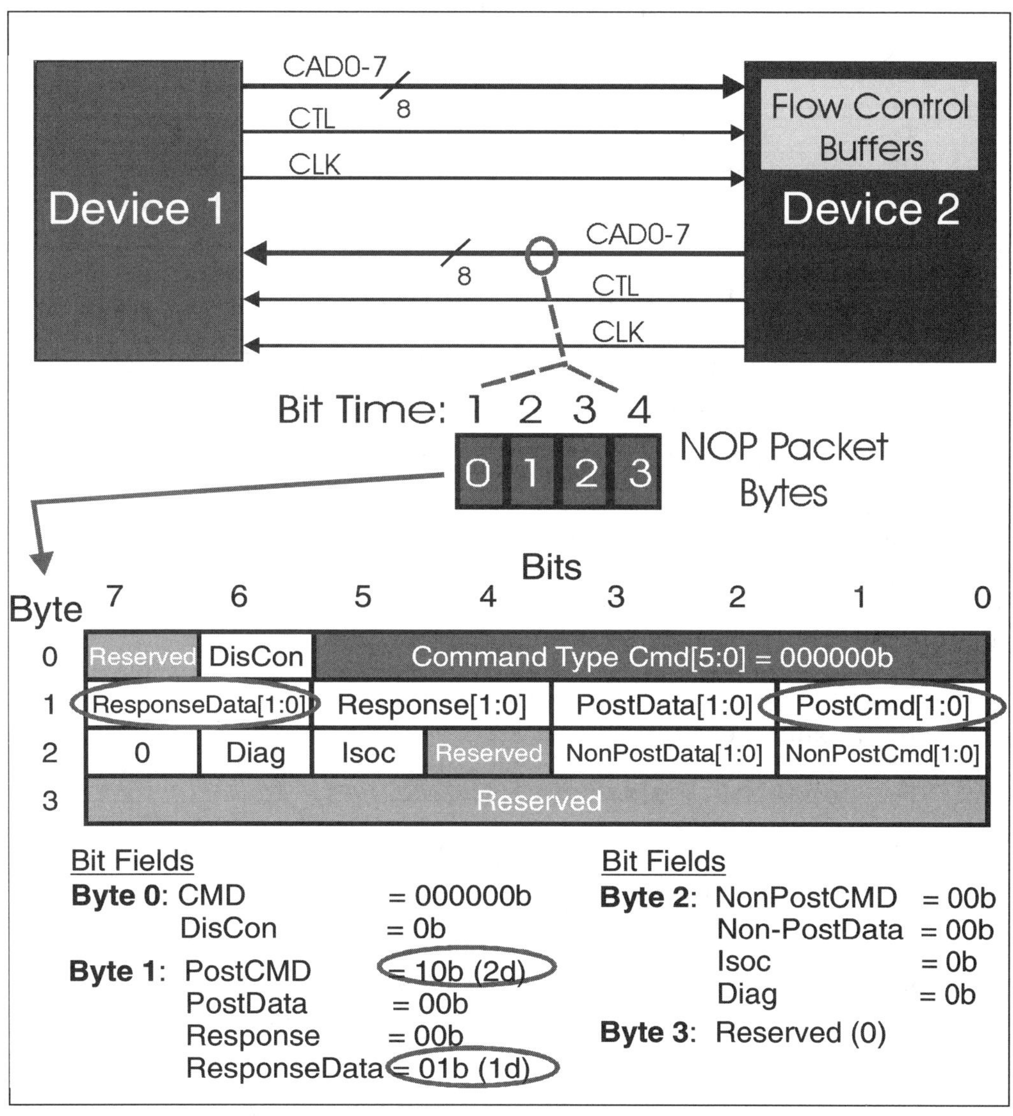

Example 1: NOP Packet Setup

(Refer to Figure 7-1 on page 145.)

Command[5:0] Field (Byte 0, Bit 5:0)

The NOP information packet command code. There are no option bits within this field. **For this example, field = 000000b.**

DisCon Bit Field (Byte 0, Bit 6)

This bit is set to indicate an LDTSTOP# sequence is beginning. Not active for this example. Refer to "HT Link Disconnect/Reconnect Sequence" on page 223 for a discussion of the DisCon bit and the LDTSTOP# sequence. **For this example, field = 0b.**

PostCMD[1:0] Field (Byte 1, Bits 1:0)

This NOP field is used by all devices to dynamically report the number of new entries (0-3) in the receiver's *Posted Request (CMD)* flow control buffers which have become available since the last NOP reporting this information was sent. In our example, assume that two new entries have become available. **For this example, field = 10b.**

PostData[1:0] Field (Byte 1, Bits 3:2)

This NOP field is used by all devices to dynamically report the number of new entries (0-3) in the receiver's *Posted Request (Data)* flow control buffers which have become available since the last NOP reporting this information was sent. In our example, assume no new entries in this buffer have become available. **For this example, field = 00b.**

Response[1:0] Field (Byte 1, Bits 5:4)

This NOP field dynamically reports the number of new entries (0-3) in the receiver's *Response (Data)* flow control buffers which have become available since the last NOP reporting this information was sent. In our example, assume no new entries in this buffer have become available. **For this example, field = 00b.**

ResponseData[1:0] Field (Byte 1, Bits 3:2)

This NOP field is used by all devices to dynamically report the number of new entries (0-3) in the receiver's *Response (Data)* flow control buffers which have become available since the last NOP reporting this information was sent. In our example, assume one new entry in this buffer has become available. **For this example, field = 01b.**

Non-PostCMD[1:0] Field (Byte 2, Bits 1:0)

This NOP field dynamically reports the number of new entries (0-3) in the receiver's *Non-Posted Request (CMD)* flow control buffers which have become available since the last NOP reporting this information was sent. In our example, assume that no new entries have become available. **For this example, field = 00b.**

Non-PostData[1:0] Field (Byte 2, Bits 3:2)

This NOP field dynamically reports the number of new entries (0-3) in the receiver's *Non-Posted Request (Data)* flow control buffers which have become available since the last NOP reporting this information was sent. In our example, assume no new entries in this buffer have become available. **For this example, field = 00b.**

Isoc Bit Field (Byte 2, Bit 5)

If set, this NOP field indicates that all six flow control buffer updates contained within this NOP should be applied to the transmitter's isochronous flow control buffer counter set. In our example, assume that the update applies to the standard flow control buffer entries. **For this example, field = 0b.**

Diag Bit Field (Byte 2, Bit 6)

If set, this NOP field indicates the transmitter is entering the CRC diagnostic test mode. In this example, this bit is cleared. **For this example, field = 0b.**

Bits Not Mentioned

Reserved bits in bytes 0,1; these should be driven = 0.

Example 1: NOP Sequence Of Events On The Link

(Refer to Figure 7-1 on page 145)

NOP packets are sent continuously during idle times on each link. Because they aren't subject to flow control and don't have data or response packets associated with them, the sequence of events is simple. For the example in Figure 7-1 on page 145:

1. Device 2 has processed two packets received previously from Device 1 and is about to report the new entries which have suddenly become available in its corresponding receiver flow control buffers.
2. Device 2 prepares the next NOP packet, filling in the six fields used to report buffer updates and clears the Isoc bit because this update is for the standard flow control buffer set. Flow control buffer fields in the response are cleared to "0" for buffers which haven't changed since the last NOP. Other bits carried in NOP packets are not enabled in this example (e.g. *DisCon, Diag*).
3. The transmitter interface in Device 2 drives the four byte NOP packet byte out onto the 8-bit link, least significant byte first. After four bit times (2 clock periods), the NOP packet has been sent; Device 2 clears its internal receiver flow control update counters — indicating the update is complete.
4. Upon receiving the NOP packet update, Device 1 adds the indicated "credits" to its corresponding flow control counters. It also checks the status of the other bits in the packet and detects them clear (no *DisCon, Diag*, etc.).
5. Device 2 returns to sending "null packet" NOPs (all flow control update fields = 0) during idle times on the link until another buffer entry becomes available.

Generic Request And Response Packet Formats

Many of the following examples are based on variants on the sized read (RdSized) and sized write (WrSized) request control packets and the Target Done and Read response packets associated with them. While the use of field bits varies with the specific request and response, the formats of the generic WrSized/RdSized request packet and the Read/Target Done response packet are shown in Figure 7-2 and Figure 7-3 on page 149 for reference.

Figure 7-2: Generic RdSized And WrSized Request Packet Format

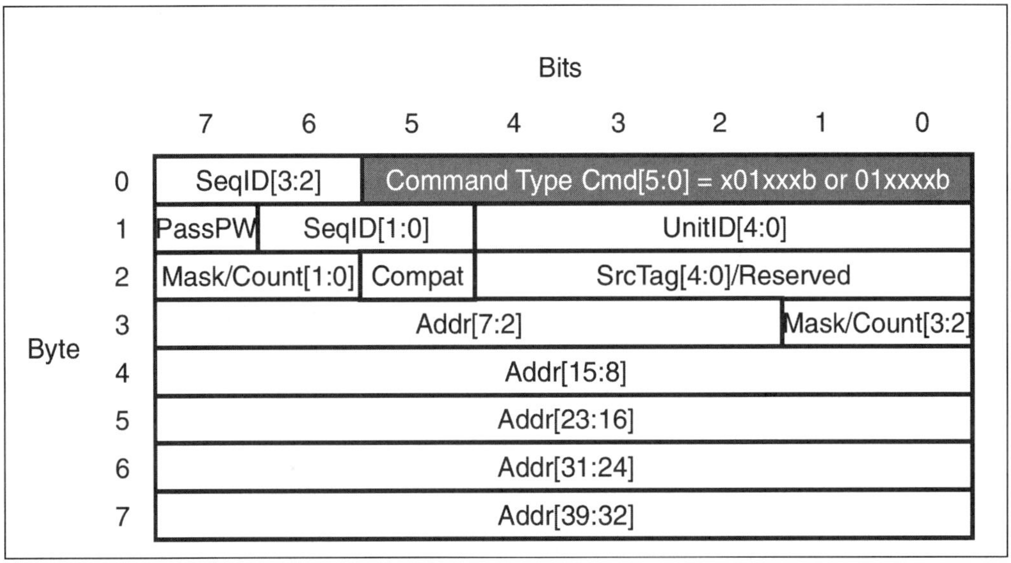

		Bits							
		7	6	5	4	3	2	1	0
Byte	0	SeqID[3:2]		Command Type Cmd[5:0] = x01xxxb or 01xxxxb					
	1	PassPW	SeqID[1:0]		UnitID[4:0]				
	2	Mask/Count[1:0]		Compat	SrcTag[4:0]/Reserved				
	3	Addr[7:2]						Mask/Count[3:2]	
	4	Addr[15:8]							
	5	Addr[23:16]							
	6	Addr[31:24]							
	7	Addr[39:32]							

Figure 7-3: Generic Read/Target Done Response Packet Format

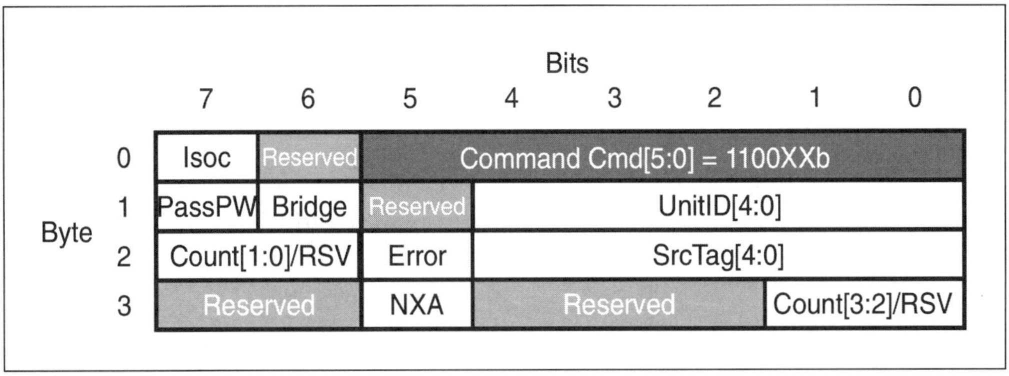

		Bits							
		7	6	5	4	3	2	1	0
Byte	0	Isoc	Reserved	Command Cmd[5:0] = 1100XXb					
	1	PassPW	Bridge	Reserved	UnitID[4:0]				
	2	Count[1:0]/RSV		Error	SrcTag[4:0]				
	3	Reserved		NXA	Reserved			Count[3:2]/RSV	

Example 2: Non-Posted WrSized (Dword) Transaction

Problem: Device 2 (UnitID2) in Figure 7-4 on page 150 targets main memory with a non-posted, isochronous, sized (dword) write of two dwords.

Figure 7-4: DMA Non-Posted Write Targeting Main Memory

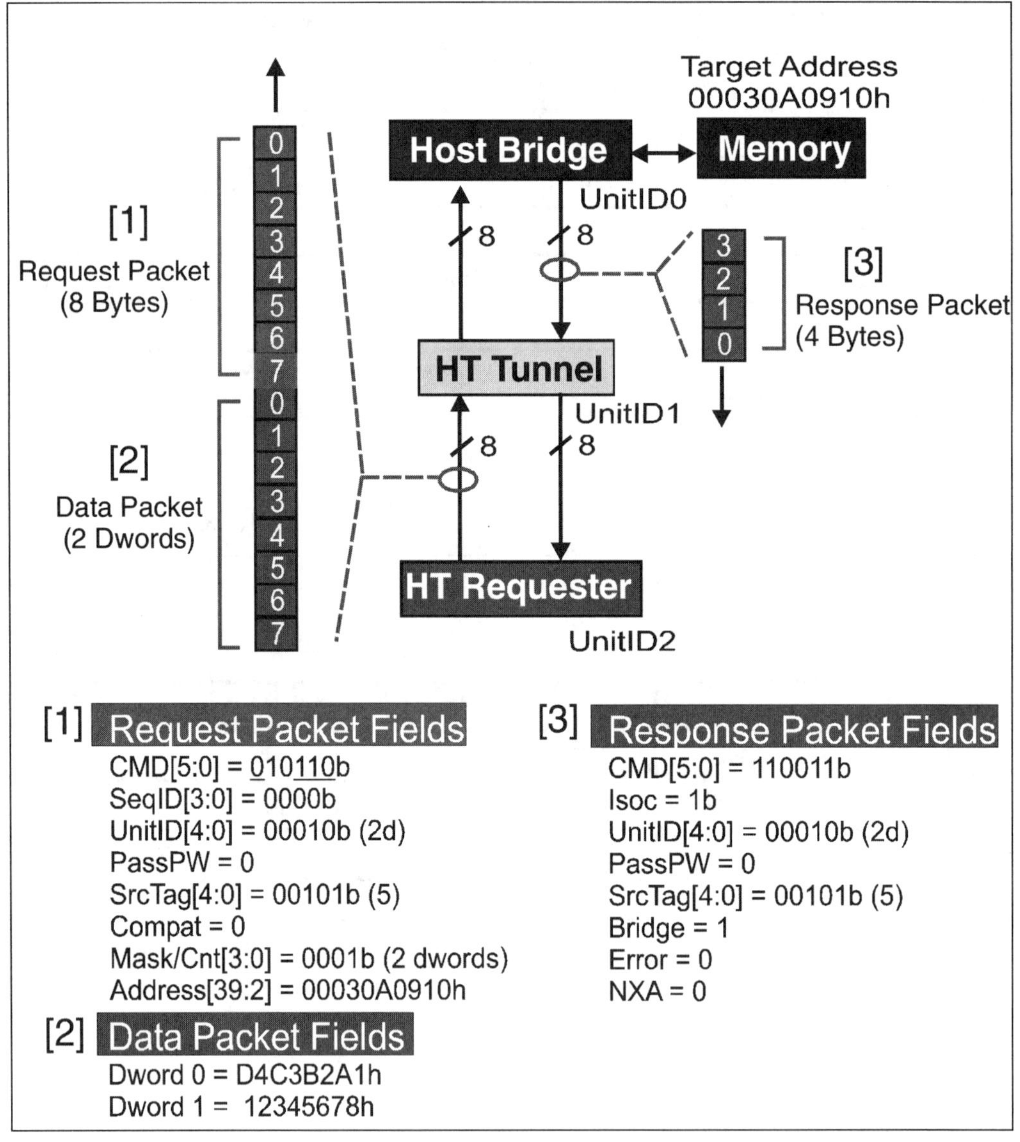

[1] Request Packet Fields
CMD[5:0] = 010110b
SeqID[3:0] = 0000b
UnitID[4:0] = 00010b (2d)
PassPW = 0
SrcTag[4:0] = 00101b (5)
Compat = 0
Mask/Cnt[3:0] = 0001b (2 dwords)
Address[39:2] = 00030A0910h

[2] Data Packet Fields
Dword 0 = D4C3B2A1h
Dword 1 = 12345678h

[3] Response Packet Fields
CMD[5:0] = 110011b
Isoc = 1b
UnitID[4:0] = 00010b (2d)
PassPW = 0
SrcTag[4:0] = 00101b (5)
Bridge = 1
Error = 0
NXA = 0

Example 2: WrSized (Dword) Request Packet Setup

(Refer to [1] in Figure 7-4 on page 150)

Command[5:0] Field (Byte 0, Bit 5:0)

This is the sized write (WrSized) command code. There are several option bits within this field (see underlined bits below). Assume *dword* (bit 2) and *Isoc* bits (bit 1) are enabled; the *posted* (bit 5) and *coherency* (bit 0) bits are disabled (cleared = 0). This indicates an isochronous, non-posted sized (dword) write with no cache coherency requirement. **For this example, field = 010110b.**

SeqID[3:0] Field (Byte 0, Bit 7:6) and (Byte 1, Bit 6:5)

This field tags groups of requests that were issued as part of an ordered sequence from a single UnitID. This field is cleared to indicate a request is not part of a sequence. **For this example, field = 0000b.**

UnitID[4:0] Field (Byte 1, Bits 4:0)

For both upstream and downstream requests, this field is programmed with the UnitID of the requester. UnitID is assigned during initialization by configuration software; a device may consume more than one UnitID — meaning it can own more than one transaction stream. In this example, the request originates at Device 2 (UnitID2). **For this example, field = 00010b.**

PassPW Bit Field (Byte 1, Bit 7)

This bit indicates whether this packet may pass packets in the posted request channel. Cleared for full producer/consumer ordering model. **For this example, field = 0b.**

SrcTag[4:0] Field (Byte 2, Bits 4:0)

This field tags the 32 outstanding non-posted transactions allowed to each UnitID. All reads, non-posted writes, Atomic RMW, and Flush requests in a particular transaction stream must have a unique source tag; the value is assigned by the requester from a pool of available tags. Assume that tag 5 is the one to be used for this request. **For this example, field = 00101b.**

Compat Bit Field (Byte 2, Bit 5)

Used by bridges to tag downstream requests targeting the compatibility chain (where subtractive decoder is found). Reserved field for upstream requests. **For this example, field = 0b.**

Mask/Count[3:0] Field (Byte 2, Bits 7:6) & (Byte 3, Bits 1:0)

Interpretation of this field depends on whether *byte or dword* data size option was selected in Command field (bit 2). In this case, a two <u>dword</u> transfer is being done and the value programed into this field should be *dword count -1*. **For this example, field = 0001b.**

Start Address Field (Bytes 4-7, Bit 7:0) & (Byte 3, Bit 7:2)

This field sets up the 40-bit target start address. Note that the lower bits [1:0] are not in the field, and are assumed to be = 00b. This means that the start address is dword-aligned. For this example, the target address is in the main memory portion of the system memory map. **For this example, the 40-bit address = 00030A0910h.**

Example 2: WrSized (Dword) Transaction Data

(Refer to [2] in Figure 7-4 on page 150)

Because this is a WrSized (dword) command, 1-16 dwords may be transferred. In this example, two dwords are to be written to memory. This is reflected in the transfer count in the Mask/Count[3:0] field above. In a sized (dword) write transaction, data transfer immediately follows the request, starting with least significant dword. Two restrictions on dword data packets:

1. All transfers must be multiples of whole dwords (1-16).
2. All addresses within the range defined by the *start address* to the *start address plus the transfer count* are considered valid for the transfer, so there is no byte mask with a sized (dword) write data packet.

For this example, assume the following dword values are to be written:

- Dword 0 = D4C3B2A1h
- Dword 1 = 12345678h

Example 2: WrSized (Dword) Request Sequence Of Events

(Refer to Figure 7-4 on page 150)

1. UnitID2 checks its transmitter *Isochronous Non-Posted Request (CMD)* and *Isochronous Non-Posted Request (Data)* flow control counters to make sure it has "credits" in both before issuing the request.
2. The transmitter starts sending out the request packet information (see [1] in Figure 7-4). On this eight bit interface, the transmitter sends one byte of request packet info during each successive bit time (there are two bit times per clock). During the request packet transmission, the CTL signal is asserted and the least significant bytes are sent first. After 8 bit times, it has finished sending the request and commences sending the data packet (see [2] in Figure 7-4). It will take 8 bit times to send the 2 dwords of data over this 8 bit interface, and CTL will be deasserted during this time.
3. When the tunnel receives this request and data, it will check its own transmitter flow control counters for the upstream link and prepare to reissue the request towards the host bridge. If the tunnel supports isochronous traffic, it will give the transaction priority and use the isochronous non-posted virtual channels for both the request and data as well as the subsequent response; if not, it uses the standard virtual channels. The packet formats will remain the same in either case.
4. When the host bridge receives the request and data, it will check the target address and command in the request. Because the *coherency* bit (Command field, bit 0) is not set, the bridge is allowed to schedule the write to main memory without checking for coherency with caches located in the CPU or elsewhere. If the host supports isochronous traffic, it will give this traffic priority over other HyperTransport traffic it may be processing.
5. Once the write is complete to memory, the host bridge prepares to initiate a *Target Done* response to the original requester (UnitID2).

Example 2: The WrSized (Dword) Response Packet

(Refer to [3] in Figure 7-4 on page 150)

Because this is a non-posted sized (dword) write transaction, the target owes the original requester a confirmation response indicating the transfer is complete. This information is returned in a four-byte *Target Done* response packet initiated by the original target if the request arrived properly. In the event the request was not claimed by the original target, the Target Done response will be sourced by the end-of-chain device which unexpectedly decodes the unclaimed request. Two bits in the response packet described below, *Error* and *NXA*, are used to indicate whether an error occurred and if it is being reported by the original target or the end-of-chain device. In setting up a response packet, the sender uses several of the fields which were encoded in the original request; this is mainly to assure that the response finds its way back to the original requester and travels in the proper virtual channel and transaction stream as it returns.

Command[5:0] Field (Byte 0, Bit 5:0)

This is the command code for the Target Done response. There are no option bits in this field. **For this example, field = 110011b.**

Isoc Bit Field (Byte 0, Bit 7)

This bit indicates whether this response should travel in the isochronous response virtual channel as it moves back to the original requester. Because the request which caused this response was isochronous, this bit will automatically be set in the response. Any devices in the path of the response which support isochronous traffic will give this packet priority; any that don't support isochronous traffic will route it through standard virtual channels. **For this example, field = 1b.**

UnitID[4:0] Field (Byte 1, Bits 4:0)

For responses, there are two ways this field is handled. If the response is traveling upstream, it is always programmed with the UnitID of the bridge (UnitID0). If the response is traveling downstream (as this one is), the UnitID field contains the Unit ID of the <u>original Requester</u> and is used much like an address to help route the response to the proper device. **For this example, field = 00010b.**

Bridge Bit Field (Byte 1, Bit 6)

Set by host bridges in responses they issue downstream (such as this example). Devices use this information to decide if they may attempt to decode and claim a response; only downstream responses may be claimed by non-bridges. Upstream responses must be forwarded by tunnels in the direction they are already moving. **For this example, field = 1b.**

PassPW Bit Field (Byte 1, Bit 7)

This bit indicates whether this packet may pass packets in the posted request channel (for same transaction stream). Because the PassPW bit was not set in the original request, it won't be set here either. **For this example, field = 0b.**

SrcTag[4:0] Field (Byte 2, Bits 4:0)

Used to tag the 32 outstanding non-posted transactions allowed to each requester UnitID. Whatever tag was assigned in the original request will also be inserted in this response field. In our example, the Source Tag was 5. **For this example, field = 00101b.**

Error Bit Field (Byte 2, Bit 5)

In a non-posted write transaction, this bit in the response will be set to indicate an error was encountered in the delivery of data. When the response finds its way back to the original requester, this bit means that the data was not delivered; action taken by the requester is design-specific, but may range from an interrupt to a sync flood. Refer to "Response Errors" on page 248 for a discussion of how response errors are handled in HyperTransport. **For this example, field = 0b.**

NXA Bit Field (Byte 3, Bit 5)

This response bit is only valid if the Error bit (Byte 2, Bit 5) is set. IF NXA and Error are both set, the original request did not find the proper target at all, and the response is being returned by a device at the end of the chain. If NXA is clear and Error set, the error occurred at the target device. **For this example, field = 0b.**

Example 2: WrSized (Dword) Response, Sequence Of Events

(Refer to Figure 7-4 on page 150)

1. Because the request that caused this response was isochronous, the Host Bridge also uses the isochronous virtual channel to send the response back (if it supports isochronous channels). Otherwise it uses the non-isochronous response virtual channel.
2. The Host Bridge transmitter sends one byte of the four byte response packet during each successive bit time on its 8-bit interface. The CTL signal is asserted during the transmission of the response, and the least significant byte is sent first (see [3] in Figure 7-4).
3. When the tunnel receives the response from the Host Bridge, it checks the Bridge bit and UnitID to determine if the response belongs to it; a non-bridge is only allowed to claim responses moving downstream (Bridge bit set = 1) and carrying the proper UnitID. In this case, the UnitID = 2, so the tunnel forwards the response downstream intact — again using the virtual channel if it is supported.
4. When the response reaches Device 2 (UnitID (2) it is decoded and claimed. The SrcTag[4:0] field is used to identify the particular outstanding request this response is associated with (in this case, transaction #5).

Example 3: Posted Byte Write Request

The Host Bridge in Figure 7-5 on page 157 is targeting I/O address space in Device 2 (UnitID2) with a posted, sized (byte) write of three bytes.

Example 3: WrSized (Byte) Request Packet Setup

(Refer to [1] in Figure 7-5 on page 157)

Command[5:0] Field (Byte 0, Bit 5:0)

This is the sized write (WrSized) control packet command code. The option bits within this field: the *posted* (bit 5), *dword* (bit 2), *Isoc* (Bit 1), and *coherency* (bit 0) are all (cleared = 0), meaning the request is non-posted, uses byte data size, travels in the standard virtual channels, and coherence is not required. **For this example, field = 010000b.**

Figure 7-5: Posted WrSized (Byte) Write Targeting A Downstream Device

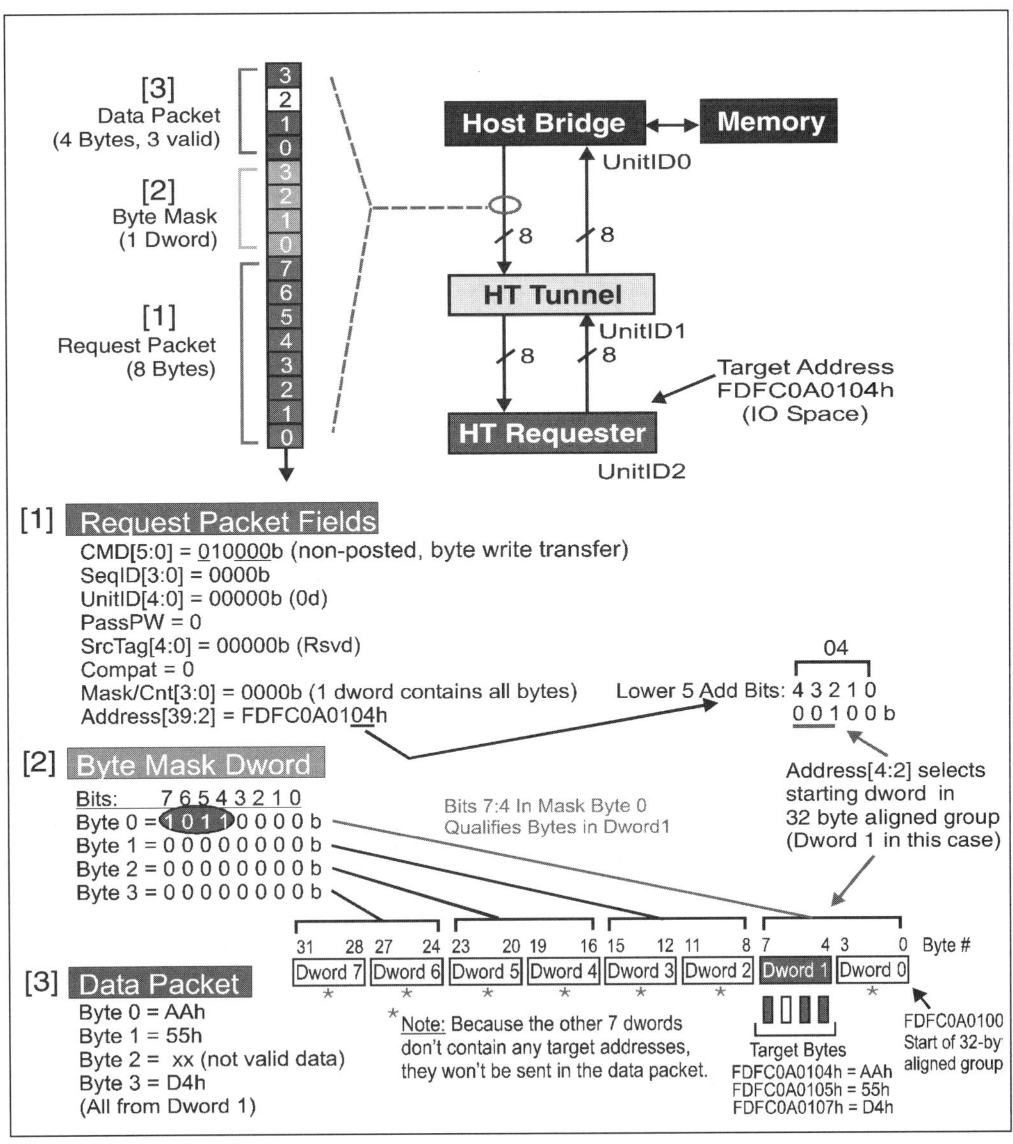

SeqID[3:0] Field (Byte 0, Bit 7:6) and (Byte 1, Bit 6:5)

This field tags groups of requests that were issued as part of an ordered sequence. This field is cleared to indicate a request is not part of a sequence. **For this example, field = 0000b.**

UnitID[4:0] Field (Byte 1, Bits 4:0)

For all requests, this field is programed with the UnitID of the requester. In this example, the request originates at the Host Bridge (UnitID 0) **For this example, field = 00000b.**

PassPW Bit Field (Byte 1, Bit 7)

This bit indicates whether this packet may pass packets in the posted request channel (for same transaction stream). For full producer/consumer ordering model, this bit is cleared. **For this example, field = 0b.**

SrcTag[4:0] Field (Byte 2, Bits 4:0)

This field tags the 32 outstanding non-posted transactions allowed to each UnitID. This field is reserved and set = 0 for posted requests. **For this example, field = 00000b.**

Compat Bit Field (Byte 2, Bit 5)

Used by bridges to tag downstream requests targeting the compatibility chain (where a subtractive decoder may be found). Cleared in this case because target is not on the compatibility chain. **For this example, field = 0b.**

Mask/Count[3:0] Field (Byte 2, Bits 7:6 and Byte 3, Bits 1:0)

The sized (byte) write command can transfer any combination of bytes within the 32-byte address aligned group which contains the starting address. The value programmed in the Mask/Count field represents the dwords <u>actually sent</u> in the data packet; the value is "the span of dwords in a 32-byte address aligned group from the first one containing a valid byte to the last dword in the group containing a valid byte". This value ranges from 1 (1 dword data plus the 1 dword mask) to 8 (8 dwords plus the 1 dword mask). In our example, all three valid bytes reside in the same dword, the Mask/Count[3:0] is set = 1 (1 dword plus the dword mask). **For this example, field = 0001b.**

Start Address Field (Bytes 4-7, Bit 7:0) and Byte 3, Bit 7:2)

This field sets up the 40-bit target start address of the dword containing the first byte to be written. Note that the lower bits [1:0] are not in the field, and are assumed to be = 00b. For this example, the target address is in the I/O address portion of the system memory map. **For this example, the 40-bit start address = FDFC0A0104h. The specific addresses bytes to be written are: FDFC0A0104h, FDFC0A0105h and FDFC0A0107h (note the skipped byte).**

Example 3: Sized (Byte) Write Data Packet And Mask

(Refer to [2] and [3] in Figure 7-5 on page 157)

All Sized (Byte) Writes include a 32 bit (1 dword) mask which precedes the data packet and qualifies each of the 32 bytes (contained in 8 dwords) allowed to be written in one of these transactions. There is a one-to-one correspondence between the dword mask bits and the 32 bytes which may be transferred (mask bit 0 qualifies byte 0, etc.). In the event that less than 32 bytes are to be written, it isn't necessary to send dwords below the first dword with a valid byte or above the last dword with a valid byte. This eliminates the need to send 31 bytes of "don't care" data to send one valid byte. The start address, transfer count (in dwords), and the byte mask are used by the target of the request to determine the bytes of interest within the address-aligned, 32 byte group.

In this example, the WrSized (byte) write is updating three bytes within a single dword. Because of this, only one dword of data (plus the dword mask) must be sent. The starting address bits 4:2 determine the dword containing the three valid bytes (dword 1 in this example). Mask bits 7:4 qualify the bytes within dword 1, and the three mask bits corresponding to the target addresses are enabled: bits 4, 5, and 7. The mask bit for the one invalid byte in dword 1 and the mask bits for all of the unsent dwords are all cleared. (Refer to [2] in Figure 7-5 on page 157). *Note: Because the start address does not contain bits 1:0 (is always dword aligned), it is only the mask bits which provide the byte resolution needed. Basically, the two least significant address bits(1:0) are represented by the first valid mask bit in the target dword.* The data values written to the three target addresses in this example transaction are: **FDFC0A0104h = AAh; FDFC0A0105h = 55h; FDFC0A0107h = D4h.**

Example 3: WrSized (Byte) Request, Sequence Of Events

(Refer to Figure 7-5 on page 157)

1. The Host Bridge checks its transmitter *posted request (CMD)* and *posted request (Data)* flow control counters to make sure it has "credits" before issuing the request.
2. The Host Bridge transmitter sends the request packet information. On this eight bit interface, the transmitter sends one byte of request packet info during each successive bit time and the CTL signal is asserted during the request transfer. After 8 bit times, it deasserts CTL and commences sending the dword byte mask over the next 4 bit times, followed by the single dword of data (containing three valid bytes) during the final 4 bit times.
3. When the tunnel receives this request and mask/data, it will first decode the request to make sure it is not the intended target. In this case, the address does not match its own and it will prepare to reissue the packets downstream, unchanged.
4. When UnitID2 receives the request and mask/data, it will decode the command and target address in the request. It then uses the byte mask information and the start address it received to determine the actual three bytes it should update internally.

A Couple Of Notes About WrSized (Byte)

The main use of the WrSized (byte) command is to provide compatibility with PCI write cycle *byte enables* which allows:

- writes to individual bytes in the system address map
- burst writes to <u>discontinuous memory mapped I/O locations</u>.

The HyperTransport WrSized (Byte) byte mask permits writing any combination of bytes within a 32 byte address-aligned group.

Most of the bulk write transfers in HyperTransport would be done using the WrSized (dword) commands instead of this one because larger transfers are possible (16 dwords) and because most writes can be done in blocks with all target addresses in the block being valid. This eliminates the need for a "write mask" such as the sized (byte) write transaction uses, but is restricted to dword data objects and to writing all locations in the block which is sent.

Example 4: Dword Read Request

Problem: Device 3 (UnitID3) in Figure 7-6 on page 161 targets main memory with a sized (dword) read of two dwords. Start address is 0020400020h; this request is part of an ordered sequence (#9).

Figure 7-6: DMA Dword Read Targeting Main Memory

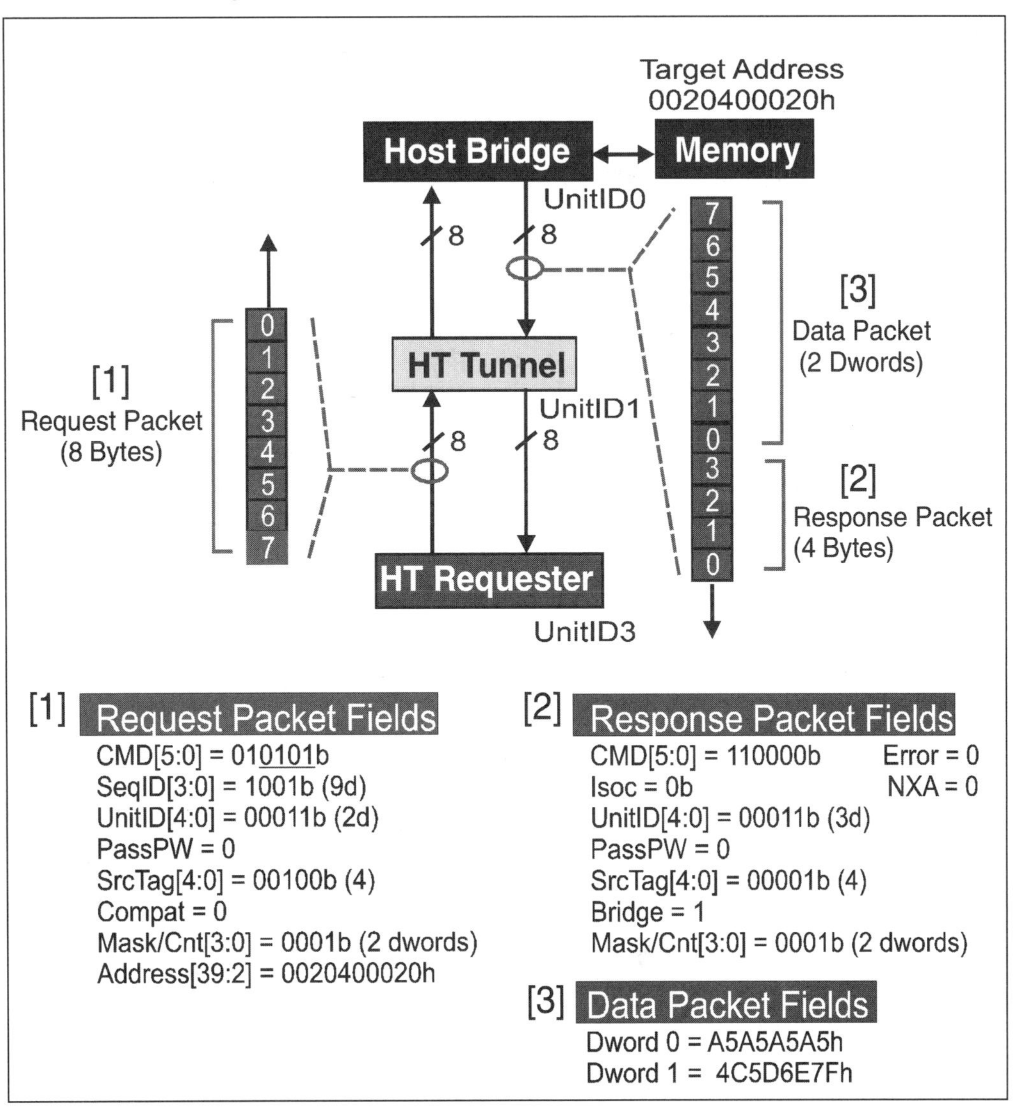

Example 4: RdSized (Dword) Request Packet Setup

(Refer to [1] in Figure 7-6 on page 161)

Command[5:0] Field (Byte 0, Bit 5:0)

This is the sized read (RdSized) command code. There are several option bits within this field (see underlined bits below). Assume *dword* (bit 2) and *coherency* (bit 0) bits are enabled; the *Isoc* (bit 1) and *response may pass posted requests* (bit 3) bits are disabled. This encoding describes a RdSized (dword) request which must assure coherency with CPU caches. **For this example, field = 010101b.**

SeqID[3:0] Field (Byte 0, Bit 7:6) and (Byte 1, Bit 6:5)

This field tags groups of requests that were issued as part of an ordered sequence. In this example, this request is part of ordered sequence #9 for this transaction stream. Tunnel devices are required to maintain the ordering of each request from a single transaction stream that carries the same non-zero sequence ID. **For this example, field = 1001b.**

UnitID[4:0] Field (Byte 1, Bits 4:0)

This field is programmed with the UnitID of the requester. In this example, the request originates at Device 3 (UnitID3). **For this example, field = 00011b.**

PassPW Bit Field (Byte 1, Bit 7)

This bit is carried with read requests and isn't used until the response is generated. At that time the PassPW bit in the response will be set to match this bit. **For this example, field = 0b.**

SrcTag[4:0] Field (Byte 2, Bits 4:0)

This field tags the 32 outstanding non-posted transactions allowed to each UnitID. The code is assigned by the requester from a pool of available tags. Assume that the tag for this read request is 4. **For this example, field = 00100b.**

Compat Bit Field (Byte 2, Bit 5)

Used by bridges to tag downstream requests targeting the compatibility chain (where subtractive decoder is found). Reserved field for upstream requests. **For this example, field = 0b.**

Mask/Count[3:0] Field (Byte 2, Bits 7:6) and (Byte 3, Bits 1:0)

For a *dword* read request, the value programed into this field should be *dword count-1*. In this example, the transfer is two dwords. **For this example, field = 0001b.**

Start Address Field (Bytes 4-7, Bit 7:0) and (Byte 3, Bit 7:2)

This field is used to set up the 40-bit target start address. For this example, the target address in the main memory portion of the system memory map. **For this example, the 40-bit address = 0020400020h.**

Example 4: RdSized (Dword) Request, Sequence Of Events

(Refer to Figure 7-6 on page 161)

1. UnitID3 checks its transmitter *non-Posted Request (CMD)* flow control counter to make sure it has at least one "credit" before issuing the request.
2. The transmitter asserts CTL and starts sending out the request packet information. On this eight bit interface, the transmitter sends one byte of request packet info during each successive bit time, least significant byte first.
3. When the tunnel receives this read request, it will check its own transmitter non-posted request (CMD) flow control counter for the upstream link and prepare to reissue the request towards the host bridge.
4. When the host bridge receives the request, it will check the target address and command in the request. Because the <u>*coherency* bit is set</u> (Command field, bit 0) and the address targets main memory, the bridge is required to take any action required to guarantee that the latest copy of data is returned, even if main memory is stale. How exactly this is done is system specific. After the coherency action is taken, the sequence of steps taken by the Host Bridge in returning data is the same.
5. Having obtained all requested data, the Host Bridge prepares a read response followed by the data.

Example 4: Dword Read Response Packet Setup

(Refer to [2] in Figure 7-6 on page 161)

When the Host Bridge has completed the read of memory on behalf of the requester, it prepares to initiate a *read* response to the original requester (UnitID3). The purpose of the response is to inform devices in the topology that read data follows, and identifies who the recipient should be. Many of the fields in the Host Bridge read response reflect the setting of similiar fields in the request which caused it.

Command[5:0] Field (Byte 0, Bit 5:0)

This is the command code for the Read Response. There are no option bits in this field. **For this example, field = 110000b.**

Isoc Bit Field (Byte 0, Bit 7)

This bit indicates whether this response should travel in the isochronous response virtual channel as it moves back to the original requester. Because the request which caused this response was not isochronous, this bit will not be set. **For this example, field = 0b.**

UnitID[4:0] Field (Byte 1, Bits 4:0)

For responses moving downstream, (as this one is), the UnitID field contains the Unit ID of the <u>original Requester</u> and is used much like an address to help route the response to the proper device. **For this example, field = 00011b.**

Bridge Bit Field (Byte 1, Bit 6)

Set by host bridges in responses they issue downstream (such as this example). Only downstream responses (Bridge = 1) may be claimed by non-bridges. **For this example, field = 1b.**

PassPW Bit Field (Byte 1, Bit 7)

This bit indicates whether this packet may pass packets in the posted request channel (for same transaction stream). Because the PassPW bit was not set in the original request, it won't be set here either. **For this example, field = 0b.**

SrcTag[4:0] Field (Byte 2, Bits 4:0)

The field is set to the same value as seen in the request. In our example, the Source Tag was 4. **For this example, field = 00100b.**

Error Bit Field (Byte 2, Bit 5)

In a read transaction, this bit in the response will be set to indicate an error was encountered in obtaining requested data. When the response finds its way back to the original requester, this bit means that the accompanying data is not valid. **For this example, field = 0b.**

Count[3:0] Field (Byte 2, Bits 7:6) and (Byte 3, Bits 1:0)

For dword reads, the response count field is programmed to match the Mask/ Count[3:0] field in the original request. **For this example, field = 0001b.**

NXA Bit Field (Byte 3, Bit 5)

This response bit indicates a non-existent address error and is only valid if the Error bit (Byte 2, Bit 5) is set. **For this example, field = 0b.**

Example 4: Sized (Dword) Read Data Packet

(Refer to [3] in Figure 7-6 on page 161)

The sized (dword) read request which caused this data packet to be sent specified a start address of 0020400020h and a transfer count of two dwords.

For this example, the two dwords being read from memory are:

- Address 0020400020h (Dword 0) = A5A5A5A5h
- Address 0020400024   (Dword 1) = 4C5D6E7Fh

Example 4: RdSized (Dword) Response, Sequence Of Events

(Refer to Figure 7-6 on page 161)

1. The Host Bridge fetches the two dwords of data from the target addresses in memory. It prepares the four byte read response for transmission down-stream. The *Bridge* bit is set = 1 and if there were no errors performing the read, the *Error* and *NXA* bits will also be clear. As the target does not reside on the compatibility chain in this example, the *Compat* bit is also clear. The *Isoc* bit is clear and the *UnitID* (3) and *SrcTag (4)* fields are set up to match the read request.

2. The Host Bridge transmitter sends one byte of the four byte response packet during each successive bit time on its 8-bit interface. The CTL signal is asserted during the transmission of the response, and the least significant byte is sent first.

3. When the response packet has been sent, the Host Bridge deasserts CTL and sends the two dword data packet.

4. When the tunnel receives the response and data from the Host Bridge, it checks the Bridge bit to see if it is set (it is) and UnitID to determine if the response belongs to it (it doesn't). Unable to claim the response and data, the tunnel forwards the packets as-is downstream to the next device.

5. When the response reaches Device 3 (UnitID 3) it is decoded and claimed. The SrcTag[4:0] field is used to identify the particular outstanding request this response is associated with (in this case, transaction #4).

Example 5: Byte Read Request

Problem: Device 2 (UnitID2) in Figure 7-7 on page 167 targets main memory with a sized (byte) read of two bytes. Start dword address is 0010060204h; bytes of interest are at 0010060206h and 0010060207h.

Figure 7-7: DMA Byte Read Targeting Main Memory

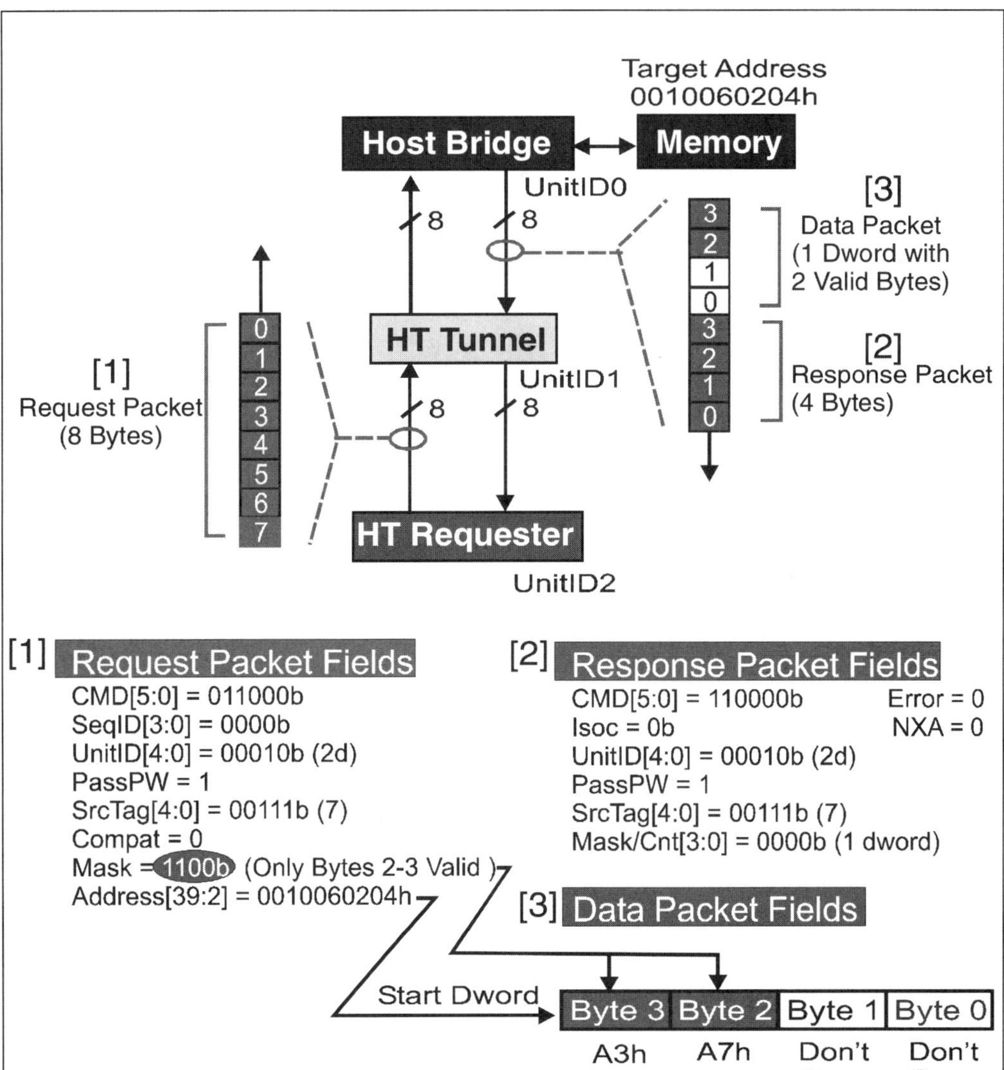

Example 5: RdSized (Byte) Request Packet Setup

(Refer to [1] in Figure 7-7 on page 167)

Command[5:0] Field (Byte 0, Bit 5:0)

This is the sized read (RdSized) command code. The option bits within this field are to be set up for byte transfer size, standard virtual channels, no coherency, and for relaxed ordering of response with respect to posted requests. So, *response may pass posted requests* (bit 3) bit is set; *dword* (bit 2), *coherency* (bit 0), and the *Isoc* (bit 1) bits are all disabled. **For this example, field = 011000b.**

SeqID[3:0] Field (Byte 0, Bit 7:6) and (Byte 1, Bit 6:5)

This field is used to tag groups of requests that were issued as part of an ordered sequence. In this example, this request is not part of an ordered sequence. **For this example, field = 0000b.**

UnitID[4:0] Field (Byte 1, Bits 4:0)

For both upstream and downstream requests, this field is programmed with the UnitID of the requester. In this example, the request originates at Device 2 (UnitID2). **For this example, field = 00010b.**

PassPW Bit Field (Byte 1, Bit 7)

This bit is carried with read requests and isn't used until the response is generated. At that time the PassPW bit in the response will be set to match this bit. For this example, the response/data will allowed to pass posted requests in the same transaction stream. **For this example, field = 1b.**

SrcTag[4:0] Field (Byte 2, Bits 4:0)

This field tags the 32 outstanding non-posted transactions allowed to each UnitID. The code is assigned by requester from a pool of available tags. Assume that the tag for this read request is 7. **For this example, field = 00111b.**

Compat Bit Field (Byte 2, Bit 5)

Used by bridges to tag downstream requests targeting the compatibility chain (where subtractive decoder is found). Reserved field for upstream requests. **For this example, field = 0b.**

Mask/Count[3:0] Field (Byte 2, Bits 7:6) and (Byte 3, Bits 1:0)

For a sized (byte) read request, the value programmed into this field is a 4-bit mask used to indicate which of the four bytes in the target dword are of interest. This is comparable with the *byte enable* scheme used on buses such as PCI to get byte resolution on a dword-based bus. For this example, we are interested in accessing locations 0010060206h and 010060207h; both are contained in the dword located at start address 0010060204h. This means that the starting address will be 0010060204h and mask bits 2,3 will be enabled (set = 1), while mask bits 0,1 will be cleared (= 0). **For this example, field = 1100b.**

Start Address Field (Bytes 4-7, Bit 7:0) and Byte 3, Bit 7:2)

This field carries the 40-bit target start address. Again, the start address is always dword aligned and in sized (byte) reads the Mask/Count[3:0] bits are used to indicate the target bytes within the starting dword (see previous field). For this example, the target dword address for the request is in main memory. **For this example, the 40-bit address = 0010060204h.**

Example 5: RdSized (Byte) Request, Sequence Of Events

(Refer to Figure 7-7 on page 167)

1. After UnitID2 checks its transmitter *non-Posted Request (CMD)* flow control counter for credits, the transmitter interface asserts CTL and starts sending out the request packet information. On this eight bit interface, the transmitter sends one byte of request packet info during each successive bit time, least significant byte first.
2. When the tunnel receives this read request, it will check its own transmitter non-posted request (CMD) flow control counter for the upstream link and prepare to reissue the request towards the host bridge.
3. When the host bridge receives the request, it will check the target address and command in the request. Because the _coherency bit is not set_ (Command field, bit 0) and the address targets main memory, the bridge is not required to take any coherency action with respect to caches before accessing the target location in memory. Note that a device is allowed to disregard the mask bits defining the bytes of interest within the target dword if it can be guaranteed that accessing all four bytes in the dword will never cause a problem. It has to return a dword data packet anyway, and in this case it simply sends all four bytes of the dword. On the other hand, there are some devices

(e.g. FIFOs) which are not compatible with reading extra data; in those cases, the byte mask must be applied explicitly.

4. Having obtained all requested data, the Host Bridge prepares initiate a read response followed by the sending of the data.

Example 5: Byte Read Response Packet Setup

(Refer to [2] in Figure 7-7 on page 167)

When the Host Bridge has completed the dword read of memory (or a subset of a dword if the mask requires this), it prepares to return a *Read* response and the single dword to the original requester (UnitID2). The purpose of the response is to inform devices in the topology that read data follows and it identifies who the recipient should be. Many of the fields in the Host Bridge read response reflect the setting of similiar fields in the request which caused it.

Command[5:0] Field (Byte 0, Bit 5:0)

This is the command code for the Read Response. There are no option bits in this field. **For this example, field = 110000b.**

Isoc Bit Field (Byte 0, Bit 7)

This bit indicates whether this response should travel in the isochronous response virtual channel as it moves back to the original requester. Because the request which caused this response was not isochronous, this bit will not be set. **For this example, field = 0b.**

UnitID[4:0] Field (Byte 1, Bits 4:0)

Because this response is being issued by a Host Bridge and traveling downstream, the UnitID field carries the identity of the original requester (UnitID2). **For this example, field = 00010b.**

Bridge Bit Field (Byte 1, Bit 6)

Set by host bridges in responses they issue downstream (as in this example) so devices may distinguish upstream and downstream responses. **For this example, field = 1b.**

PassPW Bit Field (Byte 1, Bit 7)

This bit indicates whether this packet may pass packets in the posted request channel (for same transaction stream). Because the PassPW bit was set in the original request, it will also be set in the response. **For this example, field = 1b.**

SrcTag[4:0] Field (Byte 2, Bits 4:0)

The field is set to the same value as seen in the request. In our example, the Source Tag was 7. **For this example, field = 00111b.**

Error Bit Field (Byte 2, Bit 5)

In a read transaction, this bit in the response will be set to indicate an error was encountered in obtaining requested data. In this example, there were no errors. Refer to "Response Errors" on page 248 for a discussion of how response errors are handled in HyperTransport. **For this example, field = 0b.**

Count[3:0] Field (Byte 2, Bits 7:6) and (Byte 3, Bits 1:0)

For dword reads, the response count field is programmed to match the Mask/Count[3:0] field in the original request. For byte read transfers, the count is always set = 0 (1 dword). The count field in a read response informs tunnels and other devices in the path to the target how much data is to follow. **For this example, field = 0000b.**

NXA Bit Field (Byte 3, Bit 5)

This response bit is only valid if the Error bit (Byte 2, Bit 5) is set. IF NXA and Error are both set, the original request did not find the proper target at all, and the read response is being returned by a device at the end of the chain. If NXA is clear and Error set, then a read error occurred at the target device. **For this example, field = 0b.**

Example 5: Sized (Byte) Read Data Packet

(Refer to [3] in Figure 7-7 on page 167)

The sized (byte) read request which caused this data packet to be sent specified a start address of 0010060204h. A single dword data packet is always returned for this type of request. The bytes within the dword data packet will be valid if their corresponding Mask/Count[3:0] bit was set =1; bytes within the target dword with a Mask/Count[3:0] bit not set will be driven to whatever value the agent sending the response decides. For our example, after the read only the two upper bytes (corresponding to addresses 0010060206h and 001060207h) are valid. (Two data values are arbitrary).

- Address 0010060204h: Data value = (Don't Care)
- Address 0010060205h: Data value = (Don't Care)
- Address 0010060206h: Data value = A7h
- Address 0010060207h: Data value = A3h

Example 5: RdSized (Byte) Response, Sequence Of Events

(Refer to Figure 7-7 on page 167)

1. The Host Bridge fetches the requested data from memory using the target start address and the Mask/Count[3:0] bit pattern. It prepares the dword read response for transmission downstream. The *Bridge* bit is set = 1 and if there were no errors performing the read, the *Error* and *NXA* bits will also be clear. The *Compat* and *Isoc* bits are cleared and the *UnitID* (2) and *SrcTag* (7) fields are set up to match the original read request.
2. The Host Bridge transmitter sends one byte of the four byte response packet during each successive bit time on its 8-bit interface. The CTL signal is asserted during the transmission of the response, and the least significant byte is sent first.
3. When the response packet has been sent, the Host Bridge deasserts CTL and sends the dword data packet; any bytes tagged invalid by the Mask/Count[3:0] bit pattern are padded to a value selected by the Host Bridge.

4. When the tunnel receives the response and data from the Host Bridge, it checks the Bridge bit to see if it is set (it is) and UnitID to determine if the response belongs to it (it doesn't). Unable to claim the response and data, the tunnel forwards the packets as-is downstream to the next device.
5. When the response reaches Device 2 (UnitID 2) it is decoded and claimed. The SrcTag[4:0] field is used to identify the particular outstanding request this response is associated with (in this case, transaction #7). Invalid bits in the data dword are ignored.

SrcTag[4:0] Field (Byte 2, Bits 4:0)

This field is used to tag the 32 outstanding non-posted transactions allowed to each UnitID. The code is assigned by requester from a pool of available tags. Flush requests are always non-posted, so must have a source tag assigned to them. Assume that the tag for this Flush request is 6. **For this example, field = 00110b.**

Example 6: Flush Request

Problem: Device 2 (UnitID2) in Figure 7-8 on page 174, issues a Flush request which causes all previous posted writes <u>in the same transaction stream</u> (sourced by UnitID2) to be forced to host memory.

Figure 7-8: A Flush Request Issued By UnitID 2

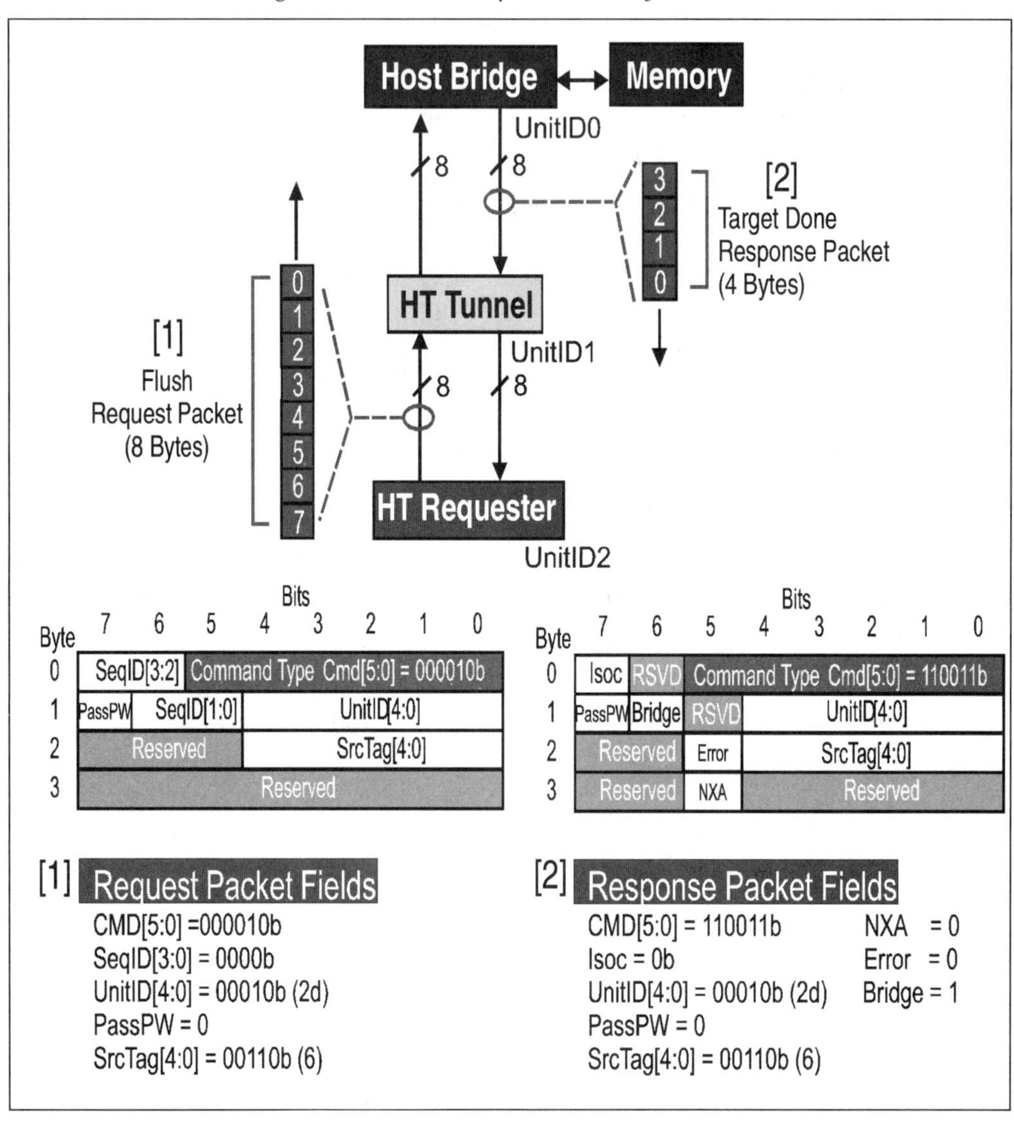

Example 6: Flush Request Packet Setup

(Refer to [1] in Figure 7-8 on page 174)

Command[5:0] Field (Byte 0, Bit 5:0)

This is the Flush request command code. There are no option bits. **For this example, field = 000010b.**

SeqID[3:0] Field (Byte 0, Bit 7:6) and (Byte 1, Bit 6:5)

This field is used to tag groups of requests that were issued as part of an ordered sequence. Flush requests are never part of an ordered sequence, so this bit should be cleared. Setting SeqID[3:0] to a non-zero value may have indeterminate results. **For this example, field = 0000b.**

UnitID[4:0] Field (Byte 1, Bits 4:0)

For both upstream and downstream requests, this field is programmed with the UnitID of the requester. UnitID is assigned during initialization by configuration software. In this example, the request originates at Device 2 (UnitID2). **For this example, field = 00010b.**

PassPW Bit Field (Byte 1, Bit 7)

This bit is cleared in a Flush request so that the request will perform its intended function: pushing earlier posted writes ahead of it. Setting this bit = 1 may have indeterminate results. **For this example, field = 0b.**

Example 6: Flush Request, Sequence Of Events

(Refer to Figure 7-8 on page 174)

1. After UnitID2 checks its transmitter *non-Posted Request (CMD)* flow control counter for credits, the transmitter interface asserts CTL and starts sending out the Flush request packet information. On this eight bit interface, the transmitter sends one byte of request packet info during each successive bit time, least significant byte first.
2. When the tunnel receives this Flush request, it will check its own transmitter non-posted request (CMD) flow control counter for the upstream link and prepare to reissue the request towards the host bridge.
3. When the host bridge receives the Flush request, it will check the UnitID in the Flush request against the UnitID of each posted write request it has pending. Any matches indicate a posted write request which must be completed (they are in the same transaction stream).
4. Once the Flush is done, the Host Bridge prepares initiate a Target Done response to the requester.

Example 6: Flush Response Packet Setup

(Refer to [2] in Figure 7-8 on page 174)

When the Host Bridge has completed the Flush to memory (or downstream for peer-to-peer traffic) it prepares to return a *Target Done* response to the original requester (UnitID2). The purpose of the response is to inform the requester the operation is done and that it may take any action that may have been pending its completion.

Command[5:0] Field (Byte 0, Bit 5:0)

This is the command code for the Target Done Response. There are no option bits in this field. **For this example, field = 110011b.**

UnitID[4:0] Field (Byte 1, Bits 4:0)

Because this response is being issued by a Host Bridge and traveling downstream, the UnitID field carries the identity of the original requester (UnitID2). **For this example, field = 00010b.**

Bridge Bit Field (Byte 1, Bit 6)

Set by host bridges in responses they issue downstream (as in this example) so devices may distinguish upstream and downstream responses. **For this example, field = 1b.**

PassPW Bit Field (Byte 1, Bit 7)

This bit indicates whether this packet may pass packets in the posted request channel (for same transaction stream). This bit should be clear in Flush request and the subsequent Target Done response. **For this example, field = 0b.**

SrcTag[4:0] Field (Byte 2, Bits 4:0)

The field is set to the same value as seen in the Flush request. In our example, the Source Tag was 6. **For this example, field = 00110b.**

Error Bit Field (Byte 2, Bit 5)

In a read transaction, this bit in the response is set to indicate an error was encountered in obtaining requested data. In this example, there were no errors. See Chapter 10, entitled "Error Detection And Handling," on page 229 for a discussion of how response errors are handled in HyperTransport. **For this example, field = 0b.**

NXA Bit Field (Byte 3, Bit 5)

This response bit is only valid if the Error bit (Byte 2, Bit 5) is set. IF NXA and Error are both set, the original Flush request did not find the proper target at all, and the Target Done response is being returned by a device at the end of the chain. **For this example, field = 0b.**

Example 6: Flush Response, Sequence Of Events

(Refer to Figure 7-8 on page 174)

1. When the Host Bridge completes the forwarding of all posted write data for the transaction stream being Flushed, it prepares a Target Done response packet which will be routed back to the requester as a confirmation that the Flush is complete. There is no data returned with a Target Done response. The Host bridge programs the UnitID[4:0] and SrcTag[4:0] fields with the same information that was contained in the Flush request. It also sets the Bridge bit indicating a downstream response, and clears the error bits.
2. Assuming it has credits in its Response flow control buffer counter, the Host Bridge transmitter then sends one byte of the four byte Target Done response packet during each successive bit time on its 8-bit interface. The CTL signal is asserted during the transmission of the response, and the least significant byte is sent first.
3. When the tunnel receives the Target Done response from the Host Bridge, it checks the Bridge bit to see if it is set (it is) and the UnitID to determine if the response belongs to it (it doesn't). Unable to claim the response, the tunnel forwards the packet as-is downstream to the next device.
4. When the response reaches Device 2 (UnitID 2) it is decoded and claimed. The SrcTag[4:0] field identifies the particular outstanding request this response is associated with (in this case, transaction #6).

A Few Notes About Flush Operations

- Flush requests are intended to make previous write data "globally visible" in the host. To do this, any posted write data still in buffers between the sender and memory must be forwarded. The Target Done response at the end of a Flush is confirmation that the operation is complete.
- There is no Isoc bit associated with a Flush because these requests do not affect traffic in the isochronous virtual channels.
- If one or more of the posted writes which preceded the Flush operation targeted another device (called a peer-to-peer transaction) instead of main memory, then when the Target Done response returns confirming the completion of the Flush operation, it only guarantees that those posted writes were chased to the Host Bridge; they may not have reached the target yet.
- Flush requests are only sent from devices to host bridges, or from one host bridge to another. Tunnels are not required to take any special action with a Flush other than to forward it in the direction it is already moving.

- If a Flush request mistakenly reaches an end-of-chain device, it is required to decode the command and return the required Target Done response with the Error and NXA bits set.

Example 7: Fence Request

Problem: Device 3 (UnitID3) in Figure 7-9, issues a Fence request to provide a barrier between previous and subsequent posted writes; it applies to all I/O streams and virtual channels (except isochronous).

Figure 7-9: A Fence Request Issued By UnitID 3

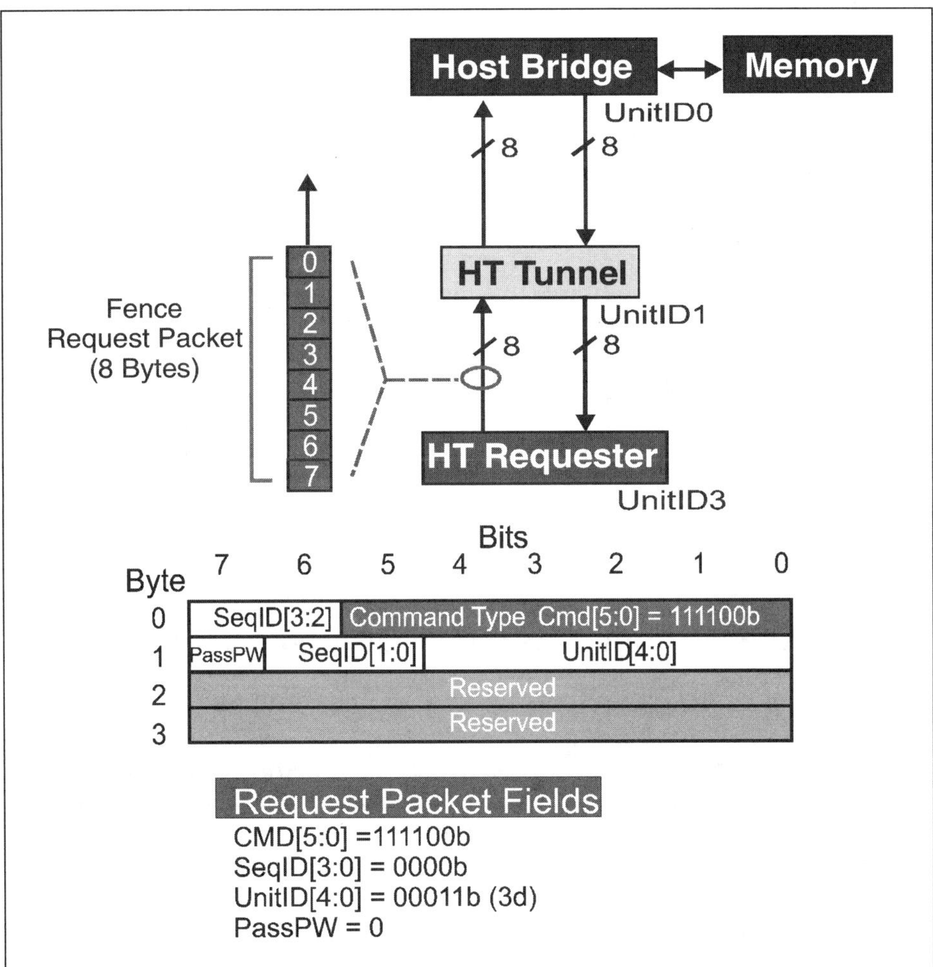

Example 7: Fence Request Packet Setup

(Refer to Figure 7-9 on page 179)

Command[5:0] Field (Byte 0, Bit 5:0)

This is the Fence request command code. There are no option bits within this field. **For this example, field = 111100b.**

SeqID[3:0] Field (Byte 0, Bit 7:6) and (Byte 1, Bit 6:5)

This field tags groups of requests that were issued as part of an ordered sequence. Fence requests are never part of an ordered sequence, so this bit should be cleared. Setting SeqID[3:0] to a non-zero value may have indeterminate results. **For this example, field = 0000b.**

UnitID[4:0] Field (Byte 1, Bits 4:0)

For both upstream and downstream requests, this field is programmed with the UnitID of the requester. UnitID is assigned during initialization by configuration software. In this example, the request originates at Device 3 (UnitID3). **For this example, field = 00011b.**

PassPW Bit Field (Byte 1, Bit 7)

This bit should be cleared in a Fence request so that the request will perform its intended function: pushing earlier posted writes for all transaction streams ahead of it. Setting this bit = 1 may have indeterminate results. **For this example, field = 0b.**

Example 7: Fence Request, Sequence Of Events

(Refer to Figure 7-9 on page 179)

1. After UnitID3 checks its transmitter *Posted Request (CMD)* flow control counter for credits, the transmitter interface asserts CTL and starts sending out the Fence request packet information. On this eight bit interface, the transmitter sends one byte of request packet info during each successive bit time, least significant byte first.
2. When the tunnel receives this Fence request, it will check its own transmitter Posted Request (CMD) flow control counter for the upstream link and

prepare to reissue the request towards the host bridge.

3. When the host bridge receives the Fence request, it creates an "internal barrier" which guarantees that any subsequent non-isochronous posted writes (from any transaction stream) will not be able to pass any posted writes which arrived prior to the Fence request. Note: normally, there are no ordering rules between transactions belonging to different transaction streams; Fence creates an exception.

A Few Notes About Fence Operations

- Fence requests are intended to act as a barrier between posted writes. It is different from Flush in that it applies to all transaction streams (all UnitIDs). It also travels in the posted virtual channel, meaning that, unlike Flush, there will not be a response to this request. It also means that there is no source tag (SrcTag[4:0]) field required for Fence.
- If isochronous flow control mode is not supported on a link, isoc packets travel in the standard virtual channels and <u>are</u> affected by the Fence command.
- Fence requests are only sent from devices to host bridges, or from one host bridge to another. Tunnels are not required to take any special action with a Fence other than to forward it in the direction it is already moving.
- If a Fence request mistakenly reaches an end-of-chain device, it is required to decode the command and then drop it (no response is required). This may be logged as an end-of-chain error if the end-of-chain device is programmed to handle it that way. Refer to "End-Of-Chain Errors" on page 243 for a discussion of error handling by end-of-chain devices.

HypterTransport System Architecture

Example 8: Atomic Read-Modify-Write

Problem: Device 2 (UnitID2) in Figure 7-10 targets main memory with an Atomic RMW transaction. Memory start address is 0082402020h.

Figure 7-10: Atomic Read-Modify-Write Targeting Main Memory

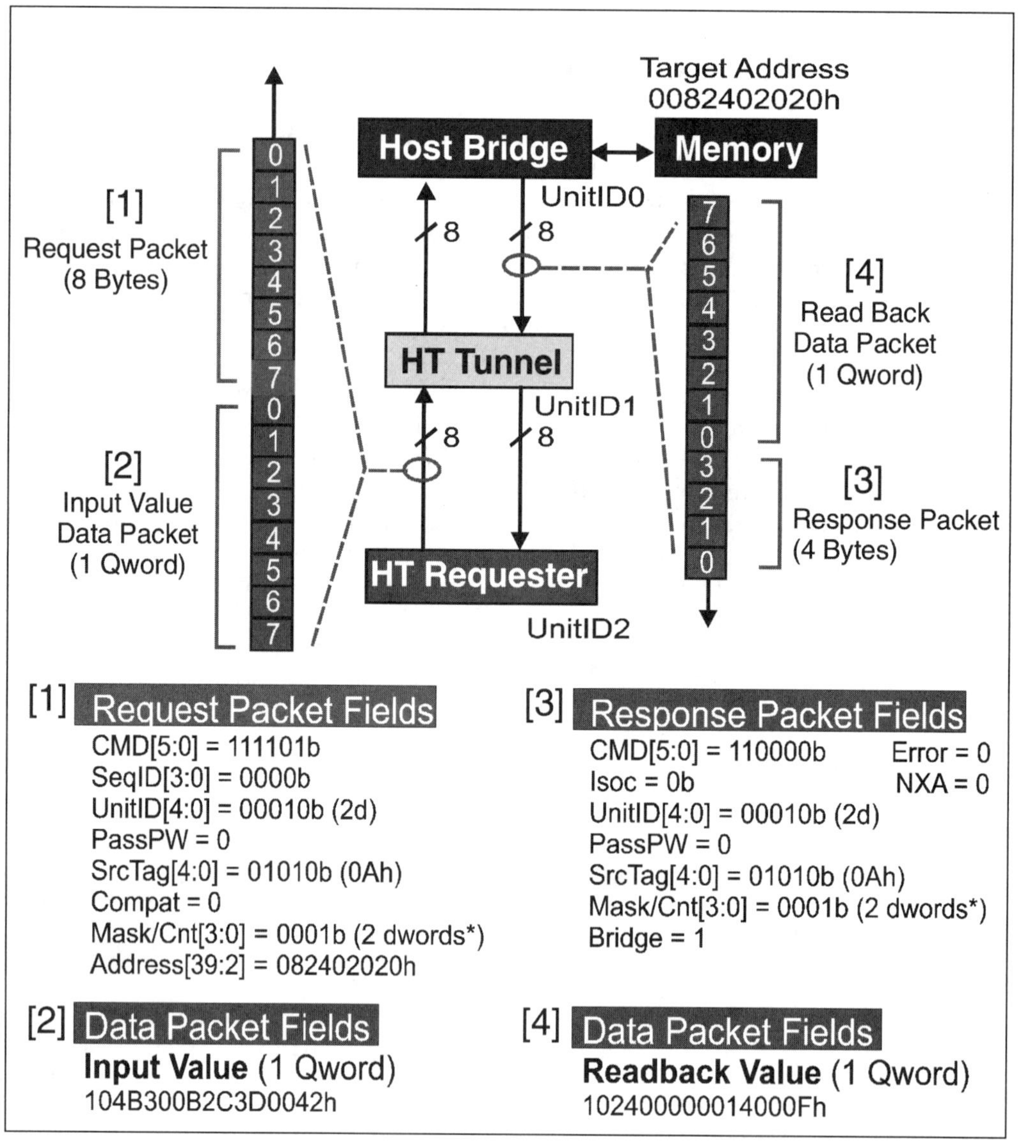

Example 8: Atomic RMW Request Packet Setup

(Refer to [1] in Figure 7-10 on page 182)

Command[5:0] Field (Byte 0, Bit 5:0)

This is the Atomic RMW command code. There are no option bits within this field and the following limits are imposed: isochronous Atomic RMW is not allowed, target location may be cached, the response to an Atomic RMW may not pass posted writes. **For this example, field = 111101b.**

SeqID[3:0] Field (Byte 0, Bit 7:6) and (Byte 1, Bit 6:5)

This field tags groups of requests that were issued as part of an ordered sequence; the field is cleared if a request is not part of a sequence. In this example, this request is not part of an ordered sequence. **For this example, field = 0000b.**

UnitID[4:0] Field (Byte 1, Bits 4:0)

For both upstream and downstream requests, this field is programmed with the UnitID of the requester. UnitID is assigned during initialization by configuration software. In this example, the request originates at Device 2 (UnitID2). **For this example, field = 00010b.**

PassPW Bit Field (Byte 1, Bit 7)

This bit is reserved for Atomic RMW requests, and should be cleared. **For this example, field = 0b.**

SrcTag[4:0] Field (Byte 2, Bits 4:0)

This field is used to tag the 32 outstanding non-posted transactions allowed to each UnitID. The code is assigned by requester from a pool of available tags. Assume that the tag for this read request is 10d (Ah). **For this example, field = 01010b.**

Compat Bit Field (Byte 2, Bit 5)

Used by bridges to tag downstream requests targeting the compatibility chain (where subtractive decoder is found). Atomic RMW requests target main memory, so this bit should be clear. **For this example, field = 0b.**

Mask/Count[3:0] Field (Byte 2, Bits 7:6) and (Byte 3, Bits 1:0)

The value programmed in this field depends on which of the two Atomic RMW variants is being performed. For the *Fetch and Add* operation, the request is accompanied by one qword (2 dwords) of data. In this case, the Mask/Count would be programmed with <u>one</u> (*dword count -1*). For the *Compare and Swap* operation, the request is accompanied by two qwords (4 dwords) of data. In this case, the Mask/Count would be programmed with <u>three</u> (again, *dword count -1*). *Note: the value in this field is used by the Host Bridge to determine which variant is being requested.* In this Fetch and Add example, the transfer is one qword (two dwords). **For this example, field = 0001b.**

Start Address Field (Bytes 4-7, Bit 7:0) and (Byte 3, Bit 7:2)

This field carries the 40-bit target start address. Note that the lower bits [1:0] are not in the field, and are assumed to be = 00b. This means that the start address is dword-aligned. For Atomic RMW, the target address is also expected to be in the main memory portion of the system memory map. **For this example, the 40-bit address = 0082402020h.**

Example 8: Atomic RMW Request Data Packet

(Refer to [2] in Figure 7-10 on page 182)

The Atomic RMW Fetch and Add request is accompanied by a one qword Input value carried in the data packet. This is the value which will be added to the value currently in the target memory location. For this example, the Input value is: 104B300B2C3D0042h.

Example 8: Atomic RMW Request, Sequence Of Events

(Refer to Figure 7-10 on page 182)

1. UnitID2 checks its transmitter *non-Posted Request (CMD)* and *non-Posted Request (Data)* flow control counters to make sure it has at least one "credit" in each before issuing the request.
2. The transmitter asserts CTL and starts sending out the request packet information. On this eight bit interface, the transmitter sends one byte of request packet info during each successive bit time, least significant byte first. Fol-

lowing the request, it will de-assert CTL and drive the qword (2 dwords) data packet containing the Input value across the interface.

3. When the tunnel receives this read request and data, it checks its own transmitter non-posted flow control counters for the upstream link and prepares to reissue the packets towards the host bridge.
4. When the host bridge receives the request and data, it will lock the memory location which prevents access by other agents.
5. The Host Bridge reads the current data from target memory address.
6. The Host Bridge adds the value just read to the Input value which accompanied the request and writes the sum back to the target memory location.
7. The Host Bridge unlocks the memory location and prepares to send the read response and (readback) data to the original requester.

Example 8: Atomic RMW Response Packet Setup

(Refer to [3] in Figure 7-10 on page 182)

When the Host Bridge has completed the read/modify/write of the target memory location, it still must return the "original" value read from memory to the requester. This is always a qword (2 dword) data object. The Host Bridge prepares to initiate a *Read* response to accompany the "readback" data, using the original UnitID and Source Tag information provided in the request.

Command[5:0] Field (Byte 0, Bit 5:0)

This is the command code for the Read Response. There are no option bits in this field. **For this example, field = 110000b.**

Isoc Bit Field (Byte 0, Bit 7)

This bit indicates whether this response should travel in the isochronous response virtual channel as it moves back to the original requester. Atomic RMW transactions are not allowed to use the isochronous virtual channels so this bit should be cleared. **For this example, field = 0b.**

UnitID[4:0] Field (Byte 1, Bits 4:0)

For responses, there are two ways this field is handled. If the response is traveling upstream, it is always programmed with the UnitID of the bridge (UnitID0). If the response is traveling downstream (as this one is), the UnitID field contains the Unit ID of the <u>original Requester</u>. **For this example, field = 00010b.**

Bridge Bit Field (Byte 1, Bit 6)

Set by host bridges in responses they issue downstream (such as this example). Devices use this information to decide if they may attempt to decode and claim a response; only downstream responses (Bridge = 1) may be claimed by non-bridges. **For this example, field = 1b.**

PassPW Bit Field (Byte 1, Bit 7)

This bit should be cleared because Atomic RMW responses are not allowed to pass posted writes. **For this example, field = 0b.**

SrcTag[4:0] Field (Byte 2, Bits 4:0)

The field is set to the same value as seen in the request. In our Atomic RMW example, the Source Tag was 0Ah. **For this example, field = 01010b.**

Error Bit Field (Byte 2, Bit 5)

In a read transaction, this bit in the response will be set to indicate an error was encountered in obtaining requested data. When the response finds its way back to the original requester, this bit means that the accompanying data is not valid; action taken by the requester is design-specific, but may range from an interrupt to a sync flood. Refer to "Response Errors" on page 248 for a discussion of how response errors are handled in HyperTransport. **For this example, field = 0b.**

Count[3:0] Field (Byte 2, Bits 7:6) and (Byte 3, Bits 1:0)

For both variants of Atomic RMW, this field should be set = one (2 dwords = 1 qword). **For this example, field = 0001b.**

NXA Bit Field (Byte 3, Bit 5)

This response bit is only valid if the Error bit (Byte 2, Bit 5) is set. IF NXA and Error are both set, the original request did not find the proper target at all, and the read response is being returned by a device at the end of the chain. If NXA is clear and Error set, then a read error occurred at the target device. **For this example, field = 0b.**

Example 8: Atomic RMW Response Data Packet

(Refer to [4] in Figure 7-10 on page 182)

The Atomic RMW request that caused this data packet to be sent always results in a return of 1 qword (2 dwords) of data. The response data packet contains the "original" data value read from memory at the beginning of the Fetch and Add operation. For this example, the qword of "original" data at the target start address was: 102400000014000Fh.

Example 8: Atomic RMW Response, Sequence Of Events

(Refer to Figure 7-10 on page 182)

1. After the Host Bridge performs the memory fetch, adds the input value to it and writes the result back to memory, it prepares to send the read response and "original" read data back to the requester. The *UnitID* (2) and *SrcTag* *(0Ah)* fields are set up in the response to match the request, the *Bridge* bit is set = 1 and if there were no errors performing the operation, the *Error* and *NXA* bits will also be clear. The *Compat* and *Isoc* bits are also cleared.
2. The Host Bridge transmitter sends one byte of the four byte response packet during each successive bit time on its 8-bit interface. The CTL signal is asserted during the transmission of the response, and the least significant byte is sent first.
3. When the response packet has been sent, the Host Bridge deasserts CTL and sends the single qword (two dword) data packet.
4. When the tunnel receives the response and data from the Host Bridge, it checks the Bridge bit to see if it is set (it is) and UnitID to determine if the response belongs to it (it doesn't). Unable to claim the response and data, the tunnel forwards the packets as-is downstream to the next device.
5. When the response reaches Device 2 (UnitID2) it is decoded and claimed. The SrcTag[4:0] field is used to identify the particular outstanding request this response is associated with, in this case the transaction is #10d (0Ah).

Some Notes About Atomic RMW Operations

(Refer to Figure 7-10 on page 182)

- The purpose of this request type is to allow an atomic read, modify, and write of a memory location <u>without the possibility another agent will access the location while any part of the hybrid operation is in progress</u>. The memory controller is required to lock <u>at least the one qword</u> of address range which contains the target start address until it completes.
- The Fetch and Add variant of Atomic RMW takes the Input Value (1 qword of data) and adds it to the value currently at the target location and writes it back to memory. The "original" value read from memory (before the addition) is returned to the requester with the read response. This means that both the request and the response packets are accompanied by data packets.
- The Compare and Swap variant of Atomic RMW request is accompanied by a two qword data packet: the Input Value and a Compare value. When the memory controller detects a Compare and Swap (based on the value in the Mask/Count[3:0] field), it reads the value currently at the target memory location and checks it against the Compare value; if the Compare value is greater than the memory location, the Input value is over-written to memory (a swap has occurred). If the Compare value is less than the "original" value in memory, the Input value is discarded (no swap). In either case, the "original" qword value in memory is returned to the requester with the read response.
- The value in the Atomic RMW request Mask/Count[3:0] field is expected to be either one (Fetch and Add) or three (Compare and Swap). Behavior resulting from programming a value other than these two into the request Mask/Count field is undefined. Note that the Response returned for either variant of Atomic RMW request always carries a Mask/Count[3:0] field of one (1 qword).
- Atomic RMW requests move upstream towards the Host bridge. The *Compat* in the response should be clear.
- No target, other than a host, is required to support atomic operations. Non-hosts which decode Atomic RMW requests may either return a single qword of data with the Error bit set in the response or they may allow the read, modify, and write in a non-atomic manner.
- A host is not required to support Atomic RMW requests which target address outside of main memory. If they see one, they may abort it or reflect it as a peer-to-peer operation. Peer-to-peer Atomic RMW operations are to be supported in future revisions of the HyperTransPort Specification.

Example 9: WrSized Request Crosses A Bridge

Problem: Device 2 (UnitID2) in Figure 7-11 on page 189 targets main memory with a non-posted sized (dword) write of one dword. The non-isochronous transaction must cross a HyperTransport-HyperTransport bridge; there is no coherency requirement.

Figure 7-11: Sized (Dword) Write Transaction Must Cross A Bridge

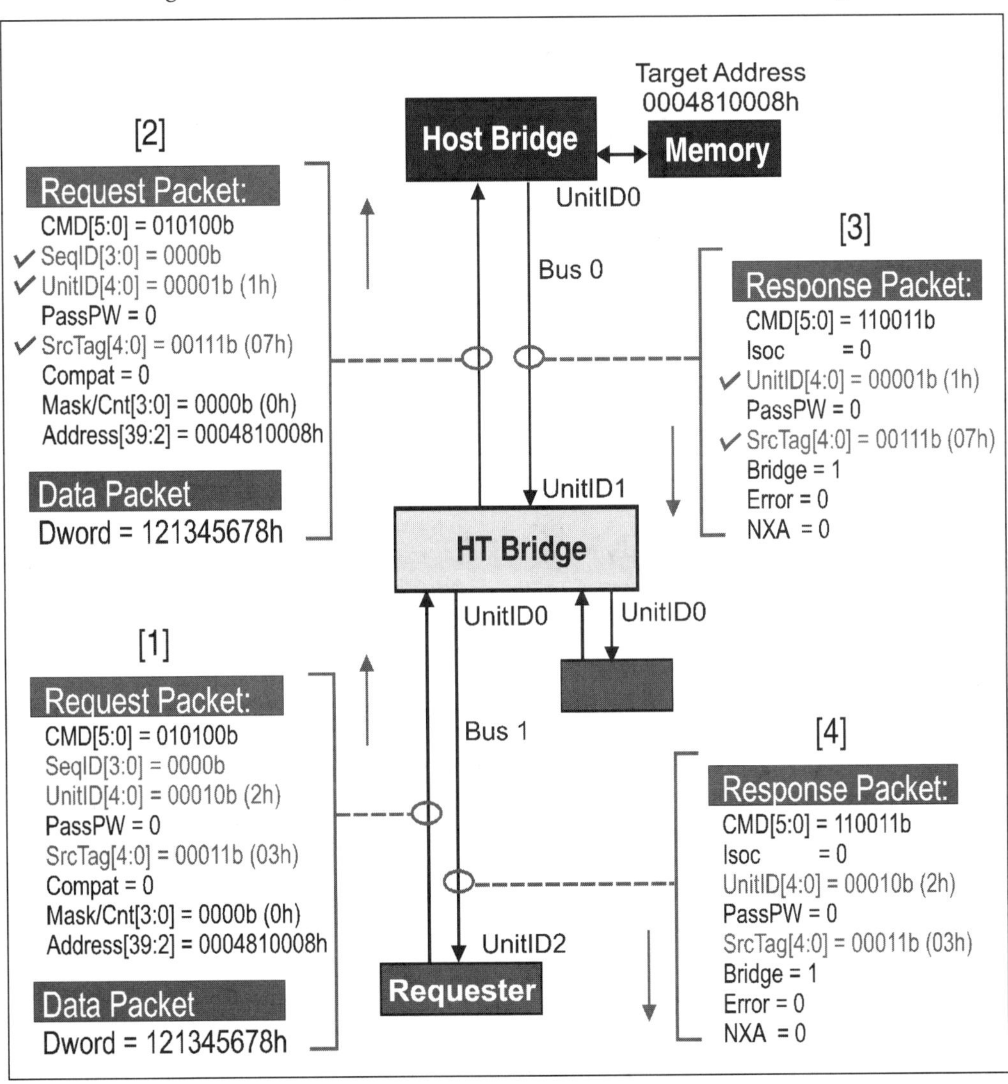

Example 9: Request Packet On Bus 1

(Refer to [1] in Figure 7-11 on page 189)

Command[5:0] Field (Byte 0, Bit 5:0)

This is the sized write (WrSized) command code. Assume *dword* (bit 2) is enabled; the *posted* (bit 5), *Isoc* (bit 1), and *coherency* (bit 0) bits are all disabled (0). This indicates a non-isochronous, non-posted sized (dword) write with no cache coherency requirement. **For this example, field = 010100b.**

SeqID[3:0] Field (Byte 0, Bit 7:6) and (Byte 1, Bit 6:5)

This field tags groups of requests that were issued as part of an ordered sequence from a single UnitID. This field is cleared to indicate a request is not part of a sequence. **For this example, field = 0000b.**

UnitID[4:0] Field (Byte 1, Bits 4:0)

This field is programmed with the UnitID of the requester. For transactions which move through a bridge, the UnitID and SrcTag fields in an incoming request are tracked internally and are replaced by those of the bridge as it reissues the request on the next bus. For bus 1, the UnitID of the requester is two. **For this example, field = 00010b.**

PassPW Bit Field (Byte 1, Bit 7)

This bit indicates whether this packet may pass packets in the posted request channel (for same transaction stream). **For this example, field = 0b.**

SrcTag[4:0] Field (Byte 2, Bits 4:0)

All non-posted requests require a source tag (0-31d); the value is assigned by UnitID2 from a pool of available tags. Assume that tag 3 is the one to be used for this request. **For this example, field = 00011b.**

Compat Bit Field (Byte 2, Bit 5)

Used by bridges to tag downstream requests. Cleared here. **For this example, field = 0b.**

Mask/Count[3:0] Field (Byte 2, Bits 7:6) & (Byte 3, Bits 1:0)

In this example, a single <u>dword</u> transfer is being done and the value programmed into this field should be *dword count -1*, which is zero. **For this example, field = 0000b.**

Start Address Field (Bytes 4-7, Bit 7:0) & (Byte 3, Bit 7:2)

This field carries the 40-bit target start address. For this example, the target address is in the main memory portion of the system memory map. **For this example, the 40-bit address = 0004810008h.**

Example 9: Sized (Dword) Write Data Packet: Bus 1

In this example, one dword is to be written to memory. This is reflected in the transfer count in the Mask/Count[3:0] field above (0). In a sized (dword) write transaction, data transfer immediately follows the request.

For this example, assume the following dword value is to be written:

Dword = 12345678h

Example 9: Request/Data Sequence Of Events On Bus 1

(Refer to Figure 7-11 on page 189)

1. UnitID2 on bus 1 checks its transmitter *Non-Posted Request (CMD)* and *Non-Posted Request (Data)* flow control counters to make sure it has "credits" in both before issuing the request.
2. The transmitter asserts the CTL signal and starts sending out the request packet information on its eight bit interface. After 8 bit times, it has finished sending the request and commences sending the dword data packet; the CTL signal will be deasserted during this time.
3. When the bridge receives this request and data, it checks the target address and recognizes it is upstream; it prepares to reissue the request on bus 0.

Example 9: Bridge Reissues Request Packet: Bus 0

(Refer to [2] in Figure 7-11 on page 189)

Note: When the bridge re-issues a request packet onto a new bus, it substitutes its own transaction stream information into the request's UnitID and Source Tag fields. It does this because transaction stream ordering is on a "per-bus" basis; the bridge is host (UnitID 0) on bus 1, but is simply UnitID 1 on the upstream bus (bus 0). If a response is required, the bridge will "remember" the transaction stream information it replaced in the request so it can be restored when the response is later driven onto the source bus. Except for UnitID and Source Tag, the bus 0 request packet fields are the same as on bus 1.

Command[5:0] Field (Byte 0, Bit 5:0)

The bridge issues the same command (with option bits) as it received in the request from bus 1. This is a non-isochronous, non-posted sized (dword) write with no cache coherency requirement. **For this example, field = 010100b.**

SeqID[3:0] Field (Byte 0, Bit 7:6) and (Byte 1, Bit 6:5)

This field is used to tag groups of requests that were issued as part of an ordered sequence from a single UnitID. This field is cleared to indicate a request is not part of a sequence. **For this example, field = 0000b.**

UnitID[4:0] Field (Byte 1, Bits 4:0)

This field is programmed with the UnitID of the requester, which on bus 0 is UnitID 1. **For this example, field = 00001b.**

PassPW Bit Field (Byte 1, Bit 7)

This bit indicates whether this packet may pass packets in the posted request channel (for same transaction stream). **For this example, field = 0b.**

SrcTag[4:0] Field (Byte 2, Bits 4:0)

This value is assigned by the bridge from its pool of available tags. Assume that the bridge's next available tag for non-posted requests it issues onto bus 0 is seven. **For this example, field = 00111b.**

Compat Bit Field (Byte 2, Bit 5)

Used by bridges to tag downstream requests. Cleared here. **For this example, field = 0b.**

Mask/Count[3:0] Field (Byte 2, Bits 7:6) & (Byte 3, Bits 1:0)

The count field is copied from the original; this example is a single <u>dword</u> transfer and the counter is loaded with *dword count* - 1. **For this example, field = 0000b.**

Start Address Field (Bytes 4-7, Bit 7:0) & (Byte 3, Bit 7:2)

This field is also copied over from the original request. **For this example, the 40-bit address = 0004810008h.**

Example 9: Sized (Dword) Write Data Packet: Bus 0

(Refer to [2] in Figure 7-11 on page 189)

In this example, one dword is to be written to memory. The single dword data packet that accompanied the request on bus 1 is simply repeated on bus 0.

For this example, assume the following dword values are to be written:

- Dword 0 = 12345678h

Example 9: Request/Data Sequence Of Events: Bus 0

(Refer to Figure 7-11 on page 189)

1. The Host Bridge (UnitID1 on bus 0) checks its transmitter *Isochronous* Non *Posted Request (CMD)* and *Isochronous Non-Posted Request (Data)* flow control counters to make sure it has "credits" in both before issuing the request.
2. The Host Bridge bus 0 transmitter interface asserts the CTL signal and starts sending out the request packet information on its eight bit interface. After 8 bit times, it has finished sending the request and commences sending the dword data packet; the CTL signal will be deasserted during this time.
3. When the bridge receives this request and data, it checks the target address and performs the non-posted write of the single data dword to memory. It then prepares to send a Target Done Response downstream.

Example 9: Response Packet On Bus 0

(Refer to [3] Figure 7-11 on page 189)

Because this is a non-posted sized (dword) write transaction, the Host Bridge owes the requester a confirmation response indicating the transfer is complete. This information is returned in a four-byte *Target Done* response packet. The UnitID and Source Tag information contained in the bus 0 response packet will belong to the HT-to-HT bridge bus 0 interface. It will be the HT-to-HT bridge's responsibility to restore the original transaction stream information before forwarding the response packet to the next bus (bus 1).

Command[5:0] Field (Byte 0, Bit 5:0)

This is the command code for the Target Done response. There are no option bits in this field. **For this example, field = 110011b.**

Isoc Bit Field (Byte 0, Bit 7)

This bit will be cleared because isochronous traffic was not enabled in the request. **For this example, field = 1b.**

UnitID[4:0] Field (Byte 1, Bits 4:0)

For downstream responses, this field contains the UnitID if the requester. In this example, it will contain UnitID 1 (the HT-to-HT bridge bus 0 interface). **For this example, field = 00001b.**

Bridge Bit Field (Byte 1, Bit 6)

Set by host bridges in responses they issue downstream (such as this example). **For this example, field = 1b.**

PassPW Bit Field (Byte 1, Bit 7)

This bit will be cleared because passing was not enabled in the request. **For this example, field = 0b.**

SrcTag[4:0] Field (Byte 2, Bits 4:0)

Programmed with the Source Tag provided by the HT-to-HT bridge (7). **For this example, field = 00111b.**

Error Bit Field (Byte 2, Bit 5)

Cleared in this example. **For this example, field = 0b.**

NXA Bit Field (Byte 3, Bit 5)

Cleared in this example. **For this example, field = 0b.**

Example 9: Response, Sequence Of Events On Bus 0

(Refer to Figure 7-11 on page 189)

4. Once it has the Response (CMD) flow control credits, the Host Bridge transmitter asserts the CTL signal and sends the four byte Target Done response onto Bus 0.
5. When the HT-to-HT bridge receives the response from the Host Bridge, it checks the Bridge bit and UnitID (1) to determine if the response belongs to it (it does). The HT-to-HT bridge then attempts to match the Source Tag field in the response to one of the entries in a table of outstanding requests it has been tracking.
6. Using its Source Tag, it locates the original transaction stream information for the device on bus 1 which made the request. It then prepares to forward the response onto bus 1 with the original transaction stream information (UnitID and Source Tag) restored.

Example 9: Response Packet On Bus 1

(Refer to [4] Figure 7-11 on page 189)

The HT-to-HT bridge forwards the response packet onto bus 1. With the Bridge bit set and the original UnitID (2) and Source Tag (3) restored, the response will be claimed by the original requester.

Command[5:0] Field (Byte 0, Bit 5:0)

This is the command code for the Target Done response. There are no option bits in this field. **For this example, field = 110011b.**

Isoc Bit Field (Byte 0, Bit 7)

This bit will be cleared because isochronous traffic was not enabled in the request. **For this example, field = 1b.**

UnitID[4:0] Field (Byte 1, Bits 4:0)

Restored by the HT-to-HT Bridge to the value provided by the original requester (2). **For this example, field = 00010b.**

Bridge Bit Field (Byte 1, Bit 6)

Set by host bridges in responses they issue downstream (such as this example). **For this example, field = 1b.**

PassPW Bit Field (Byte 1, Bit 7)

This bit will be cleared because passing was not enabled in the request. **For this example, field = 0b.**

SrcTag[4:0] Field (Byte 2, Bits 4:0)

Restored by the HT-to-HT bridge with the Source Tag provided by the UnitID2 (3). **For this example, field = 00011b.**

Error Bit Field (Byte 2, Bit 5)

Cleared in this example. **For this example, field = 0b.**

NXA Bit Field (Byte 3, Bit 5)

Cleared in this example. **For this example, field = 0b.**

Example 9: Response, Sequence Of Events On Bus 1

(Refer to Figure 7-11 on page 189)

1. Once the HT-to-HT bridge has prepared the response to be driven onto bus 1 (with the original UnitID and Source Tag restored), it asserts the CTL signal and sends the four byte Target Done response onto Bus 1.
2. When UnitID 2 receives the response from the HT-to-HT Bridge, it checks the Bridge bit and UnitID to determine if the response belongs to it (it does).

UnitID 2 then attempts to match the Source Tag field in the response (this is the source tag it sent with its request) to one of the entries in a table of outstanding requests it has been tracking.

3. Using its Source Tag, it locates the pending transaction and retires it. If there had been an error on either bus, the Error bit (and possibly the NXA bit) in the response would have ben set.

8   *HT Interrupts*

The Previous Chapter

To review the principles of HT transactions and to provide a more comprehensive understanding, the previous chapter presented examples of complex system transactions, including reads, posted and non-posted writes, and atomic read-modify-write.

This Chapter

HT uses an interrupt signaling scheme very similar to PCI's Message Signaled Interrupts. This chapter defines how HT delivers interrupts to the Host Bridge via posted memory writes. This chapter also defines an End of Interrupt message and details the mechanism that HT uses for configuring and setting up interrupt transactions (which is different from the PCI-defined mechanisms).

The Next Chapter

Rather than requiring HT devices to incorporate additional pins for signaling system-related items such as power management events, System Management messages are defined that permit signaling via transactions, thereby creating virtual wires. The next chapter defines the mechanism used to send these messages, by detailing the System Management packets and protocol. Some System Management functions such as changing the operating frequency of an HT link require that transactions be stopped. The chapter also introduces the ability to temporarily disconnect the links to allow these types of state changes, and to save power.

Introduction

HT, unlike most legacy I/O bus implementations, does not define the use of interrupt pins, nor an interrupt controller. Instead, interrupt delivery is distributed to the HT devices themselves. Each device delivers interrupts by performing memory writes to memory address locations reserved for that purpose. The data written to these locations provides information that historically comes from or is handled by an interrupt controller (such as interrupt priority and vector information that specifies the location of the interrupt service routine). This method of interrupt delivery is commonly referred to as Message Signaled Interrupts.

HT supports message signaled interrupts via two message types:

- **Interrupt Request message** — Interrupt requests are forwarded upstream as sized write transactions that target a reserved interrupt request address range. The host bridge receives these packets and based on the target address recognizes the transaction as an interrupt request. The specific actions taken by the bridge to process the interrupt request is platform-specific and not specified.
- **End of Interrupt message** — HT also supports an End Of Interrupt (EOI) message that may be used by devices that require confirmation that their interrupt service routine has completed. These messages originate at the host and are forwarded downstream as a broadcast. Like the interrupt request message, the EOI request packet address must also fall within the reserved address range.

Discovering a Device's Interrupt Requirements

HT defines an interrupt capability block as illustrated in Figure 8-1 on page 201. The presence of the capability block in configuration space indicates that the HT device uses interrupts. This block, named the Interrupt Discovery and Configuration Capability block, defines the number of interrupt sources each HyperTransport technology function can generate. This block also allows software to configure each interrupt independently.

Figure 8-1: Interrupt Capability Block Indicates that the Device Supports Interrupts

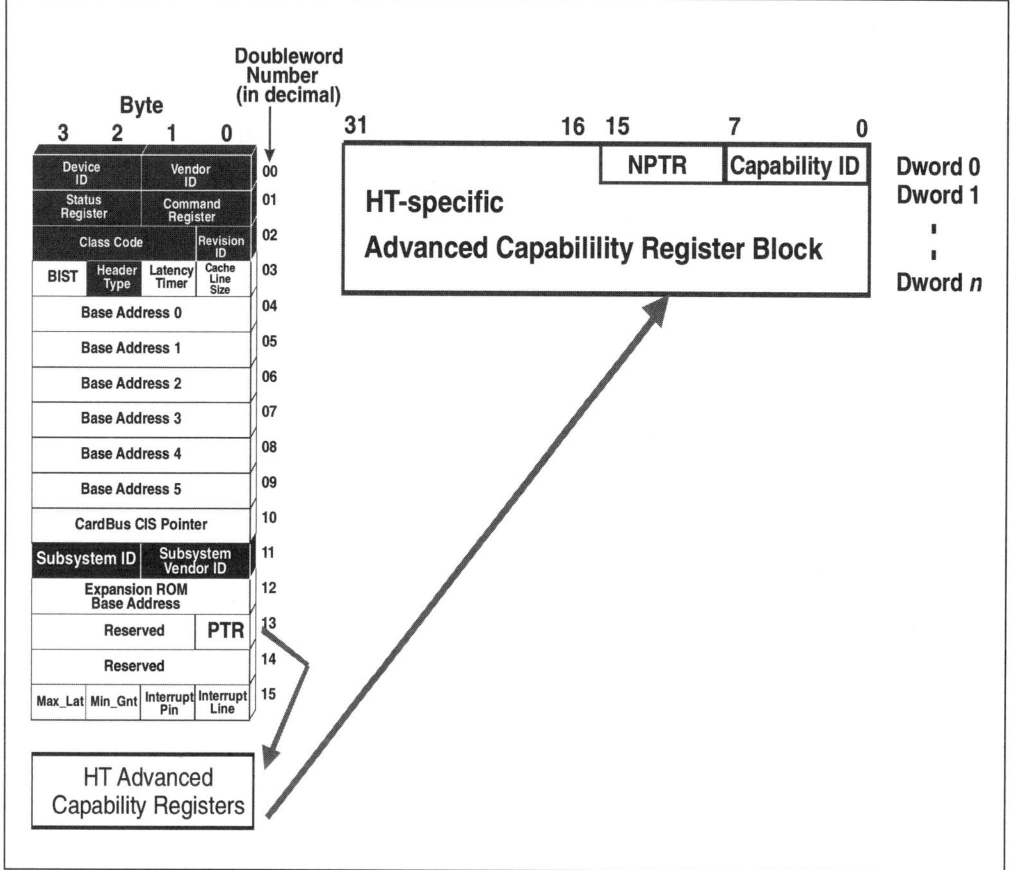

The Interrupt Message Address Range

The reserved address range used by the Interrupt Request and EOI messages is pictured in Figure 8-2 on page 202. This 3984MB address range is mapped from location FD_0000_0000h to FD_F8FF_FFFFh.

The specified range reserved for the interrupt request packets seems straight forward until one looks at the interrupt request packet definition (Figure 8-2 on page 202). Note that the address field (byte 7) defined by the specification includes only Addr[39:32]. These upper eight address bits identify a 4GB address range starting at FD_0000_0000h. When reviewing the address map in Figure 8-2, it can be seen that several reserved address blocks fall within this 4GB address range — from Legacy PCI Ack through Configuration.

Figure 8-2: Interrupt Request and EOI Message Reserved Address Range

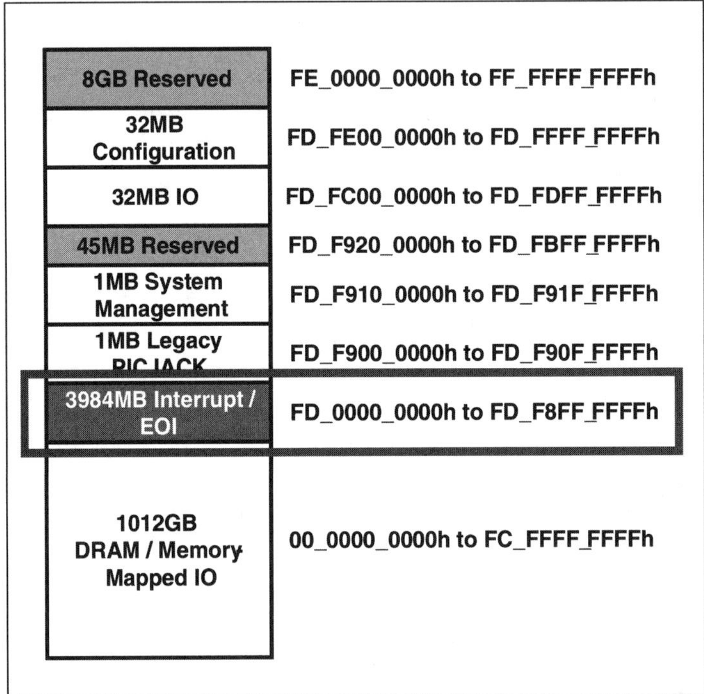

Figure 8-3 on page 203 depicts the format of the Interrupt Request packet. Note that the specification defines only Addr[39-32] to identify the interrupt packet address. If the interrupt request and EOI packets were limited to Addr[39:32], the host could not differentiate the interrupt packets from packets associated with other address ranges and functions.

The address associated with the interrupt packets must include additional address bits to distinguish between the difference address ranges. If the Host Bridge is to resolve the address to within the specified interrupt address range (FD_0000_0000h to FD_F8FF_FFFFh), then Addr[31:24] must be included within the interrupt packets. Verification of the specification's intent can be found in the x86 compatibility definitions, which specify Addr[31:24] be delivered in the IntrInfo[31:24] field of the interrupt packet. To maintain compatibility with earlier versions of HT implementations, the specification sets a default value of F8h for IntrInfo[31:24].

Figure 8-3: Interrupt Request Packet Address Field

Bytes \ Bits	7	6	5	4	3	2	1	0
0	SeqID[3:2]		Cmd[5:0] = 101000 (posted, Wr(sized), Byte					
1	PassPW	SeqID[1:0]		UnitID[4:0]				
2	Count[1:0]		Reserved					
3	Interrupt Information[7:2]						Count[3:2]	
4	Interrupt Information[15:8]							
5	Interrupt Information[23:16]							
6	Interrupt Information[31:24]							
7	Addr[39:32] = FDh							
8	Interrupt Information[39:32]							
9	Interrupt Information[47:40]							
10	Interrupt Information[55:48]							
11	Reserved							

For system platform implementations other than x86, the specification leaves open the possibility of the interrupt range being extended, but does not explicitly state that the interrupt range can be extended in the absence of the PIC IACK, System Management, and IO mappings. For example, some platforms may only need support for the interrupt and configuration packets. This would require the use of Addr[39:26], thereby permitting the Host Bridge to distinguish between the interrupt and configuration requests.

Interrupt Requests

Interrupt request messages originate within HT I/O devices and are sent upstream using the posted-write virtual channel. This assures that any posted write transactions that preceded the interrupt request are pushed ahead of it to memory before the host bridge receives the interrupt.

HT uses message-signaled interrupts that behave much like HT Sized Write (byte) transactions. The interrupt request packet format is defined, but how bit fields are used is implementation-specific

While interrupt information contained in a request varies with the implementation, basic content might include:

- Type of interrupt
- Target address or CPU ID of the recipient
- Interrupting device's vector
- Whether an End-Of- Interrupt (EOI) acknowledgement is required.

The specification defines a generic interrupt request packet format, thereby providing flexibility for supporting the interrupt protocols used in different platforms. For example:

- An interrupt vector might be defined as 8 bits (e.g. x86 machines), while it may be 32 bits in other architectures.
- Interrupt requests may come from devices attached directly to an HT link, and from devices residing on a legacy bus (e.g., PCI), where interrupt requests are gathered by an interrupt controller and delivered by a HT-to-PCI bridge to the HT bus and transported to the host.
- Interrupt requests may all be directed to a single processor for handling, or particular interrupt requests may be directed to different processors in a multi-processor system. Etc.

Interrupt Request Packet

The interrupt request packet specifies a posted byte sized-write command with an address that falls within the reserved interrupt address range. The request packet is immediately followed by a single 4 byte data packet which carries interrupt information. Figure 8-4 on page 205 illustrates the format of the interrupt request packet.

Figure 8-4: Format of Interrupt Request Packet

Bytes \ Bits	7	6	5	4	3	2	1	0
0	SeqID[3:2]		Cmd[5:0] = 101000 (posted, Wr(sized), Byte					
1	PassPW	SeqID[1:0]		UnitID[4:0]				
2	Count[1:0]		Reserved					
3	Interrupt Information[7:2]						Count[3:2]	
4	Interrupt Information[15:8]							
5	Interrupt Information[23:16]							
6	IntrInfo[31:24] (Addr[31:24] = F8h)*							
7	Addr[39:32] = FDh							
8	Interrupt Information[39:32]							
9	Interrupt Information[47:40]							
10	Interrupt Information[55:48]							
11	Reserved							

* Default value for Interrupt Information[31:24].

The specification requires specific values in some fields within the interrupt request packet:

- **Command Type (Cmd)** — This field must specify a posted, byte, sized write packet with a value of 101000b.
- **Count [3:0]** — The count field must contain a value of zero to specify that the data packet size is 4 bytes. There is no "byte mask"; instead, the single dword contains additional interrupt information.
- **Interrupt Information [31:2]** — The interrupt information fields within the request packet are defined for interrupt-specific information as required by a given platform. The specification defines the contents of these fields for x86-based platforms only. The following Interrupt info fields have optional definitions:
  - o **Interrupt Type (IntrInfo [4:2])** — This 3-bit field defines the type of

interrupt request. All values except 111b are defined as implementation specific. The only definition provided by the specification (111b) defines the interrupt packet as an EOI packet (See "EOI Packet Format" on page 208). Refer to "HT Method of Handling APIC Interrupts" on page 498 for examples of other message types (NMI, SMI, etc.) used in x86-based platforms.

o **Request EOI (IntrInfo [5])** — This bit is set by an HT device to request that an EOI request packet be returned after the interrupt service has finished.

o **IntrInfo [31:8]** — The specification states that contents of these bit location during the request "will be retuned in the EOI message, although some hosts may not support use of all bits."

o **IntrInfo[31:24]** — Note that byte 6 of Figure 8-4 is defined as Addr[31:24] that has a default value of F8h, rather than simply IntrInfo[31:24]. See Addr description below.

- **Addr[39:24]** — This address represents the upper address bits (FD_F8xx_xxxxh) of the reserved interrupt address range. This address permits the Host Controller to differentiate interrupt packets from the other reserved packet types. (See "The Interrupt Message Address Range" on page 201.) Additional address lines may be required depending on the platform implementation. Also see "X86 Interrupt Support" on page 494 for definition of the address field in x86 platforms.

Interrupt Request Data Packet

Figure 8-5, illustrates the 4-byte data packet, which is concatenated directly to the end of the Interrupt Request packet. The data packet fields are defined as:

- **IntrInfo [55:32]** = The upper 24 bits of interrupt information are defined as platform-specific. Figure 22-7 on page 502 defines these fields for x86 platforms.
- **Reserved** — The last byte of the data packet is reserved.

Figure 8-5: Format of the Interrupt Request Data Packet

Bytes \ Bits	7	6	5	4	3	2	1	0
0	SeqID[3:2]		Cmd[5:0] = 101000 (posted, Wr(sized), Byte					
1	PassPW	SeqID[1:0]		UnitID[4:0]				
2	Count[1:0]		Reserved					
3	Interrupt Information[7:2]						Count[3:2]	
4	Interrupt Information[15:8]							
5	Interrupt Information[23:16]							
6	IntrInfo[31:24] (Addr[31:24] = F8h)*							
7	Addr[39:32] = FDh							
8	Interrupt Information[39:32]							
9	Interrupt Information[47:40]							
10	Interrupt Information[55:48]							
11	Reserved							

* Default value for Interrupt Information[31:24].

The End of Interrupt (EOI) Message

The HT specification defines the mechanism used to notify an interested party that an interrupt service routine has completed execution. The EOI is used to notify an Advanced Programmable Interrupt Controller (APIC) that an interrupt request has been processed. This is needed when more than one device is sharing the same interrupt line via level triggering. In such cases, the APIC needs confirmation that an interrupt has been serviced prior to sending another interrupt request. {See IA32 Processor Architecture book for more details.)

The EOI request is sent downstream through the HT fabric as a broadcast message that travels in the posted channel. Also, the EOI message targets the same reserved address ranges that interrupt requests use. When the broadcast EOI message reaches the end of a chain, it is simply dropped by the last device (no response is expected or sent because EOI travels in the posted channel)

EOI Packet Format

The EOI packet (Figure 8-6 on page 209) is structured like other broadcast message requests. The specification defines the use or behavior of some fields within the EOI packet, as described below:

- **Command Type (Cmd)** — This field must specify a posted, broadcast packet with a value of 111010b.
- **Interrupt Type (IntrInfo [4:2])** — This 3-bit field defines the type of interrupt request. The specification defines a value of 111b, identifying the packet as EOI.
- **IntrInfo [31:08]** = These fields are duplicated from the corresponding interrupt request packet. Other requirements may also exist as discussed below:
    - o IntrInfo[15:08] may optionally also be 00h and used as a "wild card" that can match any value of IntrInfo[15:08].
    - o **IntrInfo[31:24]** — Note that byte 6 of Figure 8-6 on page 209 is defined as Addr[31:26]. See Addr description below for additional information.
- **Addr[39:32]** — This address represents the upper address bits of the reserved interrupt address range (FD_0000_0000 to FD_F800_0000h). This address permits the Host Controller to differentiate interrupt packets from the other reserved packet types. (See "The Interrupt Message Address Range" on page 201. for more detail.) Additional address lines may be required depending on the platform implementation.

Figure 8-6: EOI Packet Format

Bytes \ Bits	7	6	5	4	3	2	1	0
0	SeqID[3:2]		Cmd[5:0] = 111010 (posted, broadcast)					
1	PassPW	SeqID[1:0]		UnitID[4:0]				
2	Reserved							
3	Reserved			MT[2:0] = 111b			Reserved	
4	Interrupt Information[15:8]							
5	Interrupt Information[23:16]							
6	IntrInfo[31:26] = Addr[31:26]*						IntrInfo[25:24]	
7	Addr[39:32] = FDh							

* Note that the specification defines byte 6 as Interrupt Information[31:24]

Interrupt Discovery and Configuration Capability Block

Each function can have its own capability block, facilitating a mapping of interrupts to functions. The capability block not only defines the number of interrupts the function is designed to use, but also provides a way for system software to define the contents of the Interrupt Information fields that will be delivered to the host during each Interrupt Request.

Interrupt Capability Block Format

Figure 8-7 on page 210 illustrates the fields within the capability block and further illustrates the relationship between each interrupt supported and its associated Interrupt Definition Registers. Each field has the following purpose:

- Capability ID (08h) — This field identifies this as an HT Capability Block
- Capability Type (80h) — This field specifies the type of HT Capability Block as the Interrupt Discovery and Configuration Capability Block
- Capabilities Pointer — Pointer to the next Capability Block
- Index — The value selects the entry within the register stack to be accessed.

- Dataport — Configuration transactions targeting this dataport permit access to the register selected by the index register value.

The index field of the Interrupt Discovery and Configuration Capability Block selects 32-bit entries within a register stack (or data structure) that defines the number of interrupts the function uses and includes a corresponding Interrupt Definition Register for each interrupt.

Figure 8-7: Format of Interrupt Discovery and Configuration Capability Block

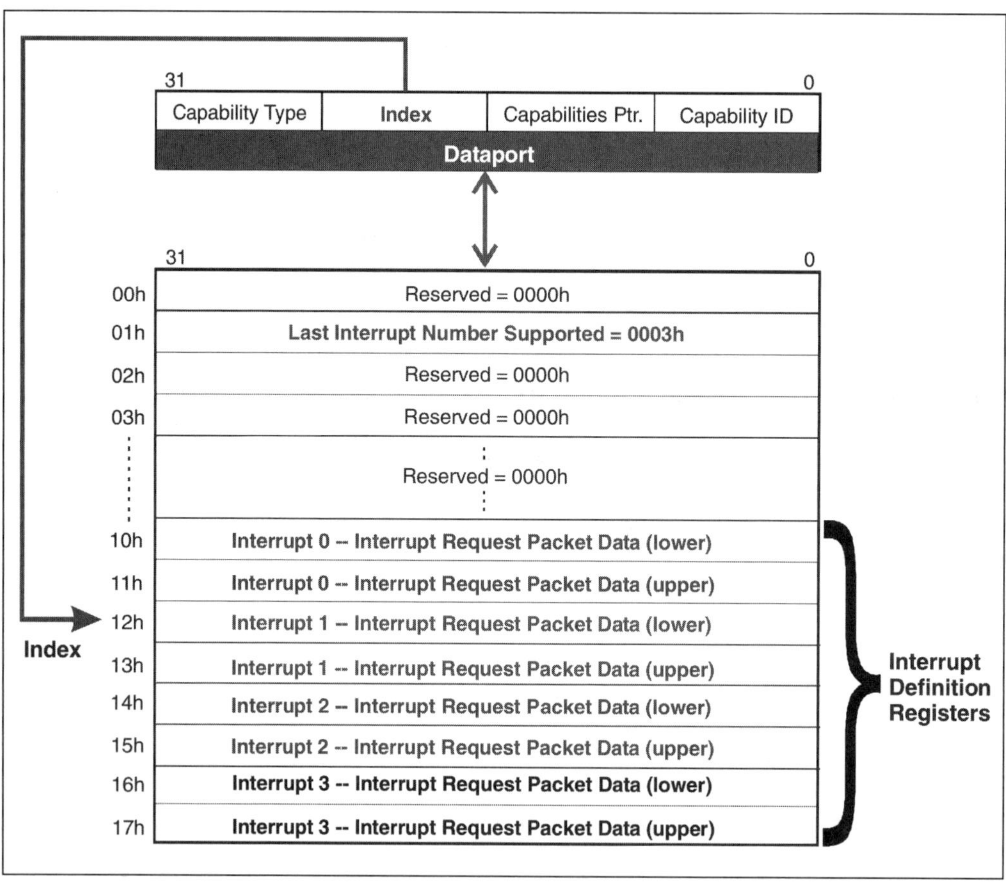

Last Interrupt Supported

Figure 8-7 on page 210 illustrates the Last Interrupt Supported register. This register is accessed through the dataport field of the capability block when selected with an index value of 01h. The register specifies the highest interrupt number used in conjunction with the function or ReqID. Interrupt numbers start at zero and are assigned sequentially. The example chosen in Figure 8-7 shows a value of 3 in the Last Interrupt Supported register, indicating there are 4 interrupts used by this function.

Interrupt Definition Registers

Each interrupt message specified for this function has a corresponding Interrupt Definition register (See Figure 8-7). These 64-bit registers are accessed 32-bits at a time via the Interrupt Discovery and Configuration Capability Block index field. The specification defines the relationships between the index and interrupt numbers as listed below:

- Index 10h points to bits [31:0] of Interrupt 0 Interrupt Definition register.
- Index 11h points to bits [63:32] of Interrupt 0 Interrupt Definition register.
- Index 12h points to bits [31:0] of Interrupt 1 Interrupt Definition register.
- Index 13h points to bits [63:32] of Interrupt 1 Interrupt Definition register.
- Index 14h points to bits [31:0] of Interrupt 2 Interrupt Definition register.
- Index 15h points to bits [63:32] of Interrupt 2 Interrupt Definition register.
- etc.

Table 8-1 on page 212 defines the contents of the Interrupt Definition register.

Table 8-1: Contents of the Interrupt Definition Registers

Bit	R/W	Reset	Description
63	R/C	0	Waiting for EOI — This bit only has meaning if Request EOI (bit 5) is set (1). This bit is set by hardware when an interrupt request is sent and it is cleared when the corresponding EOI is returned. Software may also clear this bit by writing a 1 to it.
62	R/W	0	Pass PW — This bit determines the state of the PassPW bit within the interrupt request packet. When set (1) no ordering of this message is guaranteed with respect to other upstream transactions. When cleared (0) ordering of this message is maintained within the posted channel. This bit may be read-only if single behavior is supported.
61:56	R/O	0	Reserved
55:32	R/W	0	IntrInfo[55:32] — value is implementation specific.
31:24	R/W	F8h	IntrInfo[31:24] — This field is implemented as part of the interrupt address (Addr[31:24]). The default value is F8h for backward compatibility with 1.01 and earlier HT devices. The actual number of address bits used within a device is implementation specific.
23:6	R/W	0	IntrInfo[23:6] — value is implementation specific.
5	R/W	0	IntrInfo[5] (Request EOI) — This bit permits a device to request an EOI packet be sent after the interrupt service has completed. When set (1) the device waits for an EOI packet (by monitoring the "Waiting for EOI" bit to be cleared) before sending the next interrupt.
4:2	R/W	0	IntrInfo[4:2] Message Type — value is implementation specific. See Chapter 22, entitled "X86 CPU Compatibility," on page 491 for an example use in x86 platforms.
1	R/W	0	Polarity — Specifies the polarity of the interrupt trigger in HT devices with external interrupts (e.g. PCI bridge). 1=active low and 0=active high.

Table 8-1: Contents of the Interrupt Definition Registers

Bit	R/W	Reset	Description
0	R/W	1	Mask — When set (1) interrupt messages are not sent from this source.

9 *System Management*

The Previous Chapter

HT uses an interrupt signaling scheme very similar to PCI's Message Signaled Interrupts. The previous chapter defines how HT delivers interrupts to the Host Bridge via posted memory writes. The chapter also defined an End of Interrupt Message and details the mechanism that HT uses for configuring and setting up interrupt transactions (which is different from the PCI-defined mechanisms).

This Chapter

Rather than requiring HT devices to incorporate additional pins for signaling system-related items such as power management events, System Management messages are defined that permit signaling via transactions, thereby creating virtual wires. This chapter defines the mechanism used to send these messages, by detailing the System Management packets and protocol. Some System Management functions such as changing the operating frequency of an HT link require that transactions be stopped. This chapter also introduces the ability to temporarily disconnect the links to allow these types of state change and to save power.

The Next Chapter

There are two aspects in dealing with link or internal errors in HyperTransport: detection and handling. The next chapter describes the error types defined by the specification and what devices may do about them. Some devices may choose to detect and handle some errors but not others. The PCI configuration space registers, used to program the error strategy and log errors, are described here — as are the reporting mechanisms: error response, fatal and non-fatal interrupts, and Sync flood.

System Management Transactions

HT provides a message passing mechanism between the Host Bridge and the System Management Controller (SMC). One of the primary purposes of HT messages is to eliminate dedicated pins and traces that would otherwise be required to signal various events, reducing pin count and cost. These System Management (SM) messages are delivered via packets that support a wide variety of functions including:

- HT Power Management
- X86 Power Management
- X86 Legacy CPU Signalling (e.g. A20M, FERR#, and IGNNE#)

HT System Management messages in conjunction with LDTSTOP# may be used to support operations such as changes in operating frequency and link width, or to disable the links to save power. It is also through System Management (SM) requests that many of the x86 compatibility mechanisms are accomplished as indicated above. Further, x86 platforms are required to support SM and LDT-STOP# for power management. Power Management support for HT devices is optional in non-x86 platforms; however, many non-x86 systems do support power management. Note also that the specification requires all HT devices to forward SM packets in both directions.

Sources of SM Request

System Management requests may be either sent in the upstream or downstream direction, as illustrated in Figure 9-1 on page 217. All SM requests moving upstream originate at the System Management Controller (SMC) and downstream requests originate at the Host Bridge. Note that the SMC typically resides in the south bridge (or I/O Controller Hub) where the legacy signals typically originate and where power management registers reside.

Figure 9-1: SM Request Sources

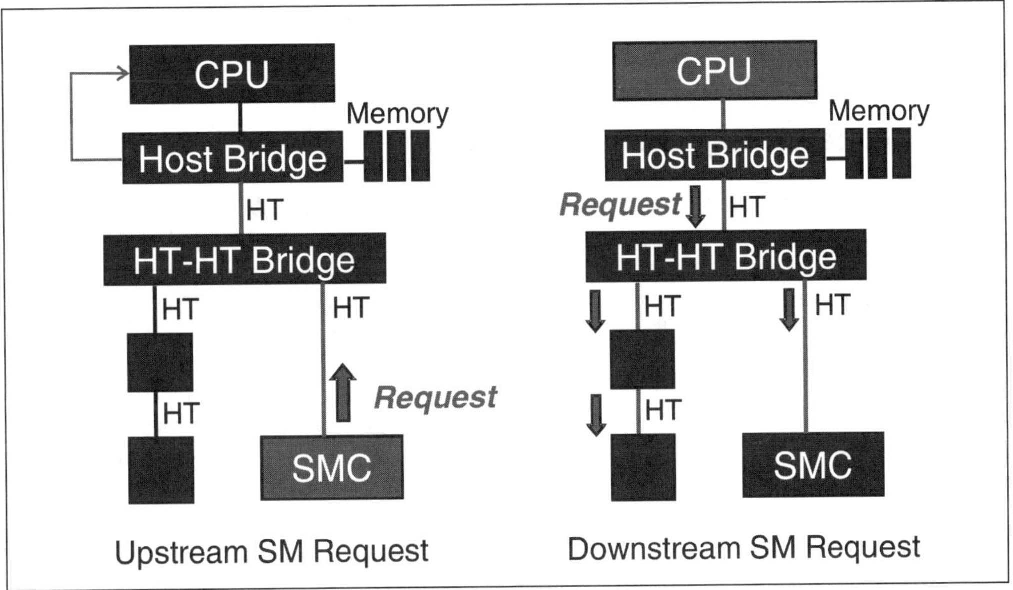

System Management Address Range

System Management transactions are recognized by their assigned address range. The HT specification reserves a 1MB address range for system management transactions from FD_F910_0000h to FD_F91F_FFFFh. In reality, only the upper address bits are needed to identify that the transaction falls within the assigned 1MB range. SM request packets include only the upper 20 bits (A39:A20) of the HT address for identifying the SM range (FD_F91h). Note that the lower 5 nibbles (or 20 bits) of the address are not defined and could theoretically be any value between 0_0000h and F_FFFF. The 1MB block of SM address space serves only to identify SM transactions and does not actually target any memory locations.

The SMC & Upstream Request Packets

The System Management Controller generates SM requests in response to both software initiated events (i.e., writes to registers within the south bridge) and hardware events (e.g. inactivity timeouts).

Upstream Request Packet Format

SMC-originated messages are delivered as posted Sized Write transactions, consisting of a SM request packet followed by a 4 byte data packet. An SM transaction is identified by both the Sized Write command and the assigned SM address range. The format of the upstream moving SM request packet is illustrated in Figure 9-2. The specification requires the following field values:

- Byte 0 — Cmd defined as posted, sized, byte write = 101000d
- Bytes 2 & 3 — Count [3:0] = 0000b (specifies 4 byte data packet)
- Byte 4 — Defines types of upstream SM Request. (See "System Management Commands — Upstream" on page 219 for details.)
- Bytes 5 & 6 — Address[39:20] = FDF91h

An interesting aspect of the upstream SM transactions is that the Host Bridge reflects them all back downstream across all links as broadcast SM messages.

Figure 9-2: Format of SM Request Packet Issued by the System Management Controller

Bytes \ Bits	7	6	5	4	3	2	1	0
0	SeqID[3:2]		Cmd[5:0] = 101000 (posted, Wr(sized), byte)					
1		SeqID[1:0]		UnitID[4:0]				
2	Count[1:0]		Reserved					
3	Reserved						Count[3:2]	
4	SysMgtCmd[7:0]							
5	Addr[23:20] = 1h				Reserved			
6	Addr[31:24] = F9h							
7	Addr[39:32] = FDh							

System Management Commands — Upstream. The System Management Command field (SysMgtCmd) defines the type of SM request being issued. Table 9-1 on page 219 summarizes the SM command options for the upstream direction. All of the upstream messages represent state changes of various signals as indicated in Table 9-1. (Chapter 22, entitled "X86 CPU Compatibility," on page 491 details each signal type.)

Three distinct pieces of information may be included in the upstream SysMgtCmd:

- The Base Command Type — The upper nibble defines the primary command type (processor input signals and STPCLK) and is always present.
- Signal State Bit Map — Defines new state of the signal (1 = signal asserted, 0 = signal deasserted). For example, STPCLK may be asserted by the SMC (bit 0 of SysCmd = 1) to indicate a power management request. Subsequently, the SMC would need to deassert STPCLK by sending another message, with bit 0 of SysCmd = 0. In addition, when an SM message is sent to the Host Bridge, the bridge will send the packet back downstream using its SM request packet.
- System Management Action Field (SMAF) — In the upstream direction this field is only defined only for the STPCLK message. SMAF qualifies the nature of the power management event being signalled. Note that the definition of the SMAF is platform specific.

Table 9-1: Summary of Upstream SysMgtCmd Encodings

SysMgtCmd 7:4 3:0	Command Type
0000 xxxx	Reserved
0001 xxss	x86 legacy inputs to the processor. Bits [3:0], labeled "s," are bit maps that correspond to each input and the new logic state of the specified signal. (1 = signal asserted and 0 = signal deasserted) Bit 0 = IGNNE state Bit 1 = A20M state Bits 2 and 3 = Reserved

Table 9-1: Summary of Upstream SysMgtCmd Encodings

SysMgtCmd 7:4 3:0	Command Type
0011 aaas	STPCLK (Stop Clock) used to communicate a request to change power management states and can also be an input to the processor. Bit [0], labeled "s" specifies the new state of the STPCLK signal. Bits [3:1] are defined as system management action fields (SMAFs). (These values define the nature of the power management request.)

The Host Bridge & Downstream Request Packets

The Host Bridge generates all SM packets moving in the downstream direction. These transactions are delivered via Broadcast request packets that have no associated data packet.

Downstream Request Packet Format

The SM request packet format delivered by the Host Bridge is illustrated in Figure 9-3. The packet format for downstream SM messages contains the following required data field values:

- Byte 0 — Cmd defined as posted, broadcast = 111010d
- Byte 4 — Defines types of downstream SM Request. (See "System Management Commands" on page 221 for details.)
- Bytes 5 & 6 — Address[39:20] = FDF91h

Figure 9-3: Format of SM Request Packet Issued by the Host Bridge

Bytes \ Bits	7	6	5	4	3	2	1	0
0	SeqID[3:2]		Cmd[5:0] = 111010 (posted, Broadcast)					
1		SeqID[1:0]		UnitID[4:0]				
2	Reserved							
3	Reserved							
4	SysMgtCmd[7:0]							
5	Addr[23:20] = 1h				Reserved			
6	Addr[31:24] = F9h							
7	Addr[39:32] = FDh							

System Management Commands. Table 9-2 on page 221 summarizes the SM command options. The highlighted entries in Table 9-2 indicate commands that originate at the Host Bridge, while the other entries originate from the SMC and are reflected back downstream by the Host Bridge.

Table 9-2: Summary of SysMgtCmd Encodings

SysMgtCmd 7:4 3:0	Command Type
0000 xxxx	Reserved
0001 xxss	x86 legacy inputs to the processor — Originates at the SMC and is reflected downstream across all links. Bit 0 = IGNNE state Bit 1 = A20M state Bits 2 and 3 = Reserved
0010 xxxs	*x86 legacy outputs from the processor. The bit map [3:0] corresponds to each output and the new logic state of the signal. The specification currently defines only FERR. (1 = signal asserted and 0 = signal deasserted)* *Bit 0 = FERR#* *Bits [3:1] = Reserved*

HyperTransport System Architecture

Table 9-2: Summary of SysMgtCmd Encodings

SysMgtCmd 7:4 3:0	Command Type
0011 aaas	STPCLK (Stop Clock) — Originates at the SMC and is reflected downstream across all links.
0100 xxxx	*x86 CPU Halt Special Cycle (xxxx = Reserved)*
0101 xxxx	*x86 CPU Shutdown Special Cycle (xxxx = Reserved)*
0110 aaax	*STOP_GRANT Special Cycle. The SMAF values define the type of STOP_GRANT request.* *Bit 0 = reserved* *Bits [3:1] = system management action fields (SMAFs).*
0111 xxxx	*VID/FID Change (xxxx = Reserved)*
1000 xxxx	*WBINVD Special Cycle (xxxx = Reserved)*
1001 xxxx	*INVD Special Cycle (xxxx = Reserved)*
1010 xxxs	*SMIACK Special Cycle* *Bit 0 = new logic state of SMIACK signal* *Bits [3:1] = Reserved*
1011 xxxx	Reserved for x86 platform-specific functions — can originate at either the host or SMC.
11xx xxxx	Reserved

Detailed descriptions of each of the virtual wires and special cycles can be found in Chapter 22.

HT Link Disconnect/Reconnect Sequence

The specification defines the ability of the HT bus to disconnect all of its links simultaneously. This mechanism uses the LDTSTOP# signal and NOP packet (with disconnect bit set) to gracefully disconnect and subsequently reconnect all links within the HT fabric. This feature has five specified uses:

- Disconnecting HT links to conserve power
- An alternate and faster method of changing link frequency and width during initialization, when compared to Soft Reset
- Support for Host Voltage ID and Frequency ID (VID/FID) change
- Processor is entering certain ACPI-specified states
- System is entering certain ACPI-specified states

Regardless of the system motivation for asserting LDTSTOP# the sequence of events associated with one of the five previous features is initiated in one of the following ways:

- Host software accesses a register within the I/O Controller Hub (or South Bridge)
- I/O Controller Hub (or South Bridge) logic initiates the sequence.
- The host sends a VID/FID SM request.

Reference Information: LDTSTOP# Procedures

The specification defines the following procedures associated with the disconnect and reconnect sequence.

1. Once LDTSTOP# is asserted, it must remain asserted for at least 1 us. LDTSTOP# assertion must not occur while new link frequency and width values are being assigned by link-sizing software, or undefined operation may occur. (This is because both sides of a link must have link width and frequency programmed, and if one side has been programmed with new values and the other has not yet been programmed, the width and/or frequency of the two sides will not match.)
2. PWROK and RESET# assertions have priority over LDTSTOP# assertion, and LDTSTOP# must be deasserted before RESET# is deasserted.
3. A transmitter that recognizes the assertion of LDTSTOP# finishes sending any control packet that is in progress and sends a disconnect NOP packet. After sending this packet, the transmitter continues to send disconnect NOP packets through the end of the current CRC window (if the window is

incomplete) and continuing through the transmission of the CRC bits for the current window. After sending the CRC bits for the current window, the transmitter continues to drive disconnect NOP packets on the link for no less than 64 bit-times, after which the transmitter waits for the corresponding receiver on the same device to complete its disconnect sequence, and disables its drivers (if enabled by the LDTSTOP# Tristate Enable bit). No CRC bits are transmitted for the last (partial) CRC window, which only contains disconnect NOP packets. Since the HyperTransport protocol allows control packets to be inserted in the middle of data packets, and since transmitters react to the assertion of LDTSTOP# on control packet boundaries, a given data packet could be distributed amongst two or more devices after the disconnect sequence is complete.

4. A receiver that detects the disconnect NOP packet continues to operate through the end of the current CRC window and into the next CRC window until it receives the CRC bits for the current window. After sampling the CRC bits for the current window, the receiver disables its input receivers to the extent required by the LDTSTOP# Tristate Enable bit.

5. Note that LDTSTOP# can deassert either before or after the link disconnection sequence is complete. A link transmitter is not sensitive to the deassertion of LDTSTOP# until both its disconnect sequence as described in step 3 is complete, and the disconnect sequence for the associated receiver on the same device is complete. A link receiver is not sensitive to the deassertion of LDTSTOP# until both its disconnect sequence is complete and the disconnect sequence for the associated transmitter on the same device is complete.

6. A transmitter that perceives and is sensitive to the deassertion of LDTSTOP# enables its drivers as soon as the implementation allows, begins toggling the CLK with a minimum frequency of 2MHz and places the link in the state associated with the beginning of the initialization sequence (CTL = 0, CAD = 1s, CLK toggling). The transmitter is required to have CLK running within 1 us (to assure that the receive logic has a clock source). The clock frequency does not have to match the currently programmed frequency before CTL is asserted. A receiver that perceives and is sensitive to the deassertion of LDTSTOP# waits at least 1 us before enabling its inputs. This 1-us delay is required to prevent a device from enabling its input receivers while the signals are invalid before the transmitter on the other side of the link has perceived and reacted to the deassertion of LDTSTOP#. When a transmitter's corresponding receiver on the same device has been enabled, it is free to begin the initialization sequence.

7. After reconnecting to the link, the first transmitted packet after the initialization sequence must be a control packet, as implied by the state transitions of the CTL signal during link initialization. This is true even if the link was disconnected in the middle of a data packet transmission.

8. The CRC logic on either side of the link should be re-initialized after a disconnect sequence in exactly the same way as for a reset sequence.
9. Link disconnect and reconnect sequences do not cause flow control buffers to be flushed, nor do they cause flow control buffer counts to be reset.
10. LDTSTOP# should not be reasserted until all links have reconnected to avoid invalid link states. The means to ensure this is beyond the scope of this specification, although it is expected that this will be under software control.

Example SM Sequence: Link Initialization Disconnect

This example illustrates a link initialization disconnect sequence; that is, the events that would occur when BIOS software uses disconnect (LDTSTOP#) to change the link frequency and width during initialization.

Background

During initialization, devices that share a HT link must determine the maximum clock frequency and maximum link width supported by both devices. The first step in this process involves a device procedure that establishes a safe but not necessarily optimum clock frequency and link width immediately following reset. Initialization software (BIOS) must tune the link width and frequency. Software simply reads the maximum capability registers within each device, determines the maximum values that each device supports, and loads the link control registers with these values. However, these values do not take effect until BIOS software initiates either a soft reset or LDTSTOP disconnect. This example assumes that the system is designed to perform the disconnect rather than a soft reset. The disconnect method may be chosen because it completes more quickly than a soft reset. Details regarding link initialization can be found in Chapter 12, entitled "Reset & Initialization," on page 275.

Setup and Assumptions

This example explains the relationships between the various messages, responses, and signals involved in the disconnect sequence. Figure 9-4 on page 226 illustrates an HT-based system with an I/O Controller Hub (ICH) that contains a System Management Controller. The LDTSTOP# signal is an input only to all devices except the ICH in an x86 system.

Figure 9-4: Theoretical System with LDTSTOP# Support

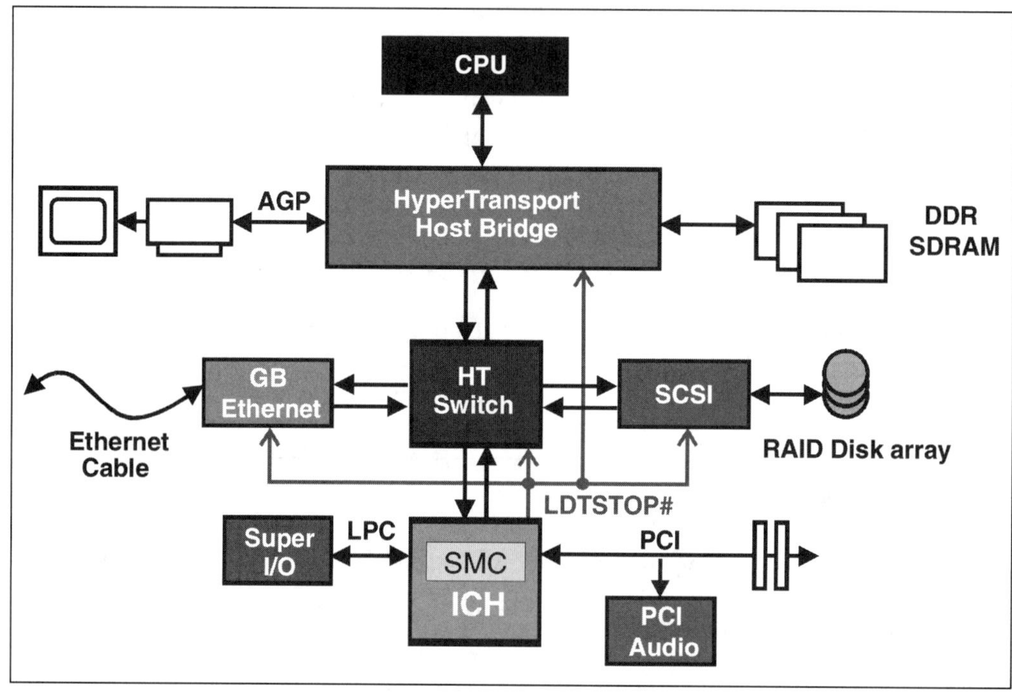

The Host Bridge and ICH must be designed specifically to perform the link initialization disconnect. This requires support of several key features including:

- A trigger mechanism to initiate link initialization disconnect — The ICH must include a mechanism (e.g. a register) that permits BIOS software to initiate the link initialization disconnect sequence for changing the link frequency and width.
- STPCLK system management message — The ICH must be able to assert and deassert STPCLK in response to a software request to initiate the link initialization disconnect. The platform must also define an SM Action Field (SMAF) code that identifies the reason for STPCLK being signaled. The host bridge must be designed to detect the message and respond appropriately.
- STOP_GRANT system management message — The system must support the delivery of the STOP_GRANT message and the associated SMAF code that identifies the reason for sending the message. The ICH must be designed to detect the message and respond appropriately.
- LDTSTOP# signal — The ICH must be able to assert and deassert LDT-STOP# using the specified sequence and timing required.

The Link Initialization Disconnect Sequence

The following steps define the sequence of events beginning with BIOS initiation of the sequence to return to normal operation.

Step 1: The BIOS code initiates the link-initialization disconnect sequence by writing to register within the ICH, in this example implementation. The write transaction is assumed to be a non-posted operation, which requires a TargetDone response.

Step 2: The ICH responds by sending a STPCLK assertion SM message to the host with a UnitID that matches the UnitID of the TargetDone response pending for step 1. The STPCLK assertion message contains a SMAF value that defines the reason for STPCLK assertion in this example.

Step 3: After the STPCLK assertion message is sent to the host, the ICH is allowed to send the response to the initiating transaction from step 1. Note that this sequence is important to ensure correct ordering of events based on some OS implementations. The response must follow the STPCLK SM message to guarantee that the host does not execute any additional instructions after the initiating command of step 1.

Step 4: When the STPCLK assertion message reaches the host, the host reflects the message downstream to all links in the fabric. Reflecting STPCLK assertion downstream has no specific purpose, it is simply earier for the host to reflect all SM messages rather than just some.

Step 5: In addition to reflecting the STPCLK assertion message in the downstream direction, the host must also respond to the STPCLK assertion message by broadcasting a STOP_GRANT SM message across all downstream links. This is intended to indicate that the host is ready for the next step in the state transition, and notifies all devices of the power state being entered.

Step 6: The ICH asserts LDTSTOP# in response to receiving and decoding the STOP_GRANT system management message and SMAF code. The ICH must delay signaling LDTSTOP# after receiving STOP_GRANT to allow time for STOP_GRANT to reach all other devices in the system. All devices upon detecting LDTSTOP# perform the disconnect sequence that includes updating the link width and frequency based on new values loaded into the Link Control Register.

Step 7: LDTSTOP# is deasserted under control of the system management logic. Recall that LDTSTOP# can be deasserted either before or after the link disconnection sequence is complete as described in item 5 on page 224.

Step 8: When a device completes the disconnect sequence and has detected

LDTSTOP# deasserted, it enters its reconnect sequence.

Step 9: After LDTSTOP# is deasserted, the ICH must send the STPCLK deassertion system management message to the host to notify the host that it can resume normal operation (i.e. to exit the STOP_GRANT state). The Host Bridge in turn reflects the STPCLK deassertion message downstream to all chains.

10 *Error Detection And Handling*

The Previous Chapter

Rather than requiring HT devices to incorporate additional pins for signaling system-related items such as power management events, System Management messages are defined that permit signaling via transactions, thereby creating virtual wires. The previous chapter defined the mechanism used to send these messages, by detailing the System Management packets and protocol. Some System Management functions such as changing the operating frequency of an HT link require that transactions be stopped. This chapter also introduced the ability to temporarily disconnect the links to allow these types of state change and to save power.

This Chapter

As indicated in the chapter title, there are two aspects in dealing with link or internal errors in HyperTransport: detection and handling. This chapter describes the error types defined by the specification and what devices may do about them. Some devices may choose to detect and handle some errors and not others. The PCI configuration space registers used to program the error strategy and log errors are described here, as are the reporting mechanisms: error response, fatal and non-fatal interrupts, and Sync flood.

The Next Chapter

Reset signalling and timing along with actions taken by the system and devices during reset are the primary topics discussed in the next chapter. It also discusses the software initiated reset and why it's required. The process of determining the default speed and link width and the subsequent software tuning of bus speed and link width are also detailed.

Introduction

HyperTransport defines six types of errors, and three basic ways they may be reported to the system.

Types Of Errors

The error types which may be detected, logged, and reported are:

1. CRC (Cycle Redundancy Code) Errors
2. Protocol Errors
3. Receive Buffer Overflow Errors
4. End Of Chain Errors
5. Chain Down Errors
6. Response Errors

Reporting Methods

Once an error is detected, it can be conveyed to other devices in the system in the following ways:

1. Error Responses
2. Error Interrupts (fatal and non-fatal)
3. Sync Flooding

The Role Of PCI Configuration Space

The PCI Configuration Space required of each HyperTransport device performs several roles in error handling. The *Command* and *Status* registers in the header and the *Link Error* and *Error Handling* registers in the HyperTransport Advanced Capability Register block are used to report error handling capabilities, program the error reporting mechanism to be used if an error occurs, and to log the errors which occur so that software can later assess the error events seen by each device.

Once the error capabilities of a device have been determined and the error reporting strategy is programmed in configuration space, any errors which occur will be handled accordingly. For example, a HyperTransport device which detects a protocol error may be programmed to set the corresponding log bit in the configuration space *Error Handling* register and generate a fatal interrupt message.

Most Types Of Error Checking Are Optional

To accommodate differences in how devices and applications may view certain types of errors, the specification only requires CRC generation/checking on each link; other aspects of error detection and handling are optional. If a particular error is not checked, the corresponding enable and logging bits in configuration space must be hardwired to 0.

System Handling Of HyperTransport Errors Varies

As in many other bus protocols, HyperTransport bus behavior during error events is well specified but the action taken by the system in response to reported errors is implementation specific. However, if Sync flood is used as a reporting mechanism, a reset is required on the affected chain(s) to restore proper protocol.

The Error Types

The following section summarizes the required *CRC generation/checking* as well as the optional *protocol, receive buffer overflow, end of chain, chain down,* and *response* error handling.

CRC Errors

The Cycle Redundancy Code (CRC) is used to detect transmission errors on all enabled byte lanes on each link. The 32 bit CRC value is calculated and sent at prescribed intervals by each transmitter, then checked against the CRC value calculated by the corresponding receiver as packets arrive. CRC is calculated by finding the remainder when the sum of packet data (CAD bits plus CTL signal during each bit time) is divided by the CRC polynomial. The polynomial used is:

$$X^{32} + X^{26} + X^{23} + X^{22} + X^{16} + X^{12} + X^{11} + X^{10} + X^8 + X^7 + X^5 + X^4 + X^2 + X + 1$$

CRC On 8, 16, or 32 bit Interfaces

For interfaces which are 8-, 16-, or 32-bits wide, CRC is independently generated and checked for each byte of CAD width. Figure 10-1 on page 232 illustrates CRC "stuffing" into the CAD packet stream on each 8-bit CAD interface.

Figure 10-1: 8/16/32 Bit Interfaces: CRC Inserted Into CAD Stream Every 512 Bit Times

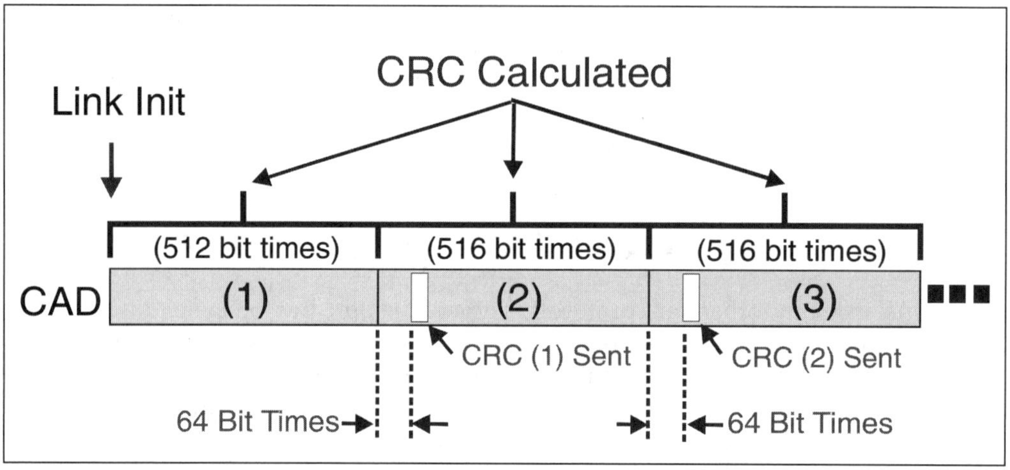

CRC Generation/Checking: 8/16/32 bit links

(Refer to Figure 10-1 on page 232)

1. After link initialization, each transmitter begins sending packets (NOP, etc.). CRC calculation is based on "raw" CAD/CTL bit patterns on each CAD byte without regard to the packet types being sent.
2. 512 bit times after initialization, the first 32-bit CRC value has been calculated for each byte lane. The window for "stuffing" the 32-bit CRC value into its CAD stream is 64 bit times <u>into the next "window"</u>. Note: because of this delay, there is no CRC sent during the first window.
3. Although each window for CRC calculation is 512 bit times, in reality all windows (after the first one) are actually 516 bit times because CRC for each window is inserted into the following one for four additional bit times. Note that the CRC value stuffed into each window is not included in the subsequent CRC calculation for that window.
4. There is no special signalling associated with CRC transmission; both devices simply count the bit times starting with link initialization and "know" where the CRC payload falls in each window.
5. CRC is calculated and sent independently for each 8 bits of CAD width. The

CTL signal itself is included in the CRC calculation for the lowest byte of CAD (bits 0-7). On a bus wider than 8 bits, the CTL signal is also factored into the CRC calculation for each of the upper CAD bytes, <u>but is assumed to be 0</u> during all bit times.

6. During the driving of the CRC value itself, the CTL signal is driven = 1 (Control) by the transmitter. The CRC bits are inverted before being transmitted onto the link.

CRC Generation/Checking: 2/4 bit links

On links narrower than 8 bits, the CRC value is generated in the same way as for 8-bit links carrying the same value. It simply takes longer to move the packets and CRC value across the link — causing the calculation window and stuffing point for the CRC value to be stretched accordingly. The extra assertions of the CTL signal (after the first bit time in each byte) are not used by the transmitter or receiver in the CRC calculation.

4 Bit CAD Width. A CAD width of four bits requires twice as many bit times as an 8 bit bus for moving information across the link. Therefore:

- The CRC window size is 1024 bit times.
- The CRC stuffing point starts128 bit times after the start of a window.
- It takes 8 bit times to transfer the 32-bit CRC value.

2 Bit CAD Width. A CAD width of two bits requires four times as many bit times as an eight bit bus for moving information across the link. Therefore:

- The CRC window size is 2048 bit times.
- The CRC stuffing point starts 256 bit times after the start of a window.
- It takes 16 bit times to transfer the 32-bit CRC value.

Logging CRC Errors

CRC errors impact both control and data information; if these errors occur on any CAD byte lane, the corresponding error bit(s) will be set in the HyperTransport Advanced Capability block *Link Control* CSR. The four bits (one for each byte lane) are illustrated in Figure 10-2 on page 234 below.

Figure 10-2: Link Control CSR: CRC Error Logging Bits

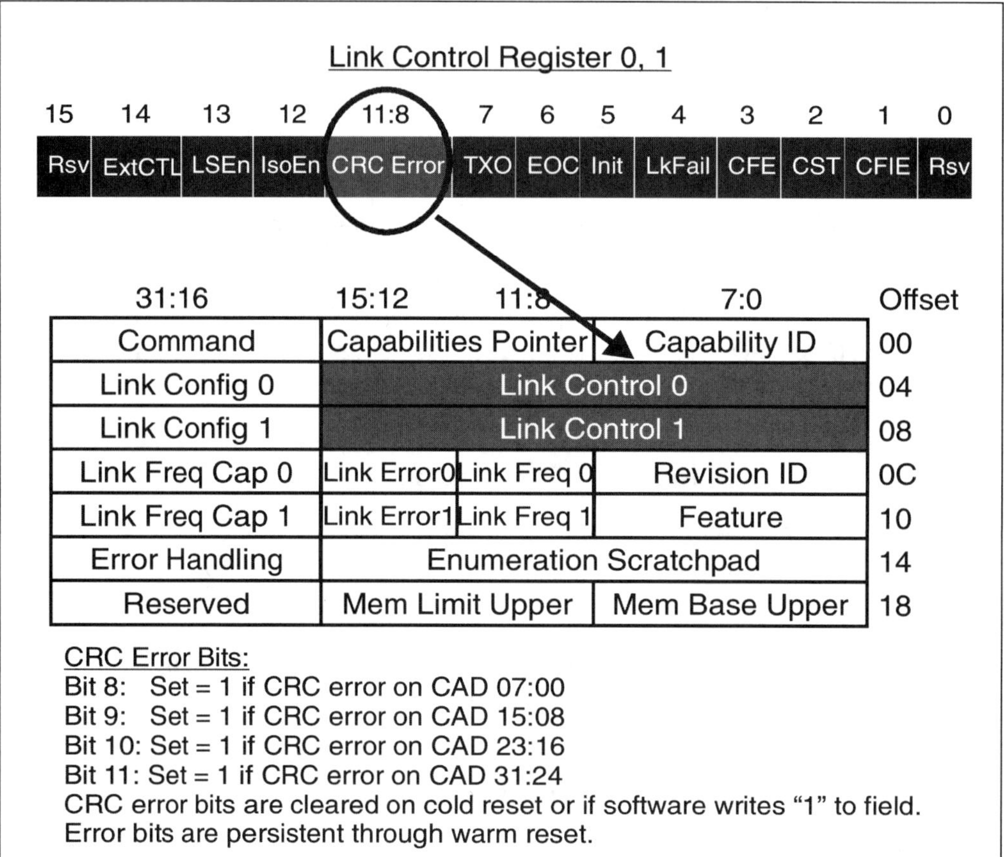

Programming The CRC Error Reporting Policy

Informing the system of a CRC error on one or more of the links is handled in the manner programmed at boot time in the Advanced Capability *Error Handling* and *Link Control* Registers. Options include sending a fatal interrupt message, non-fatal interrupt message, or initiation of a sync flood.

CRC Interrupts. Figure 10-3 on page 235 below illustrates the CRC interrupt error enable bits in the Error Handling register. If the fatal or non-fatal interrupt enable bit is set (bit 6 in byte 0 and byte 1), the device will generate a WrSized interrupt message into the address range reserved for interrupts.

Chapter 10: Error Detection And Handling

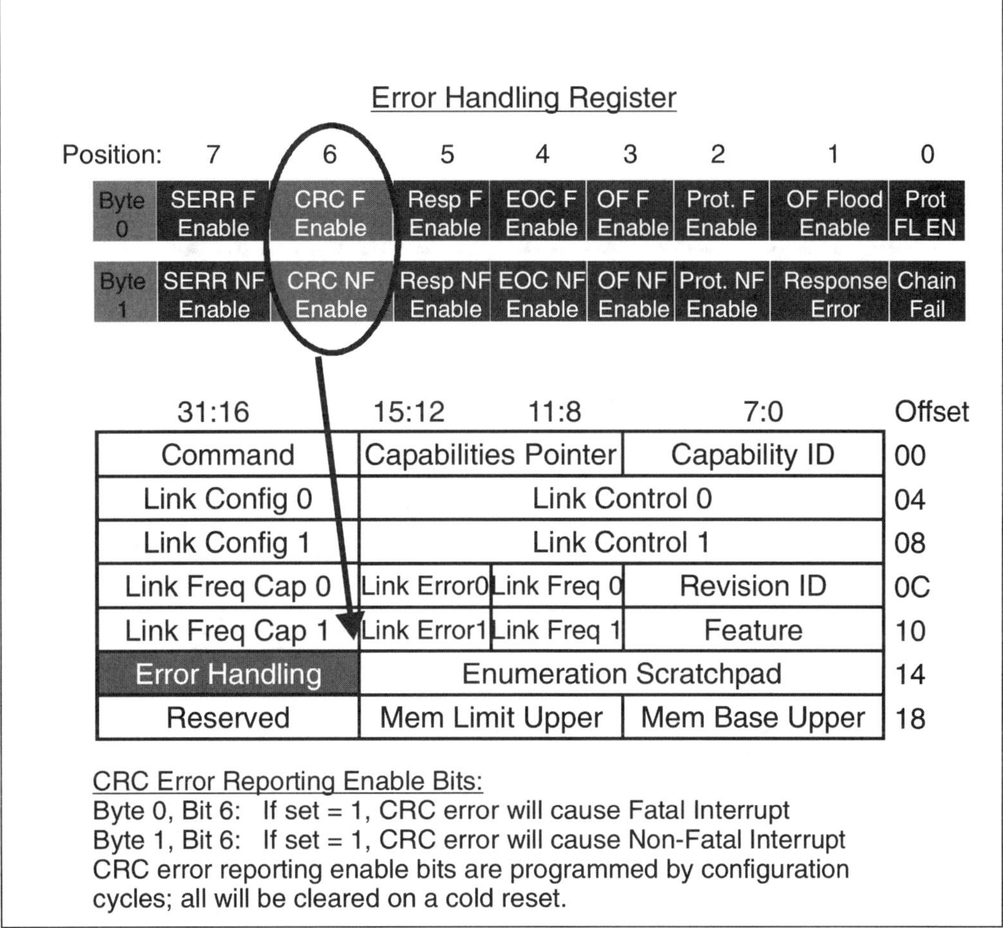

CRC Sync Flood. An alternative to sending fatal or non-fatal interrupts in response to a CRC error is the use of Sync flood. The bit to enable CRC Sync flood is contained in the Link Control Register as depicted in Figure 10-4 on page 236. When the CFIE bit is set, Sync flood will be issued on any CRC error and the Link Fail bit (bit 4) in this register will also be set.

Figure 10-4: Link Control Register: CRC Sync Flood Enable bit

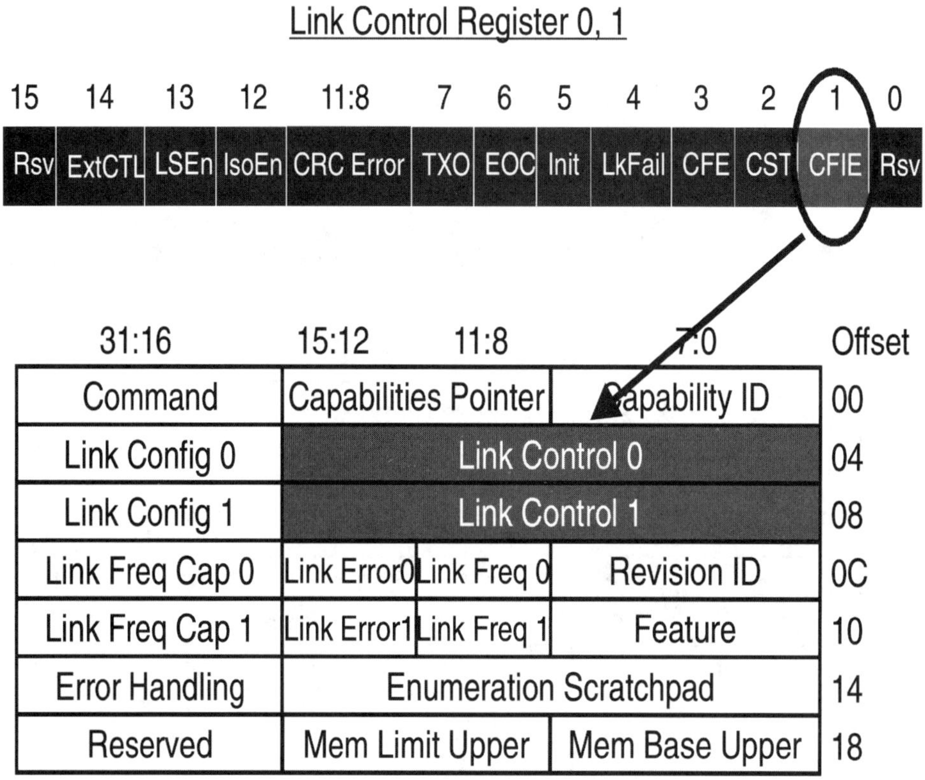

Link Control Register 0, 1

15	14	13	12	11:8	7	6	5	4	3	2	1	0
Rsv	ExtCTL	LSEn	IsoEn	CRC Error	TXO	EOC	Init	LkFail	CFE	CST	CFIE	Rsv

31:16	15:12	11:8	7:0	Offset
Command	Capabilities Pointer		Capability ID	00
Link Config 0	Link Control 0			04
Link Config 1	Link Control 1			08
Link Freq Cap 0	Link Error0	Link Freq 0	Revision ID	0C
Link Freq Cap 1	Link Error1	Link Freq 1	Feature	10
Error Handling	Enumeration Scratchpad			14
Reserved	Mem Limit Upper		Mem Base Upper	18

CRC Sync Flood Enable Bit:
If set = 1, CRC error on any enabled byte lane will result in Sync flood.
When detected, Sync flood is repeated on all links in chain. A reset is
required on the chain to recover from a Sync flood. Regardless of state
of this bit, CRC errors will still be logged (See bits 11:08).

CRC Test Mode

If both devices on a link support the CRC diagnostic testing mode (determined by checking bit 2 in the Feature Capability register for each device), then software may enable a test sequence that allows stress tests of CRC generation and checking. The basic events involved in link CRC testing include:

1. Software writes a "1" to the CRC Start Test bit of the Link Control register (refer to Figure 10-4 on page 236). Setting this bit informs the transmitter interface that it should enter the CRC diagnostic mode for the following 512 bit times on each enabled byte lane. For 4- or 2-bit CAD widths, this time is stretched to 1024 or 2048 bit times, respectively.
2. The transmitter sends a NOP packet with the *Diag* bit set; this informs the receiver that it should ignore CAD and CTL signals for the next 512 bit times but still is required to check CRC. Again, for 4- or 2-bit CAD widths, this time is stretched to 1024 or 2048 bit times, respectively.
3. With the normal buffers suspended, the transmitter may generate any test pattern it wants; CRC is still stuffed into the CAD test pattern stream in the normal way.
4. CRC errors detected during this time will be logged normally, and if the Sync flood is enabled, it will be performed. All data content is "don't care" during this time and is dropped.
5. If the CRC Force Error (CFE) bit is also set during the test (see bit 3 in Figure 10-4 on page 236), then the test pattern sent by the transmitter will contain at least one CRC error in each of the active byte lanes.
6. When the test is complete, hardware automatically clears the CRC Start Test bit. This bit may be polled by software to check completion.
7. At the end of the CRC Diagnostic test, normal packet transfer resumes.

Protocol Errors

Protocol errors are failures on the link involving low-level packet violations. These include the following:

CTL Signal Four-Byte Boundary Violation

The CTL signal may only transition between low-high on four byte boundaries. The exception to this rule is during the CRC diagnostic test mode. If an illegal transition is detected, then either the transmitter has lost track of packet start and ending boundaries or the receiver has.

CTL Deassertion Violation

Other than when CRC diagnostic test mode is in use, a transmitter only deasserts the CTL signal during data packets associated with earlier requests requiring them. Deasserting CTL when data packets are not in transit is another protocol violation.

CTL/Data Interleaving Violation

A transmitter is allowed to interleave new control packets into the data packet associated with an earlier request if the new control packet does not have any immediate data of its own. If an attempt is made to interleave a control packet with immediate data (e.g. a write request) into a data packet already in transit, this is a protocol violation.

Bad Command Code In Control Packet

Control packets (request, response, information) have a 6-bit command field in the first byte to encode the intended operation. Some codes are not used, and are reserved. Sending an illegal command code is another protocol violation.

CTL Deassertion Timeout Violation

The HyperTransport specification limits the amount of time the CTL signal may be deasserted. There are two maximum timeout options (1 millisecond or 1 second) and the one in effect is programmed in bit 15 of the Link Error Register (see Figure 10-5). If the transmitter exceeds the programmed maximum CTL deassertion timeout, it is a protocol violation.

CTL Deasserted During CRC Transmission

CTL is always asserted during the transmission of the 32-bit CRC code in each calculation window. If a receiver detects CTL deasserted during a CRC stuffing period, it is a protocol violation.

Logging Protocol Errors

Protocol error checking is optional. If protocol violations are checked, the *Link Error* register log the errors; refer to Figure 10-5 on page 239.

Figure 10-5: Link Error Register: Protocol Error Logging Bits

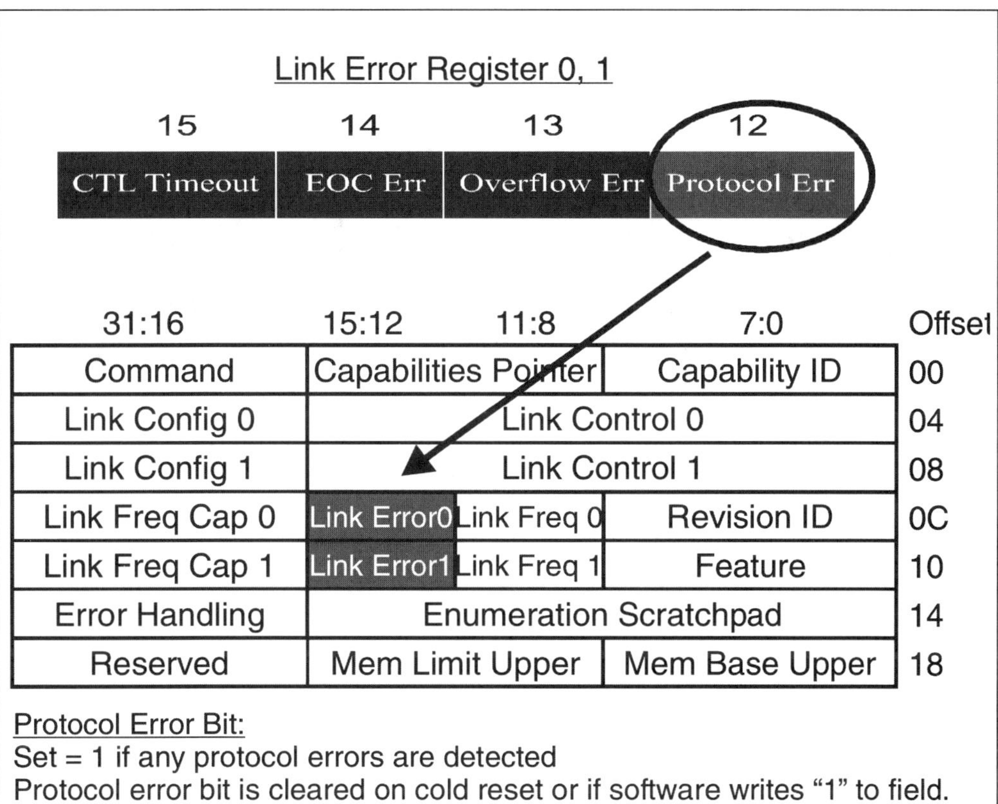

Protocol Error Bit:
Set = 1 if any protocol errors are detected
Protocol error bit is cleared on cold reset or if software writes "1" to field.
Error bit is persistent through warm reset.
Note: CTL Timeout bit (bit 15 above) sets CTL timeout threshold; if time
is exceeded, a protocol violation occurs.
  If Bit 15 = 0, CTL deassertion timeout period is 1 mS.
  If Bit 15 = 1, CTL deassertion timeout period is 1 S.

Programming The Protocol Error Reporting Policy

Informing the system of a protocol error on one or more of the links is handled
in much the same way as for CRC errors. They may be mapped to a fatal or non-
fatal interrupt message, or a sync flood. The reporting strategy is programmed
in the Error handling CSR, as shown in Figure 10-6 on page 240.

Figure 10-6: Error Handling CSR: Protocol Error Reporting Enables

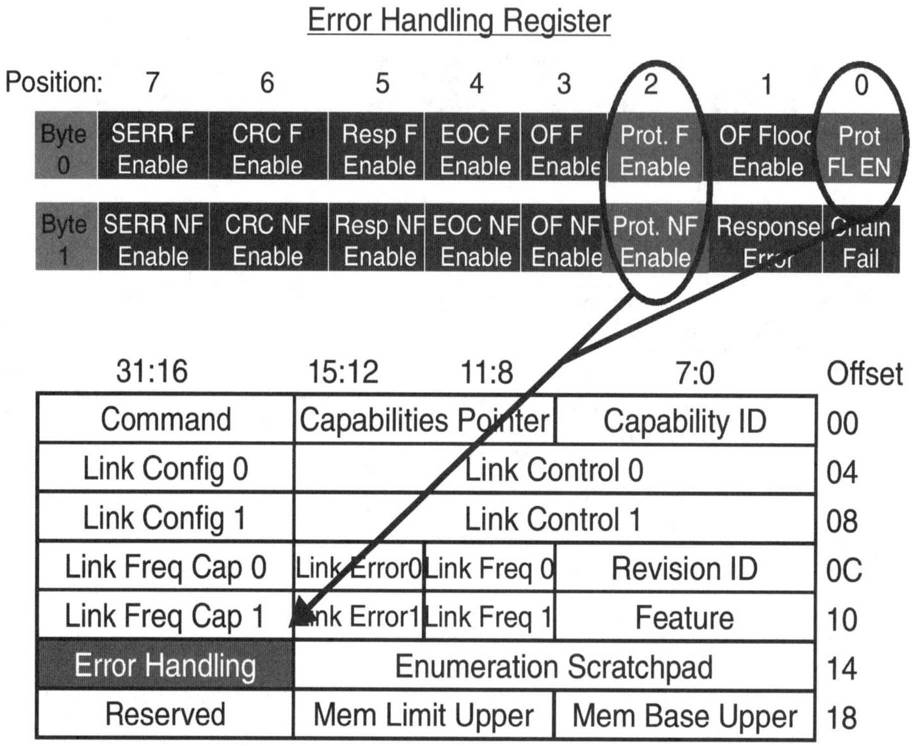

Protocol Error Reporting Enable Bits:
Byte 0, Bit 2: If set = 1, Protocol error will cause Fatal Interrupt
Byte 1, Bit 2: If set = 1, Protocol error will cause Non-Fatal Interrupt
Byte 0, Bit 0: If set = 1, Protocol error will cause a Sync Flood

Protocol error reporting enable bits are programmed by configuration cycles; all will be cleared on a cold reset.

Receive Buffer Overflow Errors

Receive buffer overflow errors can occur if a link transmitter no longer maintains an accurate count of available flow control buffers at the receiver. If a flow-controlled packet (posted request, non-posted request, or response) is sent without an available receiver flow control buffer to accept it, the packet will be lost.

Logging Receive Buffer Overflow Errors

In the event a receive buffer overflow is detected, the Overflow Error bit will be set in the Link Error CSR.

Figure 10-7: Link Error Register: Receive Buffer Overflow Error Logging Bits

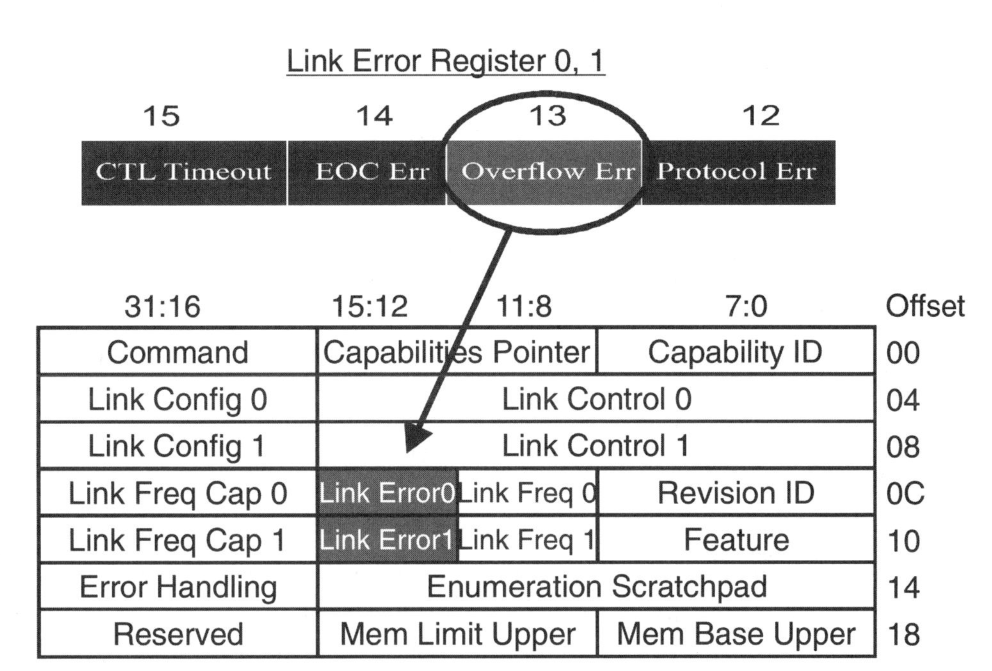

Programming The Buffer Overflow Error Reporting Policy

As in the cases of CRC and Protocol errors, buffer overflow errors may be mapped to fatal/non-fatal interrupts, or a sync flood. The reporting strategy is programmed in the Error handling CSR, as shown in Figure 10-8 on page 242.

Figure 10-8: Error Handling CSR: Receive Buffer Overflow Error Reporting Enables

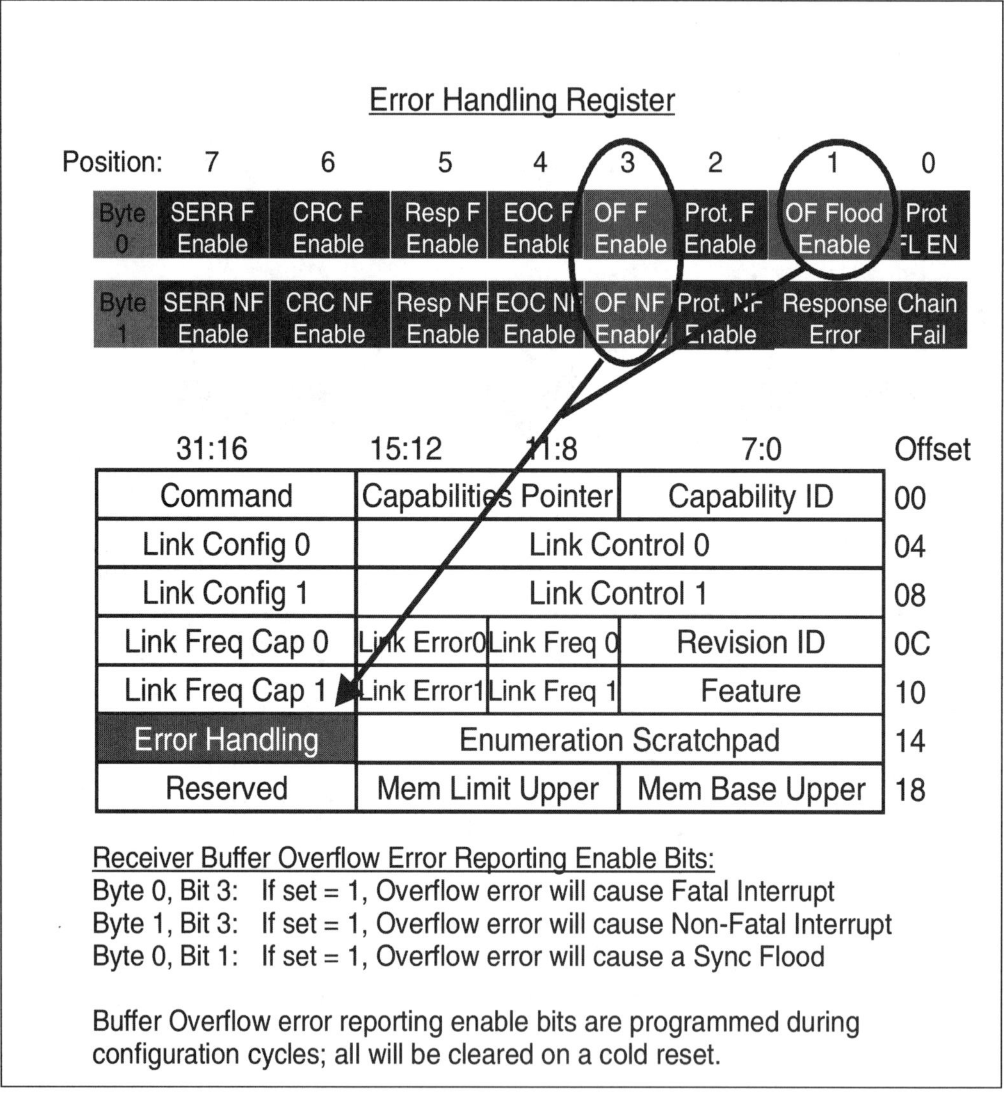

End-Of-Chain Errors

End-Of-Chain (EOC) errors result when a packet moving through HyperTransport is either not claimed by, or does not reach, the intended recipient. Other devices which see the packet forward it and eventually it reaches the device at the end of the chain, where the packet must be handled. Some of the possible reasons for EOC errors include; improper address in a request, invalid Unit ID in a response, the target device is broken, or it has not been programmed properly with UnitID or target base address range.

EOC errors are analogous to the *master abort* event in PCI. Unlike PCI, however, "misdirected" transactions must be handled by the EOC device rather than simply having the initiator of the transaction time out after a prescribed amount of time. This is important in HyperTransport because it is a series of point-to-point connections rather than a shared bus, and an initiator simply sends packets to the neighboring device and has no way of immediately "knowing" whether the ultimate recipient receives it. The EOC error handling mechanism helps with link management in two ways:

1. For posted requests and responses which inadvertently reach an EOC device, the EOC error bit and reporting mechanism may be used to let the system know a packet never reached its destination — information that otherwise would be unknown.
2. For non-posted requests which reach an EOC device in error, the error logging and reporting can also be used. In addition, the EOC device will act as a surrogate for the target and send back a *Read* or *Target Done* response to the requestor (with error bits set). For read requests, all of the requested data is also sent back by the EOC device — although it is obviously invalid (all data values are driven to FFh). Sending back the responses (and data) allows all devices in the path back to the requestor to deallocate internal buffer space and retire the outstanding transaction. The original requester examines the response, decodes the error bits, and takes whatever action is appropriate.

How A Device Knows It Is At The End Of A Chain

Single link peripherals (also known as *End* or *Cave* devices) are always end-of chain-devices. Any packets reaching these device that they are not programmed to accept (by Command type, UnitID, or Address range), are considered lost. No software programming is required for these devices to carry out their EOC function other than setting up the error reporting mechanism to be used.

Tunnel and HyperTransport bridge devices have multiple links. PC board layout determines whether they are physically at the end of a chain or not. At reset, each device receives a bit pattern from the device on the other end of the link. If one of the tunnel links or one of the bridge secondary interfaces does not have a device attached to it, all CAD inputs are seen as logic "0". After reset, if a device senses one of its links is unconnected, it sets the EOC bit in the corresponding Link Control register. This is shown in Figure 10-9 on page 244.

After initialization, links may also be enabled and disabled under software control by setting the EOC bit in the Link Control register(s).

Figure 10-9: End-Of-Chain Device Determination

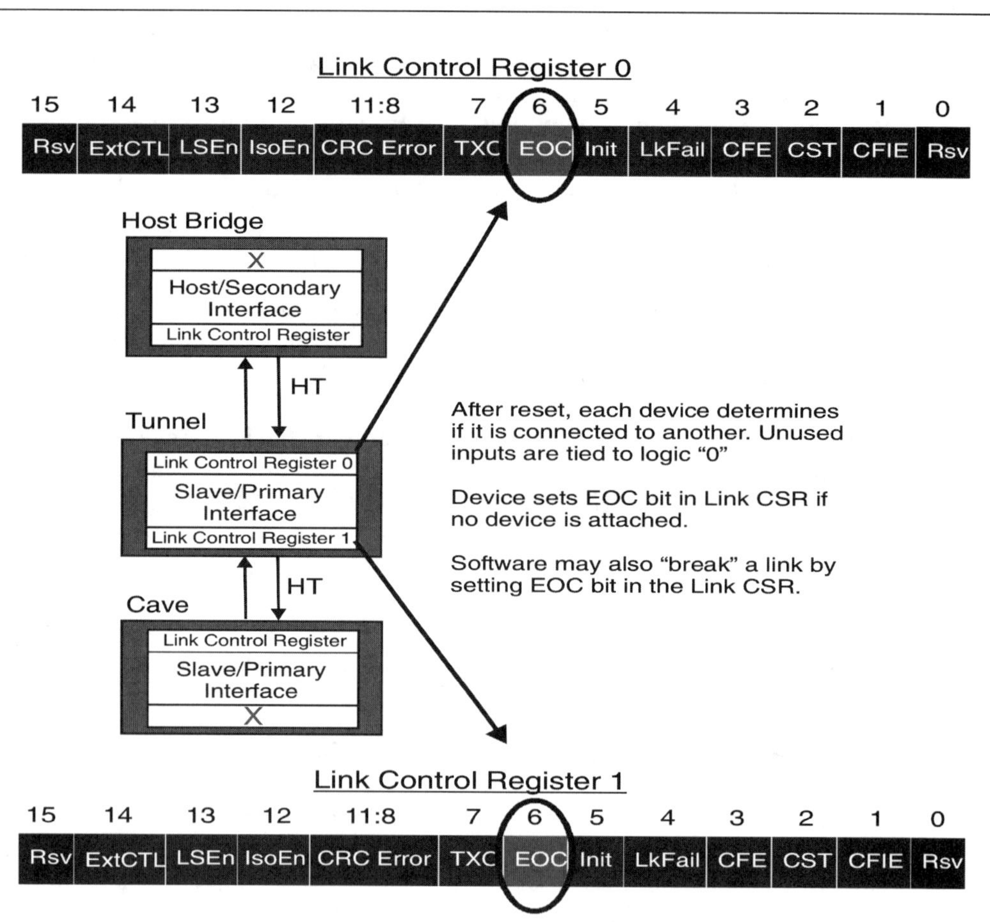

Logging End-Of-Chain Errors

In the event a packet reaches an EOC device in error, the *EOC Err* bit will be set in the Link Error CSR. Refer to Figure 10-10 on page 245. Note in this figure that this is the Advanced Capability register format for a tunnel device; it has two sets of Link Control, Link Configuration, Link Frequency, Link Error, and Link Frequency capability registers — one set for each link. An end (cave) device implements only one set of each these registers.

Figure 10-10: Link Error Register: End-Of-Chain Error Logging Bits

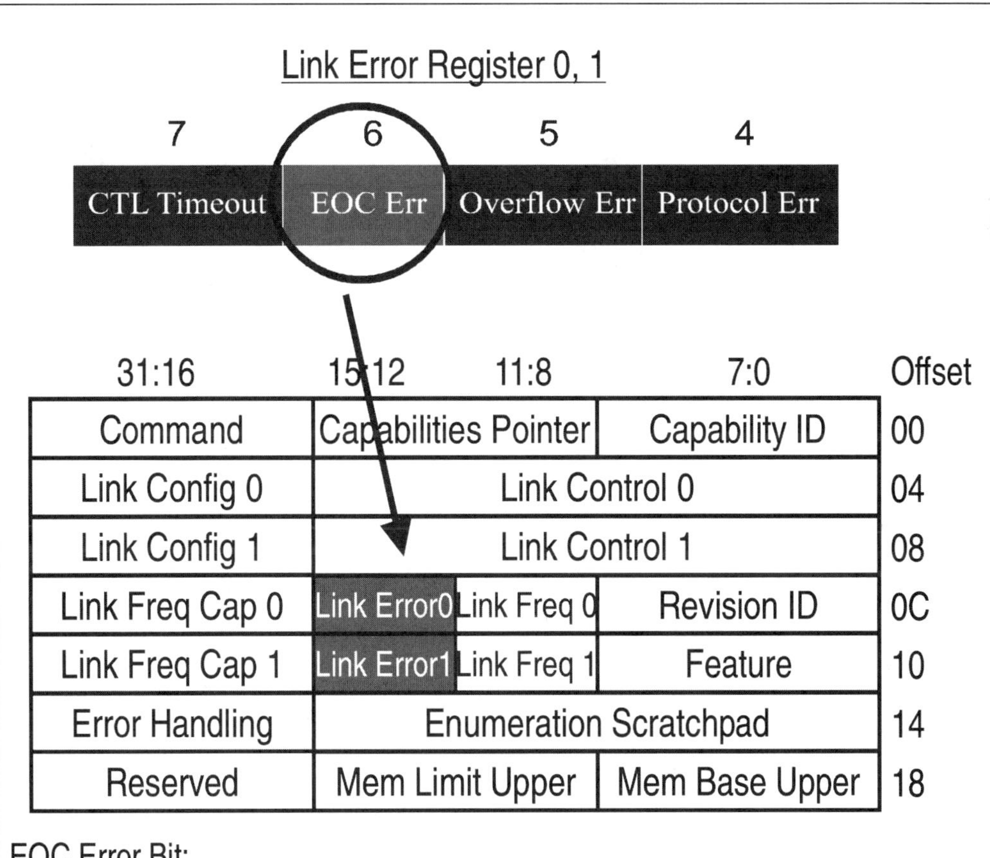

Programming The EOC Error Reporting Policy

As in the cases of CRC, protocol, and receive buffer overflow errors, EOC errors may be mapped to a fatal or non-fatal interrupts. There is no option for Sync flood reporting for EOC errors. The reporting strategy is programmed in the Error handling CSR, as shown in Figure 10-11 on page 246.

Figure 10-11: Error Handling CSR: End-Of-Chain Error Reporting Enables

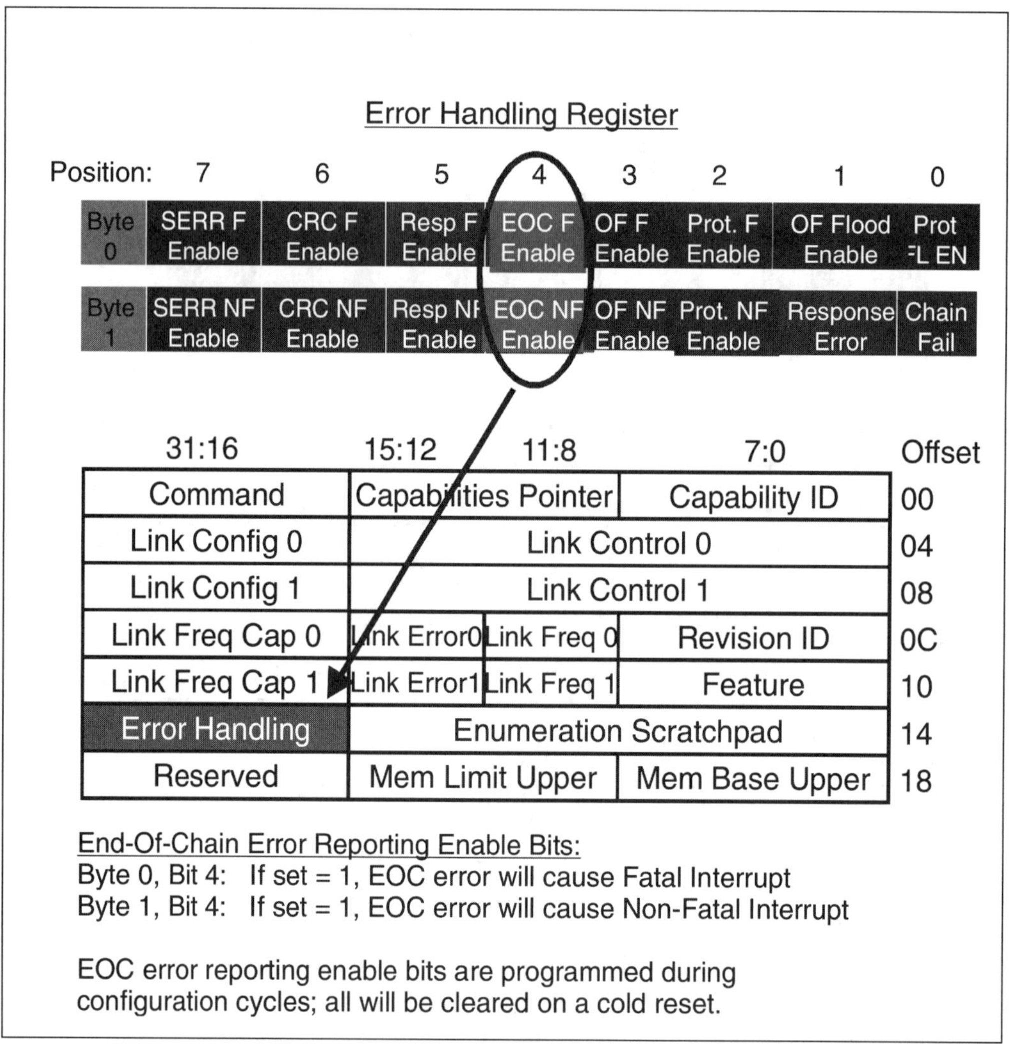

Chain Down Errors

If a device detects a Sync flood or an error that would cause a Sync flood, it sets the *Chain Fail* bit in its Error Handling register and waits for a bus reset. The action taken when the chain goes down depends on the device type:

- Host interfaces track outstanding non-posted requests for devices below them. On chain down errors, they flush the state of all internal non-posted requests and return non-NXA error responses to the requesters for each one that is pending.
- Slave devices have their internal states re-initialized when the RESET# occurs after a chain goes down; there is generally no need for a flush operation of non-posted requests by these devices. If a slave device were implemented that maintained its state through a HyperTransport RESET#, it would need to perform the non-posted request flush operation after the chain goes down as well.

Figure 10-12: Error Handling Register: Chain Fail Bit

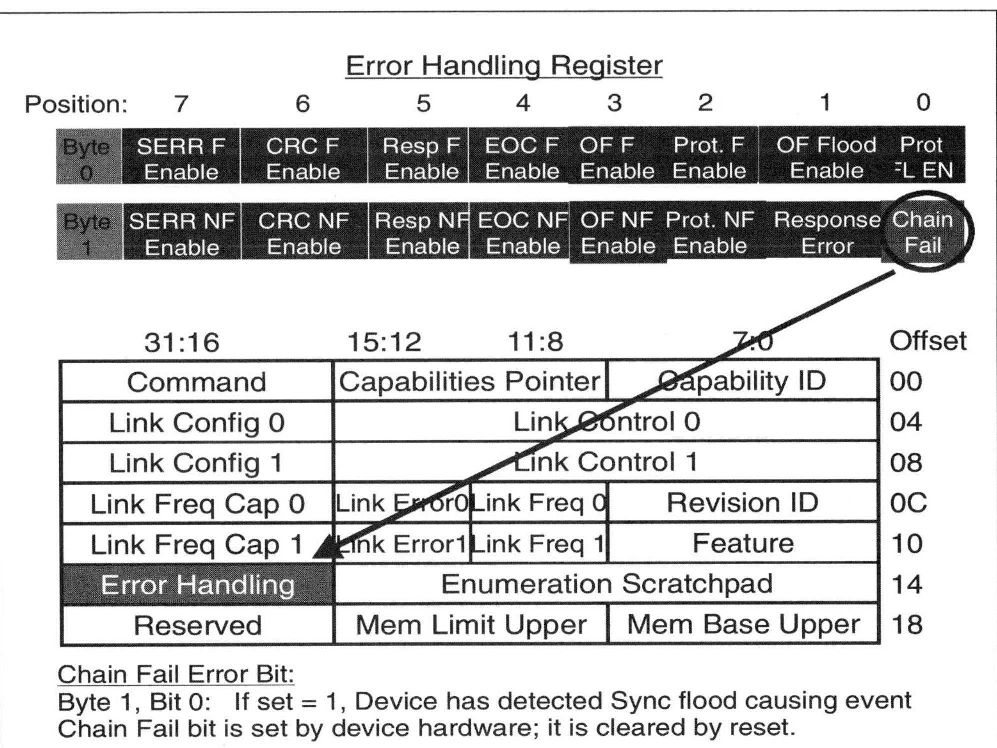

Response Errors

All non-posted requests that are issued require either a Read or Target Done response. The requester programs *UnitID* and *source tag* information into each request packet it issues so that when the response is returned it may be tagged with the same information and find its way back to the original requester. When a downstream response is detected, each device compares the UnitID to its own to see if it should claim the response; if so, it then checks the source tag to determine which of its outstanding transactions is being completed.

It is possible a response may return and be claimed by a requester (UnitID is OK), but not be recognized as being valid. Some of the reasons this might happen include:

1. A read response (RdResponse) is received by a device which carries the correct UnitID, but has an invalid source tag (SrcTag field). The recipient cannot associate the response with any of its outstanding transactions.
2. A read response (with data) is received with the correct UnitID and SrcTag fields, but the response type is incorrect (requester is expecting a Target Done response).
3. A Target Done response is received for a RdSized or Atomic RMW request.
4. A read response (with data) is received for a RdSized or Atomic RMW, but the (data) count field doesn't match what the requester originally asked for.

Response Error Logging And Reporting Policy

The logging of response errors as well as the reporting policy to be used when they occur is handled in the Error Handling CSR. These errors may be mapped into fatal or non-fatal interrupts There is no Sync flood option for response errors. Figure 10-13 on page 249 depicts the response error logging bit and the fatal and non-fatal interrupt enables.

Figure 10-13: Error Handling CSR: Response Error Logging And Reporting Policy Bits

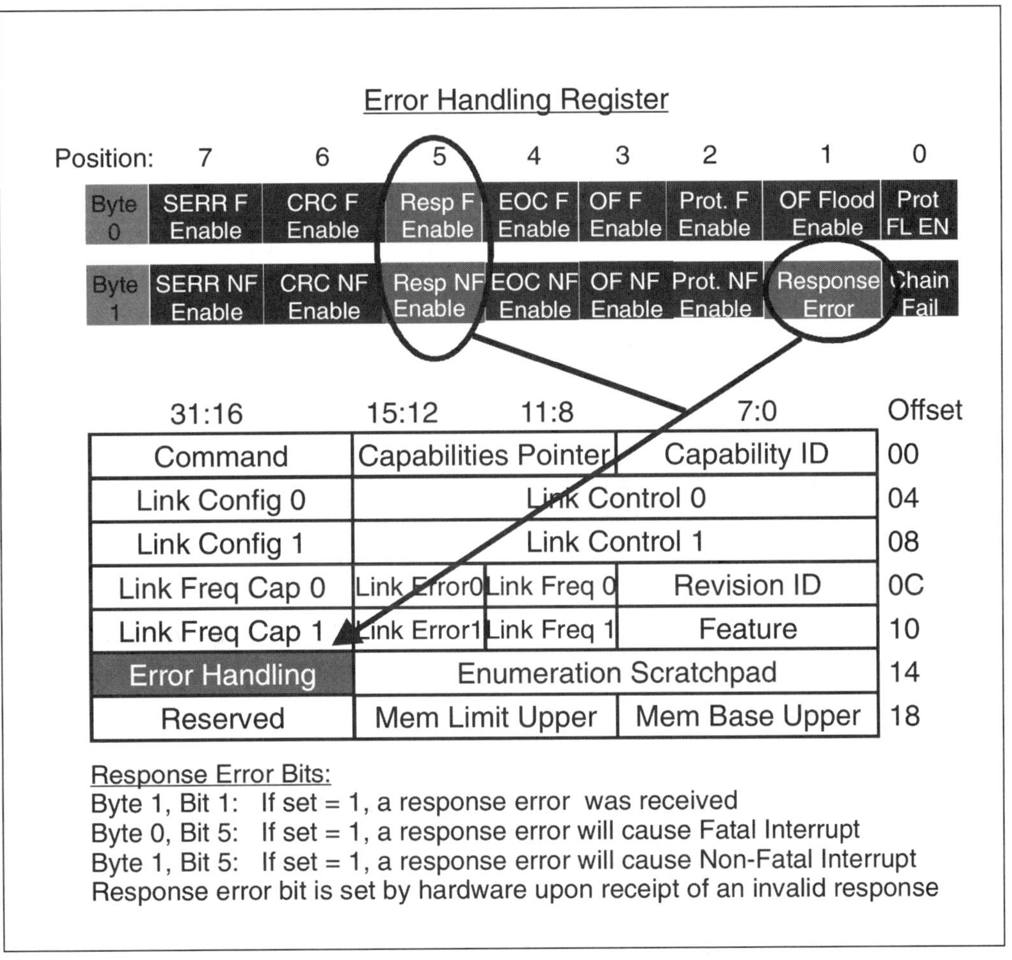

Error Reporting

The three error reporting methods, error responses, fatal and non-fatal interrupts, and Sync flood have different system implications. They are described here in order of increasing severity.

Error Responses (Non-Posted Requests Only)

The HyperTransport specification considers error responses the preferred error reporting mechanism because they are the most localized (conveyed only from target to requester). Error responses are transaction-specific and do not prevent the link from performing other transfers — even to or from the same device.

Every RdSized or Atomic Read-Modify-Write request results in the return of a *Read* response from the target, followed by all of the requested data. All non-posted WrSized and Flush requests result in the return of a *Target Done* response which confirms the completion of the operation, but is not accompanied by data. The response packet format is shown in Figure 10-14 on page 250.

Figure 10-14: Response Packet And Error Bits

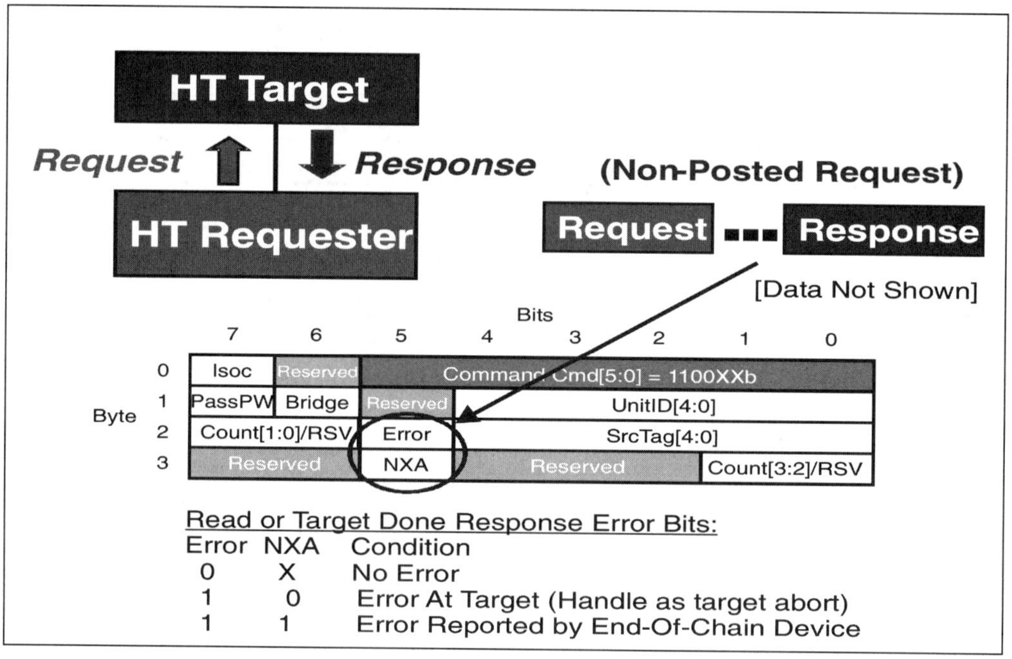

Chapter 10: Error Detection And Handling

When either a Read or Target Done response packet is returned to a requester, the requester checks the state of the two error bits — *Error* and *NXA* (Non-Existent Address) — contained in the packet to determine if the transaction completed properly. The two sources of error responses are the target device and, in the case of a non-existent address, the end-of-chain device.

Error Response Returned By The Target

If a non-posted request reaches a target, but the target cannot complete the operation (can't source or accept data, etc.), the target will return the appropriate response with the Error bit set. If the request called for the return of data (RdSized or Atomic RMW), all requested data (as indicated in the Mask/Count field of the request) will also be returned. Sending the data (even though it is invalid) allows devices in the path to deallocate buffer space and retire the outstanding transaction.

A returning response with Error set and NXA cleared is equivalent to a PCI target abort; HyperTransport requesters detecting this "non-NXA" error response set the *Received Target Abort* bit in the PCI Status register. Bridges seeing this error on a secondary bus would set the bit in the Secondary Status CSR.

Error Response Returned By An End-Of-Chain Device

If a non-posted request fails to reach the target (bad address, etc.), an end-of-chain device must send the response on its behalf. The response will have <u>both</u> the Error and NXA bits set. As in the target response above, if the request called for the return of data, all requested data (again, invalid) will be returned as FFh.

A returning response with both Error and NXA set is equivalent to a PCI master abort; HyperTransport requesters detecting the NXA error response set the *Received Master Abort* bit in the PCI Status register. Bridges seeing this error on a secondary bus would set the equivalent bit in the Secondary Status CSR.

Fatal And Non-Fatal Interrupts

Using interrupts to inform the system of errors is slightly more complex because the interrupt message must travel up through the topology to the host. Interrupts can indicate a non-fatal error (roughly analogous to INTR# in an x86 machine) which implies that the device issuing it has seen an error, but may be able to recover from it; or an interrupt can indicate a fatal error condition (analogous to NMI# in an x86 machine) which indicates that the nature of the error is such that recovery is not possible. Interrupts of either type do not prevent the link from performing other transfers. The conditions under which fatal or non-fatal interrupts are to be used are device and driver specific.

In HyperTransport, interrupts are typically sent using an interrupt message scheme rather than sideband interrupt signals as found in other buses. Devices are not prevented from using external pins as an option, although this method is beyond the scope of the HyperTransport specification.

An interrupt message transaction is actually a special case of the standard size byte write (WrSized Byte) request. Devices in the system can distinguish interrupt messages being sent from other sized writes by the following attributes of the request:

- Interrupt requests target a reserved address in the system address map (from FD_0000_0000h to FD_F8FF_FFFFh).
- The command type is WrSized (byte)
- The Count field is always programmed with a "0", indicating a single dword of data content follows. In standard byte writes, this would be the byte mask dword; in interrupt requests, the single dword data payload contains information about the interrupt.

Note: the data payload for interrupt requests is system-specific. Refer to Chapter 8, entitled "HT Interrupts," on page 199 for a detailed discussion of HyperTransport interrupts.

Sync Flood: When All Else Fails

In some cases, one or more links in HyperTransport may get into a state where ordinary packets cannot be sent reliably. For example, a device may detect a series of CRC errors which indicates to it that either the external link is broken or, more likely, it may not be synchronized with its neighbor with respect to CRC stuffing in the CAD stream. If this is the case, it can't send new packets; it also can't convey the fault using fatal/non-fatal interrupts because they travel in the same channels as other packets.

Sync flood reports errors that cannot be signalled by other methods. It is roughly analogous to the PCI SERR# (system error) event and has a serious impact on the entire chain. Sync flood packets put the chain into an inactive state pending a warm reset to restore normal packet protocol. The behavior of the device initiating the sync flood is slightly different from the other devices which propagate it. The basic rules are described below.

Device Initiating The Sync Flood

1. The device initiating sync flood must have the *SERR# Enable* bit in the configuration header Command register set = 1 before it initiates a sync flood for any reason.
2. If the device intends to initiate sync flood for CRC errors, buffer overflow errors, or protocol errors, it must first check the corresponding "flood enable" bits in the Error Handling and Link Control registers.
3. The device which initiates a sync flood sets the *Signaled System Error* bit in the configuration header Status register, *LinkFail* bit in its Link Control Register, and the *Chain Fail* bit in the Error Handling CSR. Note: if all conditions for a sync flood have been met, Link Fail is always set — even if the SERR# enable bit in the configuration header Command register is clear (preventing the sync flood packets from actually being sent.)

Devices Detecting Sync Flood

1. Devices detecting sync flood at a receiver input cease all normal packet transmission on the affected chain.
2. Each device sets the *Chain Fail* bit in its Error Handling CSR.
3. Each device drives sync packets onto all transmitter interfaces, including back to the device which initiated the flood. This assures that sync is seen on all links on the chain.

Sync Flooding And HyperTransport Bridges

1. Bridges set the *Detected System Error* bit in the Secondary Status register if they see a sync flood on the secondary bus.
2. Bridges may forward a secondary bus sync flood upstream to the primary bus if the *SERR# Enable* bit in its PCI Command register is set. This is similiar to the behavior of PCI-PCI bridges when SERR# is detected on a secondary bus. The bridge may optionally convert the secondary bus sync flood to a fatal or non-fatal interrupt on the primary bus.
3. Bridges always propagate primary bus sync floods downstream onto their secondary bus(ses).

Miscellaneous Notes

Flooding Continues Until Reset. Once a device commences the sync flood operation, it must continue until a reset is detected on the affected bus. This assures that the sync flood propagates throughout the chain.

CRC Not Checked During Sync Flood. CRC is not generated or checked on links where a sync flood event is in progress. Normal packet protocol, including CRC, resumes after a reset occurs on the chain. Refer Chapter 12, entitled "Reset & Initialization," on page 275 for a discussion of reset and initialization.

Sync Flood Example

Figure 10-15 on page 255 below illustrates the sequence of events associated with a sync flood on a single chain.

Sequence of events: (Figure 10-15 on page 255)

1. UnitID 2 detects an error condition which it cannot convey using a fatal or non-fatal interrupt.
2. If it has been enabled to generate sync flood (*SERR# Enable* bit in its configuration header Command register) <u>and</u> one or more error cases have been qualified to use sync flood (see Error Handling register in last section), it sets the Link Control register *Link Fail* bit, the Error Handling CSR *Chain Fail* bit, and the configuration header Status register *Signaled System Error* bit.
3. UnitID 2 starts driving sync packets towards the tunnel (1). It continues to do this until reset occurs, and CRC is neither generated or checked.
4. UnitID 1 (the tunnel) detects the sync flood on its downstream receiver. It terminates normal packet transmission, sets its Error Handling register

Chain Fail bit, and starts driving sync packets onto both its upstream and downstream links (2). CRC is neither generated or checked.

5. When UnitID 0 (the Host Bridge) detects the sync flood on the chain, it also ceases normal packet transfers, sets its *Chain Fail* bit, and drives sync flood packets downstream (3). CRC is neither generated or checked.

A warm reset on this chain will restore normal protocol.

Figure 10-15: Sync Flood Example

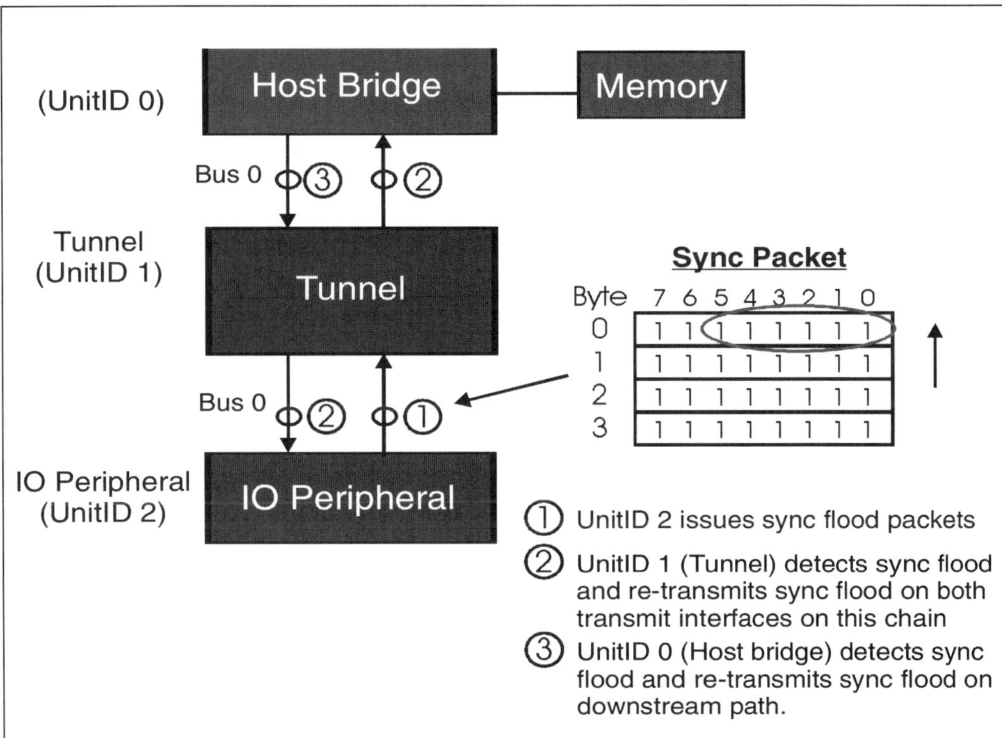

11 *Routing Packets*

The Previous Chapter

The previous chapter described HyperTransport *flow control*, used to throttle the movement of packets across each link interface. On a high-performance connection such as HyperTransport, efficient management of transaction flow is nearly as important as the raw bandwidth made possible by clock speed and data bus width. Topics covered here include background information on bus flow control and the initialization and use of the HyperTransport virtual channel flow control buffer mechanism defined for each transmitter-receiver pair.

This Chapter

This chapter describes the rules governing acceptance, forwarding, and rejection of packets seen by HyperTransport devices. Several factors come into play in routing, including the packet type, the direction it is moving, and the device type which sees it. A related topic also covered in this chapter is the fairness algorithm used by a tunnel device as it inserts its own packets into the traffic it forwards upstream on behalf of devices below it. The HyperTransport specification provides a fairness algorithm and a hardware method for tunnel management packet insertion.

The Next Chapter

The next chapter describes the ordering rules which apply to packets associated with the three types of HyperTransport I/O traffic: PIO, DMA, and Peer-to-Peer. Depending on the whether compatibility with the full producer-consumer ordering model used in PCI is required or relaxed ordering is permissible, attribute bits in request and response packets may be set or cleared. These bits are defined by the requester and are used by devices in the path to the target, and within the target, to enforce proper ordering. HyperTransport applies dedicated sets of ordering rules for upstream I/O traffic, downstream I/O traffic, and the special ordering required of host bridges and in double-hosted chains. *Refer to Chapter 20, entitled "I/O Compatibility," on page 457 for a description of the additional ordering requirements when interfacing HyperTransport to other compatible protocols (e.g. PCI, PCI-X, and AGP).*

Packet Routing: Shared Bus vs. Point-Point Topology

Routing information in a shared bus topology such as PCI or PCI-X is somewhat simpler than in a point-point topology such as HyperTransport.

Shared Bus Routing

Referring to the PCI/PCI-X shared bus example illustrated in Figure 11-1 on page 258, it should be clear that if a transaction appears on the shared bus, all devices "see it" and have an opportunity to decode the address and command and claim the cycle. Devices other than bridges have no responsibilities for routing information to their neighbors. Also note that arbitration on a shared bus is simple because a single arbiter can manage the entire bus. In PCI/PCI-X, the arbiter is typically in the bus Host Bridge; the arbiter considers requests from each master, then grants the bus to each in turn, hopefully applying a reasonable fairness algorithm.

Figure 11-1: Routing: Shared Bus vs. HyperTransport Point-Point

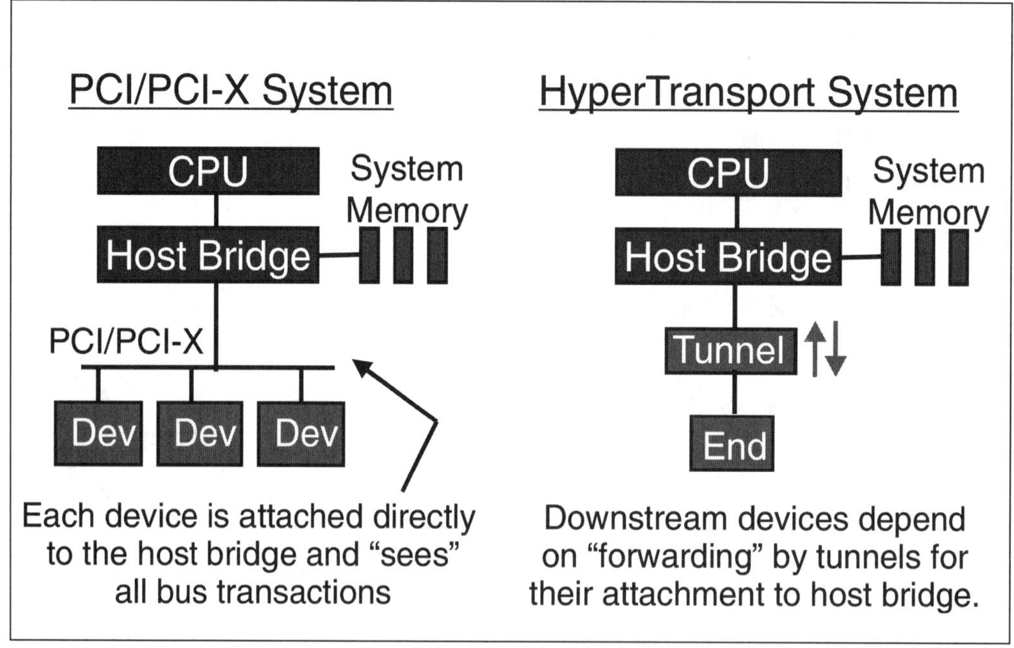

HyperTransport Point-Point Routing

In contrast to the shared bus approach, the HyperTransport topology distributes responsibility for routing and forwarding packets among all devices, with the exception of single-link end (cave) devices. For example, the tunnel peripheral device in Figure 11-1 on page 258 must observe a set of rules governing acceptance, forwarding, and rejection of packets moving both upstream and downstream. The end device in Figure 11-1 on page 258 is dependent on the tunnel to do this. Note that a benefit of a point-point bus is the elimination of shared bus arbitration. Packet transfer is subject only to flow control on each link.

Review Of Packet Types And Formats

How a packet is routed depends in large part on the type of packet it is. Each packet in HyperTransport is a multiple of four bytes in size, and the specification divides packets into two types: control and data. All control packets contain a *Command Type* field in the first byte which identifies which type of control packet it is and the format of the remaining packet fields to follow. It also indicates whether data packets follow immediately (writes), will return later (reads), or are not required.

Control Packets

Control packets are sent across a link to initiate specific tasks; they contain information fields used for several purposes: address decoding, virtual channel and transaction stream management, error reporting, and routing. Devices perform routing functions by extracting information from key fields in control packets. Control packets are further divided into three groups: information, requests, and responses.

Information Packets: No Routing Required

Information packets include *NOP* and *Sync/Error*. These four-byte packets are used for communication between two ends of a link interface. When issued by a transmitter, they are always accepted by the corresponding receiver; they are never forwarded to another link. This means that are no routing issues associated with them. These two packet types will not be discussed further in this chapter.

Request Packet Routing Information

Request packets are used to initiate various transactions and control operations. Packet format depends on the request type; four byte request packets are sent when no address field is needed; eight byte requests are sent otherwise. Figure 11-2 on page 260 depicts a generic eight byte RdSized or WrSized request packet and the key fields used in request packet routing.

Figure 11-2: Generic WrSized Or RdSized Request Packet: Key Routing Fields

Byte		

Bits

	7	6	5	4	3	2	1	0
0	SeqID[3:2]		Command Type Cmd[5:0] = x01xxxb or 01xxxxb					
1	PassPW	SeqID[1:0]		UnitID[4:0]				
2	Mask/Count[1:0]	Compat	SrcTag[4:0]/Reserved					
3	Addr [7:2]							Mask/Count[3:2]
4	Addr[15:8]							
5	Addr[23:16]							
6	Addr[31:24]							
7	Addr[39:32]							

<u>Key Fields In Routing Request Packets:</u>
Command Type (Cmd[5:0])
UnitID (UnitID[4:0])
Compat Bit
Address[39:2]* (8 byte request packets only)

Table 11-1: Definitions Of Request Packet Fields Used In Routing

Byte	Bit	Function
0	5:0	**Command Type Code.** This code indicates the request type.
1	4:0	**UnitID.** This is the UnitID (0-31d) of the requester
2	5	**Compat.** This bit is set by bridges in downstream request packets targeting the system subtractive decode device (e.g. compatibility bridge) residing on the system "compatibility chain."
3 4-7	7:2 7:0	**Start Address[39:2]** The dword-aligned, 40 bit target start address. Refer to the HyperTransport address map for address use.

Six Request Types. HyperTransport supports six request command types. Most have a number of variants. The following table summarizes each request, the number of bytes in the packet, the Command Code (Byte 0, Bits 5:0 of the request packet), and notes about its use.

Table 11-2: Request Packet Command Code Summary

Command Name	Packet Size	CMD Code	Comments
Broadcast Message [Always Posted]	4 Bytes	111010b	Originate at host bridges and travel downstream. All devices accept them and propagate them downstream onto all links. EOC device accepts message and drops it.
Sized Read (RdSized) [Always non-posted]	8 Bytes	01xxxxb	May be issued by any device. Read response is returned. Use of "xxxx" option bits: [3] = response may pass (if 1) [2] = dword/byte (1 = dword) [1] = Isoc channel (1 = Isoc) [0] = Coherency (1 = Req'd)
Sized Write (WrSized) [Posted or Non-posted]	8 Bytes	x01xxxb	May be issued by any device. Use of "xxxx" option bits: [5] = posted req (1 = posted) [2] = dword/byte (1 = dword) [1] = Isoc channel (1 = Isoc) [0] = Coherency (1 = Req'd)
Flush [Always non-posted]	4 Bytes	000010b	Forces all preceding posted writes in same transaction stream to host bridge.
Fence [Always posted]	4 Bytes	111100b	Barrier to subsequent posted writes from all streams (except Isoc) until all previous posted writes complete.
Atomic Read-Modify-Write [Always non-posted]	8 Bytes	111101b	Hybrid read and write command for modifying a memory location atomically.

Response Packet Routing Information

Response packets are used in the completion of non-posted transactions. There are two types, *Read* response and *Target Done* response. Read responses are returned by the target of an earlier read or Atomic RMW request to indicate which transaction has been serviced, how much data immediately follows, and whether an error occurred in fetching the data.

The Target Done response is returned by the target of an earlier non-posted write or Flush request. No data accompanies this response; it is returned simply to report if the requested operation completed successfully or not.

The four byte packet format is different for a Read vs. Target Done response. Figure 11-3 on page 262 shows a generic Read/Target Done response packet and the key fields used in response packet routing.

Figure 11-3: Generic Read/Target Done Response Packet: Key Routing Fields

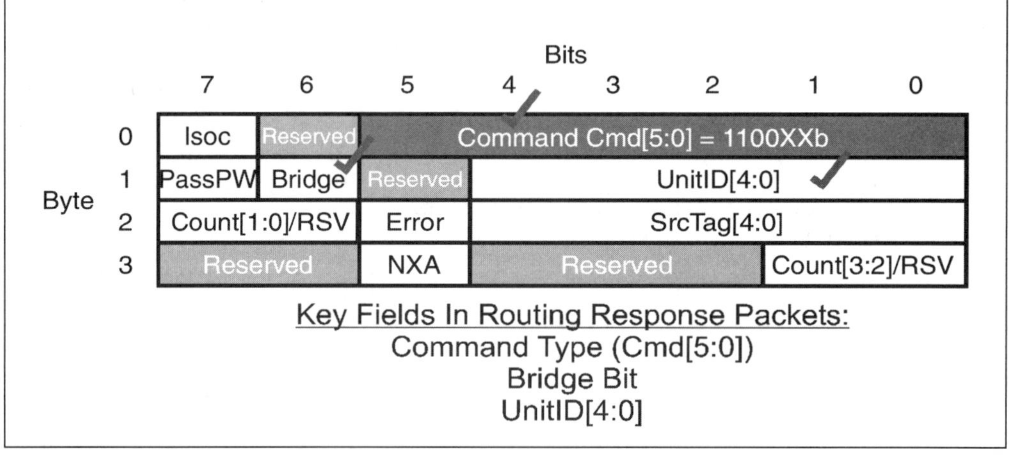

Table 11-3: Definitions Of Response Packet Fields Used In Routing

Byte	Bit	Function
0	5:0	**Command Type Code.** This code indicates the response type.
1	5	**UnitID.** In downstream responses, this field is UnitID of the original requester. In upstream responses, this is the UnitID of the bridge (0).
1	6	**Bridge.** This bit is set by bridges to indicate a response is traveling downstream. Non-bridges may only claim responses with this bit set.

Data Packet Routing Depends On Control Packets

Because data packets are always accompanied by a control packet (request or response), they do not contain any routing information of their own. The control packet indicates the size of the data packet payload, the virtual channel it travels in, whether it is bytes or dwords, where it is going, and even whether it is valid or not. For this reason, data packets are not mentioned much in the packet routing discussion that follows; they are assumed to be immediately behind the control packets which accompany them.

Directed vs. Broadcast Requests

HyperTransport defines two classes of requests: *directed* and *broadcast*. Directed requests travel in the posted or non-posted virtual channel, and always target a particular device. The request destination is indicated either by the address field (e.g. RdSIzed and WrSized requests), or is implied in the request type (Flush and Fence commands target the host bridge). Directed requests move through the chain until they reach the target device and are absorbed. Devices in the path of a directed request forward it to the next device in the direction of the target.

Broadcast requests are issued by bridges and travel in the posted virtual channel. They are accepted by each node then forwarded downstream on all links. When the broadcast request reaches an end-of-chain device, it is absorbed and dropped.

Only the *Broadcast Message* command type (see Table 11-2 on page 261) uses the broadcast request routing method. All other command types are directed requests.

Accepting Packets

A packet is accepted by a device if certain fields indicate it is the intended recipient. For directed requests or responses, this will be the end of packet travel. For broadcast requests, the packet is consumed and passed downstream to the next device.

Rules For Acceptance

A node will accept (consume) an incoming packet if <u>any</u> of the following conditions are met.

1. The packet is a broadcast request (determined from the command type)
2. The packet is a directed request which carries a UnitID = 0 (coming from a bridge), the compatibility bit is clear (only subtractive decoders may claim a request if compatibility bit is set), and the address is owned by this node.
3. The packet is a directed request which carries a UnitID = 0 (coming from a bridge), the compatibility bit is set (only the subtractive decoder may claim this request), and this node <u>is</u> the subtractive decoder or a bridge to it.
4. The packet is a response with the bridge bit set (traveling downstream from a bridge), and the UnitID field matches that of the node.
5. Because of double-hosted chain topologies, tunnels must be able to accept downstream packets from either link.

A Note About The Subtractive Decoder

In the rules for packet acceptance, an allowance is made in HyperTransport for a "compatibility chain". If implemented, this chain would host the system subtractive decode device (e.g. compatibility bridge) which is responsible for handling transactions to *legacy* devices. Often, these devices do not support plug and play addressing and the system may not be aware of the address ranges they are using. When the host bridge targets such devices, it sets the Compat bit in the request which indicates that:

1. No device, other than the subtractive decoder, is allowed to claim the request packet — regardless of the address it carries.
2. Bridges in the path to the subtractive decoder must reissue the request onto the proper downstream link.
3. The subtractive decode device must accept the request on behalf of the legacy devices it supports.

Forwarding Packets

Any node that forwards a packet sends it in the same direction it is already moving.

Rules For Forwarding

A node will forward an incoming packet to its outbound link if <u>any</u> of the following conditions are met.

1. The packet is a broadcast request (determined from the command type)
2. The packet is a directed request which carries a UnitID = 0 (coming from a bridge), the compatibility bit is clear, and the address is not owned by this node.
3. The packet is a directed request which carries a UnitID = 0 (coming from a bridge), the compatibility bit is set, and this node is not the subtractive decoder or a bridge to it.
4. The packet is a response with the bridge bit set (traveling downstream from a bridge), and the UnitID field does not match that of the node.
5. The packet is a directed request which carries a non-zero UnitID (coming from an interior node). Non-bridges are not allowed to claim requests from interior nodes (no direct Peer-to-Peer transfers).
6. The packet is a response with the bridge bit clear (coming from an interior node). Non-bridges are not allowed to claim responses which are not sourced by a bridge (bridge bit must be set).

Other Notes On Forwarding

Forwarding Into The End Of Chain

An attempt to forward a packet into the end of a chain (device has EOC bit set) will result in a rejected packet. How the rejection is handled on the link is described in the next section on packet rejection. In addition, error handling policy programmed into the end-of-chain device determines what additional action should be taken (log error, generate an interrupt, etc.). Refer to "End-Of-Chain Errors" on page 243 for a discussion about error handling by EOC devices.

Forwarding If Initialization Is Not Complete

Another aspect of forwarding involves the behavior of a device which detects a packet that should be forwarded, but the device has not yet completed its initialization (*EOC* and *Initialization Complete CSR* bits still clear). Whether the incoming packet will be dropped or held pending initialization is then determined by the *Drop on Uninitialized Link* bit in the HyperTransport advanced capability Command Register. Refer to Chapter 13, entitled "Device Configuration," on page 305 for a detailed discussion of HyperTransport configuration.

Rejecting Packets

Once system initialization is complete, packet rejection should be a rare event. The only non-error rejection after initialization would be the arrival of a broadcast message which is intended to travel to the end of the chain and then be dropped by the end-of-chain device. Prior to the completion of system initialization, other devices may have not completed link initialization (*Link Initialization* bit is clear). If they have not been programmed to hold incoming packets pending completion of initialization (*Drop on Uninitialized Link* bit is set), they will behave as an end-of-chain device and reject packets temporarily.

Actions taken when a packet is rejected by an end-of-chain device, or by an interior device which has not completed initialization and is behaving temporarily as one, depends on the type of packet.

Rules For Rejection

1. Broadcast requests which have completed the trip to the end of a chain are silently dropped. These are always posted, so no response is expected or sent.
2. Non-posted downstream directed requests (UnitID = 0) cause the return of a Target Done response (for non-posted writes or Flush) or a Data Response (for reads or Atomic RMW). The response for rejected non-posted downstream requests will have the Error and NXA error bits set, and the bridge bit clear. The UnitID field will be set to either 0 or that of the end-of-chain device. Read responses are accompanied by all requested data (driven to value of FFh)
3. Non-Posted upstream directed requests (non-zero UnitID) cause the return of a Target Done response (for non-posted writes or Flush) or a Data Response (for reads or Atomic RMW). The response for rejected non-posted

upstream requests will have the Error and NXA error bits set, and the bridge bit set. The UnitID field must be set to that of the original requester (without <u>its</u> UnitID and the bridge bit set in the response, an interior node won't accept the response). Read responses are accompanied by all requested data (driven to value of FFh)

4. Rejected response and posted request packets are dropped.

Host Bridge Behavior

Host bridges always reside at the ends of HyperTransport chains. They never forward packets, but will have occasion to accept and reject packets. There is additional complexity caused by the responsibilities host bridges may have in supporting double-hosted chains (this is optional) and peer-to-peer transfers (required). The following sections describe host bridge behavior in accepting and rejecting packets they receive.

Directed Request With UnitID = 0

Directed requests with a UnitID = 0 detected by the HyperTransport interface of a host bridge are inbound from the host at the other end of a double-hosted chain.

Accepted

If the host bridge recipient of the request has implemented internal memory or I/O space and owns the address range targeted in the request, it will respond to the request as an interior node would: accepting posted requests and returning Target Done or Read responses/data for non-posted requests.

The host bridge will similarly accept Type 0 configuration cycles carrying the proper fields if two conditions are met:

1. The host bridge supports double-hosted chains
2. The Host/Secondary Interface Command Register *Host Hide* bit is clear

Rejected

If the host bridge does not support double-hosted chains or doesn't own the address range contained in the request, it will be rejected in a similiar way to any other end-of-chain device. Posted requests are dropped (they may be reported as end-of-chain errors); non-posted requests cause the return of a Target Done or Read response with Error and NXA error bits set; for read requests, all requested data is also returned with the response (driven to value of FFh).

Response UnitID And Bridge Fields

A host bridge receiving a non-posted, directed request with UnitID = 0 would also set the response Unit ID field = 0 unless it is programmed to act as the slave bridge in a double-hosted chain. If this is the case, the Host/Secondary Interface Command Register *Act As Slave* bit would be set = 1, and the bridge would set the UnitID in the response to the value programmed into its Host/Secondary Interface *Device Number* Register.

Broadcast Request

Broadcast requests detected by a host bridge could only be coming from another host bridge on the other end of a double-hosted chain.

Always Accepted

If a broadcast message is seen by a host bridge, it has already been seen throughout the chain. The host bridge accepts it, and silently drops it. The specification indicates that a host bridge could optionally implement an internal space addressable with a broadcast message; if it does this, the message would be accepted and routed to the internal target.

Directed Request With Non-Zero UnitID

Directed requests with a non-zero UnitID are sourced by interior nodes and may be destined either for internal space within the host bridge (e.g. main memory) or for another device downstream (a peer-to-peer request). The host bridge evaluates the command type and address contained in the request and routes it accordingly. It is also possible that the request packet does not map to any internal space or device on the chain; in that case it will be rejected.

Accepted Requests

Internal Target. If the host bridge recipient of the request has implemented internal memory or I/O space and owns the address range targeted in the request, it will respond to the request as an interior node would: accepting posted requests and returning Target Done or Read responses/data for non-posted requests. This would be the typical behavior during DMA transfers from HyperTransport peripherals to and from main memory via the host bridge.

Peer-to-Peer Target. It is also possible that a host bridge examines the request packet address and determines that the address range is actually downstream in the HyperTransport topology. The target could either be on the same chain as the requester or on another chain if the host supports more than one. Hypertransport doesn't support direct peer-to-peer transfers and requires bridges to handle them as two separate requests (followed by two separate responses if non-posted). The sequence of events in accepting a peer-to-peer transactions include: (assume a non-posted peer-to-peer request)

1. An interior node issues a non-posted request. The UnitID is that of the requester; the Source tag field and Sequence ID fields are assigned by the requester from its pool of available tags.
2. The host bridge examines the request and determines it is peer-to-peer. It reissues the request downstream on the appropriate chain. When it does this, it changes the UnitID to 0 (its own) and changes the Source Tag and Sequence ID fields to values from its own pool of available tags. All other fields are passed through unchanged.
3. The host bridge <u>must</u> track outstanding transactions so that when the response is returned, the original UnitID and Source Tag values can be restored when the response is reissued downstream to the original requester.
4. Upon return of a response from the target (bridge bit is clear), the bridge sends the response downstream on the chain containing the requester with the bridge bit set = 1, and UnitID and Source Tag restored. The bridge may then retire the transaction from its outstanding request queue.
5. The response (and data, if any) is claimed by the original requester based on the UnitID field and the fact the bridge bit = 1. It uses the Source Tag to determine the specific transaction which has been serviced.

Compatibility Chain Requests

Another peer-to-peer possibility is that an inbound directed request from an interior node (non-zero UnitID) targets a legacy device on the compatibility chain. If the host bridge supports a compatibility chain, it may reissue requests that don't target any other address space to that chain. Again, it replaces the UnitID, Source Tag (for non-posted requests), and Sequence ID field (if non-zero) with its own values. It also sets the Compat bit in the request it sends onto the compatibility chain. This informs all devices which see it that it should be forwarded downstream to the subtractive decoder (compatibility bridge).

If the original request was non-posted, the host bridge will again track the outstanding request so it can restore the original UnitID and Source Tag information in the response packet it sends to the original requester.

Rejected Requests

If the host bridge does not recognize the address carried by an inbound directed request from an interior node and doesn't support a compatibility chain, the packet will be rejected in a similiar way to any other end-of-chain error. Posted requests are dropped (they may be reported as end-of-chain errors); non-posted requests cause the return of a Target Done or Read response with Error and NXA error bits set; for read requests, all requested data is returned (driven to value of FFh).

Responses Received By The Host Bridge

When a response is received by a host bridge, either the bridge bit is set or it is not.

Response With Bridge Bit = 1

A response with the bridge bit set always originates at a host bridge. If one is <u>received</u> by a host bridge, it means that another host bridge in a double-hosted chain attempted to respond to an interior node and the node failed to claim it. In this case, it continued to the end of the chain where it will be handled as an end-of-chain error by the other host bridge. The response is dropped, and the receiving host bridge may log it as an end-of-chain error. Refer to Chapter 10, entitled "Error Detection And Handling," on page 229 for a discussion of error handling.

Response With Bridge Bit = 0

Responses arriving at a host bridge with the bridge bit cleared belong to the host bridge. The Target Done or Read response/data is being returned in response to a non-posted request issued by this host. Devices are required to track all of their outstanding requests (those requiring responses), so the bridge simply uses the Source Tag field in the response to determine which transaction is being serviced. In the event a response returns with the bridge bit clear, but the response Source Tag doesn't match any outstanding transactions, the node may log the error and report it.

HyperTransport Bridges: Additional Routing Rules

HyperTransport Bridges and bridges between HyperTransport and PCI/PCI-X have routing responsibilities beyond those of interior devices and host bridges with a single HyperTransport interface. Refer to Chapter 16, entitled "HyperTransport Bridges," on page 407 for a discussion of bridge routing.

Tunnel Fairness And Forward Progress

Fairness Is Critical In A Point-Point Topology

Because the HyperTransport topology consists of a series of point-point links, the behavior of tunnel devices when forwarding packets on behalf of devices below them is critical. If a very active tunnel was permitted to insert its own packets onto the upstream link and ignore packets from below, flow control buffers would fill up and a bottleneck would result. Figure 11-4 on page 271 depicts a tunnel device inserting packets into the stream it is forwarding towards memory.

Figure 11-4: Tunnel Inserting Packets Into Upstream Traffic

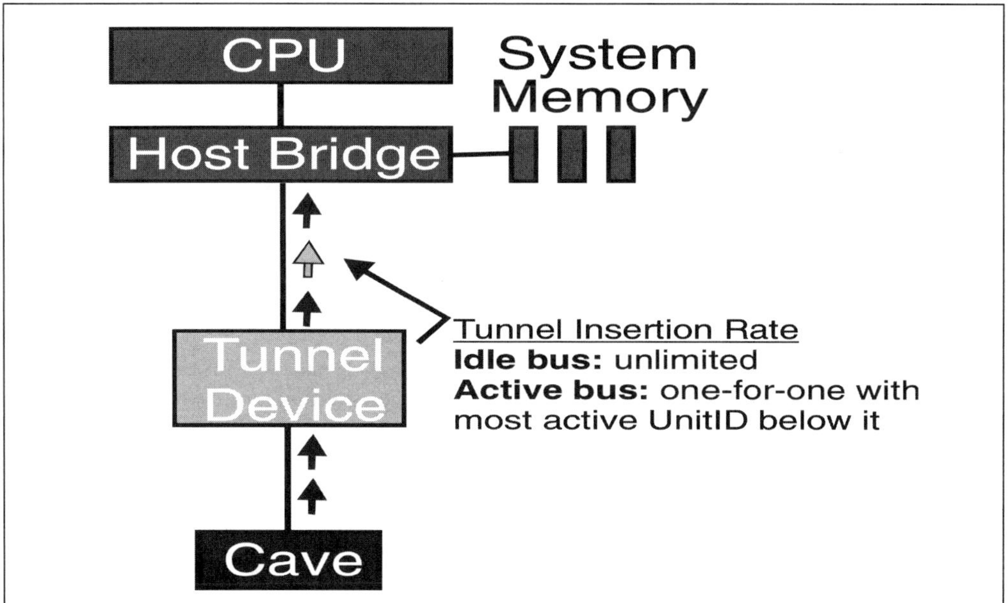

HyperTransPort Imposes A Fairness Algorithm

The Basic Policy

The fairness policy basically allows a tunnel to insert packets onto the upstream link at a rate equal to the most active device (UnitID) below it. What this means is that it must determine the most active downstream UnitID, then may insert packets one-for-one with that transaction stream. Of course, during idle times on the bus, it may insert packets at will. The goal of the specification is that there will be a continuous reassessment of the insertion rate to match changing conditions. On the subject of dynamic adjustment of the insertion rate, the specification says: "this property must be met over a window in time small enough to be responsive to the dynamic traffic patterns, yet large enough to be statistically convergent".

The Algorithm

There are two parts to the Fairness and Forward progress algorithm: calculating the insertion rate and implementing a hardware mechanism that enforces it. While a single interface is described here, tunnel devices actually must implement independent algorithms for each link. There are no programmable control/status registers associated with the algorithm; it is completely hardware based. Also note that:

1. Information packets (NOP and Sync) are issued on a per-link basis and not subject to the fairness algorithm.
2. All other packets are handled without regard to their type, virtual channel, etc. All ordering and priority issues are handled internally.

First, Calculate The Insertion Rate. This is done through the implementation of hardware registers to monitor incoming packets requiring forwarding for each transaction stream. Because HyperTransport permits a total of 32 UnitIDs per chain, the register set to manage each link consists of:

1. 32 individual 3-bit counters which are used to count the incoming packets from <u>individual transaction streams</u> (using the request and response packet UnitID field)
2. One 8-bit counter used to count the collective number of incoming packets from <u>all transaction streams</u>.

At reset, all 32 individual counters are reset to "0". The 8-bit counter is reset to "1". The sequence of events in the counters as packets arrive to be forwarded is as follows:

1. Each time a packet to be forwarded arrives, its individual 3-bit counter is indexed by 1.
2. The 8-bit counter is also indexed any time a forwarded packet arrives from any UnitID.
3. The first individual 3-bit counter to overflow (roll to "0") represents the most active UnitID.
4. The value in the eight bit counter when the first 3-bit counter overflows is used as the denominator of a fraction which is expressed as 8/denominator. This ratio is the maximum rate at which the tunnel device may insert its own packets onto the upstream link.
5. The specification recommends that the tunnel space out the interleaving of its packets, rather than bursting them all at once.

Insertion Rate Calculation Example. Assume there are three devices below a tunnel: UnitID2, UnitID4, and UnitID5. Packets targeting main memory are being issued by all three devices.

1. Incoming packets from UnitID5 cause its 3-bit counter to roll over first.
2. The 8-bit counter (denominator) was at 12 when the roll over occurred.
3. The tunnel then calculates the insertion rate: 8/denominator = 8/12 = 2/3.
4. The tunnel is allowed to insert, on average, two packets for every three which it forwards for the devices below it.

The HyperTransport specification provides additional guidelines for designers in implementing a simple priority and arbitration scheme, achieving non-integral insertion rates, and avoiding potential starvation problems.

12 Reset & Initialization

The Previous Chapter

There are two aspects in dealing with link or internal errors in HyperTransport: detection and handling. The previous chapter described all error types and discussed what devices may do about them. Some devices may chose to detect and handle some errors but not others. The PCI configuration space registers used to program the error strategy and log errors were described, as well as the reporting mechanisms: error response, fatal and non-fatal interrupts, and Sync flood.

This Chapter

Reset signalling and timing along with actions taken by the system and devices during reset are the primary topics discussed in the chapter. This chapter also discusses the software initiated reset and why it's required. The process of determining the default speed and link width and the subsequent software tuning of bus speed and link width are also detailed.

The Next Chapter

HyperTransport uses PCI configuration. The next chapter describes Hyper-Transport configuration for devices other than HyperTransport-to-HyperTransport bridges: host bridges, tunnels, and end (I/O hub) devices. HyperTransport-to-HyperTransport bridges have a different PCI header format than these devices, and are described separately in the chapter on HT Bridges. Many attributes of HyperTransport configuration are exactly the same as for generic PCI devices, although some PCI configuration space header fields are used differently in HyperTransport, and some not at all. Devices require at least one HyperTransport-specific advanced capability register block in addition to the basic PCI configuration space header fields.

General

HyperTransport defines cold reset, warm reset, and a link initialization process. Link initialization involves determining the widest link and highest clock frequency that two devices attached to a link may use. Link initialization involves a multi-step process that is initiated during cold reset.

A hardware-based handshake process determines the link width to be used immediately following cold reset. This process is performed by devices at both ends of every link. The handshake determines the smaller of the devices attached to each link with respect to maximum receiver width (up to 8 bits). For example, one receiver may have a maximum width of 4 and the other a maximum width of 8, thus the smaller CAD width (4) is used. A detailed description of this process is described in the section entitled, "Low-Level Link Width Initialization" on page 282. Also, following cold reset all devices use the default clock frequency of 200 MHz (required by all devices). The initial width negotiated and the default speed may be less than the actual capability of the devices and thus require further tuning.

Next, firmware (e.g. BIOS) optimizes all links so they use the widest path and the fastest clock that the attached devices support. This is done by reading link capability registers within the devices attached to each link. These registers report the maximum link width and maximum frequency at which each device is designed to operate. Software determines the highest common actual CAD width and clock frequency to be used.

Finally, firmware initiates a Warm RESET (or assertion of LDTSTOP#) to cause the new link width and clock values to take effect. Warm RESET or LDTSTOP# assertion is typically triggered by writing to an implementation-specific register in the I/O Controller Hub.

Details regarding the process described above are discussed in the remainder of this chapter. Note however, that portions of the RESET initialization process that involve clock and buffer initialization are detailed in the section called "Link Initialization" on page 282.

Cold Reset

Cold Reset is signaled during the power-up sequence under hardware control. This section details the sources, effects, and characteristics of a HyperTransport cold reset.

Sources of Cold Reset

In addition to the hardware generation of cold reset during the powerup sequence, platform developers may also provide hooks for generating cold reset under software control. An optional method of generating a HyperTransport cold reset is defined by the specification for the secondary bus of a HT-to-HT bridge (discussed on page 278). However, software generation of cold reset for the secondary side of the Host-to-HT bridge can be implementation specific.

Resetting the Primary HT Bus

Some implementation-specific mechanism must be defined to initiate a cold reset at powerup. The HT specification does not precisely define the source of HT cold reset for the system. It may be generated by system board logic or could be incorporated into the Host to HT bridge or other HT device residing on the system board. Figure 12-1 illustrates an example of HT RESET# generation and distribution.

Figure 12-1: Example of Reset Distribution in an HT System

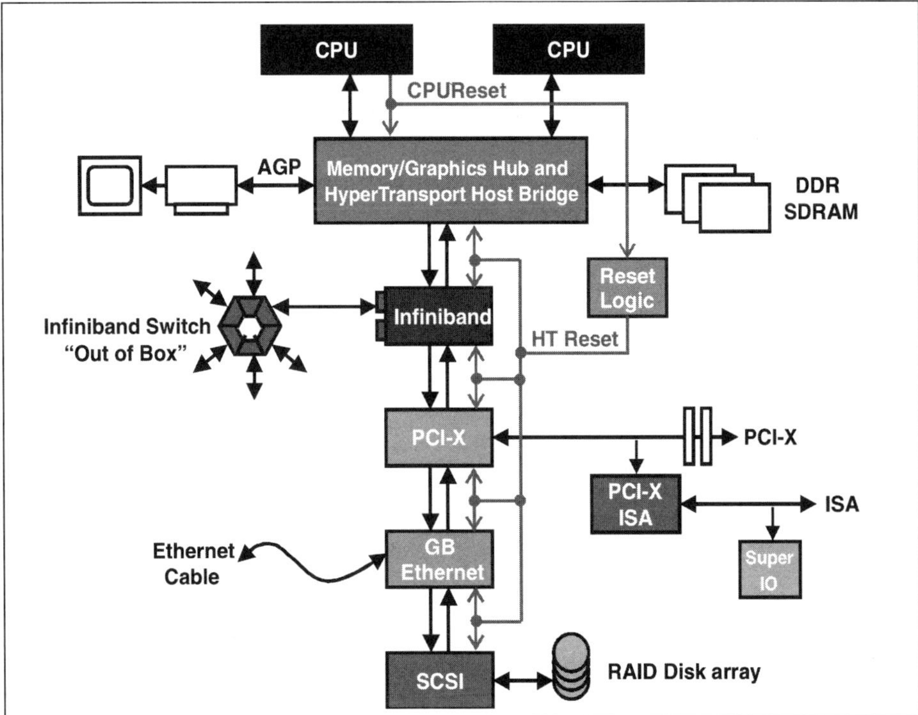

Further, the specification does not require a software controlled method of cold reset generation. However, a host bridge could optionally implement a mechanism similar to that provided by the bridge control register of an HT-to-HT bridge. (See next section.)

Once reset is signalled, any HT device has the option of extending it (via open drain signaling) to ensure the amount of time it needs to complete its internal initialization. In this way, reset remains asserted until the last HT device in the chain completes its initialization. All HT devices that signal cold reset must correctly sequence RESET# and PWROK as discussed in "Signalling and Detecting Cold Reset" on page 280.

Resetting Secondary Side of HT-to-HT Bridge

An HT Bridge is required to propagate cold reset from its primary to its secondary side, but is not allowed to propagate any form of reset from its secondary to primary side. Thus, when the HT-to-HT Bridge initiates an HT cold reset to its secondary side, it will be distributed to all devices in the downstream chain as depicted in Figure 12-2.

HT defines a optional method for HT-to-HT Bridges to generate a cold reset on the secondary bus under software control. This is done via two HT-to-HT Bridge configuration registers, the *bridge control register* in the configuration header and the *command register* located in the HT capability registers. These registers are depicted in Figure 12-3 on page 280. Each of these registers has a bit that in combination permits the generation of cold reset as follows:

- HT Command Register, Bit 0 (optional) — selects cold reset when cleared (0)
- Bridge Control Register, Bit 6 — forces a secondary bus reset when set (1)

When a cold reset is selected, the bridge will deassert PWROK as part of the reset sequence, thereby causing a cold reset. It is the responsibility of hardware to sequence PWROK and RESET# correctly, as described in "Signalling and Detecting Cold Reset" on page 280.

Figure 12-2: Example HT-to-HT Bridge Forwarding Cold Reset

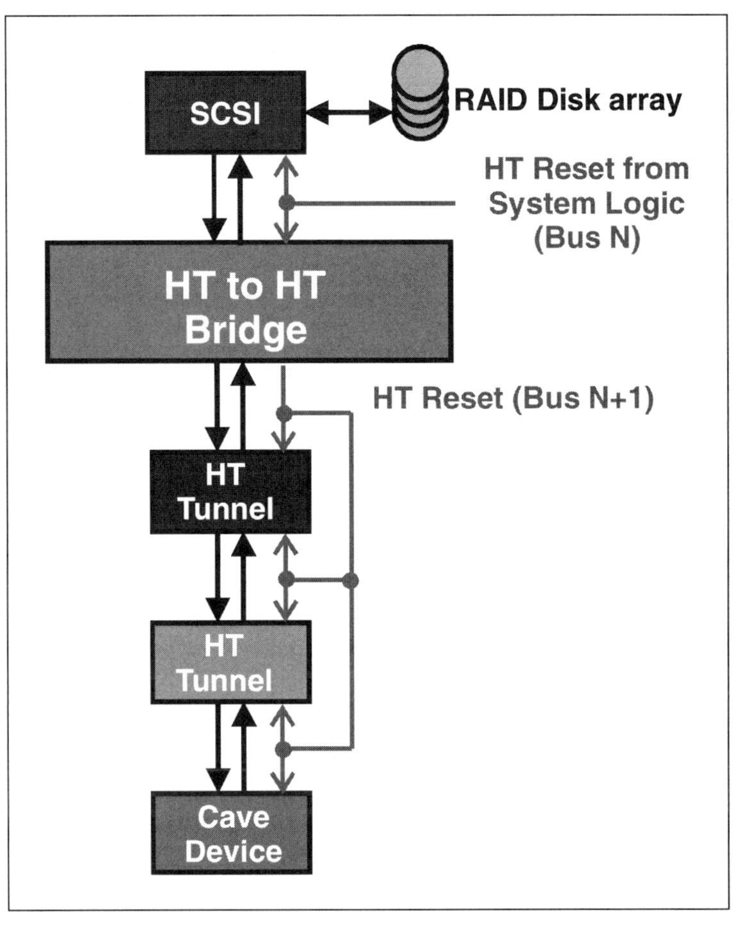

Figure 12-3: Bridge Control Register Can Force Cold Reset

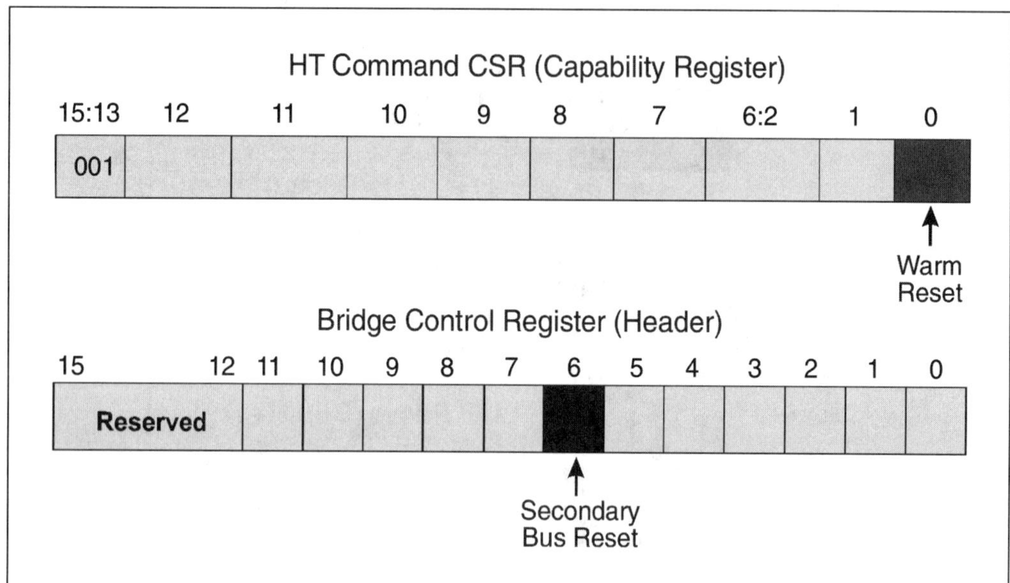

Signalling and Detecting Cold Reset

Cold reset is detected when RESET is asserted and PWROK is deasserted (i.e., power is ramping up but not yet stable) on the same clock edge as illustrated in Figure 12-4 on page 280.

Figure 12-4: Cold Reset Signalling

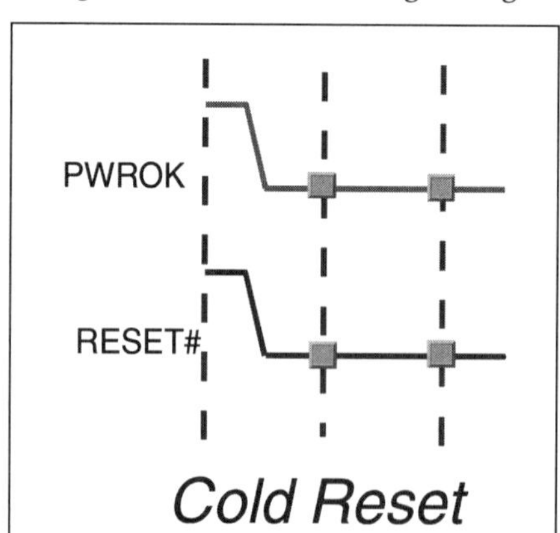

The required sequencing of PWROK and RESET# is described below and illustrated in Figure 12-5 on page 281. Valid cold reset requires that PWROK be asserted for at least 1 ms after the HT power and clocks have stabilized. Also RESET# must be asserted at least 1 ms before PWROK is valid, and remain asserted for at least 1 ms after a valid PWROK. Note that the state of RESET# is undefined during some time before PWROK is asserted. A device may require RESET# to remain low for longer than 1 ms after PWROK is asserted to stabilize its transmit clocks. This ensures that all devices are ready to transfer information prior to the deassertion of reset.

Figure 12-5: RESET# and PWROK Sequence and Timing Requirements

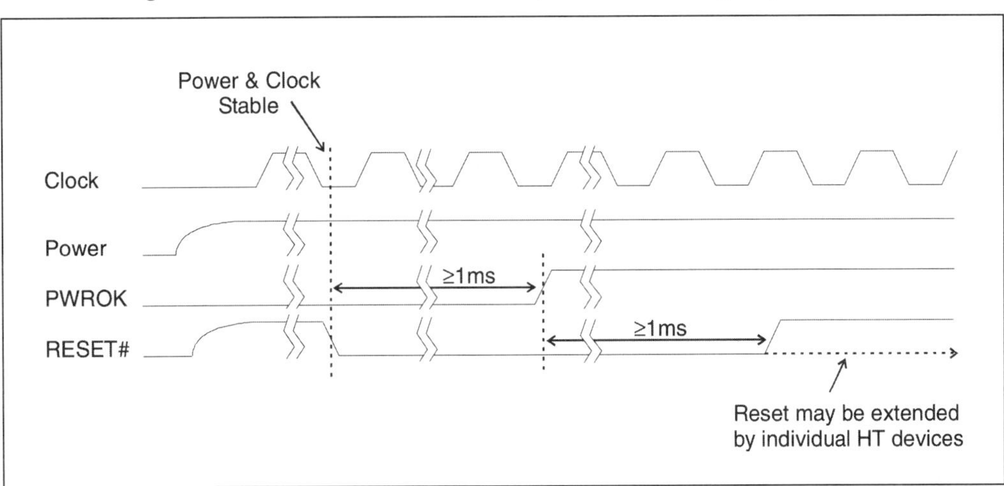

Effects of Cold Reset

Cold reset initiates the link initialization process and forces all HT devices and HT links in the fabric to their default state. The default condition includes:

- UnitID's are cleared to zero.
- all Configuration Space Registers (CSRs) are reset to their default state.
- error and status bits are cleared.

Chapter 12, entitled "Reset & Initialization," on page 275 details the required configuration registers and their default states.

Link Initialization

The process of initializing each link begins during cold reset. The complete link initialization process consists of several stages:

1. Low-level link initialization — This hardware mechanism ensures that the devices attached to a link can pass transactions safely in both directions following a cold reset. This includes:
   - Determining the link width that can be used after cold reset. This width is based on the maximum width of the smallest transmitter or receiver, but limited to 8 bits.
   - Establishing the default clock frequency of 200 MHz for all devices.
   - Synchronizing the transmit and receive clocks and setting up the receive FIFOs with the appropriate load and unload values.
   - Establishing the reference point for the beginning of packet transmission in both directions. This reference defines the beginning of 4-byte aligned packet transmission as well as the beginning of the CRC window.

2. The next stage of link initialization occurs after cold reset and is driven primarily by system firmware. This stage is needed because the low-level link initialization does not guarantee that the link is operating at maximum clock frequency and link width. The process involves:
   - Reading the maximum link-width fields from the Link Configuration register and loading the link-width control registers with the maximum common width (done for both upstream & downstream directions of a link).
   - Reading the Link Frequency Capability registers and loading the maximum common frequency into the Link Frequency control registers (done for both upstream and downstream directions).
   - Initiating a warm reset (or LDTSTOP# disconnect/connect sequence) to force the updated values to take effect.

Low-Level Link Width Initialization

Low-Level initialization of the link width is performed as a hardware sequenced point-to-point handshake between the two devices attached to each link. Once completed, the devices at each end of the link will be ready to perform transactions using either 2-, 4-, or 8-bits. This link-width negotiation sequence may not result in links operating at their maximum width. For example, since the maximum width following the negotiation is 8 bits, 16-bit, 32-bit, and asymmetrically-sized operations are not possible until enabled by software, which is the second stage of link-width initialization.

Determining Low-Level Link Width

HT permits devices with different link widths to be directly connected. This results in unused receiver and transmitter pins on the wider device. Logic within a device of course has no knowledge of the width of devices to which it connects. Consequently, a hardware handshake process is defined at powerup to ensure that all devices can determine a safe link width over which devices can communicate.

During reset all devices drive a pattern onto the link(s) to which they connect. This pattern defines the width of the transmitting device's receiver. (See Table 12-1 on page 283) The value received by each device defines the width at which they will communicate following cold reset. While this negotiated width may not be optimized for best performance, this method ensures that all devices can communicate safely. Subsequently, firmware-driven link width initialization maximizes the link width. (See "Tuning the Link Width (Firmware Initialization)" on page 295).

Table 12-1: Transmitter Value Driven to Indicate Receiver Width

Receiver Width (bits)	Transmit Values CAD[31:0] (hex)
2	0000 0003h
4	0000 000Fh
8	FFFF FFFFh
16	FFFF FFFFh
32	FFFF FFFFh

Comments regarding Table 12-1 on page 283:

- The transmitter width can be wider than the receiver, thus the values listed in column 2, are shown to be 32 bits wide (maximum possible width).
- The receiver width may be wider than the transmitter width. In this event, the transmitter cannot report the correct receiver size and is required to drive all CAD lines to 1's.
- Rows 3 and 4 list the transmit values for specifying 8-bit and 16-bit receiver widths, respectively. Note that transmit values seem to represent receiver widths that are much wider than the actual receiver size. (i.e., 32 bits of all 1's reflect a receiver width of 4 bytes). However, because the low-level link

initialization process limits the maximum link width to CAD[7:0], a value beyond FFh has no meaning. The upper lines are driven to ensure backward compatibility with the early versions of LDT.

- Row 5 defines the transmit value for a 32-bit receiver width. While this value seems to define precisely a 32-bit receiver width, the low-level receiver width is limited to FFh as described in the previous bullet.

The CAD values received during low-level initialization are interpreted by the device as shown in Table 12-2 on page 284.

Table 12-2: Interpretation of Value Received on the CAD Lines to Determine Receiver Width

CAD Value Received	Receiver Width Interpretation (result used for transmitter & receiver)
0000 0000h	No Device Attached
0000 0003h	2-bit width
0000 000Fh	4-bit width
0000 00FFh	8-bit width
0000 FFFFh	8-bit width (16-bit not supported)
FFFF FFFFh	8-bit width (32-bit not supported)

The following examples are offered to clarify the low-level receiver width detection process with different transmitter and receiver sizes. Patterns that represent 16- and 32-bit link widths are not supported by this low-level initialization process.

Example 1: 4-bit device connected to 8-bit Device. Refer to Figure 12-6 on page 285 during the following discussion. In this example an HT device with a 4-bit receiver and transmitter is connected to an HT device with an 8-bit receiver and transmitter. The specification requires that receivers that are wider than the transmitters to which they connect must connect the unused receiver inputs to logical zero as illustrated in Figure 12-6. This prevents the inputs from floating. Also, since the 8-bit device's transmitter connects to a 4-bit receiver, the upper four transmitter pins (CAD 7:4) are not connected.

During reset both devices deliver a pattern that represents the size of their receiver according to Table 12-1 on page 283, as follows:

- The 8-bit link delivers a value of FFh (logic doesn't know the receiver on the other end of the link is only 4-bits wide).
- The 4-bit link delivers a value of Fh.

The receivers then detect the pattern driven, and each device learns the link width to use when transmitting packets to the other.

- The 4-bit link device sees only Fh on CAD[3:0], and interprets the size of the remote receiver to be 4-bits wide.
- The 8-bit link device has its CAD[7:4] pins tied to differential logic 0 and detects the value Fh on CAD[3:0], and also interprets the size of the remote receiver to be 4-bits wide.

Figure 12-6: Low-Level Link Width, Example 1

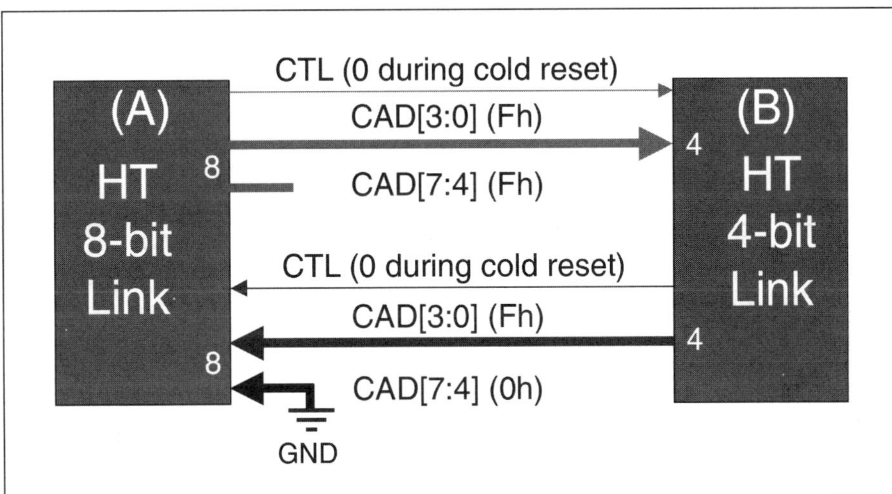

Example 2: 8-Bit Device Connected to 4/8-Bit Device. Refer to Figure 12-7 on page 286 during the following discussion. Device A with 8-bit transmitter and receiver is connected to device B with a 4-bit receiver and 8-bit transmitter. The 8-bit transmitter of device B connects to the 8-bit receiver of device A, and the 8-bit transmitter of device A connects to the 4-bit receiver of device B. CAD [7:4] of A's transmitter are not connected, because the receiver in device B is only 4-bits wide (CAD[3:0]).

During reset both devices deliver a pattern that represents the size of their receiver according to Table 12-1 on page 283, as follows:

- Device A delivers FFh from its transmitter to advertise the width of its receiver (8 bits); however, only the lower CAD line will be seen by device B.
- Device B transmits 0Fh to advertise the width of its receiver (4 bits).

The receivers then detect the pattern driven, and each device learns the link with to use when transmitting packets to the remote device.

- Device B's 4-bit receiver sees only half of the value transmitted by device A, and interprets the width of A's receiver to be 4-bits wide. In reality, device A has an 8-bit receiver, but B could not detect the correct width because its receiver is narrower than device A's transmitter. Thus, device B will only use the lower CAD lines when transmitting packets to device A even though 8-bit transfers could be made.
- Device A detects a 0Fh at its 8-bit receiver and determines that device B has a receiver that is only 4-bits wide. Thus, packets sent from A to B will also use 4 bits.

Figure 12-7: Low-Level Link Width, Example 2

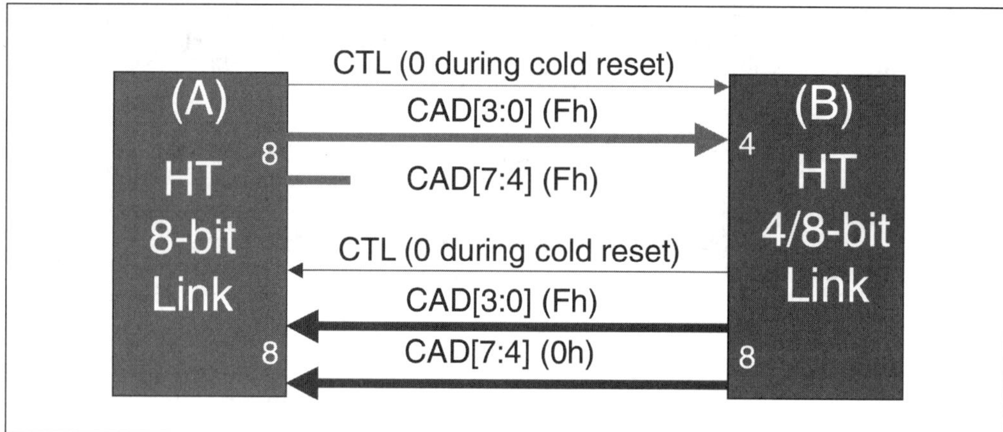

Example 3: 32-bit Upstream and 16-bit Downstream. Both devices attached to this link in this example support 16-bit widths in the downstream direction and 32-bit widths in the upstream direction.

During reset both devices deliver a pattern that represents the size of their receiver according to Table 12-1 on page 283, as follows:

- Device B delivers FFFFh from its downstream transmitter to advertise the width of its receiver. Note that device B's transmitter is only half the size of its receiver, so it cannot accurately report its receiver size. However, because the maximum width of the low-level initialization is 8-bits, there is no consequence of the transmitter not being able to report the actual receiver width.
- Device A transmits FFFF FFFFh to advertise the width of its receiver (16-bits) per Table 12-1.

Because the pattern delivered in both cases represents receivers wider than the minimum 8-bit width supported during low-level initialization, the receivers interpret the values as the maximum width (8-bits). Note that this low-level method of defining the link width to use following cold reset always results in the same width being used for both upstream and downstream transfers.

Figure 12-8: Low-Level Link Width, Example 3

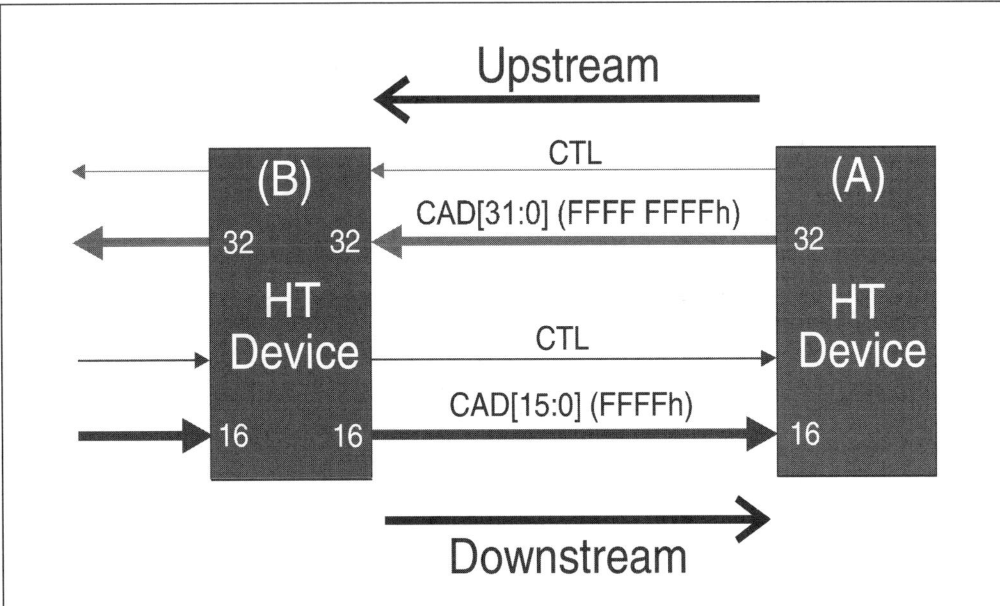

Negotiated Link Width Stored in Link Config Registers

When the low-level link initialization process has completed, each device updates its HT Link Configuration register to reflect the negotiated link width. Figure 12-9 illustrates the Link Configuration register and the link-width fields that are updated to reflect the negotiated width of the link. Two link configuration registers are defined:

1.   Link Config 0 — defines the link width for the primary interface
2.   Link Config 1— defines the link width for the secondary interface

Each of these registers contains two fields that define the link width used for the transmit and receive connections of the link. These 3-bit fields are named **LinkWidthIn** and **LinkWidthOut.** The low-level link width value loaded into these registers following cold reset becomes the default value and will always be the same for LinkWidthIn and LinkWidthOut. These fields may be updated later by software to reflect the optimum link width. Table 12-3 on page 289 lists the encoded link-width values.

Figure 12-9: Link Configuration Register

31:16	15:12	11:8	7:0	Offset
Command	Capabilities Pointer		Capability ID	00
Link Config 0	Link Control 0			04
Link Config 1	Link Control 1			08
Link Freq Cap 0	Link Error 0	Link Freq 0	Revision ID	0C
Link Freq Cap 1	Link Error 1	Link Freq 1	Feature	10
Error Handling	Enumeration Scratchpad			14
Reserved	Mem Limit Upper		Mem Base Upper	18

Link Configuration Register

15	14:12	11	10:8	7	6:4	3	2:0
DW FC Out EN	LinkWidthOut	DW FC In EN	LinkWidthIn	DW FC Out	Max Link Width Out	DW FC In	Max Link Width In

Table 12-3: Encoded Link-Width Values used in the Link Configuration Registers

Encoding	Link Width
000	8 Bits
001	16 Bits (not used for low-level init.)
010	Reserved
011	32 Bit (not used for low-level init.)
100	2 Bit
101	4 Bit
110	Reserved
111	Link Not Connected

Low-Level Clock Initialization

HT devices use a transmit clock (Tx Clk) to clock out a packet from the transmitter interface and a receive clock (Rx Clk) to receive an incoming packet. Figure 12-10 illustrates the relationship between the transmit and receive clock between two devices A and B. Device A transmits (upper half of link) a packet from the transmit FIFO to device B which receives the packet in the receiver FIFO. Similarly device B transmits (lower half of link) a packet from the transmit FIFO to device A which receives it in the receiver FIFO.

Figure 12-10: Link Interface and Clocking

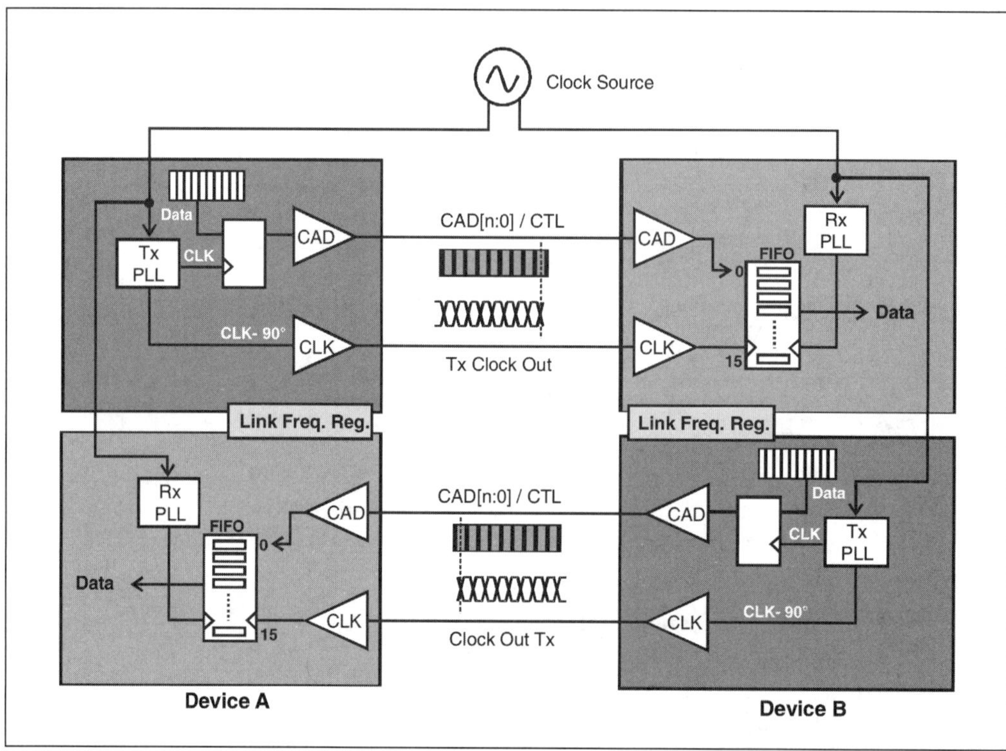

In a receiver device, the load pointer which is synchronous to Tx Clock Out (transmit time domain) points to the location within the receive FIFO where the incoming packets are stored. The unload pointer which is synchronous to Rx Clk (receiver time domain) points to the location from which data in the receive FIFO is unloaded into the core logic. The multiplexer sizes the data path width of the FIFO appropriately to match the data path width of the core logic.

The clock source supplies a clock to both devices. In this example, the clock source (source PLL) feeds the transmit and receive PLLs of both devices. A packet is clocked out synchronously to Tx Clk generated by the Transmit PLLs. The transmit PLLs also generates the Tx Clock Out signal that is visible on the link. Tx Clock Out lags Tx Clk by 90 degrees to center the clock transition within the bit time. CAD[n:0]/CTL data is loaded into the receive FIFO using Tx Clock Out clock received at a receiver device. Data is unloaded from the receive FIFO into the core logic using Rx Clk generated by the receive PLLs.

For additional detail regarding clock and FIFO initialization refer to Chapter 15, entitled "Clocking," on page 387.

The Default Clock Frequency

Following cold reset all HyperTransport I/O link transmitter and receivers must operate at the default HyperTransport clock frequency of 200 MHz. Cold reset initializes HyperTransport I/O link transmitters to this default link clock frequency.

Control and CAD Sequence after Reset is Removed

Following the deassertion of RESET#, the CTL and CAD signals go through a series of transitions that trigger various stages of low-level link initialization. These transitions and the related initialization events are listed in Table 12-4 on page 291 and detailed in the following sections.

Table 12-4: CTL/CAD Sequence Following Deassertion of RESET#

CTL	CAD	Duration (bit times) (8-, 16-, 32-bit links)	Duration (bit times) (2-, 4-bit links)	Description of Initialization Event
0	1	N/A	N/A	Values during Reset assertion
1	1	16	64, 32	CTL & CAD asserted after RESET# deasserts (device-specific)
0	0	512=4N	2048+16N,1024+8N	Initializes FIFO Pointers
0	1	4	16, 8	Transition of CAD frames incoming packets on 4-byte boundaries
1	don't care	N/A	N/A	Start of 1st control packet & beginning of CRC window

Clock Synchronization (CTL=0 & CAD=0)

Variation between the transmit clock and receive clocks can cause underrun or overrun without the appropriate FIFO size and load and unload pointer values. These pointers must be set up prior to beginning packet transmission and are triggered by CTL and CAD transition from high to low. Details regarding clock initialization, including FIFO size and separation between the FIFO write and read pointers, can be found in Chapter 15, entitled "Clocking," on page 387.

Figure 12-11 on page 293 illustrates the initialization sequence and the point at which the receive FIFO setup is referenced. To simplify the illustration, a single clock is illustrated rather than the multiple source synchronous clocks actually implemented. The specification defines the FIFO load and unload pointer setup as follows:

- The deassertion of the incoming CTL/CAD signals across a rising CLK edge is used in the transmit clock domain within each receiver to initialize the load pointer.
- The deassertion of the incoming CTL and CAD signals is synchronized to the core clock domain and used to initialize the unload pointer within each receiver.

Note that the point at which the pointers are set up may be different between the downstream and upstream transmitters and receivers. Figure 12-11 on page 293 depicts a link interface and the CTL and CAD timing associated with FIFO pointer setup. This example presumes that the link is symmetrical (i.e. 8-bits wide in both directions). Note also that the transitions of the CTL and CAD signals are shown transitioning from high to low at the same for both directions; however, this may not always be the case.

Figure 12-11 illustrates the point at which the devices at each end of the link detect that the other has asserted CTL on a rising clock edge. From this point a minimum delay must be honored before each device deasserts its CTL and CAD lines:

- 16 bit times (link widths of 8, 16, or 32 bits)
- 32 bit times (link width of 4 bits)
- 64 bit times (link width of 2 bits)

Because these timing parameters are specified as minimum values, the devices at each end of the link may implement different delays. The example shows both devices deasserting CTL and CAD at the same time, indicating that they use the same delay. The delay is extended when LDTSTOP# is asserted (>50µs) to permit time for devices to recover from a link power-down condition. (See "LDTSTOP# Disconnect Sequence" on page 304 for details).

Figure 12-11: Clock Synchronization and FIFO Load and Unload Pointer Setup

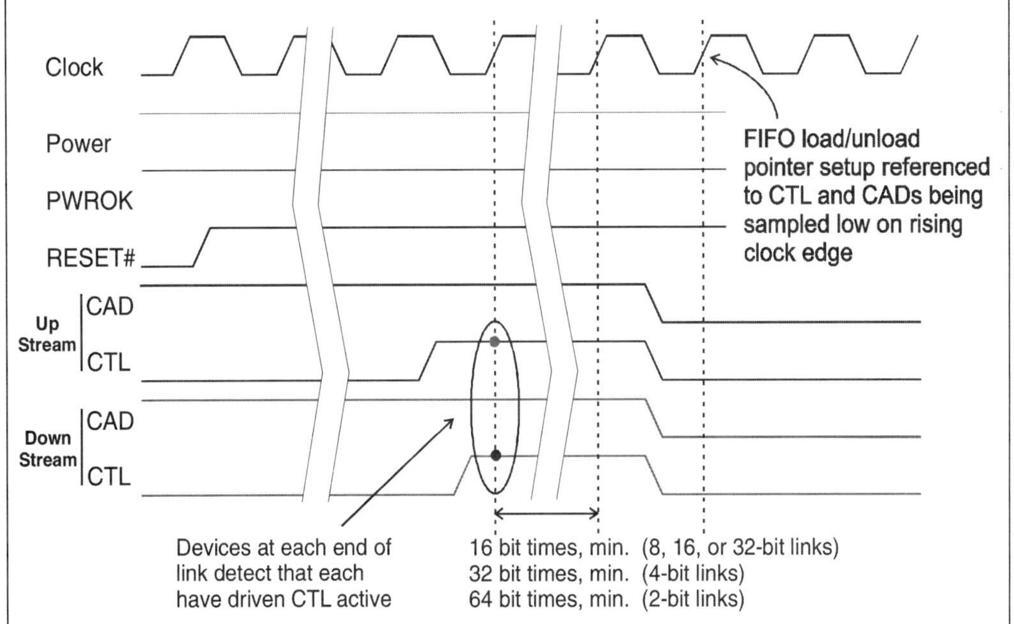

Duration of CTL & CAD Driven Low

CTL and CAD remain low for a specified minimum duration depending on CAD width as listed below and as illustrated in Figure 12-12:

- 8, 16, or 32-bit width — 512 bit times + 4N
- 4-bit width — 1024 bit times + 8N
- 2-bit width — 2048 bit times + 16N

These values ensure that the next transition is equivalent to a 4-byte boundary, and from this point, all transitions of CTL or the CAD lines must occur on 4-byte boundaries until the end of Cold RESET.

Figure 12-12: Duration of CTL & CAD Deassertion

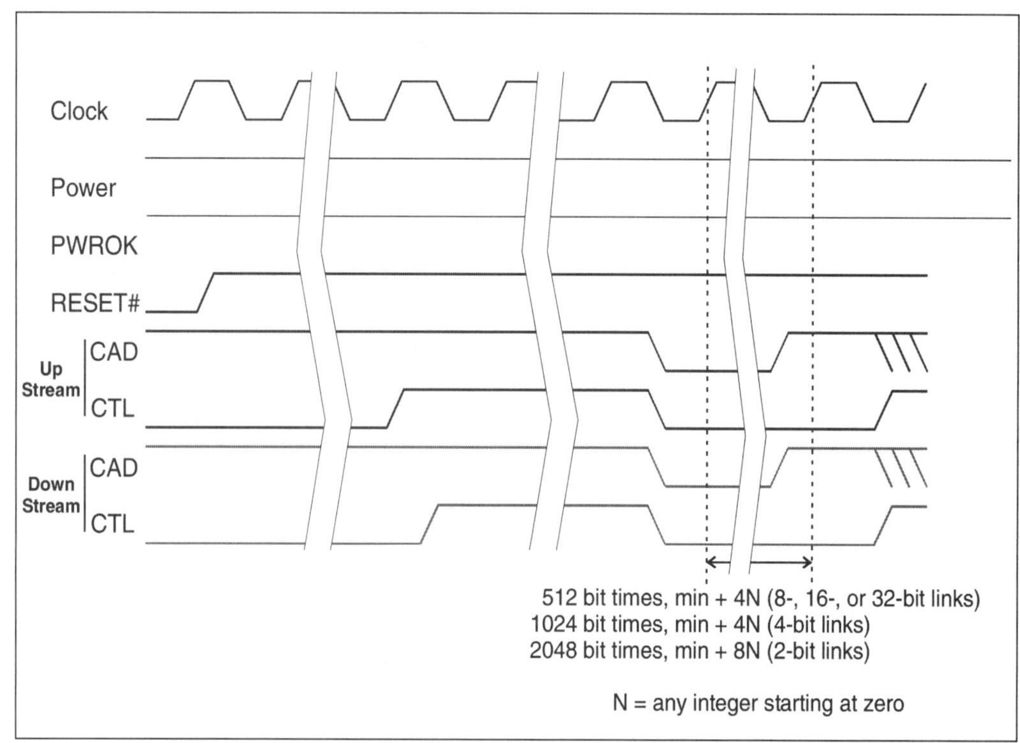

Packet Framing and Initializing the CRC Window

Once the low-level link width is determined, the transmitter and receiver pair that is used for a given direction must be synchronized as they begin packet transmission and reception. Note that each direction of the link must perform this synchronization but that they are independent of each other. This reference point specifies the alignment of packets on a 4-byte boundary, as well as the beginning of the CRC window. CRC generation is based on 512-byte intervals that begin at the end of low-level initialization on each link.

Figure 12-13 on page 295 illustrates the Reset timing that "frames" packets on 4-byte boundaries and initiates packet transmission and the beginning of the CRC window. Each device drives the CAD signals to a logic 1 across a rising CLK edge for exactly four bit-times, while leaving the CTL signal deasserted. The transition from all CAD signals deasserted to all CAD signals asserted serves to frame incoming packets. The first bit-time after these four must have CTL asserted, and is both the first bit-time of a new command packet and the first bit-time of the first CRC window. It also occurs across a rising CLK edge.

Figure 12-13: Framing and CRC Window Sequence

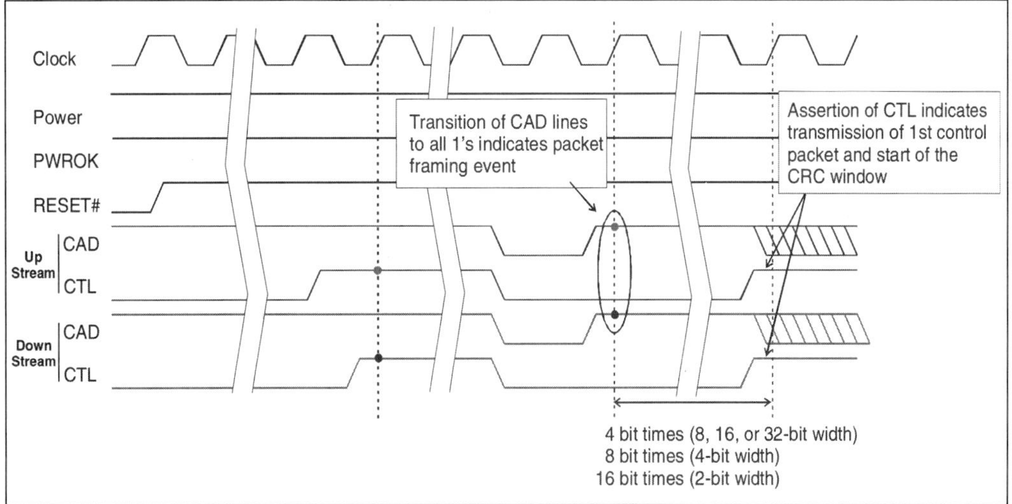

As a reminder, the examples presented above show the CTL and CAD transitions occurring at the same time for both directions of the link. However, after the receivers detect that the other has asserted its CTL signal, the synchronization for each direction of the link is independent.

Once low-level initialization is complete, the *Initialization Complete* bit in each active Link Control register is set.

Tuning the Link Width (Firmware Initialization)

Low-level link width negotiation does not guarantee that the maximum link width is used following cold reset. Therefore, system firmware or perhaps the O.S. must check the link width capability of devices at ends of each link to determine if the maximum possible link width is being used.

As discussed in "Negotiated Link Width Stored in Link Config Registers" on page 288, the Link Configuration Register contains the LinkWidthIn and Link-WidthOut fields that control the actual width used by each device's receiver and transmitter. The default value after cold reset is the negotiated value, which is always the same for upstream and downstream transfers.

Tuning the link width involves firmware that determines whether the devices attached to each link can operate at a width greater than the default setting. This is accomplished by reading the MaxLinkWidthIn and MaxLinkWidthOut fields. The largest common value for each transmitter/receiver pair is selected and stored in the corresponding LinkWidthIn and LinkWidthOut fields. Figure 12-14 on page 296 illustrates this process for one pair.

Figure 12-14: Link Width Update

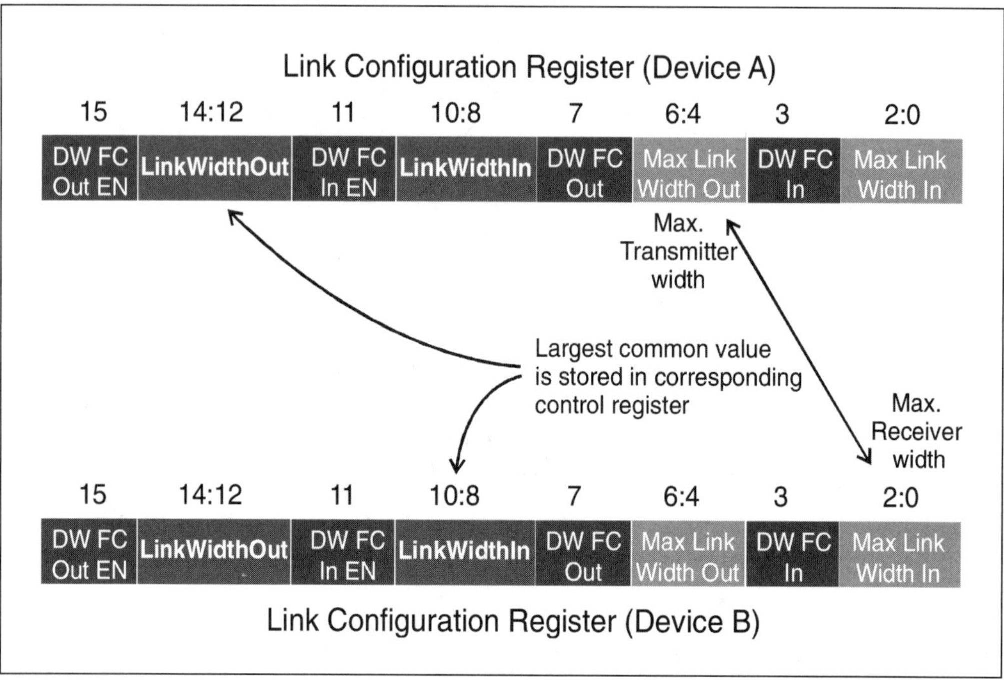

When software makes changes to the LinkWidthIn and LinkWidthOut fields, the new values loaded into these fields have no immediate effect on link operation. The updated width values take effect only after warm reset (See "Warm Reset" on page 302) or a disconnect/reconnect sequence via the LDTSTOP# protocol (See "HT Link Disconnect/Reconnect Sequence" on page 223).

The following sections describe the result of link-width tuning on the previous low-level link width examples.

Tuning Example 1: 4-bit device connected to 8-bit Device. During the low-level link width process a default link width of 4 bits was established (for both directions). Figure 12-15 on page 297 shows that for each direction the maximum link values are 4 bits and 8 bits, thus the largest common width is 4 bits in each direction. In this example, software is unable to increase the link width beyond the negotiated width.

Figure 12-15: Maximum Link Values, Example 1

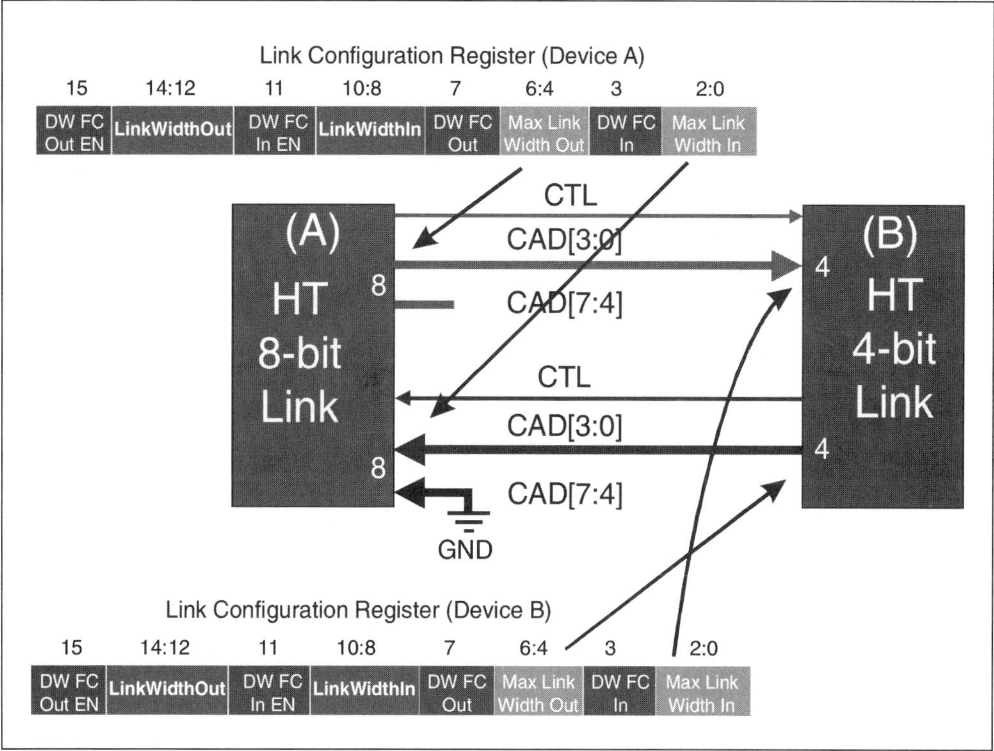

Tuning Example 2: 8-bit device connected to 4/8-bit Device. Low-level width determination in this example (described in "Example 2: 8-Bit Device Connected to 4/8-Bit Device" on page 285) results in a link width of 4 bits in both directions. A firmware check of the maximum link-width would reveal that the top portion of the link (see Figure 12-16 on page 298) can operate at a maximum width of 4 bits (the same as the negotiated width); however, the lower portion of the link has a transmitter and receiver that are both capable of transferring 8 bits at a time.

Figure 12-16: Maximum Link Values, Example 2

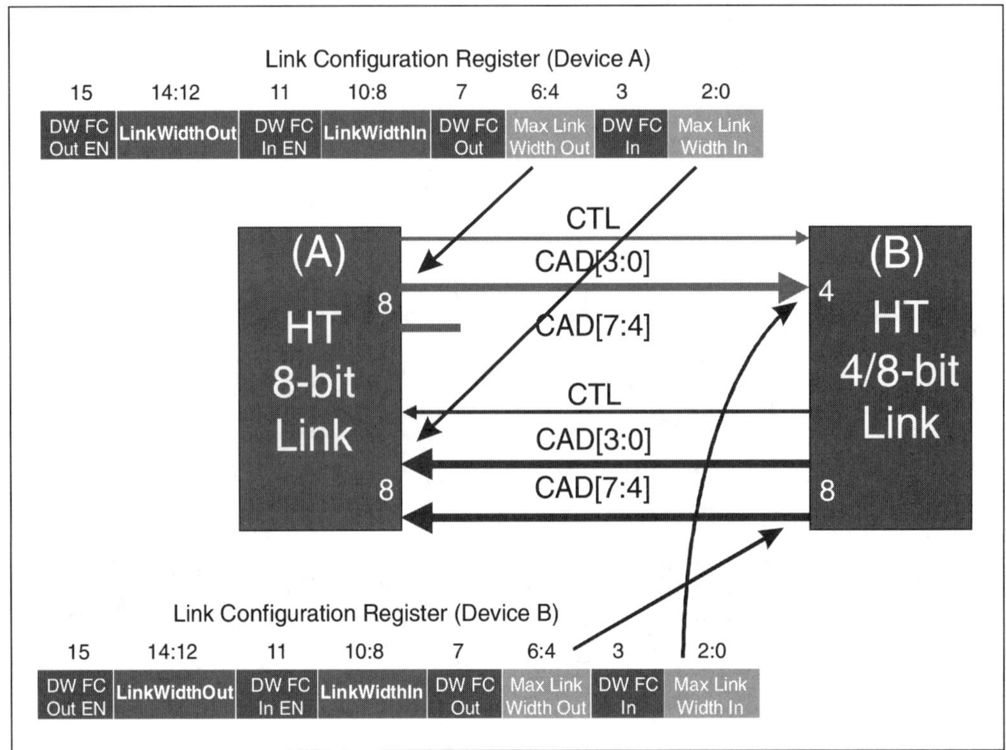

Tuning Example 3: 32-bit Upstream and 16-bit Downstream. The third example has devices with widths greater than 8 bits. Since the negotiated value was limited to 8 bits, tuning will result in far better performance than the low-level width negotiation. Figure 12-17 on page 299 illustrates that the maximum link width for the upstream link is 32 bits and the downstream link is 16 bits.

Figure 12-17: Maximum Link Values, Example 3

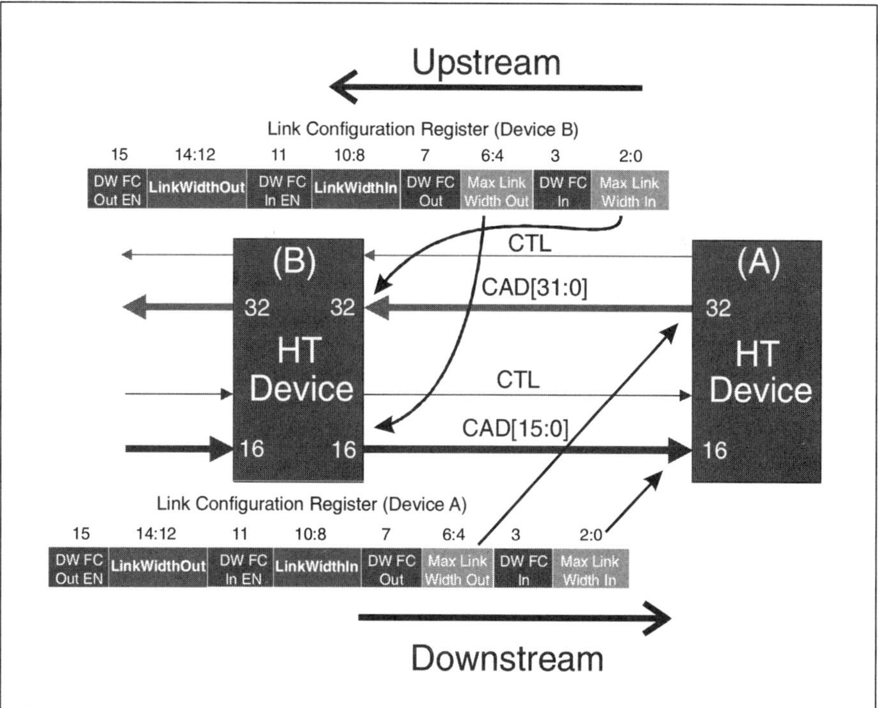

Tuning the Clock Frequency

Because the devices at the ends of each link may support a clock frequency higher than the default, initialization firmware reads the Link Frequency capability register to determine the highest common clock speed supported by the devices at each end of the link. The Frequency Capability Register is a member of the HT capability register set as illustrated in Figure 12-18 on page 300. This read-only register defines which clock frequencies a given HT device supports. Each bit corresponds to one of the specified clock frequencies, and Bit 0 must always be set to indicate support for 200 MHz operation. Note that two Frequency Capability registers (labeled Link Freq Cap 0 and 1) are defined in the capability registers to support devices such as tunnels that have two link interfaces to define. If a given device has a single link interface then only the first register is used. System firmware selects the highest common clock frequency supported by both devices connected by the link.

Figure 12-18: Tunnel Example — Link Frequency Capability Registers

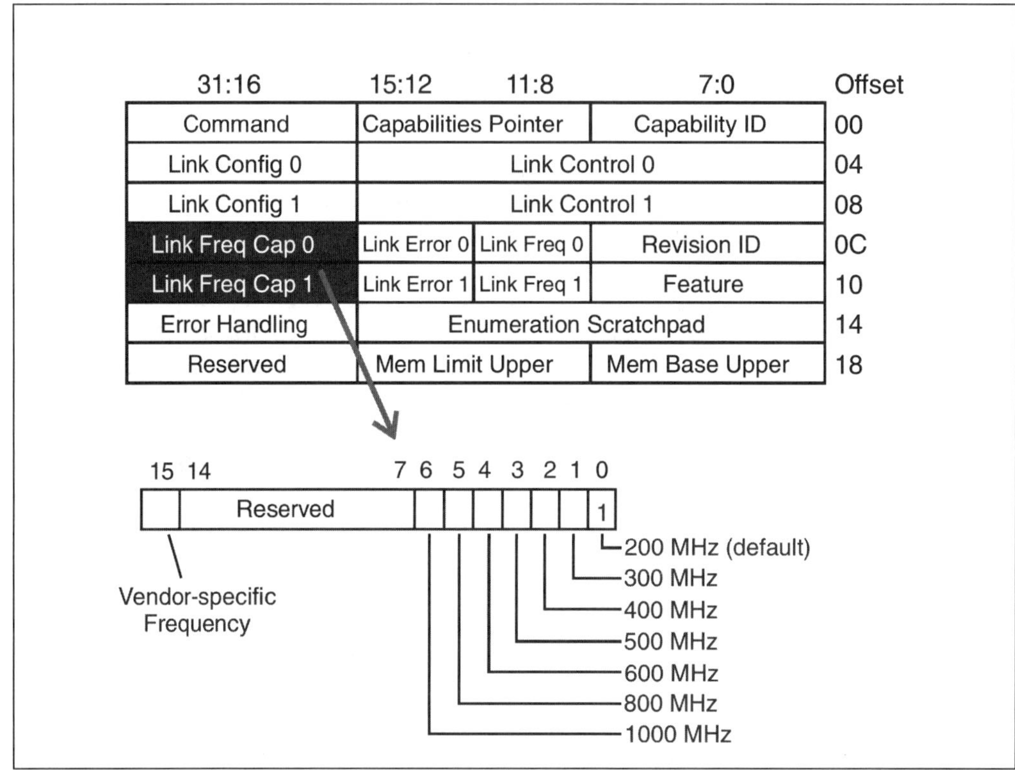

The maximum clock frequency supported is loaded into the Link Frequency register so that the link can run at the fastest speed. Figure 12-19 on page 301 illustrates the locations of the Link Frequency register within the HT capability register set, and Table 12-5 on page 301 shows the encoding for the different frequencies. Once the new maximum clock frequency value is loaded, system firmware must initiate a warm reset or LDTSTOP# disconnect sequence to invoke the changes to the link clock frequencies.

Figure 12-19: Link Frequency Register Location within the HT Capability Register Set

31:16	15:12	11:8	7:0	Offset
Command	Capabilities Pointer		Capability ID	00
Link Config 0	Link Control 0			04
Link Config 1	Link Control 1			08
Link Freq Cap 0	Link Error 0	Link Freq 0	Revision ID	0C
Link Freq Cap 1	Link Error 1	Link Freq 1	Feature	10
Error Handling	Enumeration Scratchpad			14
Reserved	Mem Limit Upper		Mem Base Upper	18

Table 12-5: Encodings for Link Frequency Field of Link Configuration Register

Link Frequency Encoding	Transmitter Clock Frequency (MHz)
0000	200 (default)
0001	300
0010	400
0011	500
0100	600
0101	800
0110	1000
0111 - 1110	Reserved
1111	Vendor Specific

Warm Reset

Warm reset can be triggered by a hardware event, or system software may initiate a warm reset. Note that the system must not generate a warm reset when changing the frequency or width of a link. Specifically, if one side of the link has had its width or frequency changed, but not the other side, following warm reset, one side of the link would think the other is faster and/or wider than what it is actually programmed to be.

Warm reset is detected when PWROK is asserted (logic high) and RESET# is asserted (logic low). Warm reset differs from a cold reset in following ways:

- The LinkWidthIn and LinkWidthOut values are not affected by a warm reset (i.e., these values persistent across a warm reset) and take effect following warm reset.
- The LinkFrequency registers are persistent across a warm reset, and take effect after warm reset.
- CAD values are driven to represent the updated link width, rather than the low-level values defined for cold reset.

Table 12-6 lists the state of the HT signals during a warm reset.

Table 12-6: Signal States During Warm Reset

Signal	State During Warm Reset
CLK	Toggling
CTL	Logic 0
CAD[n-1:0] (Updated width)	Logic 1's
CAD[31:n] (if present)	Logically undefined within electrical spec.

Following warm reset the same synchronization sequence (i.e. same as cold reset) is performed to initialize the transmit and receive clocks.

Warm Reset Generated by Software

The specification defines a bit in the HT bridge's *HyperTransport Host Command CSR* that provides a way for software to initiate warm reset on the secondary bus of an HT-to-HT bridge. Refer to Figure 12-20 on page 304. Specifically, this bit defines the type of reset that will be initiated on the secondary bus when software sets the Secondary Bus Reset bit of the Bridge Control register.

The state of the *warm reset* bit only has meaning when software initiates a reset sequence. Actions taken during the reset sequence depends on the state of the warm reset bit as follows:

- If warm reset is 1, PWROK will remain high and RESET# is asserted, causing a warm reset
- If *warm reset* is 0, PWROK will be driven low along with the RESET# signal, causing a cold reset.
- If not implemented, this bit is read-only and hardwired to 1.

It is the responsibility of the hardware to sequence PWROK and RESET# correctly.

Changing the state of the Warm Reset bit while the Secondary Bus Reset bit is asserted results in undefined behavior. Defined changes include:

- Previously assigned UnitID's are forgotten and all UnitID's default back to zero.
- Configuration Space Registers (CSR's) that are defined to be persistent through a warm reset do not go to their default state. Instead, their current state is maintained.

RESET# must be asserted for at least 1ms during a warm reset sequence.

Figure 12-20: HyperTransport Host Command CSR

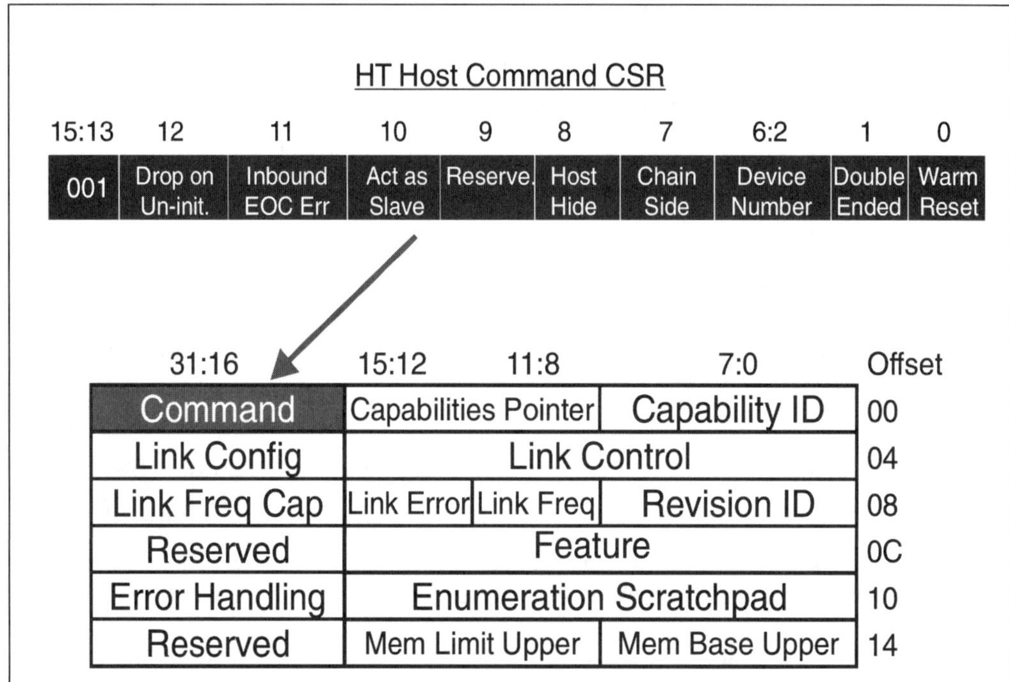

LDTSTOP# Disconnect Sequence

After a HyperTransport technology disconnect/reconnect sequence, devices that implement the LDTSTOP# protocol are required to update their link widths in exactly the same way as they do after a warm reset sequence. This allows initialization software for systems built from such devices to use the LDTSTOP# protocol rather than warm reset to invoke link width changes. The following requirements exist when during Reset with LDTSTOP#.

- The LDTSTOP# pin must be deasserted for at least 1 microsecond before RESET# is deasserted.
- LDTSTOP# must remain deasserted until the link initialization sequence, described later in this section, has completed.

13 *Device Configuration*

The Previous Chapter

Reset signalling and timing, along with actions taken by the system and devices during reset, are the primary topics discussed in the previous chapter. The chapter also discussed software initiated reset and why it's required. The process of determining the default speed and link width and the subsequent software tuning of bus speed and link width are also detailed.

This Chapter

HyperTransport uses PCI configuration. This chapter describes HyperTransport technology configuration for host bridges, tunnels, and end (cave) devices. These devices use the *type 0* configuration header format, while HyperTransport-to-HyperTransport bridges and bridges between HyperTransport and other PCI compatible protocols (e.g. PCI and PCI-X) use the *type 1* header format and are described separately in Chapter 16, entitled "HyperTransport Bridges," on page 407. Many aspects of HyperTransport device configuration are exactly the same as for generic PCI devices, although some header fields are used differently in HyperTransport, and some not at all. Devices also require at least one HyperTransport-specific advanced capability register block in addition to the basic PCI configuration space header fields.

The Next Chapter

The high speed signaling performed by HT devices is based on point-to-point differential signaling and source synchronous clocking. Details associated with link power requirements and the driver and receiver characteristics are discussed in the next chapter. Also, the characteristics of the system-related signals, including RESET#, PWROK, LDTSTOP#, and LDTREQ# are discussed.

HyperTransport Uses PCI Configuration

Many current generation computers use the PCI configuration method and the 256 byte PCI *configuration space* memory required of all PCI-compliant devices to help set up and manage system chipsets and I/O peripherals. Using PCI configuration for a bus protocol such as HyperTransport goes a long way toward promoting software compatibility with the millions of systems already supporting buses employing PCI-based configuration, including PCI, AGP, PCI-X, USB, etc. HyperTransport is designed for PCI plug-and-play configuration and to minimize impact on existing BIOS and driver software.

What PCI Configuration Accomplishes

During system initialization, low level BIOS or other system software uses configuration transaction cycles to "walk" each PCI-compatible bus (PCI, PCI-X, HyperTransport, AGP, etc.) and read the PCI configuration space of each device function it finds. Once discovered, basic and advanced capability features of each device are set up as appropriate. Collectively, PCI configuration cycles may be used for many aspects of device management, including:

- **Assignment of system resources.** Unlike earlier bus protocols, including the *Industry Standard Architecture (ISA)*, PCI compatible plug-and-play devices are not allowed to establish their own base addresses and interrupt levels using fixed schemes or through user manipulation of jumpers and switches. Instead, the designer of a PCI compatible device "hard codes" information in selected PCI Configuration Space fields describing the fixed requirements of the device with respect to memory and I/O addresses needed, whether system interrupt support is required, arbitration needs, etc. Once the system address maps and interrupt routing are determined, software then returns to programmable fields in the PCI Configuration Space of each device and programs address ranges, interrupt routing, etc.
- **Enabling of device capabilities and options.** In addition to assignment of system resources to PCI compatible devices, software also uses the PCI Configuration Space to select device options, enable bus mastering and target decoding of memory and I/O transactions, program error response strategy, and set up other basic PCI and advanced capability protocol features.
- **Checking of dynamic (error) status.** Finally, the PCI configuration space is used to log errors resulting from attempted transactions. These logged errors, if checked by software, provide a picture of the nature of the error,

which device(s) detected it, etc. The Status register in the configuration space header is used for generic PCI-type error logging; in addition, advanced capability register blocks also contain logging fields for errors related to a specific capability (e.g. HyperTransport CRC errors, buffer overflow errors, etc.).

HyperTransport System Limits

HyperTransport shares PCI terminology in describing a system in terms of the number of buses, devices, functions, and configuration space.

256 Buses In A System

PCI permits 256 buses in a system and each PCI host bridge or PCI-to-PCI bridge secondary interface is host to a new bus with a unique bus number. Unlike PCI, a HyperTransport bus may not end with a single electrical connection. Tunnel devices enable the construction of device chains which are still viewed as a single logical bus. The 256 bus limit in HyperTransport, then, is actually 256 chains.

32 UnitIDs Per Bus

PCI permits a maximum of 32 physical devices per bus. In HyperTransport, each functional device can request multiple device numbers, called UnitIDs. The reason for this is because HyperTransport ordering rules consider the transactions from each UnitID to be a unique *transaction stream*; owning multiple UnitIDs enables a device to source more than one transaction stream (e.g. a standard transaction stream and an isochronous transaction stream for its high priority traffic). The 32 device per bus limit in PCI is a 32 UnitID per bus limit in HyperTransport.

One To Eight Functions Per Device

As in PCI, HyperTransport allows 1-8 logical functions in a physical device package. Each function has its own 256 byte configuration space, and will be assigned unique UnitID(s).

256 Bytes Of Configuration Space

Just as in other PCI devices, each function of a HyperTransport device must implement a 256 byte configuration space memory. The first one-fourth of the configuration space is the header. In addition to the header, devices also must implement at least one set of HyperTransport advanced capability registers.

Configuration Accesses: Reaching All Devices

The process of HyperTransport device configuration depends on software being able to access the 256 byte configuration space of each function in each device on each bus in the system. Configuration cycles originate at the CPU that executes the configuration software; the cycles then move in the direction of the target. This section compares the PCI and HyperTransport methods used to reach the configuration space of a device which may reside on a bus many levels deep in the topology.

Implied in plug-and-play address assignment on buses such as PCI and Hyper-Transport is the fact that until it is discovered and assigned an address range by low-level software, a device can't claim normal memory or I/O transactions. Furthermore, whenever a bus reset occurs, each device "forgets" its address ranges and other information programmed in configuration space and can no longer be targeted with transactions which depend on assigned addresses. So, how can a device's configuration space be set up if it doesn't know its target address?

In addition to the problem of simple devices recognizing their own configuration cycles in an uninitialized system, the complex topologies permitted in PCI, PCI-X, and HyperTransport require that bridges be programmed to forward configuration transactions to the proper bus before a device can even consider claiming it.

Before looking at how HyperTransport differs from PCI in its handling of system-wide configuration accesses, here is a quick review of how PCI handles them.

Review: How PCI Handles Configuration Accesses

With the exception of chipsets, PCI devices generally power up (or come out of reset) disabled with respect to either generating transactions as bus master or decoding memory or I/O transactions as targets. This is because they are not aware of either their own plug-and-play addresses or those of other devices. The *Configuration Read* and *Configuration Write* transactions are the only ones a PCI device may decode following reset. Configuration cycles originate at the CPU, and instead of carrying conventional address information (which would be useless), these cycles start downstream carrying the following attributes about the target in the 32-bit address of the configuration read or write transaction:

- Bus number the target resides on (0-255 decimal)
- Device number of the target (0-31 decimal)
- Function number inside the target (0-7 decimal)
- Double Word Offset in target's configuration space (0-63decimal)

Note that while addresses are not known after reset, bus number and device number are functions of the board layout and ARE known.

Two Configuration Cycle Types

As PCI configuration cycles travel downstream, there are two variants: type 0 and type 1. The type is indicated in the lowest two bits of the 32-bit PCI address. Having two types is necessary because PCI devices don't know their bus number or device numbers and must depend on upstream bridges to help select them.

Type 1 Cycle Until Target Bus Is Reached

Starting at the host bridge, a type 1 configuration cycle is propagated downstream until it reaches the bridge with a secondary bus number equal to that of the configuration cycle *bus number* field. Type 1 configuration cycles are ignored by all devices except bridges which will claim them and pass them on to the next downstream bus if the *bus number* field of the configuration cycle is between the values programmed in the bridge's secondary and subordinate bus number registers.

Target Bus Bridge: Convert To Type 0; Assert IDSEL

The bridge owning the target secondary bus (based on the value programmed in its secondary bus number register) will convert the type 1 configuration cycle to a type 0. It will also check the *device number* field and assert the corresponding PCI *IDSEL* signal to the intended target; IDSEL acts as an explicit target device "chip select". There is a separate IDSEL signal for each device on a PCI bus; a target which detects IDSEL asserted at the same time a Configuration Read or Write cycle (type 0) occurs claims the transaction and uses the remaining information (*Function number* and *Dword offset*) to access its configuration space.

Note: In many systems, the IDSEL signals routed to each device are actually upper bits on the AD bus which are otherwise unused in type 0 configuration cycle address phases.

An Example: A PCI Configuration Space Access

Figure 13-1 on page 310 illustrates the use of type 0 and type 1 cycles; in this example, assume the CPU is accessing *Dword 1* in the configuration space of *Function 0, Device 2,* on PCI *Bus 1*.

Figure 13-1: PCI Type 1 and Type 0 Configuration Cycles

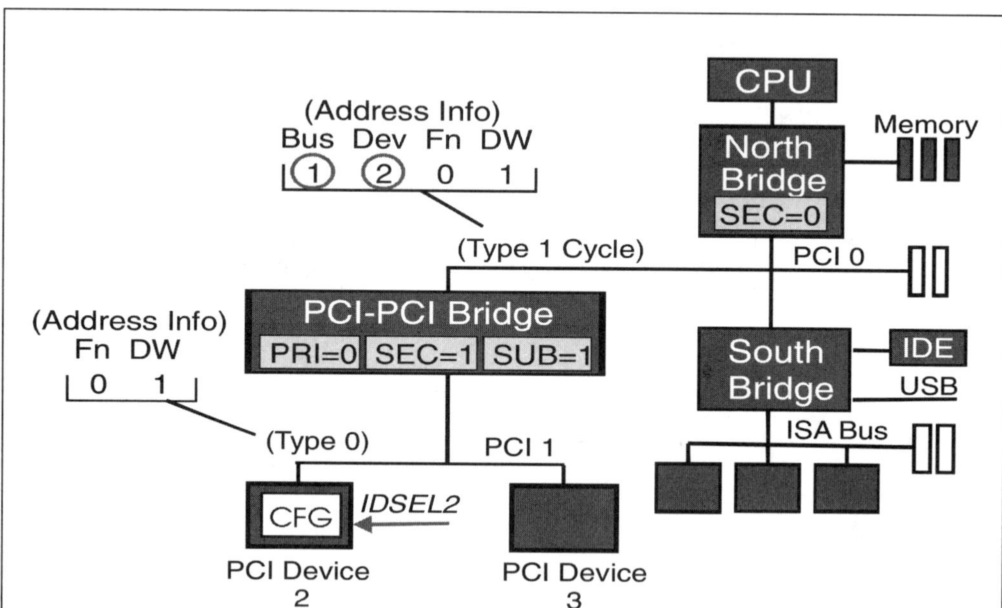

Events In PCI Configuration Space Example (see Figure 13-1)

1. Because the target bus is not bus 0, the North Bridge sends a type 1 PCI configuration cycle out on bus 0. The configuration type 1 cycle address phase includes target bus, device number, function number, and Dword offset.
2. Only bridges are allowed to respond to type 1 configuration cycles. The PCI-PCI bridge claims the cycle because the *bus number* field (bus 1) is in its range of secondary-subordinate bus numbers.
3. The PCI-PCI bridge also converts the cycle to a type 0 on bus 1 because its Secondary Bus Number register matches the *bus number* field (1). This, therefore, is the bus the target resides on.
4. The PCI-PCI bridge also asserts IDSEL2 to the target during the address phase of the configuration cycle because the *device number* field indicates Device 2.
5. Device 2 claims the type 0 configuration cycle based on the command type (configuration read/write, type 0) <u>and</u> the fact that IDSEL2 is asserted.
6. Device 2 then uses the *function number* and *Dword offset* fields in the Configuration read/write address to internally target the specific function and configuration space offset.

How HyperTransport Handles Configuration Accesses

While HyperTransport configuration cycles are compatible with PCI, there are some important differences.

Configuration Cycles Are Memory Mapped

To generate a configuration space read or write, a HyperTransport bridge simply sends a RdSized or non-posted WrSized request using a reserved address range in the 40-bit HyperTransport memory map. This 32MB range, recognized by all devices, is shown in Figure 13-2 on page 312.

Figure 13-2: Configuration Space In The HyperTransport Address Map

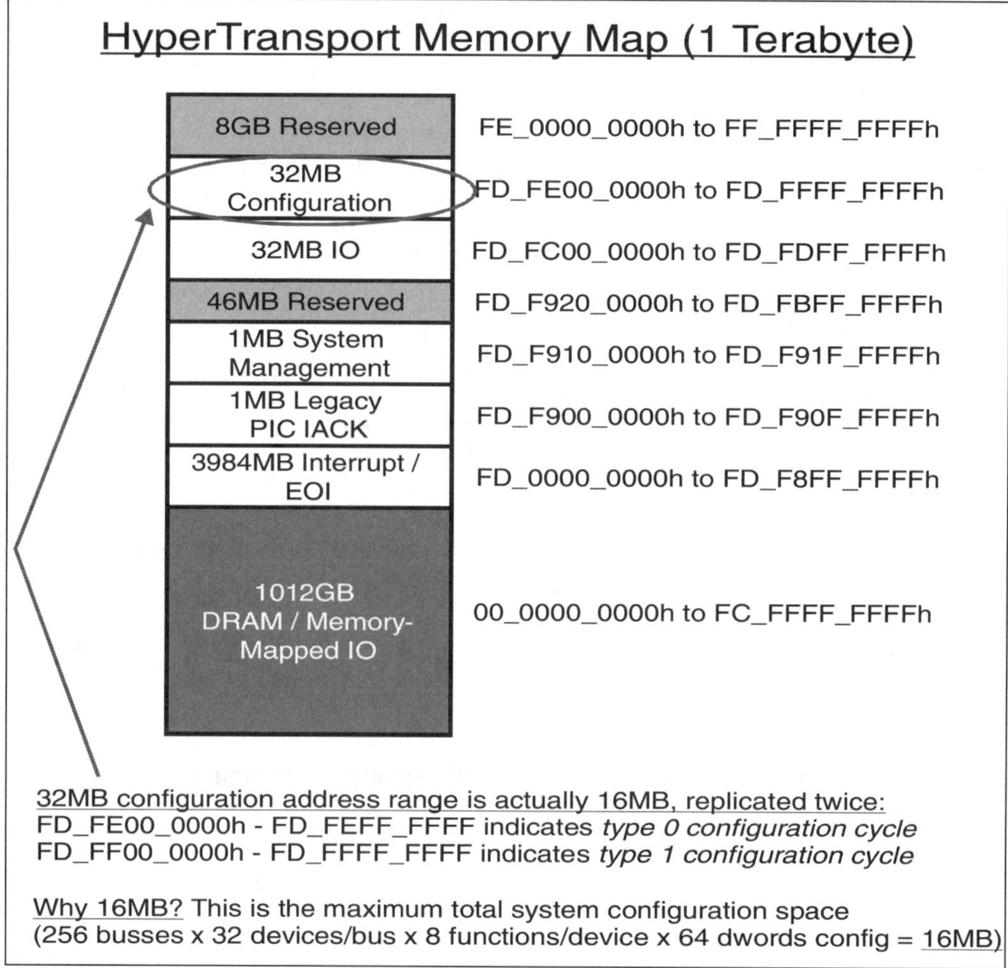

How The 32MB Configuration Area Is Used

The 32MB HyperTransport memory map address space reserved for configuration cycles is used to access the 256 byte configuration space of each function in each device on each bus. How the address range is interpreted and how a particular device can recognize configuration cycles it should claim vs. those it must forward is illustrated in Figure 13-3 on page 313 and described below.

Figure 13-3: Configuration Type 0 And Type 1 Request Packet Format

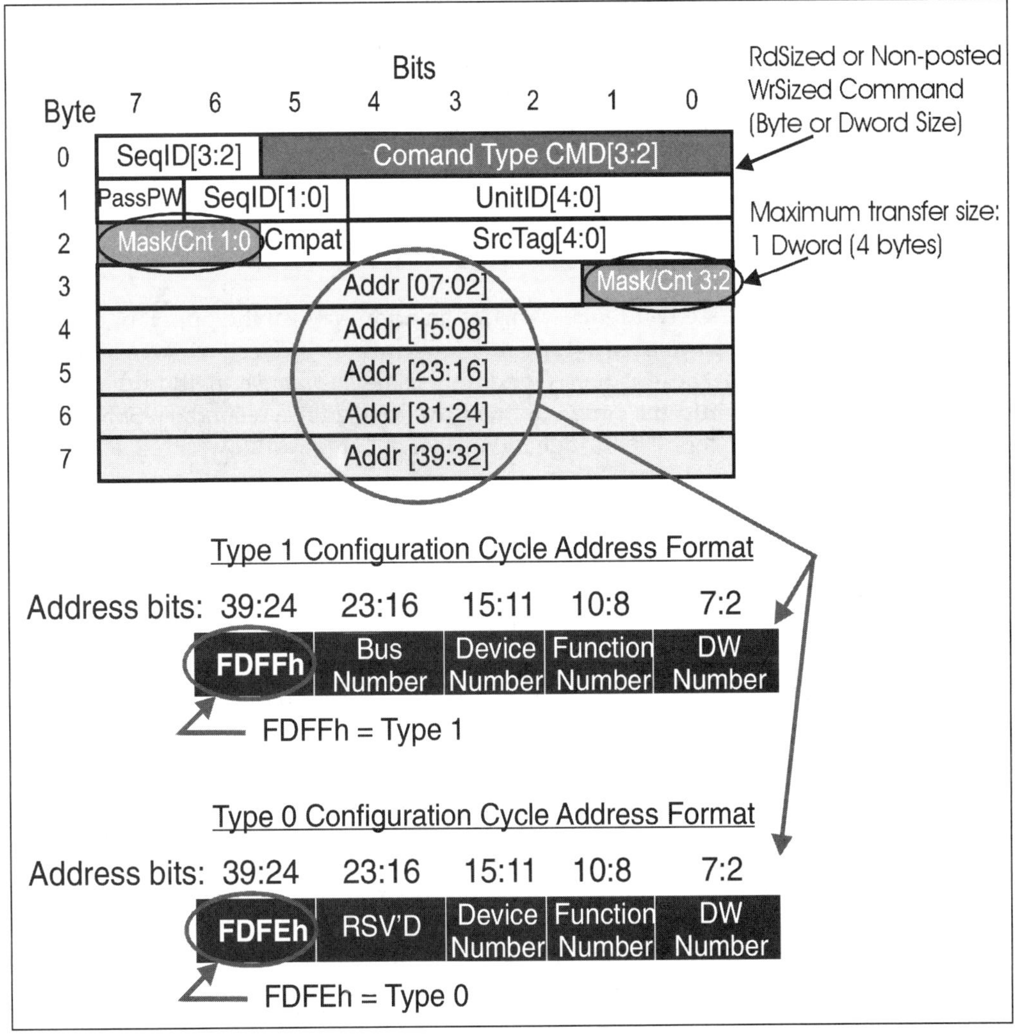

Upper 16 Address Bits Indicate Type 0 And Type 1 Cycle

As in PCI configuration cycles, HyperTransport requires two variants of configuration read/write cycles, type 1 and type 0. The type 0 configuration is generated by a bridge when the cycle has reached the target bus (chain) where the device being accessed resides; the type 1 cycle is in transit to the target bus and should be forwarded by bridges or tunnels in the target path. The bridge to the destination bus will convert it to type 0.

Because HyperTransport configuration cycles are distinguished from other read/write requests only by the fact they target the 32MB reserved configuration address range, the first problem is how to distinguish type 1 from type 0 cycles. The 32MB configuration address range is further divided into two parts: request packets carrying addresses in the upper 16MB of the range are type 1 cycles; requests with addresses in the lower 16MB are type 0 cycles. Referring to Figure 13-3 on page 313, note the following:

HyperTransport Type 1 Configuration Cycle (See Figure 13-3)

If a SizedRD or SizedWt request carries an address with the upper 16 bits set = FDFFh, then the cycle is a type 1 configuration request. Only bridges are allowed to accept these requests, and only if the *bus number* field in the address (bits 23:16) falls into the range defined by the bridges Secondary-Subordinate bus number registers. The bridge then passes the request downstream.

HyperTransport Type 0 Configuration Cycle (See Figure 13-3)

If a SizedRD or SizedWt request carries an address with the upper 16 bits set = FDF8h, then the cycle is a type 0 configuration request. This will be claimed by the device that also has a match when the *device number* field (bits 15:11) in the address matches one of its UnitIDs. It then uses the *function number* and *Dword* fields to target the particular internal function and configuration space offset.

No IDSEL Signal Needed In HyperTransport

Finally, there is no IDSEL signal to accompany a type 0 configuration cycle in the HyperTransport protocol. The need for this signal has been eliminated because a *Base UnitID* field has been included in the HyperTransport advanced capability register block so that a device is programmed to "know" its UnitID number(s). This allows the device to decode its own configuration cycles rather than depending on the upstream bridge to do it with IDSEL.

Example: HT Configuration Space Access

The following diagram, Figure 13-4 on page 315, depicts the movement of a configuration request through HyperTransport chains. In the following example, assume the CPU is targeting the configuration space in HyperTransport UnitID 2 on bus 1.

Figure 13-4: HyperTransport Type 1 And Type 0 Configuration Cycles

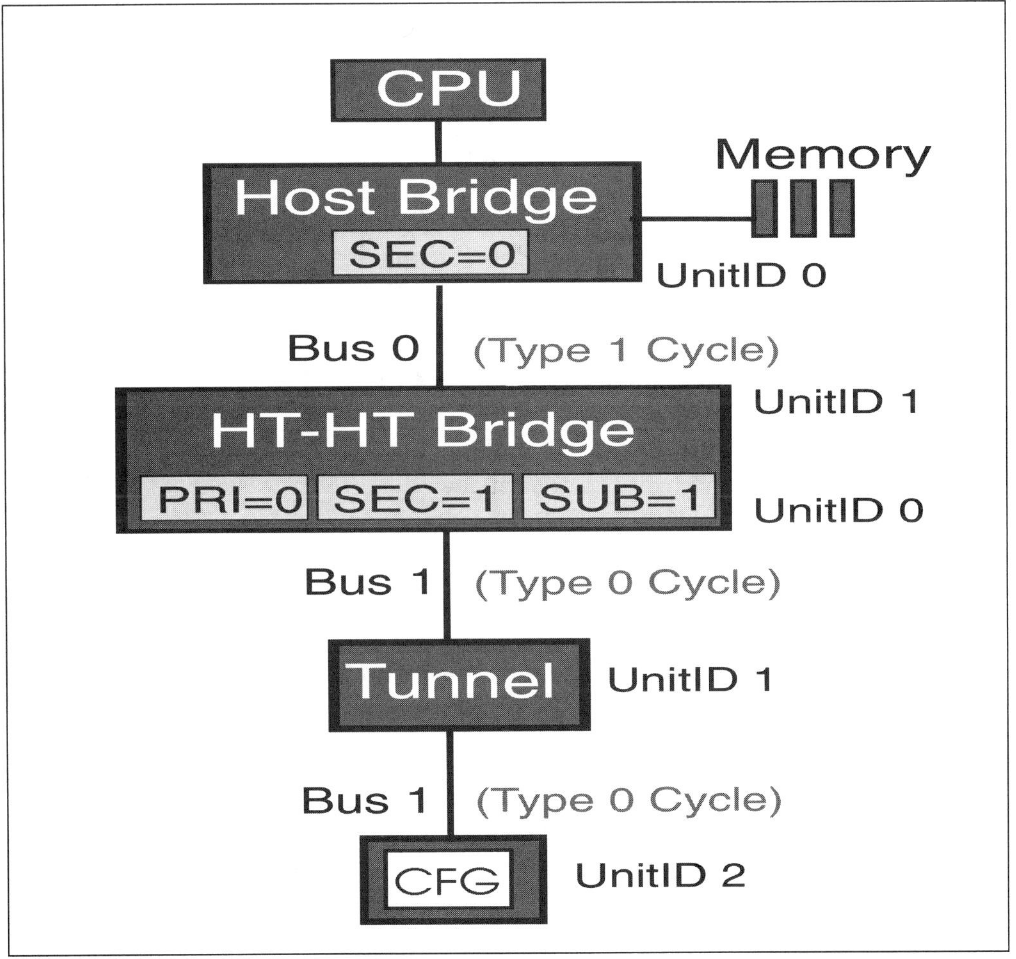

Events In HT Configuration Example (see Figure 13-4)

1. Low level software executing on the CPU requires access of the configuration space in Device 2 on Bus (chain) number 1.
2. The Host Bridge checks its secondary bus number register, recognizes the target bus <u>is not</u> its secondary bus, and sends a request packet for type 1 configuration cycle onto bus 0 (using the upper half of configuration address range).
3. The HT-to-HT bridge on Bus 0 checks the *bus number* field in the request and compares it with its own secondary, and subordinate bus numbers. Because the target bus is below it, the HT-to-HT bridge forwards the configuration cycle onto bus 1; at the same time it converts it to a type 0 because the target bus has been reached. Conversion to type 0 simply means shifting the configuration address into the lower half of the configuration address range. Note that the bus number field is stripped off when the cycle is converted to type 0 (see Figure 13-3 on page 313).
4. Device 1 claims the cycle because it is a type 0 configuration cycle AND it carries a *device number* which matches one of its assigned UnitIDs.
5. Device 1 then uses *function number* and *dword offset* fields in the request packet to target the specific internal function and offset location in its configuration space (refer to Figure 13-3 on page 313).

Initializing Bus Numbers And Unit IDs

One of the first steps in HyperTransport configuration is the initial assignment of bus numbers and UnitIDs for each device and chain in the topology. Using a depth-first search algorithm, enumeration software assigns IDs to each device it discovers; if it finds any HyperTransport bridges, it also assigns the primary, secondary, and subordinate bus numbers so that later configuration cycles may find their way to target buses other than bus 0.

Case 1: A Single Chain With One Host Bridge

In a single chain with only one host bridge, enumeration is fairly simple:

1. Following a reset assertion on a chain, the *Base UnitID* field in the Slave Command register of each in HyperTransport device is cleared to "0".
2. In addition, reset forces the primary, secondary, and subordinate bus number registers in each HyperTransport bridge and the secondary bus number register in host bridges to "0" as well.
3. The transmitter and receiver interfaces on each link perform the low-level negotiation to determine starting bus width. They also perform the required

link initialization sequence. Once synchronization is complete, the *Initialization Complete* bit in each active Link Control register is set.

4. After link synchronization, each active transmitter issues buffer release (NOP) packets to the corresponding receiver to indicate its own input flow control buffer capacities. Once this is done, each transmitter issues NOPs until configuration starts.

5. The host bridge initializes its UnitID counter so it can start assigning Unit-IDs to slave devices it discovers (it reserves UnitID 0 for itself).

6. If the host bridge's Link Control register *Initialization Complete* and *End Of Chain* bits indicate that another device is attached to its secondary bus, the host bridge sends a series of configuration cycles to the first device in the chain. These type 0 configuration cycles target Bus 0, Device 0 (UnitID 0), Function 0. Because all devices default to UnitID 0, the first device will claim the cycles. Read cycles will target configuration space locations containing Vendor ID, Device ID, Class Code, Header Type, etc.

7. At some point, the host bridge assigns new UnitID(s) to the device by reading the *Unit Count* field in the Slave Command Register and then programming (writing) the *Base UnitID* field with the next available UnitID (1). For devices which request more than one Unit ID, this *Base UnitID* is the first in a sequential set. **Note that the act of writing the Command register causes the *Base UnitID* field to be updated and the *Master Host* bit to be set (indicating the device link which points towards the host bridge).** Thereafter, the device uses its new UnitID when claiming configuration cycles, etc. Only a reset or rewriting the Slave Command register causes the *Base UnitID* field to change.

8. Once all functions in the first device are configured, the host repeats the process to access the next device in the chain. It again uses the configuration cycle attributes of *Bus 0, Device 0 (UnitID 0), Function 0*. Now, the device which is already assigned as UnitID 1, forwards the transaction downstream because the UnitID in the request (0) does not match. The second device is then programmed as the first one was, but the UnitID(s) assigned to it start where the previous device left off (i.e., UnitID 2).

9. After programming each device, the host bridge checks the *End-Of-Chain* (*EOC*) bit set in the device's downstream Link Control Register. If this bit is set = 1, the enumeration process for the chain is complete.

Case 2: A HyperTransport Bridge Is Discovered

If the enumeration process on a chain encounters a HyperTransport-To-Hyper-Transport bridge or a bridge from HyperTransport to a compatible protocol (PCI, AGP, PCI-X), then some additional initialization is needed. A bridge is detected when a read of the *Header Type* field in the configuration header indicates that the device uses the type 1 header format. Software must program the device in accordance with the type 1 header format which includes:

1. Programming the secondary and subordinate bus number registers with the next available bus number (1). This will allow this bridge to forward and/or convert subsequent configuration cycles targeting the new bus(ses) below the bridge.
2. Setting up the Base Address Registers and other fields in the configuration header in accordance with the protocol being used on the secondary bus (HyperTransport, PCI-X, PCI, etc.).

It is permissible for a HyperTransport bridge to have more than one secondary bus and/or a tunnel interface for its primary bus. Refer to Chapter 16, entitled "HyperTransport Bridges," on page 407 for a description of type 1 bridge configuration.

A Note About Bus Numbering In HyperTransport. Bus numbering in HyperTransport systems makes no distinction between HyperTransport, PCI, AGP, or PCI-X buses. As bridges to other protocols are discovered during enumeration, bus numbers are assigned without regard to the particular protocol. This is illustrated in Figure 13-5 on page 319.

Figure 13-5: Bus Numbering In A Mixed Topology

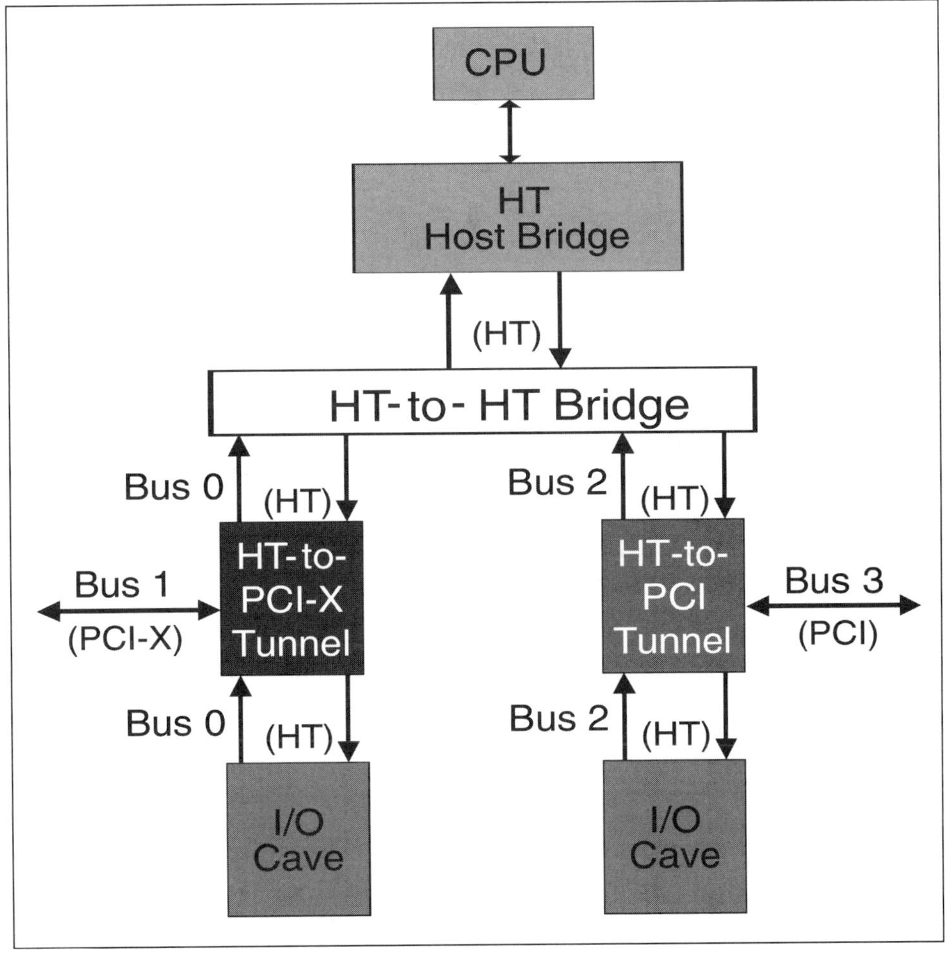

Case 3: Initializing A Double Hosted Chain

The HyperTransport I/O Link Specification makes special provision for enumeration in double-hosted chain topologies. This case does not really have an analogy in PCI or PCI-X systems. Refer to Figure 13-6 on page 320 as the initialization of a double hosted chain is described.

Figure 13-6: Bus Numbering In Double-Hosted Chains

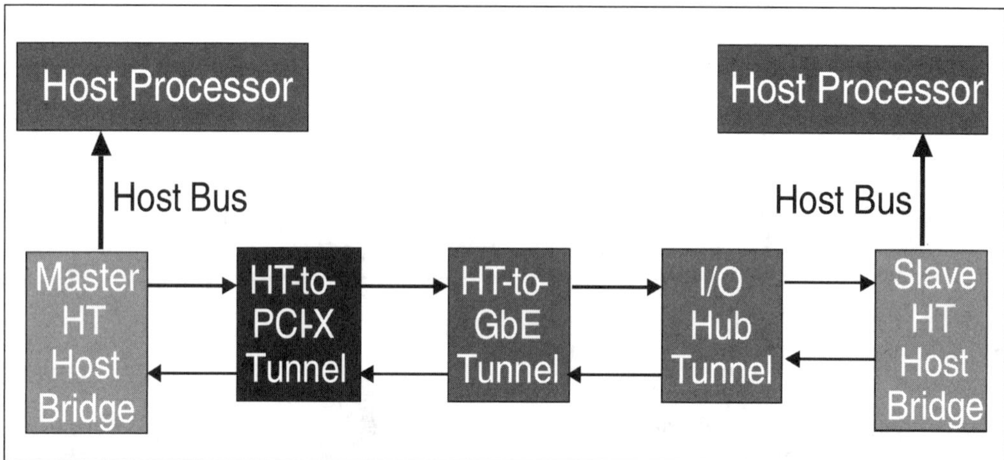

Only One Master Host Bridge. In a double-hosted chain, one host bridge is considered the master, the other is a slave. The HyperTransport I/O Link Specification does not define the method (board strapping, etc.) by which the two bridges determine their respective roles. The specification does say that the master-slave determination must be made before reset. In Figure 13-6, the master host bridge is shown at the left; the slave host bridge is at the right.

Master Bridge Initialization/Configuration Sequence. Assuming the master bridge has been designated, then the events which occur through reset and into initialization are similar to the single host bridge enumeration described above, with the following differences:

1. After reset assertion, low-level bus width negotiation, link synchronization, and NOP buffer release packets have been sent, the slave host bridge becomes inactive while the master host bridge performs configuration.
2. The master host bridge initializes its UnitID counter and starts the enumeration procedure described in the previous section. It assigns UnitIDs programs for each device it encounters on the chain, and reserves UnitID 0.
3. The slave host bridge will be discovered at the end of the chain by the mas-

ter host bridge. This occurs when the master host reads the Command register in the slave host bridge and detects the *Type* field (upper three bits in this register) set = 001b — indicating the presence of a Host Command capability block.

4. The master host bridge then programs the *Double-Ended* bit in both its own Command register and the Command register in the slave host bridge.

When the slave host bridge "wakes up", it will detect the *Double-Ended* bit set in its Command register and not initiate any configuration cycles of its own. All of the internal devices in the chain will have their Master Host bits set to point towards the master host bridge; this means that all requests they will issue will move "upstream" towards the master host bridge.

Refer to Chapter 16, entitled "HyperTransport Bridges," on page 407 for a description of double-hosted chain topologies, including the method used by software to break sharing chains into non-sharing chains.

HyperTransport Configuration Space Format

This section describes the general format of the configuration space used by a HyperTransport functional device. The discussion here focuses on two major areas:

- How a HyperTransport device is similar and different from a PCI device in its use of the generic *header* region of configuration space.
- The use of the required and optional *HyperTransport advanced capability register blocks* also located in the required 256 byte configuration space.

Two Header Formats Are Used

The first one fourth (16 dwords) of any PCI configuration space is called the *header*. As in the case of PCI devices, HyperTransport devices use two header formats: one for HT-to-HT bridges, called *header type 1*, and the other for all non-bridge devices (including tunnels and single link end (cave) devices) called *header type 0*. The lower bits in the *Header Type* field within both types of PCI configuration header is hard coded with the type code; software checks this field early in the process of device discovery to determine which of the header formats it is dealing with.

This chapter covers the format of HyperTransport header type 0 (non-bridges); refer to "HyperTransport Bridge Header Fields" on page 409 for a description of type 1 (HT-to-HT bridge) headers.

The Type 0 Header Format

Figure 13-7 on page 322 depicts the format of the type 0 (non HT-to-HT bridge) header. Basic PCI functionality is managed by having BIOS or other low level software read certain hard-coded header fields to obtain device requirements, then having it program other fields to set up plug-and-play options.

Figure 13-7: PCI Type 0 Configuration Space Header

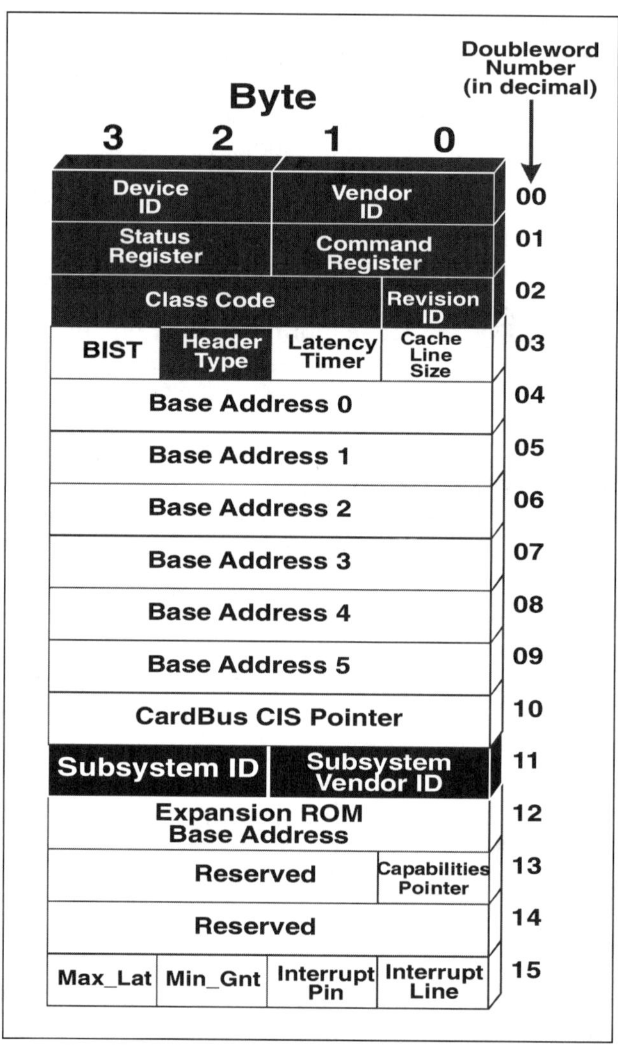

Again, note that the well-defined PCI configuration space header only occupies the first one-fourth (16 dwords) of configuration space.

PCI Advanced Capability Registers

While many early PCI devices were managed using just the register fields in the configuration space header, many additional features have been added over the years which require dedicated registers to manage them. For these devices which have capabilities beyond basic PCI compliance, the generic PCI header registers are augmented by one or more additional register sets outside of the header area, but still within the 256 byte PCI configuration space. PCI calls these *advanced capability* register blocks; refer to Figure 13-8 on page 323.

Figure 13-8: PCI Configuration Space With Advanced Capability Register Block(s)

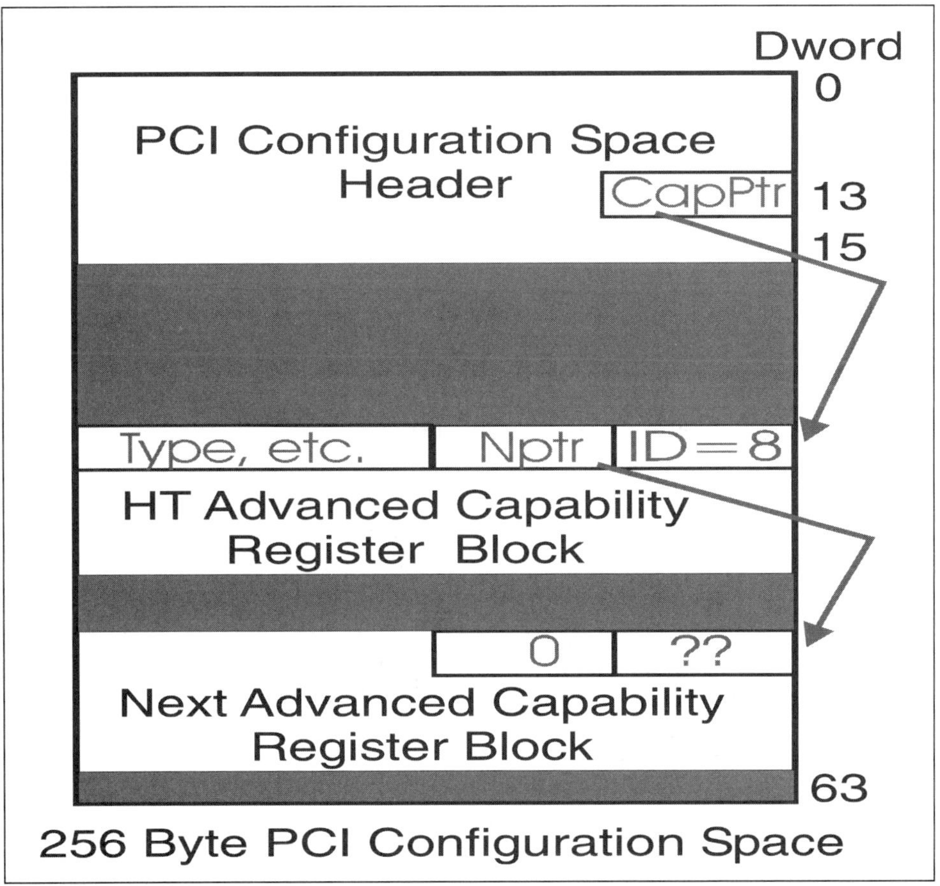

Many Advanced Capabilities Are Defined

Under the current PCI specification, advanced capability block register sets have been defined for all sorts of purposes. Two important classes are:

- Register sets for bus extensions such as HyperTransport, PCI-X, and AGP.
- Register sets for enhanced device management, including Message Signalled Interrupts (MSI), Power Management, Vital Product Data, etc.

When a PCI compatible device is designed, the basic PCI configuration space type 0 or type 1 header fields are implemented as are any additional advanced capability register blocks which may be needed. The format of an advanced capability block varies with the type, and a *Capability ID* byte at the start of each block identifies which type it is; the capability ID for HyperTransport is 08. At a minimum, a HyperTransport device must implement the 256 byte PCI configuration space memory, containing a header AND at least one HyperTransport advanced capability block (*Host/Secondary* or *Slave/Primary* Interface Block).

Discovering The Advanced Capability Blocks

If a PCI compatible device implements advanced capability blocks, low-level software must find and configure each one. Because the specific location of advanced capability blocks within the 256 byte configuration space is not specified, they must be "discovered" by executing some variation of the following software configuration process (Refer to Figure 13-8 on page 323):

1. Use the capability pointer (*CapPtr*) at dword 13 in the header to determine the configuration space offset (from the beginning of configuration space) to the first advanced capability register block. Check the first byte in the block to determine the capability *ID* (HT = 08).
2. Next, check the upper byte in the first dword to determine the HyperTransport capability block *Type*. HyperTransport supports a number of these: Host/Secondary, Slave/Primary, Interrupt Discovery & Configuration, etc.
3. Set up all of the registers in the capability block using configuration cycles.
4. Use the next pointer (*NPtr*) contained in the second byte of the first advanced capability block to determine the offset (from the beginning of configuration space) to the next capability block in the "linked list". If the ID field is "08", this is another HyperTransport capability block. Read the *Type* field, and set up the register fields as appropriate.
5. Continue the discovery and set up process until the last capability block has been located and set up. If a block is the last one, its *Nptr* field is zero — indicating the end of the linked list of advanced capability blocks.

Refer to MindShare's *PCI System Architecture, 4th Ed.* book for a complete description of configuration space advanced capability management.

HyperTransport Configuration Type 0 Header Fields

In this section, the configuration header format for non-bridge HyperTransport devices (type 0 header format) is described. For the most part, HyperTransport devices use these fields in the same way as PCI devices; the few differences are described here. Header fields not mentioned are used in the same way as in PCI devices. Refer to "HyperTransport Bridge Header Fields" on page 409 for a description of HyperTransport bridge headers.

Header Command Register

As shown in Figure 13-9, the command register occupies the lower 16 bits at dword 01. The header Command register is used by BIOS or other software to enable basic capabilities of the device <u>on the primary bus</u>, including bus mastering, target address decoding, error response capability, etc. Refer to Table 13-1 for bit definitions; bits marked "0" are tied low.

Figure 13-9: HyperTransport Technology Header Command Register Usage

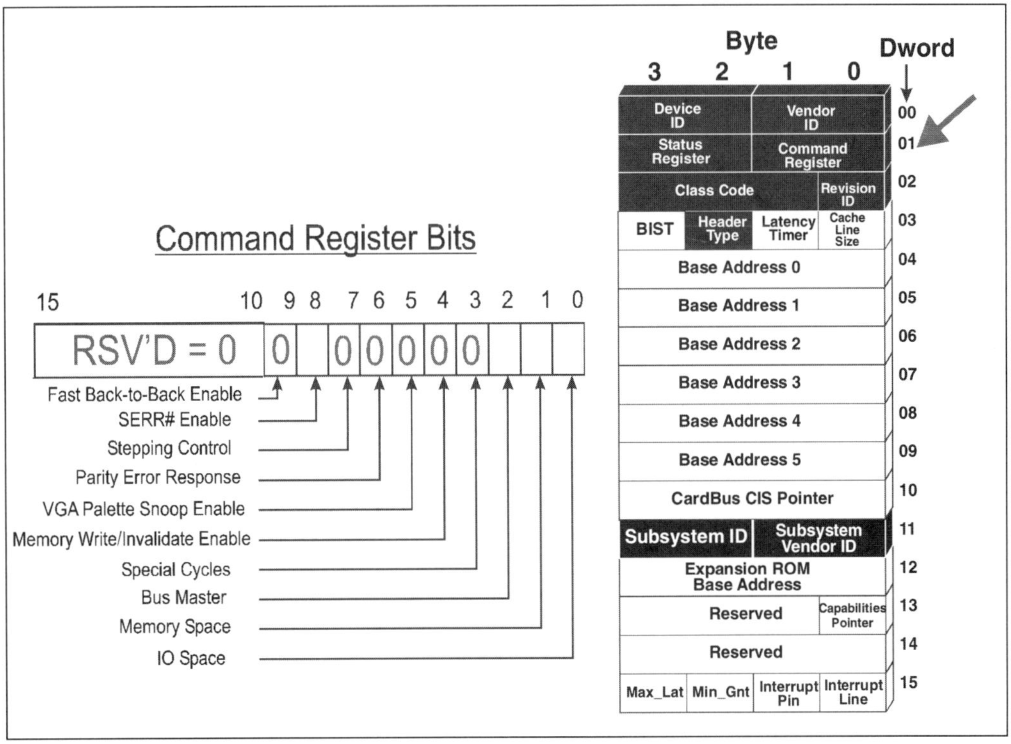

Table 13-1: HyperTransport Header Command Register Bit Assignment

Bit	Function
0	**I/O Space**. When this bit is set to a one, the device may act as a target of requests in the **I/O** portion of the HyperTransport memory map. If this is a subtractive decoder, the setting of this bit does not affect device's ability to claim requests with *Compat* bit set. Warm Reset to 0.
1	**Memory Space**. When this bit is set to a one, the device may act as a target of requests in the **Memory** portion of the HyperTransport memory map. If this is a subtractive decoder, the setting of this bit does not affect device's ability to claim requests with *Compat* bit set. Warm Reset to 0.
2	**Bus Master**. When set to a one, enables the device to issue memory or I/O (range) requests onto its HyperTransport chain. State of bit does not affect ability to forward requests on behalf of other devices. Warm Reset to 0.
8	**SERR# Enable**. When set a one, the device will flood all outgoing links with sync packets in the event of an error which has been programmed to cause a sync flood. If this bit is clear, device only generates sync packets as a part of initial link synchronization. Bit does not affect device's ability to propagate sync packets from one link to another in a chain. *Note: sync flood is similiar to SERR# assertion in PCI.* Warm Reset to 0.
Other Bits	All other Header Command Register bits are unused in HyperTransport, and should be tied to a low level.

Header Status Register

As shown in Figure 13-10 on page 327, the status register occupies the upper 16 bits at dword 01. The HyperTransport device header Status register is used by BIOS or other software to read basic capabilities and dynamic status (e.g. PCI error events) associated with the device's <u>primary bus</u>. As in the case of the header Command register, a number of these bits are used differently than PCI, or not at all. Refer to Table 13-2 on page 328 for bit definitions.

Figure 13-10: HyperTransport Technology Header Status Register Usage

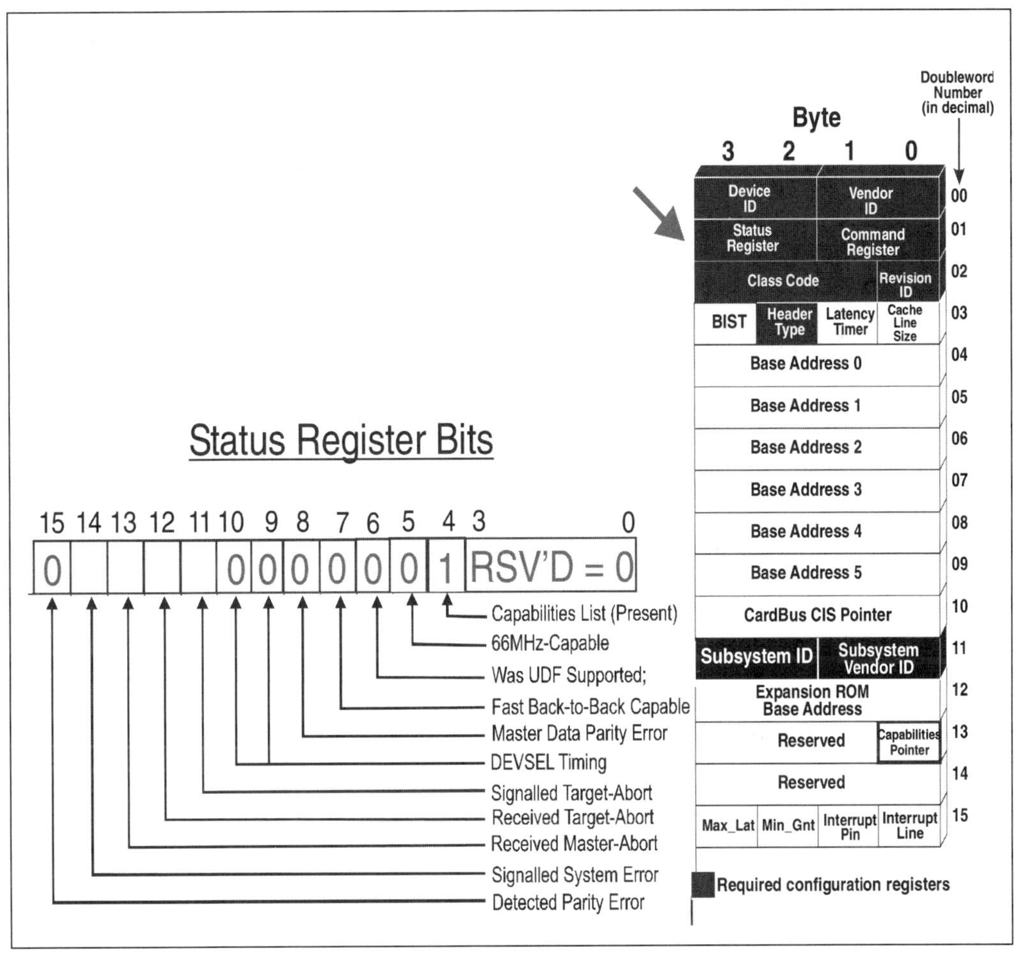

Table 13-2: HyperTransport Header Status Register Bit Assignment

Bit	Function
4	**Capabilities List**. (Introduced in the PCI 2.2 Specification) When this bit is hardcoded to a one, the device indicates there are one or more advanced capability block register sets to be discovered. All HyperTransport devices are required to have at least one advanced capability block (to support HyperTransport), so this bit is always set = 1. When software reads this bit high, it indicates that the Capabilities Pointer (at dword 13) is valid, and points to the first capability register set.
11	**Signalled Target Abort**. This status bit indicates that a HyperTransport target has aborted and has signalled a non-NXA error response to the requestor. Cold Reset = 0.
12	**Received Target Abort**. This status bit indicates that a HyperTransport requestor has received a non-NXA error response to a request it issued. Cold Reset = 0.
13	**Received Master Abort**. This status bit indicates that a HyperTransport requestor has received an NXA (non-existent address) error to a request it has issued. The NXA error response is returned by the end-of-chain device acting on behalf of the intended target. This is equivalent to the Master abort event on PCI. Cold Reset = 0.
14	**Signalled System Error**. This status bit indicates that a HyperTransport device has flooded the link with Sync packets (indicating a serious error). A device merely forwarding Sync packets from another device should not set this bit; this allows identification of the offending device(s). A link reset is required before accessing the device. Cold Reset = 0.
Other Bits	All other header Status Register bits are unused in HyperTransport, and should be tied to a low level.

Other Fields In The Header

The use of other fields in the type 0 header region of a HyperTransport device include:

Cache Line Size Register. (Offset 0Ch)

This read-only register is not implemented by HyperTransport devices. Should return 0's if read by software.

Latency Timer Register. (Offset 0Dh)

This register is not implemented by HyperTransport devices. Should return 0's if read by software.

Base Address Registers. (Offset 10h-24h)

The six Base Address Registers (BARs) are used in much the same way as for PCI devices, with the following limits:

I/O BAR. For an I/O request, a single BAR is implemented. Only the lower 25 bits of the value programmed into the BAR is used for address comparison by the target, and the upper bits of the BAR should be written to zeros by system software. Any I/O request packet sent out on a link should have the start address bits 39-25 programmed for the I/O range in the HyperTransport memory map.

Memory BAR. A request for memory using 32-bit addressing can be accomplished using a single BAR, just as in PCI. This would limit the assigned target start address for the device to the lower 4GB of the 1 TB (40 bit) HyperTransport address map.

Optionally, a HyperTransport device may support 64 bit address decoding, and use a pair of BARs to support it. If this is done, only the lower 40 bits of the 64 bit BAR memory address will be valid, and the upper bits are assumed to be zeros.

Memory windows for HyperTransport devices are always assigned in BARs on 64-byte boundaries; this assures that even the largest transfer (16 dwords/64 bytes) will never cross a device address boundary. This is important because HyperTransport does not support a disconnect mechanism (such as PCI uses) to force early transaction termination.

CardBus CIS Pointer. (Offset 28h)

This register is not implemented by HyperTransport devices. Should return 0's if read by software.

Capabilities Pointer. (Offset 34h)

This field contains a pointer to the first advanced capability block. Because all HyperTransport devices have at least one advanced capability, this register is always implemented. The pointer is an absolute byte offset from the beginning of configuration space to the first byte of the first advanced capability register block.

Interrupt Line Register. (Offset 3Ch)

The HyperTransport Specification indicates that this register should be read-writable and may be used as a software scratch pad. The information routing information programmed into this register in PCI devices isn't required in HyperTransport because interrupt messages are sent over the links and side-band interrupts are not defined.

Interrupt Pin Register. (Offset 3Dh)

This register is reserved in the HyperTransport Specification. It may optionally be implemented for compatibility with software which may expect to gather interrupt pin information from all PCI-compatible devices.

Min_Gnt and Max_Latency Registers. (Offsets 3Eh and 3Fh)

These register fields are associated with PCI shared-bus arbitration, and are not implemented by HyperTransport devices. Should return 0's if read by software.

HyperTransport Uses Advanced Capability Blocks

The generic fields in the configuration space header region of a HyperTransport device are augmented by at least one HyperTransport-specific advanced capability register block. Software discovery of capability blocks was described in an earlier section. Figure 13-11 on page 331 illustrates the format of the advanced capability block register.

Figure 13-11: HyperTransport Advanced Capability Block Types

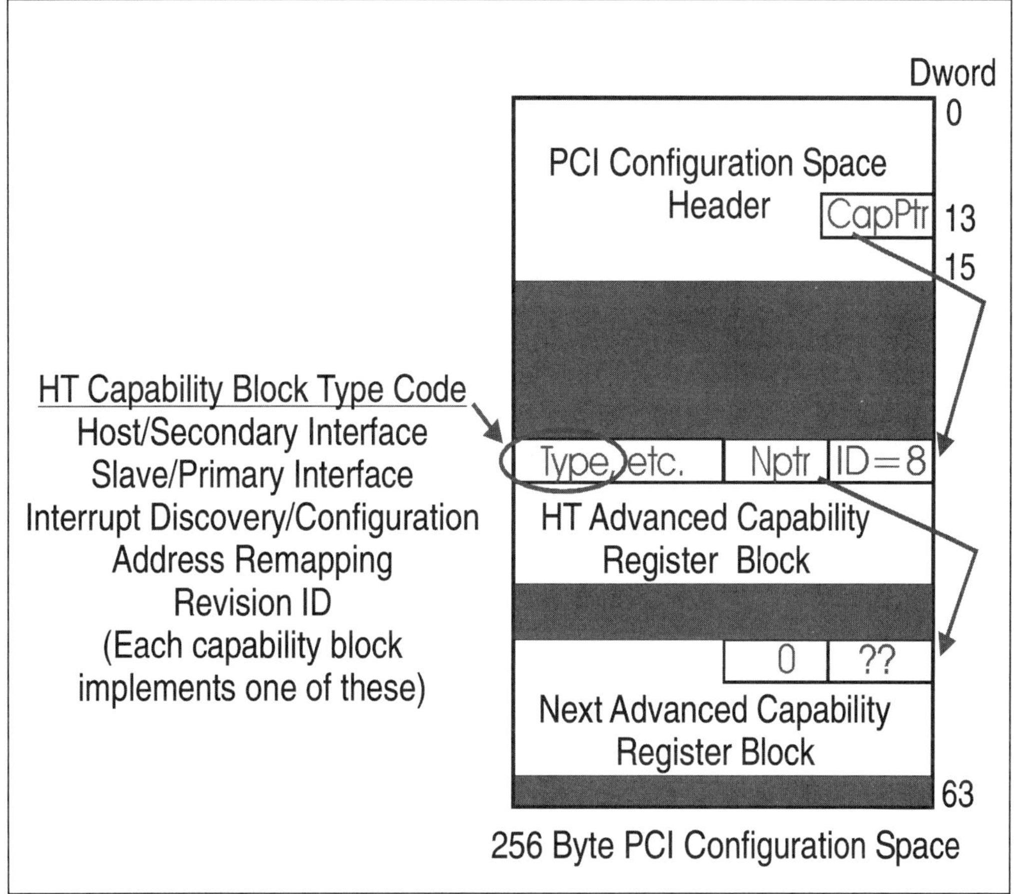

HyperTransport Block Types Currently Defined

HyperTransport is allocated a number of advanced capability blocks for various purposes. All have the same Capability ID code (08 for HyperTransport) in the first byte of the block, but carry unique *Type* field codes. Table 13-3 on page 332 summarizes the HyperTransPort advanced capabilities and their type codes.

Table 13-3: HyperTransport Advanced Capability Codes

Code	Capability Type
000xx	Slave/Primary Interface
001xx	Host/Secondary Interface
01000	Reserved (Switch)
01001-01111	Reserved
10000	Interrupt Discovery and Configuration
10001	Revision ID
10010	Reserved (UnitID Clumping)
10011	Reserved (Extended Configuration Space Access)
10100	Address Remapping
10101-10111	Reserved
11000	Reserved (Retry Mode)
11001-11111	Reserved

Block Formats Vary With Capability And Device Type

Each of the HyperTransport capability blocks has its own format. The *Type* field in the first dword of each capability block defines the format of the entire block. In addition, one of the principal capability block types (Slave/Primary Interface) also varies with the device which implements it because tunnel devices interface to two links and end (cave) devices interface to only one.

The Slave/Primary Interface Block

HyperTransport defines two principal advanced capability register block formats, Slave/Primary and Host/Secondary, which reflect the two possible roles a device interface can perform on a link. The Slave/Primary format is used by all tunnels and single-link peripheral (cave) devices. These devices never act as a host for a bus (they are slaves). In addition, because they are not bridges, they have a single primary interface to the bus and no secondary interfaces.

One complicating factor is the fact that while an end (cave) device interfaces to only one link, a tunnel must interface to two links (still only one bus, though). To accommodate this difference, each Slave/Primary interface has two sets of link management registers, one for each link. A tunnel device implements one Slave/Primary interface and both sets of link management registers; an end (cave) device also implements one Slave/Primary interface but only one set of link management registers. Figure 13-12 illustrates these interfaces.

Figure 13-12: Slave/Primary Interface For Tunnel And Cave Devices

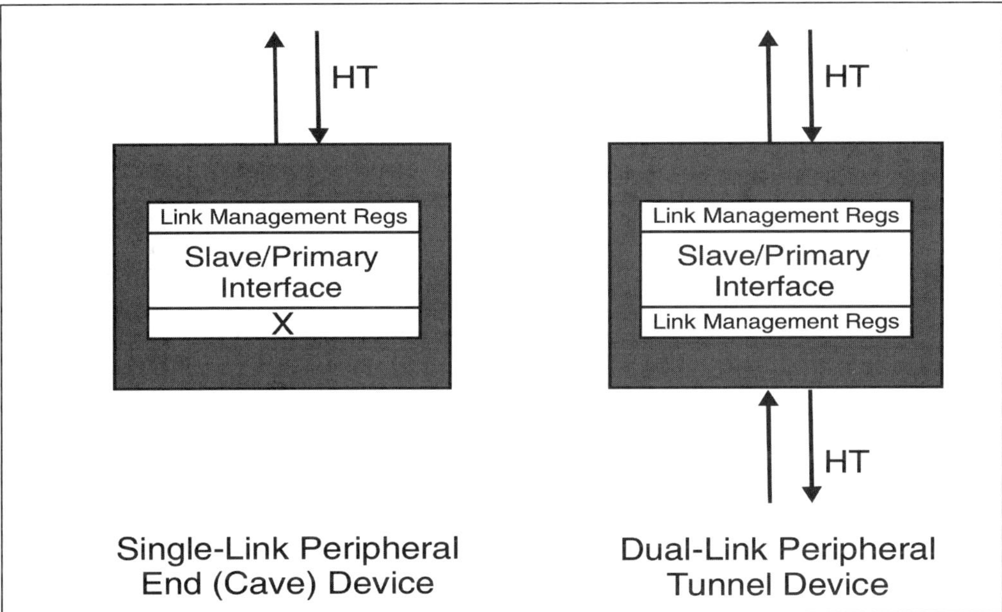

Description Of Slave/Primary Interface Fields

The following section describes the use of each field in the Slave/Primary advanced capability interface block See Figure 13-13 on page 334. Most fields are the same as those in the Host/Secondary interface block. Note that the upper part of the first dword in any HyperTransport capability register contains the *Type* field. In this block type (and the Host/Secondary block that follows), only the upper three bits are used; the code "000" in these three bits indicate this is a Slave/Primary interface block. Unimplemented bits in this register block must be tied low.

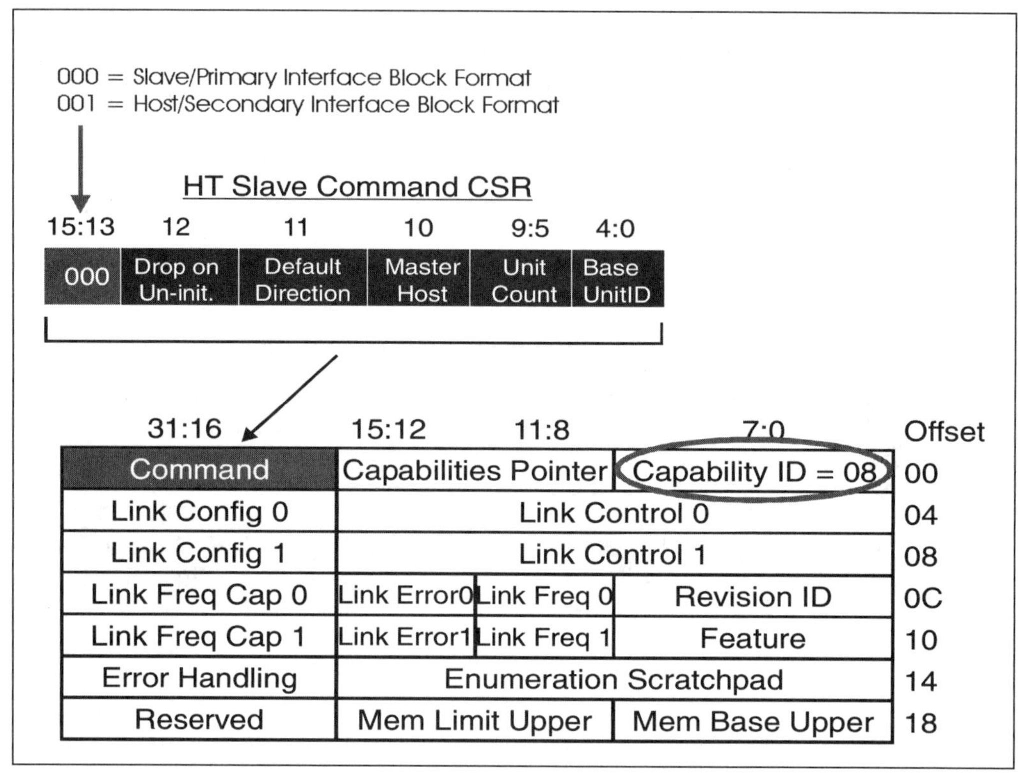

Figure 13-13: Slave/Primary Interface Block Format

Capability ID Register. (Offset 00h)

This read-only register is hard-wired to 08h for all HyperTransport advanced capability register blocks. This value is assigned by the PCISIG, and distinguishes HyperTransport advanced capability blocks from other types.

Capabilities Pointer Register. (Offset 01h)

This read-only register provides a byte offset (from the beginning of configuration space) to the next advanced capability block. If read as zero, there are no more advanced capability blocks to process.

Slave Command Register. (Offset 02h-03h)

This mixed read-only and read-write register contains fields used to set up the interface. The register is organized as shown in Figure 13-14 on page 335. Definition of the fields can be found in Table 13-4.

Figure 13-14: HyperTransport Slave Command CSR

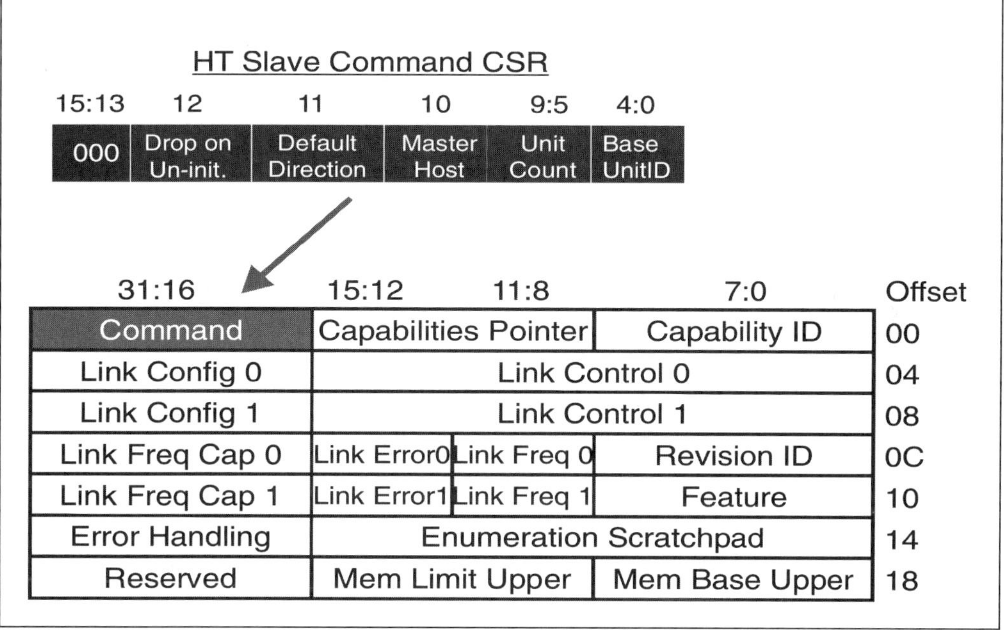

Table 13-4: Slave Interface Block Command Register Bit Assignment

Bit	Function
4:0	**Base UnitID**. Read-Write field programmed with the lowest numbered UnitID belonging to this device. Range: 0-1Fh. If multiple UnitIDs are requested (see next field), this value is the lowest numbered (base) UnitID. Warm Reset = 0.
9:5	**Unit Count**. Read only field indicating the number of UnitIDs this device requires assigned to it. Range 0-1Fh; total UnitID count per bus = 32.

Table 13-4: Slave Interface Block Command Register Bit Assignment (Continued)

Bit	Function
10	**Master Host**. This read-write bit is set by hardware automatically to indicate which link is attached to the host. Any write to the Command register will cause this bit to be set to indicate which link the write arrived on. Warm Reset = 0.
11	**Default Direction**. This read-write bit determines the default direction requests should be sent when originating with this device. A "0" in this register indicates requests should be sent in the direction of the master host (see previous bit). A "1" indicates requests should be sent in the other direction. Bit has no meaning for a single-link device. Warm Reset = 0.
12	**Drop On Un-init**. This read-write bit sets the policy an uninitialized device should use for packets it would normally issue or forward if interface initialization were complete, but can't because it doesn't know if the interface is at the end of chain. If the uninitialized device has this bit cleared, it will stall until either its end-of-chain bit becomes set or initialization completes. If the uninitialized device has this bit set, it will behave as though its end-of-chain bit is set and reject packets instead of stalling. Cold Reset = 0.
15:13	**Capability Type**. Read only field indicating the type of information in this HyperTransport advanced capability block. Codes currently supported include: 000 = Slave/Primary Interface Block 001 = Host/Secondary Interface Block 100 = Interrupt Discovery And Configuration 101= Address Mapping Others = Reserved

Link Control Registers. (Offset 04h and 08h)

Slave/Primary interfaces implement two copies of this register, one for each link. For single link devices, only one Link Control Register would be active (0); for tunnels, both Link Control Registers would be active. The Slave Interface Link Control Registers are illustrated in Figure 13-15 on page 337, and the fields are described in Table 13-5 on page 337.

Figure 13-15: Slave Interface Block Link Control Registers 0,1

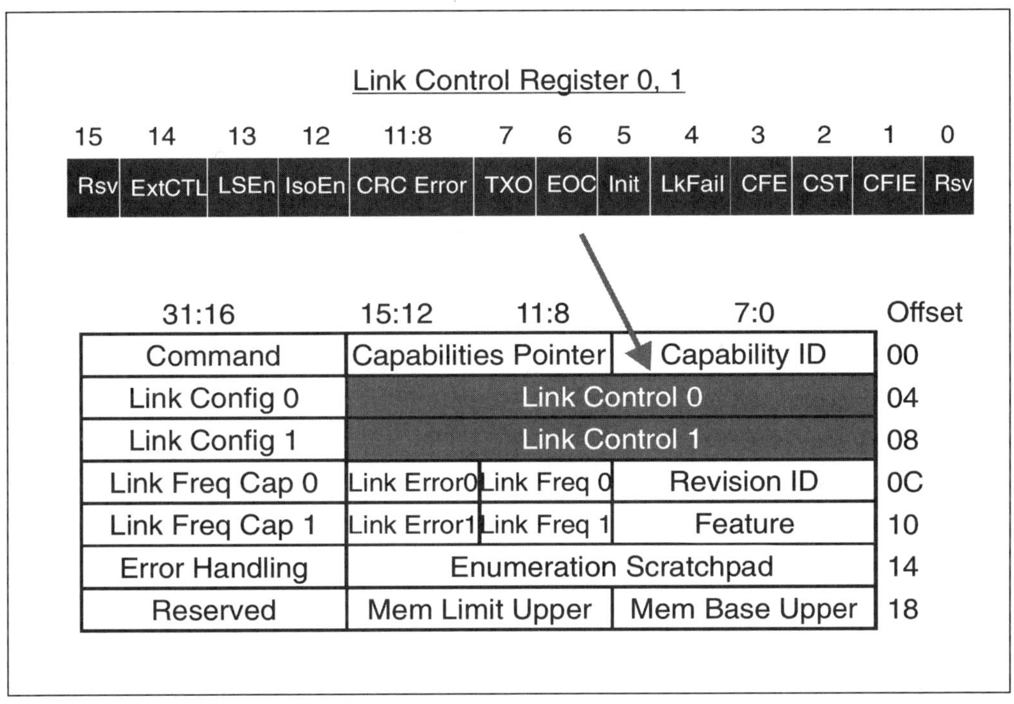

Table 13-5: Slave Interface Block Link Control Register 0,1 Bit Assignment

Bit	Function
1	**CFIE** (CRC Flood Enable). When this read-write bit is set = 1, sync flood will be initiated and the *link fail* bit (bit 4) will be set by this device whenever any CRC error bits for this link are asserted. Note that this bit has no effect on CRC error checking logic, which always is active for the entire bus width set up in the LinkWidthIn field of the corresponding Link Configuration Register. Warm Reset = 0.
2	**CST** (CRC Start Test). When this optional read-write bit is set = 1, device hardware initiates the CRC test sequence across the link. When the test is complete, hardware will clear this bit (which software can poll for completion). This feature should not be used unless the *CRC Test Mode Capability* bit has been checked for the device on the other end of the link.

Table 13-5: Slave Interface Block Link Control Register 0,1 Bit Assignment (Continued)

Bit	Function
3	**CFE** (CRC Force Error) This read-write bit is set by hardware to force bad CRC values on all data lanes enabled in the LinkWidthOut field of the corresponding Link Configuration Register. Data is not affected. Warm Reset = 0.
4	**LkFail (Link Failure)**. This read only bit is set by hardware if a sync flood resulting from bad CRC, protocol violation, or flow control buffer overflow has been generated. It is only set by device which detected the error, not those that forward resulting sync packets. Cold Reset = 0.
5	**Init (Initialization Complete)**. This read only bit is cleared by reset and set by hardware when the low-level link initialization sequence for both the transmit and receive interfaces on a link have completed successfully. If the interface is at the end of chain, the bit will not be set. Warm Reset = 0.
6	**EOC (End Of Chain)**. This bit is set to indicate that the link is not part of a chain; packets moving toward this link are either dropped or rejected with an NXA error. CRC on the link is not checked and sync flood coming from this link is ignored. This bit may be set by hardware (detecting all inputs low at reset) or by software seeking to partition a double-hosted chain. Bit can only be cleared by reset. Single-link slave devices hard-wire this bit =1 on the unused link. Warm reset = 0.
7	**TXO** (Transmitter Off). This Read/Set bit is used to disable the transmitter drivers on a link for power savings or EMI reduction. When set = 1, device outputs are steady-state, and the corresponding receiver at the other end of the link may also have to be disabled. Reset clears this bit; once set by software, it can only be cleared by a subsequent reset. Warm reset = 0.
11:8	**CRC Error.** These four bits are used to log CRC errors on a per-byte basis. bit 8 logs errors for CAD bits 0-7; bit 9 for CAD bits 8-15; bit 10 for CAD bits 16-23; bit 10 for CAD bits 24-31. Bits for unimplemented byte lanes or lanes disabled in *Max Link Width In* field of Link Configuration Register should return 0's when read. Cold Reset = 0.

Table 13-5: Slave Interface Block Link Control Register 0,1 Bit Assignment (Continued)

Bit	Function
12	**IsoEn** (Isochronous Enable). When this optional read-write bit is set = 1, isochronous mode flow control is enabled. This traffic is generally given the highest possible priority with respect to other virtual channel traffic in HyperTransport. Isochronous support is optional, and this bit should not be set unless the device on the other end of the link has been checked for isochronous support as well. Any device that detects isochronous packets but doesn't support them, simply handles them in the standard posted, non-posted, and response virtual channels. Cold reset = 0.
13	**LSEn** (LDTSTOP# Tristate Enable). When this read-write bit is set = 1, the link transmitter tristates the link (CAD, CTL, CLK) during the disconnected state of an LDTSTOP# sequence. If the bit is clear, the transmitter continues driving its outputs during these events. Cold reset = 0.
14	**ExtCTL** (Extended CTL Assertion Time). When this read-write bit is set = 1, during a link initialization sequence, CTL will be asserted for 50uS after both are true: the transmitter has asserted its CTL and it also has sampled the CTL from the other side of the link as well. If the bit is clear, both CTL assertions only have to be true for 16 bit times for 8-32 bit links, 32 bit times for 4 bit links, or 64 bit times for two bit links. The extended time is for the benefit of receivers using DLL's, and requiring additional time to lock to the input clock after a LDTSTOP# sequence. Cold reset = 0.

Link Configuration Registers. (Offset 06h and 0Ah)

Slave/Primary interfaces also implement two copies of this register, one for each link. For single link devices only one Link Configuration Register would be active; for tunnels, both Link Configuration Registers would be active. The Slave Interface Link Configuration Register is illustrated in Figure 13-16 on page 340. The field definitions can be found in Table 13-6 on page 340.

Figure 13-16: Slave Interface Link Configuration Registers 0,1

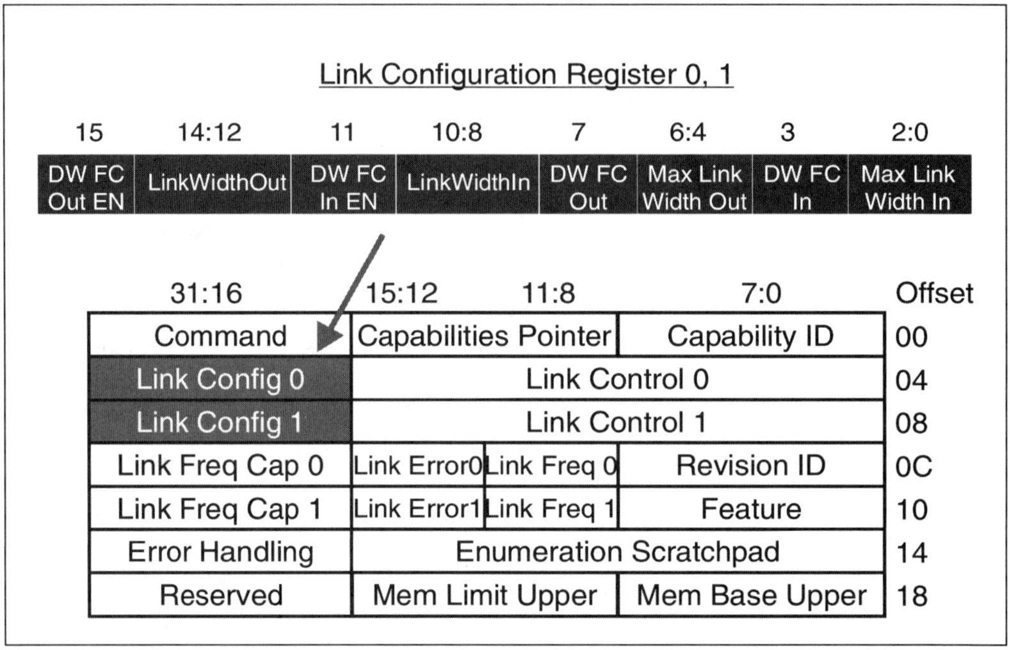

Link Configuration Register 0, 1

15	14:12	11	10:8	7	6:4	3	2:0
DW FC Out EN	LinkWidthOut	DW FC In EN	LinkWidthIn	DW FC Out	Max Link Width Out	DW FC In	Max Link Width In

31:16	15:12	11:8	7:0	Offset
Command	Capabilities Pointer		Capability ID	00
Link Config 0	Link Control 0			04
Link Config 1	Link Control 1			08
Link Freq Cap 0	Link Error0	Link Freq 0	Revision ID	0C
Link Freq Cap 1	Link Error1	Link Freq 1	Feature	10
Error Handling	Enumeration Scratchpad			14
Reserved	Mem Limit Upper		Mem Base Upper	18

Table 13-6: Slave Interface Block Link Configuration Register 0,1 Bit Assignment

Bit	Function
2:0	**Max Link Width In**. This read only register is hard coded to indicate the maximum width of the receive side of this link. HyperTransportv supports CAD bus widths of 2-32 bits in each direction. The 3 bits in this field are coded as follows: (all other codes are reserved) 000 = 8 bits wide 001 = 16 bits wide 011 = 32 bits wide 100 = 2 bits wide 101 = 4 bits wide 111 = Link Not Connected
3	**DW FC In** (Doubleword Flow Control In Capability). If this read-only bit is set = 1, the receiver is capable of double-word based flow control.

Table 13-6: Slave Interface Block Link Configuration Register 0,1 Bit Assignment (Continued)

Bit	Function
6:4	**Max Link Width Out**. This read only register is hard coded to indicate the maximum width of the transmit side of this link. HyperTransportv supports CAD bus widths of 2-32 bits in each direction. The 3 bits in this field are coded in the same way as the receive side: (all other codes are reserved) 000 = 8 bits wide 001 = 16 bits wide 011 = 32 bits wide 100 = 2 bits wide 101 = 4 bits wide 111 = Link Not Connected
7	**Dw Fc Out** (Doubleword Flow Control Out Enable). If this read-only bit is set = 1, the transmitter is capable of double-word based flow control.
10:8	**Link Width In**. This read-write field is written by software to establish the actual width to be utilized for the receive side of this link. After a cold reset, low-level device negotiation establishes the starting input width (maximum 8 bits) and presets the code in this register to reflect it. After software checks the link width capabilities of both devices, it may rewrite the value in this register to use a different width. The 3 bits in this field are coded as follows: (all other codes are reserved) 000 = 8 bits wide 001 = 16 bits wide 011 = 32 bits wide 100 = 2 bits wide 101 = 4 bits wide 111 = Link Not Connected Cold reset = 0.
11	**DW FC In EN** (Doubleword Flow Control In Enable). This read-write bit may be written = 1 to program the receiver to use double-word based flow control. Software should not program this capability before checking that devices on both sides of the link are capable of using this optional mode (see bit 3 and bit 7). Cold reset = 0.

Table 13-6: Slave Interface Block Link Configuration Register 0,1 Bit Assignment (Continued)

Bit	Function
14:12	**Link Width Out**. This read-write field is written by software to establish the actual width to be utilized for the transmit side of this link. After a cold reset, low-level device negotiation establishes the starting transmit width (maximum 8 bits) and presets the code in this register to reflect it. After software checks the link width capabilities of both devices, it may rewrite the value in this register to use a different transmit width. The 3 bits in this field are coded as follows: (all other codes are reserved) 000 = 8 bits wide 001 = 16 bits wide 011 = 32 bits wide 100 = 2 bits wide 101 = 4 bits wide 111 = Link Not Connected Cold reset = 0. Note: CAD bit width downsized through programming of Link Width Out field will result in unused drivers being shut down.
15	**DW FC Out EN** (Doubleword Flow Control Out Enable). This read-write bit may be written = 1 to program the transmitter to use double-word based flow control. Software should not program this capability before checking that devices on both sides of the link are capable of using this optional mode (see bit 3 and bit 7). Cold reset = 0.

Revision ID Register. (Offset 0Ch)

This 8 bit field indicates the Major and Minor revision of the HyperTransport I/O Link Protocol Specification to which this interface complies (See Figure 13-17). For example, if the interface was compliant with Link Protocol Specification 1.03, then:

Major Rev: 01
Minor Rev: 03

Figure 13-17: Slave Interface Revision ID Register

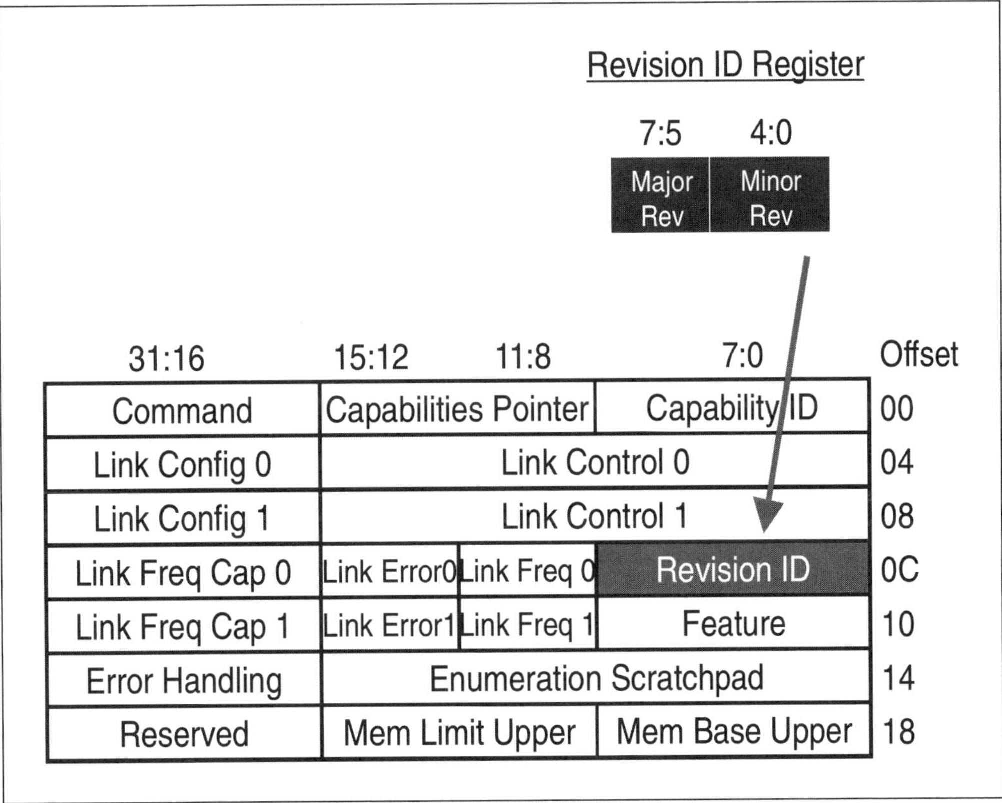

Link Frequency Registers. (Offset 0Dh and 11h)

Slave/Primary interfaces implement two copies of this register, one for each link. Figure 13-18 illustrates the slave interface Link Frequency Register, along with the possible values. For single link devices, only one Link Frequency Register would be active (0); for tunnels, both Link Frequency Registers would be active. The value in this field determines the actual clock frequency to be used on the link. After reset, the default rate of 200MHz is used and the code *000* is in this register. Software may later write a non-zero value to the register after devices on both ends of the link have been checked for clock speed capability (see Link Freq Cap Register). A warm reset or LDTSTOP# sequence is required for frequency change to take effect. Devices must support the 200MHz rate; all others are optional.

Figure 13-18: Slave Interface Link Frequency Registers 0,1

Link Frequency Encoding	Transmitter Clock Frequency (MHz)
0000	200 (default)
0001	300
0010	400
0011	500
0100	600
0101	800
0110	1000
0111 to 1110	Reserved
1111	Vendor Specific

31:16	15:12	11:8	7:0	Offset
Command	Capabilities Pointer		Capability ID	00
Link Config 0	Link Control 0			04
Link Config 1	Link Control 1			08
Link Freq Cap 0	Link Error0	Link Freq 0	Revision ID	0C
Link Freq Cap 1	Link Error1	Link Freq 1	Feature	10
Error Handling	Enumeration Scratchpad			14
Reserved	Mem Limit Upper		Mem Base Upper	18

Link Error Registers. (Offset 0Dh and 11h)

Slave/Primary interfaces implement two copies of this register, one for each link. Figure 13-19 illustrates the slave interface Link Error Register, and Table 13-7 defines the fields. For single link devices, only one Link Error Register would be active; for tunnels, both Link Error Registers would be active.

Figure 13-19: Slave Interface Link Error Registers 0,1

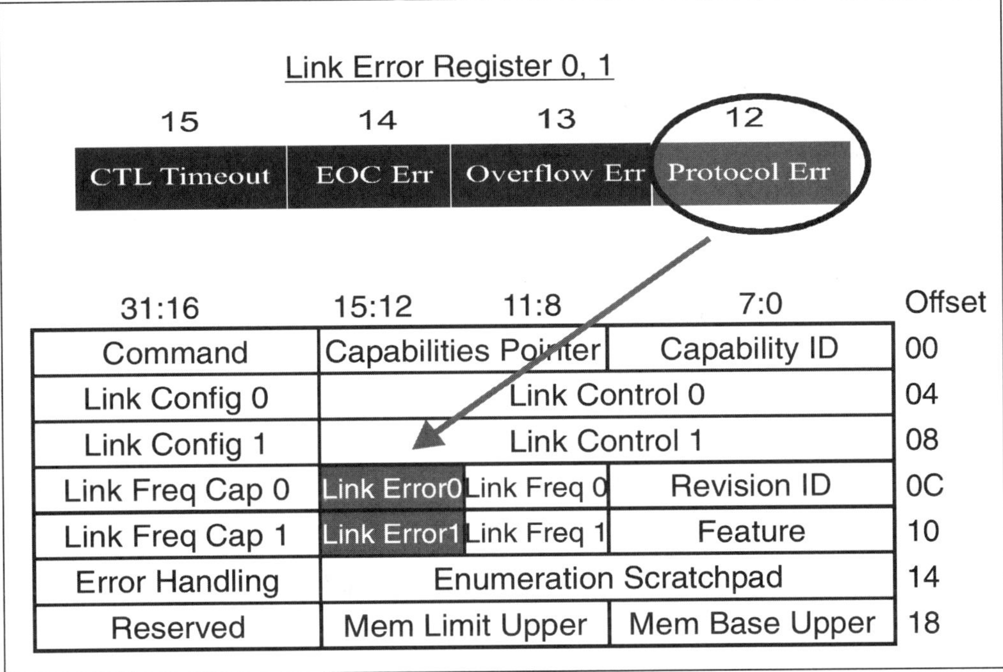

Table 13-7: Slave Interface Link Error Registers 0,1

Bit	Function
4	**Protocol Err**. This bit is set by hardware if a protocol error is detected on this link. Cold reset = 0.
5	**Overflow Err**. This bit is set by hardware if an overflow error is detected on this link. Cold reset = 0.

Table 13-7: Slave Interface Link Error Registers 0,1

Bit	Function
6	**EOC Err.** This bit is set by hardware if a protocol error is detected on this link. Cold reset = 0.
7	**CTL Time out**. This bit is hard coded to indicate the time CTL may be detected low before a protocol error is reported. If bit 7 = 1, maximum low time is 1 second; if bit 7 = 0, maximum low time is 1 millisecond

Link Frequency Capability Registers. (Offset 0Eh and 12h)

Slave/Primary interfaces implement two copies of this register, one for each link. Figure 13-20 illustrates the slave interface Link Frequency Capability Register, and Table 13-8 on page 347 defines the fields. For single link devices, only one Link Frequency Capability Register would be active (0); for tunnels, both Link Frequency Capability Registers would be active. Value hard coded is a mask of all speed capabilities.

Figure 13-20: Slave Interface Link Frequency Capability Registers 0,1

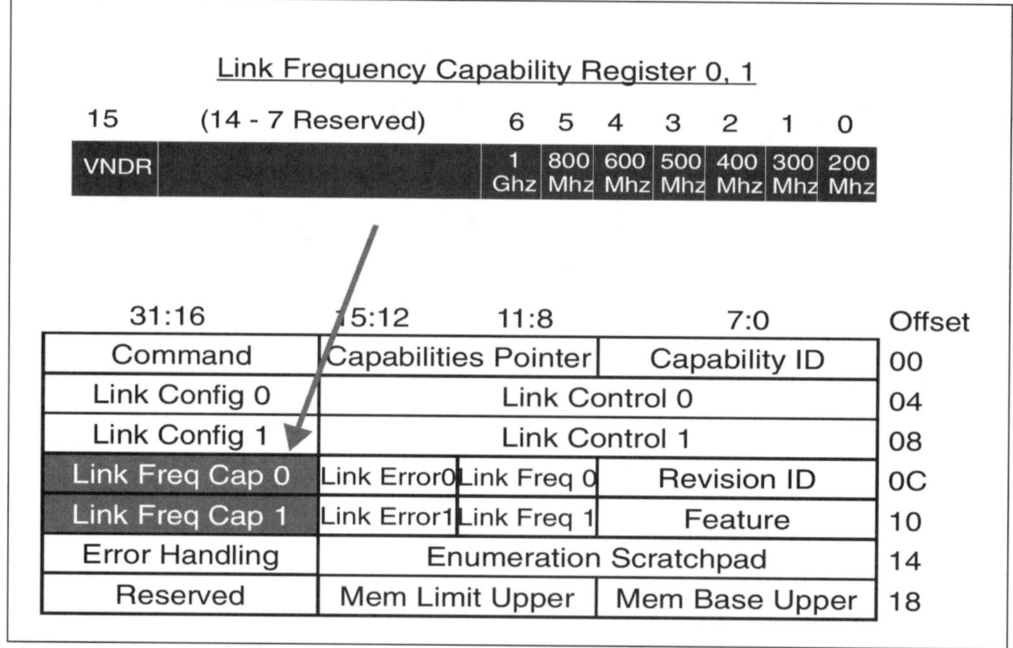

Table 13-8: Slave Interface Link Frequency Capability Registers 0,1

Bit	Function
0	**200 MHz Supported**. This read only bit (if set = 1) indicates device supports 200MHz clock speed for this link. This is the default clock speed and all devices should have this bit set for each active link.
1	**300 MHz Supported**. This read only bit (if set = 1) indicates device supports 300MHz clock speed for this link. Support is optional.
2	**400 MHz Supported**. This read only bit (if set = 1) indicates device supports 400MHz clock speed for this link. Support is optional.
3	**500 MHz Supported**. This read only bit (if set = 1) indicates device supports 500MHz clock speed for this link. Support is optional.
4	**600 MHz Supported**. This read only bit (if set = 1) indicates device supports 600MHz clock speed for this link. Support is optional.
5	**800 MHz Supported**. This read only bit (if set = 1) indicates device supports 800MHz clock speed for this link. Support is optional.
6	**1GHz Supported**. This read only bit (if set = 1) indicates device supports 1GHz clock speed for this link. Support is optional.
14:7	**Reserved**. These bits are currently reserved and unused. Devices should tie them low.
15	**Vendor Specific Speed Support**. If this bit is set, device supports vendor-specific clock frequencies. Usage is beyond scope of HyperTransport Specification.

Feature Capability Register. (Offset 10h)

This register is used to indicate to software which optional features are supported by this device. See Figure 13-21 on page 348. Table 13-9 on page 348 defines each bit field. All unspecified bits are reserved, and should read back as zero.

Figure 13-21: Slave Interface Feature Capability Register

Feature Capability Register

		7		3	2	1	0
				XTND CTL	CRC Test	LDTSTOP Support	ISOC FC

31:16	15:12	11:8	7:0	Offset
Command	Capabilities Pointer		Capability ID	00
Link Config 0	Link Control 0			04
Link Config 1	Link Control 1			08
Link Freq Cap 0	Link Error0	Link Freq 0	Revision ID	0C
Link Freq Cap 1	Link Error1	Link Freq 1	Feature	10
Error Handling	Enumeration Scratchpad			14
Reserved	Mem Limit Upper		Mem Base Upper	18

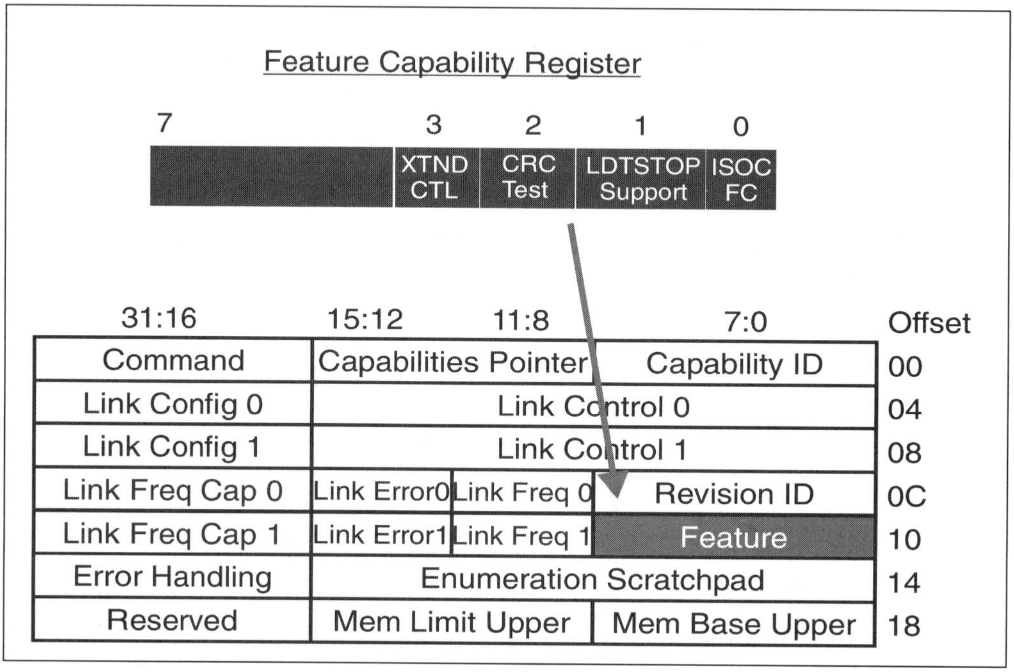

Table 13-9: Slave Interface Feature Capability Register

Bit	Function
0	**Isochronous Mode**. This read only bit (if set = 1) indicates device supports the optional Isochronous flow control mechanism. If this bit is clear, the device will handle packets with *Isoc* bit set as standard virtual channel traffic. Software should check this bit for both devices on a link before enabling isochronous traffic (see bit 12 of Link Control Register).
1	**LDTSTOP#**. This read only bit (if set = 1) indicates device supports the optional LDTSTOP# protocol.
2	**CRC Test**. This read only bit is set = 1 if device supports optional CRC test mode

Table 13-9: Slave Interface Feature Capability Register

Bit	Function
3	**Extended CTL Time Required**. This read only bit indicates whether device requires CTL signal to be asserted for the extended 50uS time following LDTSTOP# disconnect in order to allow time to reacquire lock to input clock. Bit is set = 1 if the extended time is needed; cleared otherwise.

Enumeration Scratch Pad Register. (Offset 14h)

As illustrated in Figure 13-22, this 16 bit register provides a read-writable register for software to use during the enumeration process. Cold reset = 0.

Figure 13-22: Slave Interface Enumeration Scratch Pad Register

31:16	15:12	11:8	7:0	Offset
Command	Capabilities Pointer		Capability ID	00
Link Config 0	Link Control 0			04
Link Config 1	Link Control 1			08
Link Freq Cap 0	Link Error0	Link Freq 0	Revision ID	0C
Link Freq Cap 1	Link Error1	Link Freq 1	Feature	10
Error Handling	Enumeration Scratchpad			14
Reserved	Mem Limit Upper		Mem Base Upper	18

Error Handling Register. (Offset 16h)

This register contains bits that enable the error routing strategy for the device as well as status bits used to log *chain fail* and *response error* events. See Figure 13-23. Table 13-10 defines each bit field. Devices that don't support a particular error condition should hard wire the logging and enable bits for that event to zero.

Figure 13-23: Slave Interface Error Handling Register

Error Handling Register
Position: 7 6 5 4 3 2 1 0
Byte 0
SERR F Enable
CRC F Enable
Resp F Enable
EOC F Enable
OF F Enable
Prot. F Enable
OF Flood Enable
Prot FL EN
Byte 1
SERR NF Enable
CRC NF Enable
Resp NF Enable
EOC NF Enable
OF NF Enable
Prot. NF Enable
Response Error
Chain Fail
31:16 15:12 11:8 7:0 Offset
Command
Capabilities Pointer
Capability ID
00
Link Config 0
Link Control 0
04
Link Config 1
Link Control 1
08
Link Freq Cap 0
Link Error0
Link Freq 0
Revision ID
0C
Link Freq Cap 1
Link Error1
Link Freq 1
Feature
10
Error Handling
Enumeration Scratchpad
14
Reserved
Mem Limit Upper
Mem Base Upper
18

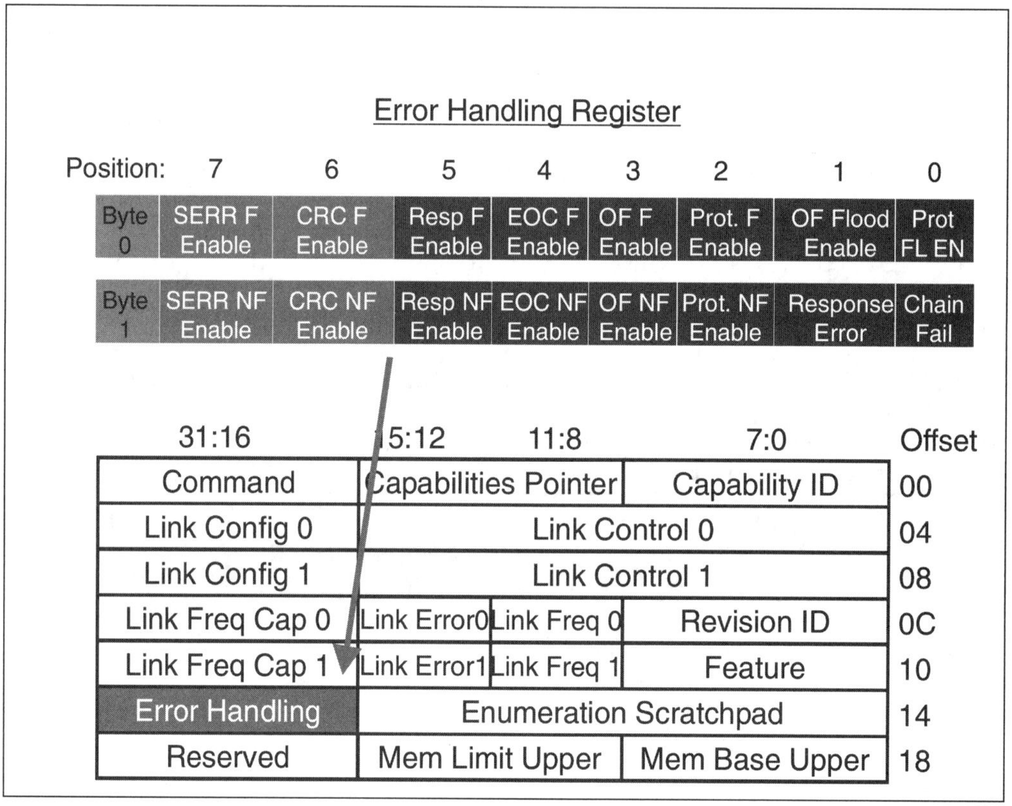

Table 13-10: Slave Interface Error Handling Register

Bit	Function
0	**Protocol Error Flood Enable.** This read-write bit, when set = 1, indicates the device should flood the link with Sync packets anytime a protocol error bit is set in one or more of its Link Error registers. Warm reset = 0.

Table 13-10: Slave Interface Error Handling Register

Bit	Function
1	**Overflow Error Flood Enable**. This read-write bit, when set = 1, indicates the device should flood the link with Sync packets anytime an overflow error bit is set in one or more of its Link Error registers. Warm reset = 0.
2	**Protocol Error Fatal Enable**. This read-write bit, when set = 1, indicates the device should issue a fatal error interrupt in the event a protocol error bit is set in one or more of its Link Error registers. If fatal interrupt not implemented, hard wire this bit = 0. Warm reset = 0.
3	**Overflow Error Fatal Enable**. This read-write bit, when set = 1, indicates the device should issue a fatal error interrupt in the event an overflow error bit is set in one or more of its Link Error registers. If fatal interrupt not implemented, hard wire this bit = 0. Warm reset = 0.
4	**End Of Chain Error Fatal Enable**. This read-write bit, when set = 1, indicates the device should issue a fatal error interrupt if the EOC error bit is set in one of its Link Error Registers. Warm reset = 0.
5	**Response Error Fatal Enable**. This read-write bit, when set = 1, indicates the device should issue a fatal error interrupt if the Response Error bit is asserted (see bit 9 in this register). Warm reset = 0.
6	**CRC Error Fatal Enable**. This read-write bit, when set = 1, indicates the device should issue a fatal error interrupt whenever any of the CRC Error bits are asserted in any of its Link Control Registers. Warm reset = 0.
7	**System Error Fatal Enable**. This bit is hardcoded = 0 for slave interfaces.
8	**Chain Fail**. This bit is used to indicate that the chain to which this device is attached has gone down. It is set by hardware whenever the device either detects a sync flood or an error which will cause it to generate a sync flood. Warm reset = 0.
9	**Response Error**. This status bit is set by the device to indicate that it has received a response error. Cold reset = 0.
10	**Protocol Error Nonfatal Enable**. This read-write bit, when set = 1, indicates the device should issue a nonfatal error interrupt in the event a protocol error bit is set in one or more of its Link Error registers. If nonfatal error interrupts are not supported, hard wire this bit = 0. Warm reset = 0.

Table 13-10: Slave Interface Error Handling Register

Bit	Function
11	**Overflow Error Nonfatal Enable**. This read-write bit, when set = 1, indicates the device should issue a nonfatal error interrupt in the event an overflow error bit is set in one or more of its Link Error registers. If nonfatal error interrupts are not supported, hard wire this bit = 0. Warm reset = 0.
12	**End Of Chain Error Nonfatal Enable**. This read-write bit, when set = 1, indicates the device should issue a nonfatal error interrupt if the EOC error bit is set in one of its Link Error Registers. If nonfatal error interrupts are not supported, hard wire this bit = 0. Warm reset = 0.
13	**Response Error Nonfatal Enable**. This read-write bit, when set = 1, indicates the device should issue a nonfatal error interrupt if the Response Error bit is asserted (see bit 9 in this register). If nonfatal error interrupts are not supported, hard wire this bit = 0. Warm reset = 0.
14	**CRC Error Nonfatal Enable**. This read-write bit, when set = 1, indicates the device should issue a nonfatal error interrupt whenever any of the CRC Error bits are asserted in any of its Link Control Registers. If nonfatal error interrupts are not supported, hard wire this bit = 0. Warm reset = 0.
15	**System Error Nonfatal Enable**. This bit is hardcoded = 0 for slave interfaces.

Memory Base Upper Register. (Offset 18h)

This 8 bit register is used only by bridges. Concatenating this 8 bits with the 32 bits programmed in to the bridges 32 bit Memory Base Register (in the bridge header region) allows extending the start (base) address for non-prefetchable addresses behind the bridge to the full 40 bits required by HyperTransport. Warm reset = 0.

Memory Limit Upper Register. (Offset 19h)

This 8 bit register is also used only by bridges. See Figure 13-24 on page 353. Concatenating this 8 bits with the 32 bits programmed in to the bridges 32 bit Memory Limit Register (in the bridge header region) allows extending the maximum address for non-prefetchable addresses behind the bridge to the full 40 bits required by HyperTransport. Warm reset = 0.

Figure 13-24: Slave Interface Memory Base/Limit Upper Registers

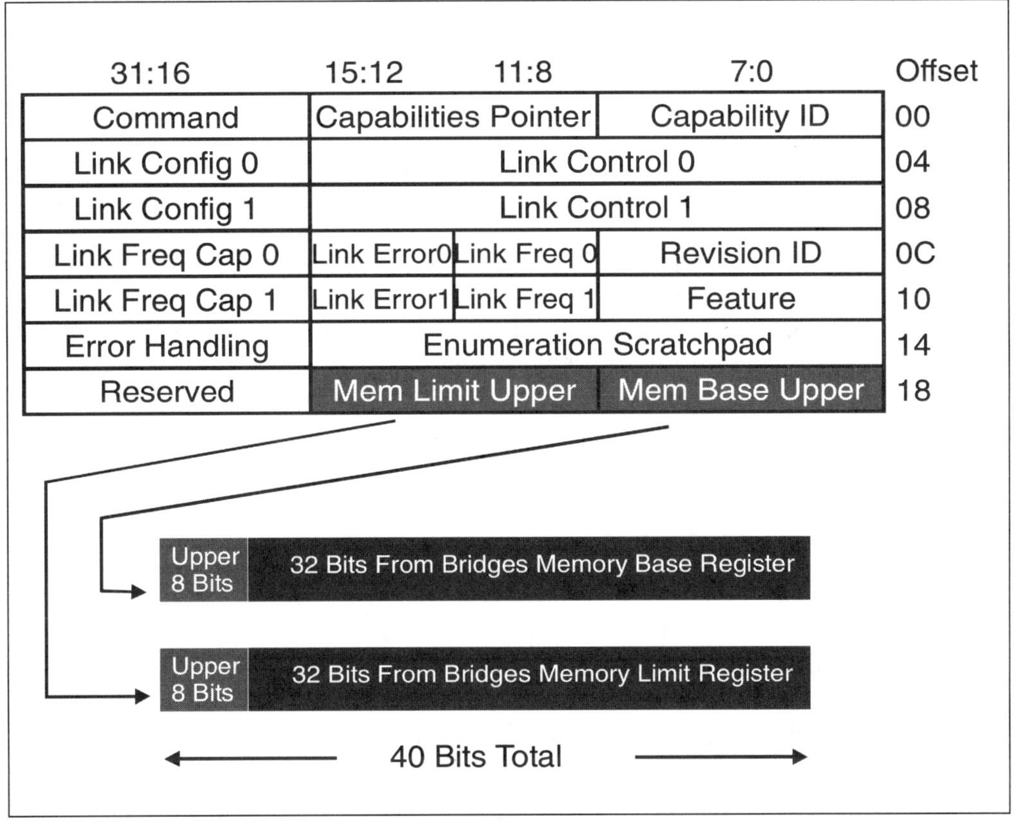

Note: The Memory Base and Memory Upper extension registers are needed because PCI compatible bridges typically don't allow accesses to addresses above 4GB (32 bits) for non-prefetchable memory, and only provide a 32 bit base-limit register pair to manage this space. Prefetchable memory is allowed above 4GB, and the bridges already have extension registers to manage this space.

The Host/Secondary Interface Block

The following section describes the Host/Secondary advanced capability interface block. Most fields are the same as those in the Slave/Primary interface block. This interface is implemented by devices which are required to act as a the host bridge for a HyperTransport chain. Such devices would include:

- Host Bridges connecting a CPU bus (or other protocol) to HyperTransport. Host bridges implement a single Host/Secondary interface with one set of link management registers.
- HyperTransport-HyperTransport bridges between chains. The bridge implements a Slave/Primary interface for its upstream connection, and one Host/Secondary interface block for each secondary bus it supports. If the bridge also implements a tunnel function, it will manage two sets of link management registers with its Slave/Primary interface.

Figure 13-25 on page 354 depicts the host bridge and HyperTransport-HyperTransport bridge (with tunnel) cases.

Figure 13-25: Host/Secondary Interface For Bridge Devices

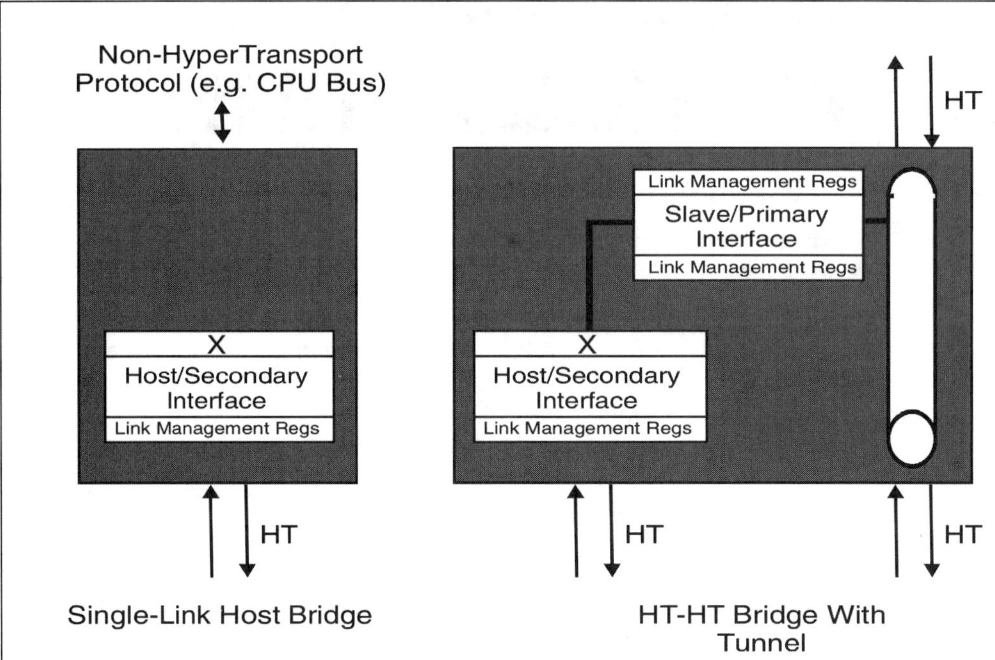

Description Of Host/Secondary Interface Fields

The following section describes the use of each field in the Host/Secondary advanced capability interface block. Most fields are the same as those in the Slave/Primary interface block. As in the case of the Slave/Primary block, the upper three bits in the Command register encodes the interface block format, as depicted in Figure 13-26. Code *001* indicates the Host/Secondary block format.

Figure 13-26: Host/Secondary Command Register Format.

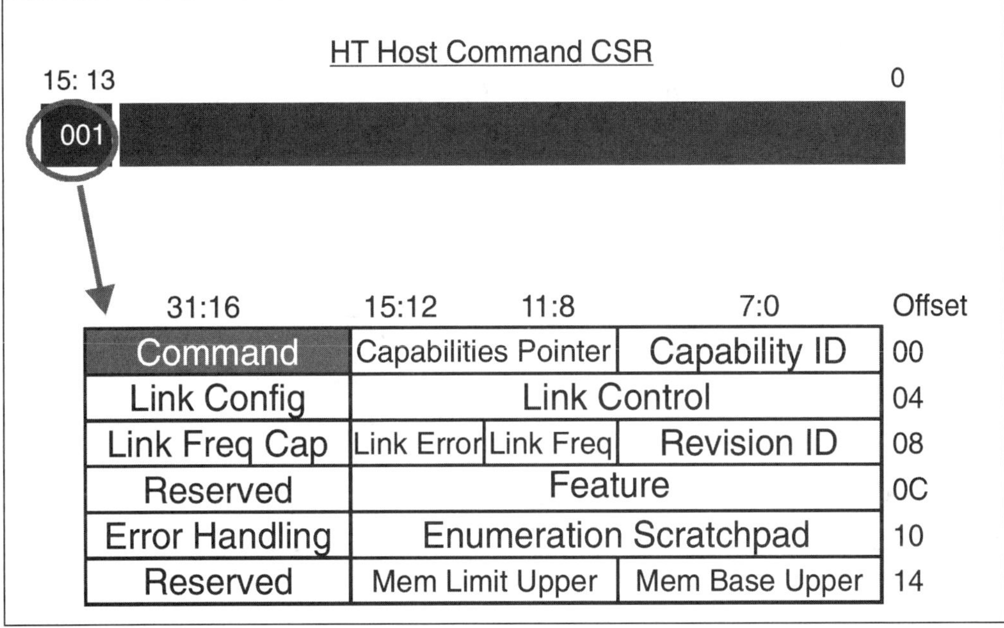

Capability ID Register. (Offset 00h)

This read-only register is hard-wired to 08h for all HyperTransport advanced capability register blocks. This value is assigned by the PCISIG, and distinguishes HyperTransport advanced capability blocks from other types.

Capabilities Pointer Register. (Offset 01h)

This read-only register provides a byte offset (from the beginning of configuration space) to the next advanced capability block. If read as zero, there are no more blocks to process.

Host Command Register. (Offset 02h-03h)

This mixed read-only and read-write register contains fields used to set up the host/secondary interface. The register is organized as shown in Figure 13-27 on page 356, and each bit field is defined in Table 13-11.

Figure 13-27: HyperTransport Host Command CSR

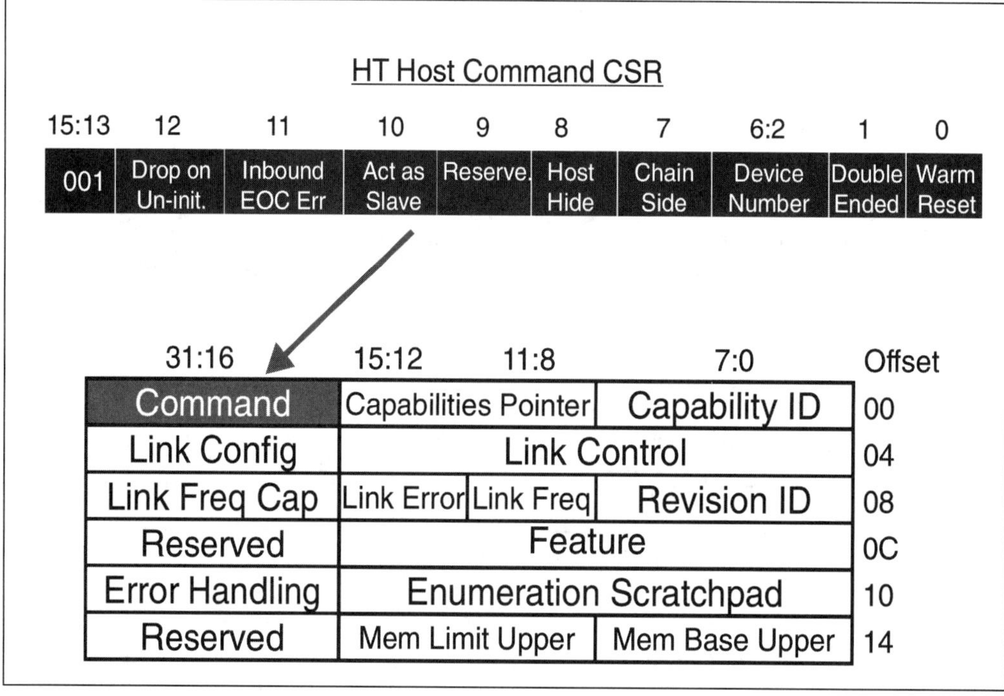

Table 13-11: Host Interface Block Command Register Bit Assignment

Bit	Function
0	**Warm Reset.** If this read-write bit is set = 1, the a secondary bus reset caused by asserting the *Secondary Bus Reset* bit in the Bridge Control Register will be a warm reset (reset with PWROK asserted). If this bit is clear, then the secondary bus reset will be a cold reset (PWROK also driven low). If not implemented, hard wire this bit = 0. If implemented, default = 1.

Table 13-11: Host Interface Block Command Register Bit Assignment (Continued)

Bit	Function
1	**Double Ended**. If this read-write bit is set = 1, there is another bridge at the other end of the chain (double-hosted chain). This bit does not affect hardware and can be used by software during initialization. If not implemented, hard wire this bit = 0. Cold reset = 0.
6:2	**Device Number**. This optional field is used to program a device number that the interface can respond to during configuration cycles initiated from the chain side. Device number is typically zero, but could be a different value for compatibility or ordering reasons. The value programmed in this register is used as the UnitID of hosts with the Act As Slave bit set (see bit 10). If the Act As Slave bit is implemented, this bit must also be implemented. If not implemented, hard wire this bit = 0. Cold reset = 0.
7	**Chain Side**. This bit is set to indicate which side of the host bridge is being accessed. A 0 in this bit indicates the read is coming from within; a 1 indicates the read is coming from the chain attached to the host interface. If double-hosted chains are not supported, this bit is 0.
8	**Host Hide**. This read-write bit is used to hide a bridges configuration space from accesses coming from the chain side. If the bit is set = 1, the host will behave as an end-of-chain device; if it is clear, configuration accesses from the chain are allowed by the device. Warm reset = 0. If the device does not support double-hosted chains, this bit is always = 1.
9	**Reserved**.
10	**Act As Slave.** This bit, when set, causes host to act as a slave, using the device number programmed in bits 6:2 as the base UnitID for requests and responses sources. In this mode, the host also won't set the Bridge bit in its responses. If this bit is clear, interface behaves as a host — using UnitID 0, setting the Bridge bit on responses it sends, etc. If host doesn't support double-hosted chains, this bit may be hard wired = 0.
11	**Inbound EOC Err.** This bit is set by hardware to indicate a mis-directed packet was received from a far host (in a double-hosted chain) and has handled it as an end-of-chain error (much as any other device handles end-of-chain errors). If the host doesn't check fro such errors, it hard wires this bit = 0. Cold reset = 0.

Table 13-11: Host Interface Block Command Register Bit Assignment (Continued)

Bit	Function
12	**Drop On Un-init**. This read-write bit sets the policy an uninitialized device should use for packets it would normally issue or forward if interface initialization were complete, but can't because it doesn't know if the interface is at the end of chain. If the uninitialized device has this bit cleared, it will stall until either its end-of-chain bit becomes set or initialization completes. If the uninitialized device has this bit set, it will behave as though its end-of-chain bit is set and reject packets instead of stalling. Cold Reset = 0.
15:13	**Capability Type**. Read only field indicating the type of information in this HyperTransport advanced capability block. Codes currently supported include: 000xxb = Slave/Primary Interface Block 001xxb = Host/Secondary Interface Block 01000b = Reserved (Switch) 10000b = Interrupt Discovery And Configuration 10001 = Revision ID 10100b= Address Mapping 11000b = Reserved (Retry Mode) Others = Reserved

Link Control Register. (Offset 04h)

Host/Secondary interfaces connect to a single link, and implement only one copy of this register to support it. The format for this register is the same as for the Slave/Primary block already described.

Link Configuration Register. (Offset 06h)

Host/Secondary interfaces connect to a single link, and implement only one copy of this register to support it. The format for this register is the same as for the Slave/Primary block already described.

Revision ID. (Offset 08h)

This 8 bit field indicates the Major and Minor revision of the HyperTransport I/O Link Protocol Specification to which this interface complies. The format for this register is the same as for the Slave/Primary block already described.

Link Frequency And Link Error Registers. (Offset 09h)

The format for these registers is the same as for the Slave/Primary block already described.

Link Frequency Capability. (Offset 0Ah)

The format for this register is the same as for the Slave/Primary block already described.

Feature Capability Register. (Offset 0Ch)

The *Feature* register in the Host/Secondary interface is slightly different than the one used for Slave/Primary interfaces. It is a 16 bit register (instead of 8), and adds one more bit field. (See Figure 13-28.) The bits are described in Table 13-12.

Figure 13-28: Host Interface Feature Capability Register

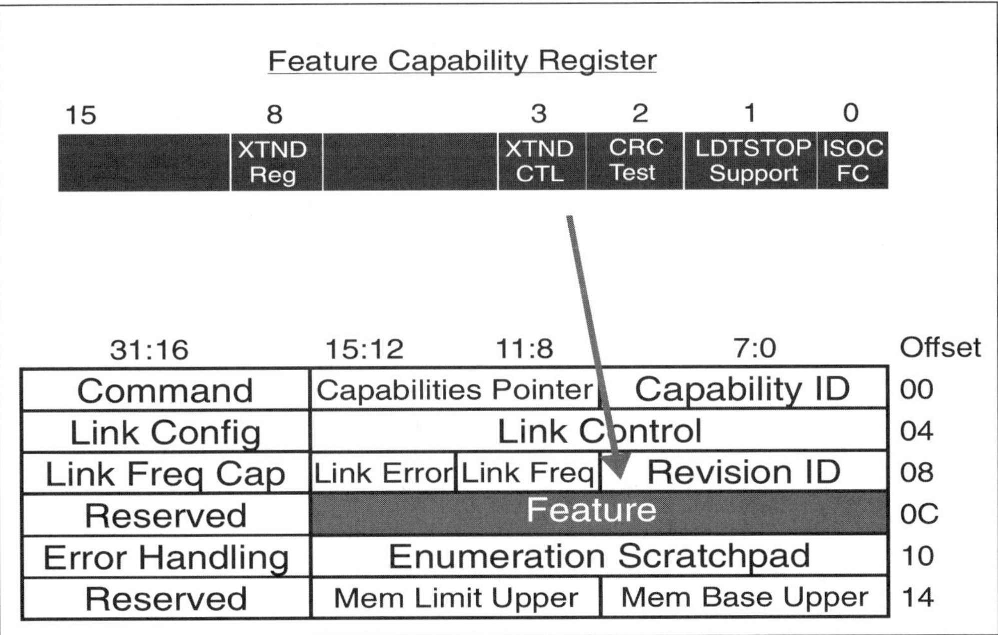

Table 13-12: Host Interface Feature Capability Register

Bit	Function
0	**Isochronous Mode**. This read only bit (if set = 1) indicates device supports the optional Isochronous flow control mechanism. If this bit is clear, the device will handle packets with *Isoc* bit set as standard virtual channel traffic. Software should check this bit for both devices on a link before enabling isochronous traffic (see bit 12 of Link Control Register).
1	**LDTSTOP#**. This read only bit (if set = 1) indicates device supports the optional LDTSTOP# protocol.
2	**CRC Test**. This read only bit is set = 1 if device supports optional CRC test mode.
3	**Extended CTL Time Required**. This read only bit indicates whether device requires CTL signal to be asserted for the extended 50uS time following LDTSTOP# disconnect in order to allow time to reacquire lock to input clock. Bit is set = 1 if the extended time is needed; cleared otherwise.
8	**Extended Register Set**. This read only bit indicates whether the following optional registers are implemented: Error Handling, and Memory Base/Limit Upper Registers. If this bit is clear, no attempt should be made to access any of these registers. Note: Implementing the extended register set is strongly recommended for host/secondary interfaces but required for slave/Primary Interface blocks. This is the reason this bit is not present in the host/secondary interface block.
Other Bits	**Reserved**. Should read back as zero.

Enumeration Scratchpad Register. (Offset 10h)

If support is indicated in bit 8, the format of this register is the same as for the Slave/Primary block already described.

Error Handling Register. (Offset 12h)

If support is indicated in bit 8, the format of this register is the same as for the Slave/Primary block already described.

Memory Base/Limits Upper Registers. (Offset 14h, 15h)

If support is indicated in bit 8, the format of these registers is the same as for the Slave/Primary block already described.

Revision ID Capability Block

Each HyperTransport function is required to report the revision of the Hyper-Transport specification to which it complies. This can be done in the *Revision ID* field of the Slave/Primary or Host/Secondary Interface block, or by implementing the Revision ID advanced capability block described here. Figure 13-29 on page 361 illustrates the format of the RevisionID capability block, and the fields are described in Table 13-13.

Figure 13-29: Revision ID Capability Block

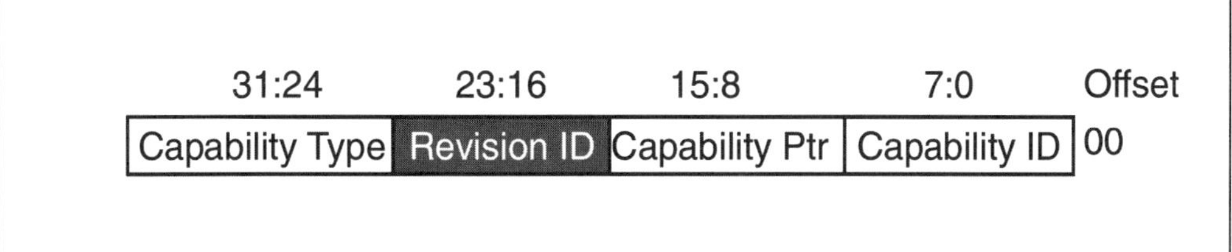

Table 13-13: Revision ID Capability Block Bit Assignment

Bits	Function
31:24	Capability Type. For the Revision ID Capability Block, the upper five bits are used and code is 10001b
23:16	Revision ID code. Bits 23:20 = Major Rev; Bits 19:16 = Minor Rev. For example, if the HyperTransport function complies with the 1.03 revision of the specification, this field would be hardcoded to 0103h.
15:8	Capability Pointer. This is the offset (in bytes) to the next advanced capability register block. Value programmed here is the offset from the beginning of configuration space. A value of zero hardcoded here indicates that this is the last advanced capability block.

Table 13-13: Revision ID Capability Block Bit Assignment (Continued)

Bits	Function
7:0	Capability ID. This identifies the protocol being supported by this block. 08h is hardcoded here for all HyperTransport capability blocks.

14 *Electrical*

The Previous Chapter

HyperTransport uses PCI configuration. The previous chapter described Hyper-Transport technology configuration for host bridges, tunnels, and end (cave) devices. These devices use the *type 0* configuration header format, while Hyper-Transport-to-HyperTransport bridges and bridges between HyperTransport and other PCI compatible protocols (e.g. PCI and PCI-X) use the *type 1* header format and are described separately in Chapter 16, entitled "HyperTransport Bridges," on page 407. Many aspects of HyperTransport device configuration are exactly the same as for generic PCI devices, although some header fields are used differently in HyperTransport, and some not at all. Devices also require at least one HyperTransport-specific advanced capability register block in addition to the basic PCI configuration space header fields.

This Chapter

The high speed signaling performed by HT devices is based on point-to-point differential signaling and source synchronous clocking. Details associated with link power requirements and the driver and receiver characteristics are discussed in this chapter. Also, the characteristics of the system-related signals, including RESET#, PWROK, LDTSTOP#, and LDTREQ# are discussed.

The Next Chapter

The next chapter focuses on the source synchronous clocking environment within HT. This involves the use of the source synchronous transmit clock to load data into a receive FIFO and the transfer of data into the receiver time domain with a receive clock that unloads data from the FIFO. Additionally, the specification defines three clocking modes that require different levels of support for passing packets between these two clock domains.

Background and Introduction

First, a brief review of the essential elements of the high-speed link is provided including an introduction to the primary aspects of the electrical signaling environment.

Each link consists of two sets of uni-directional signals (see Figure 14-1) that support concurrent data transfers in each direction. HT achieves high performance by transferring data at a maximum clock frequency of 800MHz, coupled with the use source synchronous double data rate (DDR) clocking techniques. DDR clocking permits data transfer on both the rising and falling edges of each clock. HT also relies on low voltage swing differential signaling with on-die differential termination to facilitate the high-speed data rates and to improve noise immunity.

Figure 14-1: Link Signals

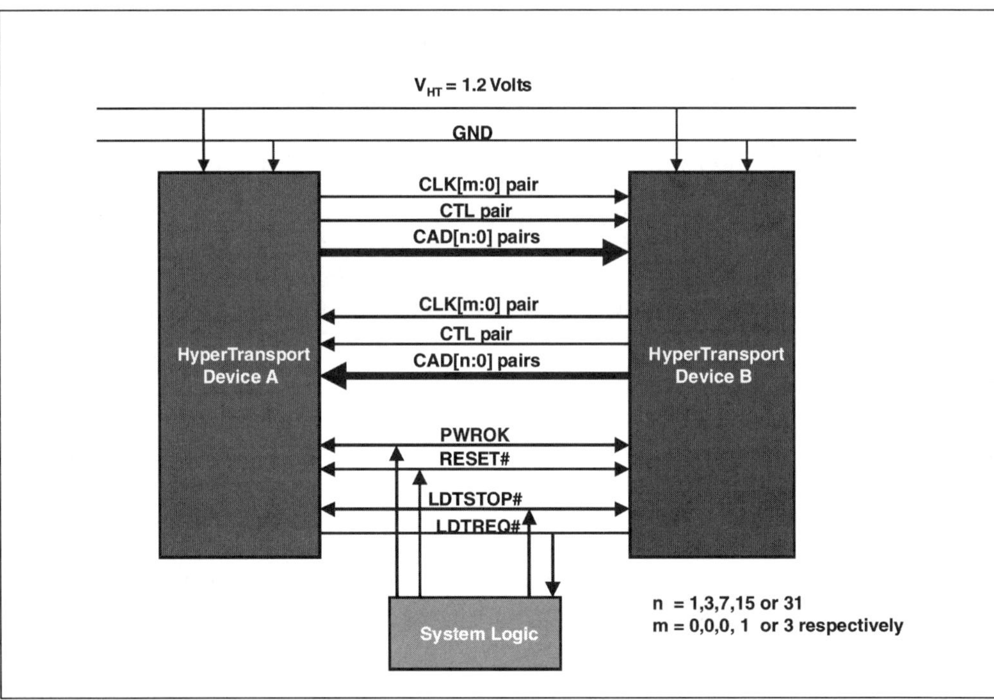

The following list reviews each of the differential high speed signals that is used when transferring data across the link.

- CAD (Command, Address and Data) — Carries HyperTransport requests, responses, data packets, and other information across the link. CAD can be different widths in each direction depending on performance needs.
- CTL (Control) — When asserted, CTL indicates that the CAD signals are carrying a control packet. When deasserted, CTL indicates that the CAD signals are carrying a data packet. There is one CTL signal for each data direction.
- CLK (Clock) — This is the source synchronous clock used when transmitting CAD and CTL signals. Each byte of CAD has its own clock. Note that the CTL signal is clocked by the same clock used for CAD[7:0].

Because the link width is scalable, the number of source clocks used also varies. Table 14-1 on page 365 below lists the transmit signals that share the same source synchronous clock.

Table 14-1: Signal Group/Source Synchronous Clock Association

Signal Group	Source Synchronous Clock
CADOUT [7:0], CTLOUT	CLKOUT (0)
CADOUT [15:8]	CLKOUT (1)
CADOUT [23:16]	CLKOUT (2)
CADOUT [31:24]	CLKOUT (3)

The four system-related signals associated with each link are implemented as single-ended LVCMOS signals and as open drain wired-OR outputs to allow multiple sources to drive them. These signals include:

- PWROK (Power OK) — Driven by system logic, this signal is a required input to each device. It may also be driven by HT devices in conjunction with RESET# to extend the reset time needed for their internal initialization.
- RESET# — This signal is driven by system logic and is a required input to each HyperTransport device. It may also be driven by HT devices in conjunction with PWROK# to extend the reset time needed for their internal initialization.
- LDTSTOP# (Lighting Data Transfer Stop) — Supports power management and other features that require a change of state on the links (e.g., it disables the link during power state transitions).

- LDTREQ# (Lightning Data Transfer Request) — An output from HT devices that permits a device to request the links be re-enabled for normal operation.

The following sections describe the power requirements, electrical, and timing characteristics of both the differential and single-ended signals, as well as an overview of the testing environment.

Power Requirements

The HT specification defines the link supply voltage requirements and the power consumption allowed by HT transmitters and receivers.

Power Supply Voltage

Single fixed power supply provides power to all the transmitter and receiver circuits. The HT link supply voltage (VLDT, or Voltage Lightning Data Transport) is rated at:

1.2 volts ± 5%

It is possible to have tight tolerance on VLDT (+-5%) on the power supply voltage output because most HyperTransport signals are differential with minimal current transients on simultaneous switching signals. The tight tolerance requirement on the supply voltage ensures that less power supply feedback noise will propagate through the system.

Differential Pair Power Consumption

The power consumption associated with the transmitter and receiver of each differential signaling pair is defined by the specification. Table 14-2 on page 367 lists these parameters. Note that the values are based on specified values of R_{ON} and R_{TT} as described in the next section.

Table 14-2: Differential Pair Power Consumption

Parameter	Min (mW)	Typical (mW)	Max (mW)
DC power per output signal pair (P_{DC})	5.9	7.2	9.0
AC power per differential pair, transmitter (P_{TAC})			53.0
AC power per differential pair, receiver (P_{RAC})			13.0
AC power per differential signal pair, total (P_{AC})			66.0

Differential Signaling Characteristics

This section describes the DC and AC characteristic of the differential transmitters and receivers. This includes impedance and voltage levels required for compliant operation of the differential transmitters and receivers.

Differential DC Characteristics

Figure 14-2 illustrates a differential link connection between a transmitter and receiver. The specification defines the impedance and subsequent voltages that will be developed under steady-state conditions (i.e., when transmitter is driving a differential 0 or 1).

Figure 14-2: HT Link Differential Driver and Receiver

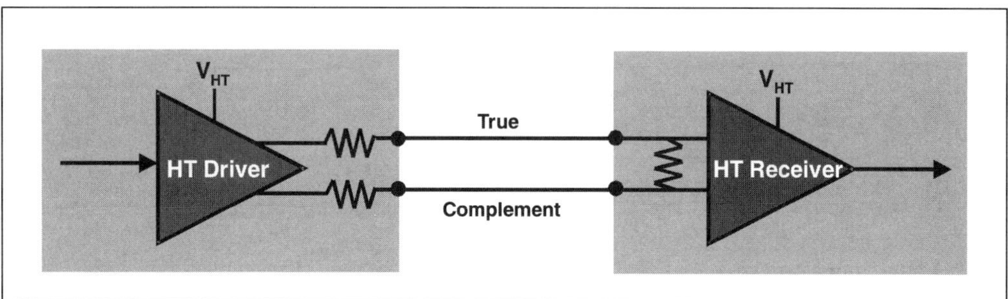

Differential DC Impedance

The DC impedance parameters are defined in the following bullet list and illustrated in Figure 14-3 on page 368 and the impedance parameter values are listed in Table 14-3 on page 369.

- **R_{TT}** is the differential input impedance under steady-state DC conditions (i.e., with a constant differential "1" or "0" driven). An on-die terminating resistor is used to adjust the receiver's input impedance to the required value. The receiver must have compensating circuitry that dynamically guarantees the impedance of the receiver stays in tolerance (see Table 14-3 on page 369) across the voltage and operating ranges of the device.
- **R_{ON}** — the driver output impedance under steady-state DC conditions. The driver must have compensating circuitry that dynamically guarantees the driver impedance remains within tolerance (see Table 14-3 on page 369) across the voltage and operating temperature ranges of the device.
- **Delta-R_{ON} (true)** — the difference in driver output impedance under DC conditions between R_{ON} (on true side) when a differential 1 is driven and when a differential 0 is driven. $[(R_{ON}+)-(R_{ON}-)]$.
- **Delta-R_{ON} (complement)** — the difference in driver output impedance under DC conditions between R_{ON} (on complement side) when a differential 1 is driven and when a differential 0 is driven. $[(R_{ON}+)-(R_{ON}-)]$.
- **Z_{Line}** — the impedance of the coupled transmission line. Notice that Z_{Line} does not exactly match ½ R_{TT}. This mismatch is intended to provide a slightly over-damped single-ended termination.

Figure 14-3: DC Impedance Values

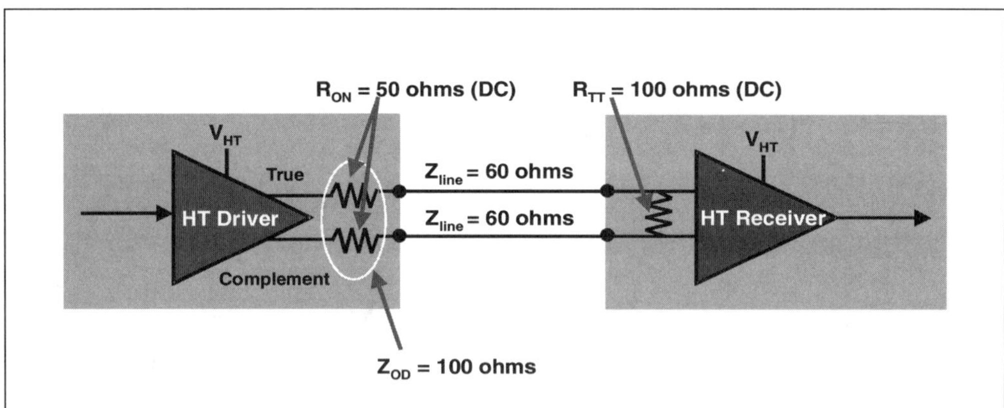

Table 14-3: DC Impedance Specification

Parameter	Min (Ohms)	Typical (Ohms)	Max (Ohms
R_{TT}	90	100	110
R_{ON}	45	50	55
Delta R_{ON} (true)	0		5%
Delta R_{ON} (complement)	0		5%
Z_{OD}		100	
Z_{line}		60	

Differential Output Voltage - DC

The specification defines the DC differential output voltage under test load conditions for compliant operation. Each parameter is described below. Figure 14-4 on page 370 illustrates the output voltage parameters, along with the test load and test points for measuring the differential voltages specified. The voltage parameter values are listed in Table 14-4 on page 370. Note that the differential output voltage and the common-mode output voltage values happen to be the same, and is not an error in the table.

- V_{OD} is the output differential voltage. It is the difference between the true signal voltage and the complement signal voltage with respect to ground (DO+)-(DO-). A Logic '0' is represented by a negative differential voltage and a Logic '1' is represented by a positive voltage difference under DC conditions.
- **Delta-V_{OD}** is the change in voltage between the differential output voltage while driving a Logic '0' and while driving a Logic '1' ($V_{OD\_0} - V_{OD\_1}$).
- V_{OCM} is the output common-mode voltage. This voltage is the average of the true signal voltage and the complement signal voltage with respect to ground under DC conditions (DO+)+(DO-)/2. This voltage is not directly measurable in an operational system, but can be derived from measuring the voltage of the true and complement voltages. However, by placing the driver in the ATE environment shown in Figure 14-4 the common mode voltage measurement can be made by hooking an oscilloscope at the V_{CM} point.

- **Delta-V$_{OCM}$** is the change in voltage between the common-mode output voltage while driving a Logic '0' and while driving a Logic '1' (V$_{OCM_0}$ – V$_{OCM_1}$).

Figure 14-4: DC Output Voltage Measurements

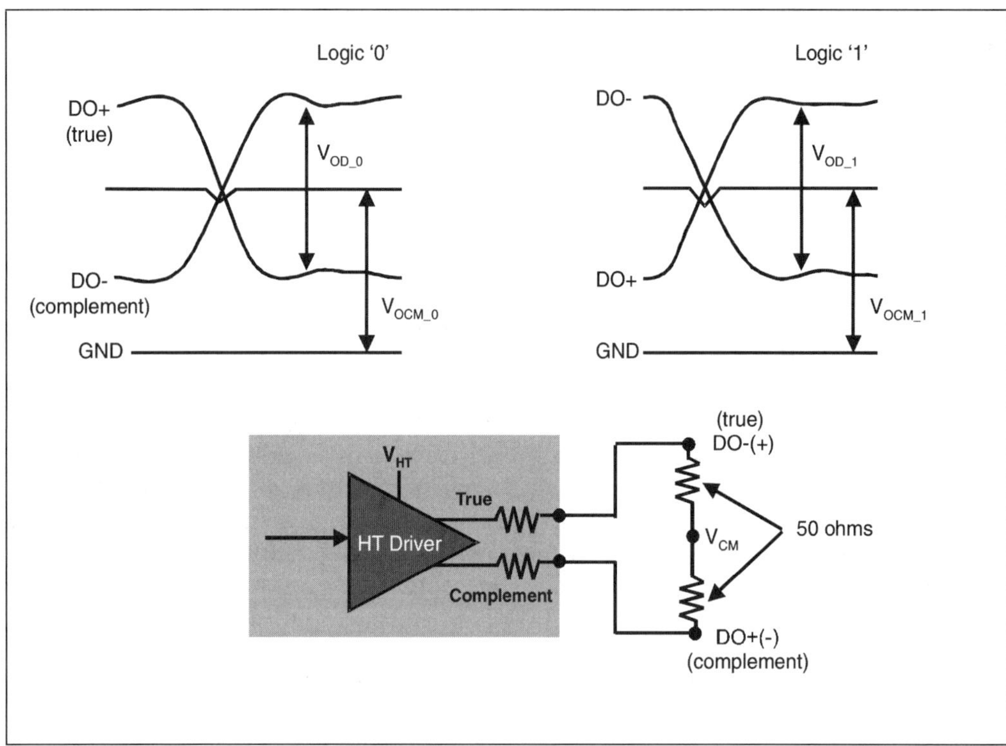

Table 14-4: Differential DC Output Voltages

Parameter	Min (mV)	Typical (mV)	Max (mV)
V$_{OD}$	495	600	715
Delta-V$_{OD}$	-15		15
V$_{OCM}$	495	600	715

Table 14-4: Differential DC Output Voltages

Parameter	Min (mV)	Typical (mV)	Max (mV)
Delta-V_{OCM}	-15		15

Differential Input Voltage - DC

The DC differential Input Voltage parameters are illustrated in Figure 14-5. The input voltage parameters are defined below and the voltage values are listed in Table 14-5 on page 372.

Figure 14-5: Differential DC Input Voltage Parameters

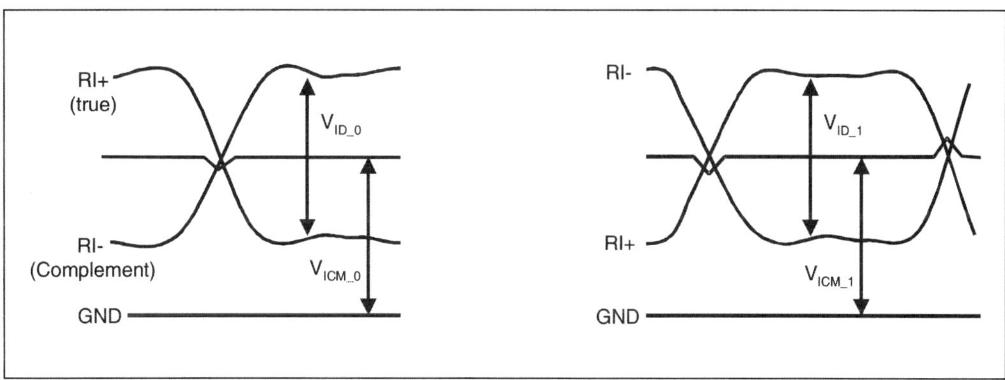

- V_{ID} is the input differential voltage at a receiver. It is the difference between the true signal voltage and the complement signal voltage with respect to ground measured at the input of a receiver.
- **Delta-V_{ID}** is the change in voltage between the differential input voltage while receiving a Logic '0' and while driving a Logic '1' ($V_{ID\_0} - V_{ID\_1}$).
- V_{ICM} is the input common-mode voltage. This voltage is the average of the true signal voltage and the complement signal voltage with respect to ground under DC conditions.
- **Delta-V_{ICM}** is the change in voltage between the common-mode input voltage while receiving a Logic '0' and while driving a Logic '1' ($V_{ICM\_0} - V_{ICM\_1}$).

Table 14-5: Differential DC Input Voltages

Parameter	Min (mV)	Typical (mV)	Max (mV)
V_{ID}	200	600	1000
Delta-V_{ID}	-15		15
V_{ICM}	440	600	780
Delta-V_{ICM}	-15		15

Differential AC Characteristics

The AC specifications are valid for a system working under normal operating conditions where signals are switching at that specified frequency and where noise is introduced due to crosstalk, reflections and inter-bit interference. A realistic system environment must be set up when performing device validation and characterization.

Differential AC Impedance

The AC impedance parameters are defined in the following bullet list and illustrated in Figure 14-6 on page 373 and the impedance parameter values are listed in Table 14-6 on page 373.

Definition of each impedance-related parameter is defined below:

- R_{TT} — the value of the differential input impedance of the receiver under AC conditions implemented with an on-die differential terminating resistor. Techniques used to compensate R_{TT} for changes due to P, V, or T fluctuations can result in R_{TT} having a non-linear I-V curve; therefore R_{TT} is specified under AC conditions and should be characterized or guaranteed over all operating ranges of voltage, and temperature.
- R_{ON} (pullup) — the driver output impedance while driving high under AC conditions. This value and tolerance must be maintained from 0.5 * VLDT_nom to VLDT_nom. R_{ON}(pulldown) is the driver output impedance while driving low under AC conditions. This value and tolerance must be maintained from 0V to 0.5 * VLDT. Techniques used to compensate the out-

put driver for changes due to P, V, or T variations can result in the driver having a non-linear I-V curve; therefore RON is specified under AC conditions and should be characterized or guaranteed over all process, voltage, and temperature operating points.

- C_{OUT} — the driver output pad capacitance and is limited to act, along with the recommended transmitter package trace single-ended impedance of 35–65 Ohms and maximum length of less than 850 mils, to create a matched impedance between the driver R_{ON} and the characteristic impedance of the package trace.

- C_{IN} — the receiver input pad capacitance and is limited to act, along with the recommended receiver package trace single-ended impedance of 35–65 Ohms and maximum length of less than 850 mils, to create a matched impedance between the interconnect transmission line and the characteristic impedance of the receiver package and input pad.

Figure 14-6: AC Impedance Values

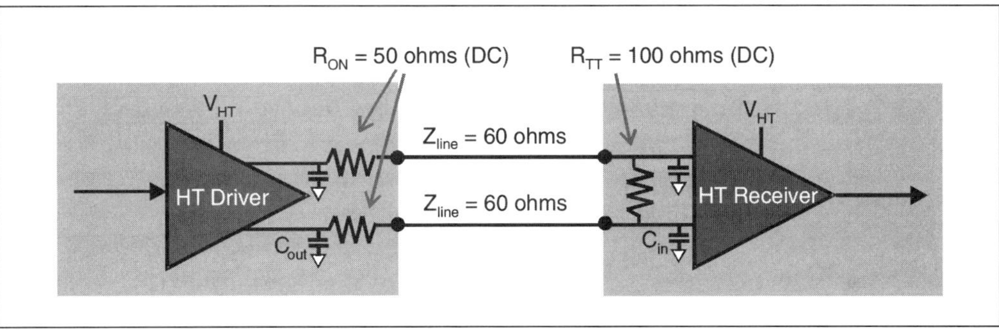

Table 14-6: AC Impedance Specification

Parameter	Min (Ohms)	Typical (Ohms)	Max (Ohms
R_{TT} - Receiver input impedance	90	100	110
R_{ON} - Driver output impedance	45	50	55
C_{out} (operation at 800MHz and above) C_{out} (operation below 800MHz)			3pF 5pF

Table 14-6: AC Impedance Specification

Parameter	Min (Ohms)	Typical (Ohms)	Max (Ohms
C_{in} (operation at 800MHz and above)			2pF
C_{in} (operation below 800MHz)			5pF

Differential Output Voltage - AC

The HT specification defines the output voltages of the driver under normal AC operating conditions. The output parameters are illustrated in Figure 14-7 on page 374. Compliant operation requires that the specified values be met when measured with the test load illustrated in Figure 14-8 on page 375. Table 14-7 on page 376 lists the specified output voltages.

- **V_{OD}** — the peak voltage difference between the true and complement signals (differential voltage) under AC conditions (RI+ – RI–).
- **Delta-V_{OD}** — the change in magnitude between the output differential voltage while driving a logic 0 and while driving a logic 1 ($V_{OD\_0} - V_{OD\_1}$).
- **V_{OCM}** — the output common-mode voltage defined as the average of the true and the complement voltage with respect to ground under AC conditions. V_{ICM} measures can be made at any moment in time and can fall at any point including crossover (RI+ + RI–) / 2.
- **Delta-V_{OCM}** — the peak change in magnitude between the output common-mode voltage while driving a logic 0 and while driving a logic 1. Delta-V_{OCM} is equal to $V_{OCM\_1} - V_{OCM\_0}$.

Figure 14-7: Differential AC Output Parameters

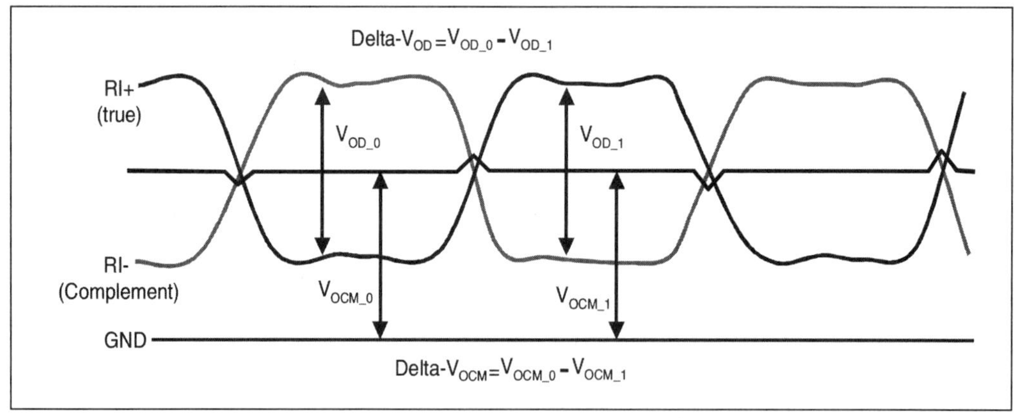

Figure 14-8: Test Setup for AC Output Voltage Measurements

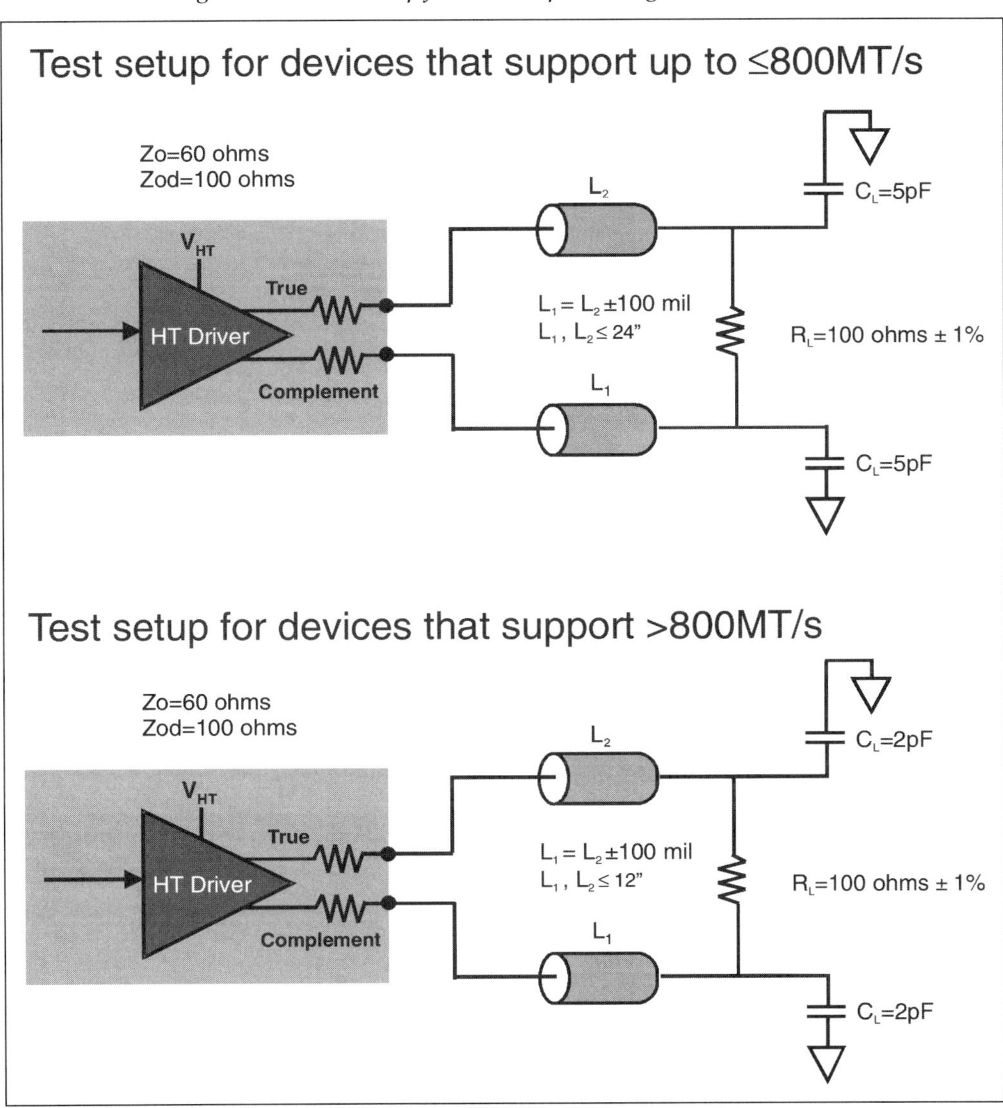

Table 14-7: AC Differential AC Output Voltages

Parameter	Min (mV)	Typical (mV)	Max (mV)
V_{OD}	400	600	820
Delta-V_{OD}	-75		75
V_{OCM}	440	600	780
Delta-V_{OCM}	-50		50

Differential Input Voltage - AC

The AC differential Input Voltage parameters are defined below and illustrated in Figure 14-9. The voltage values are listed in Table 14-8 on page 377.

- **V_{ID}** — the peak voltage difference between the true and complement signals (differential voltage) under AC conditions (RI+ − RI−).
- **Delta-V_{ID}** — the change in magnitude between the input differential voltage while receiving a logic 0 and while receiving a logic 1 (V_{ID_0} − V_{ID_1}).
- **V_{ICM}** — the input common-mode voltage defined as the average of the true and the complement voltage with respect to ground under AC conditions. V_{ICM} measures can be made at any moment in time and can fall at any point including crossover (RI+ + RI−) / 2.
- **Delta-V_{ICM}** — the peak change in magnitude between the input common-mode voltage while driving a logic 0 and while driving a logic 1. Delta-V_{ICM} is equal to V_{ICM_1} − V_{ICM_0}.

Figure 14-9: Input Voltage Parameters

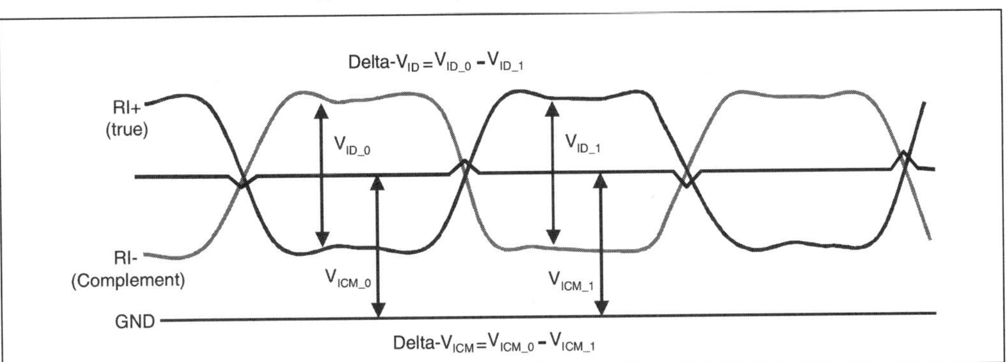

Table 14-8: Differential DC Input Voltages

Parameter	Min (mV)	Typical (mV)	Max (mV)
V_{ID}	300	600	900
Delta-V_{ID}	-75		75
V_{ICM}	385	600	845
Delta-V_{ICM}	-100		100

Input Rising and Falling Edge Rates

T_R and T_F are the rise and fall times respectively of the differential input signals. They are measured at +/-150mV point from the differential cross-over. Table 14-9 on page 378 lists the rising and falling edge rates. Note that the AC spec shows a greater range between minimum and maximum voltage parameters compared to the corresponding DC parameters. This is because the AC parameters include noise components.

Table 14-9: Differential Input Edge Rate Parameters and Values

Parameter	Min (V/ns)	Max (V/ns)
T_R Input rising edge rate	1.0	4.0
T_F Input falling edge rate	1.0	4.0

Single-Ended Signaling Characteristics

The single-ended signals must be routed to the link interface(s) of all devices. The specification states that routing "must be done in a daisy-chain fashion, with any stubs being less than 1 inch in length." The general characteristics of each of the single-ended signals are also the same:

- Output drivers must be implemented as Open Drain LVCMOS
- A pullup resistor ($\geq 1K\Omega$) is required to support the open drain drivers.
- Receivers must be implemented as single-ended 2.5V Tolerant LVCMOS

RESET# and PWROK may be implemented as input-only, depending on whether the device has been designed to extend RESET# during initialization. Table 14-10 lists the power requirements, input and output voltages, current requirements and edge rates that are specified for compliant operation.

Table 14-10: Single-Ended Signaling Characteristics

Symbol	Parameter	Test Conditions	Min	Typ	Max	Unit
V_{DD}	Supply Voltage		2.37	2.5	2.63	V
V_{IH}	Input Voltage (high)	$V_{OUT} \geq V_{VOH}(min)$	1.7		VDD+0.3	V
V_{IL}	Input Voltage (low)	$V_{OUT} \geq V_{VOL}(max)$	-0.3			V
Tr	Input Rise Time	$V_{il} < V_{in} < V_{ih}$ monotonic	0.01			V/ns

Table 14-10: Single-Ended Signaling Characteristics

Symbol	Parameter	Test Conditions		Min	Typ	Max	Unit
Tf	Input Fall Time	$V_{ih} > V_{in} > V_{il}$ non-monotonic		0.01			V/ns
V_{OL}	Output Voltage (low)	V_{DD} = min, $V_I = V_{IH}$ or V_{IL}	$I_{OL}=$ 2mA			0.7	V
I_I		V_{DD} = max, $V_I = V_{DD}$ or GND				±500	µA

Differential Timing Characteristics

This section defines the timing parameters for CAD[n:0], CTL, and CLK[m:0].

The HyperTransport link uses a simple timing methodology that accounts for simultaneous worst case combinations of uncertainties. This timing methodology attempts to cover all cases that could occur in operational systems, and is defined to provide zero additional margin. This means that board designers and designers of device transmitter and receiver interfaces must meet specification requirements over all process, voltage, and temperature corner cases.

Differential Signal Skew

The maximum skew allowed by the output driver between the true signal and its differential complement is defined as T_{ODIFF}. Figure 14-10 on page 380 illustrates this parameter, which is measured at the mid-point of the transition of the differential signal with respect to ground (i.e., the common-mode point). The maximum skew allowed is limited by Delta-V_{OCM} being within specification.

T_{IDIFF} is also defined by the specification, which is the maximum skew allowed at the input of the receiver. Specifically, T_{IDIFF} defines the max skew allowed between the true signal and its differential complement measured at the mid-point of the transition of the signal which is input to a receiver. The measurement for T_{IDIFF} is made in the same manner as for T_{ODIFF}. Table 14-11 on page 380 lists maximum skew permitted at the output of the driver and for the input of the receiver at various transmission rates.

Figure 14-10: Output Skew Measurement

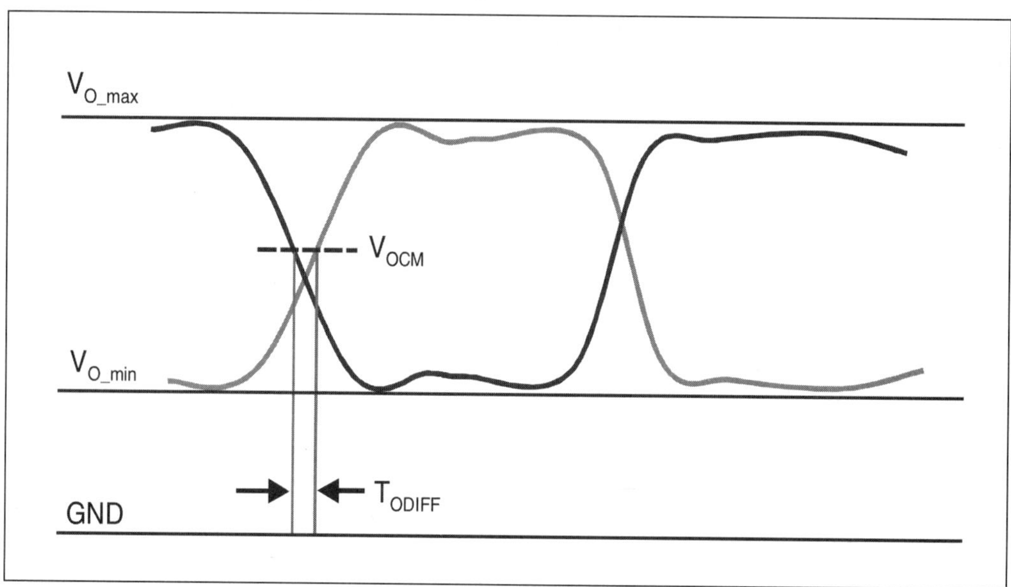

Table 14-11: Maximum Differential Output and Input Skew Values

Parameter	Description	Link Speed (Mega Transfers/s)	Maximum Skew (ps)
T$_{ODIFF}$	Output differential skew	400 MT/s	70
		600 MT/s	70
		800 MT/s	70
		1000 MT/s	60
		1200 MT/s	60
		1600 MT/s	60
T$_{IDIFF}$	Input differential skew	400 MT/s	90
		600 MT/s	90
		800 MT/s	90
		1000 MT/s	65
		1200 MT/s	65
		1600 MT/s	65

Source Synchronous Clock Skew

The source synchronous clocking employed by HT requires that CLKOUT be delayed by one-half bit time to align the clock edge in the center of the CADOUT and CTLOUT bit time. This represents a 90 degree phase shift on clock relative to the CADOUT and CTLOUT signals. Similarly, the specification defines the maximum skew permitted at the receiver between CLKIN and CADIN or CTLIN. The specification permits some tolerance in the clock delay relative to CAD and CTL for both the driver output and receiver input.

Source Synchronous Clock Skew at the Transmitter

The specified parameter for the output measurement is called CAD valid time (T_{CADV}). Figure 14-11 on page 382 illustrates the relationship between CLKOUT and CADOUT or CTLOUT. An example of this relationship is discussed below:

Assuming an 800MHz clock (which yields 1600MT/s), the clock period is 1250ps, and the bit-time period (T_{BIT}) is 625ps. When CADOUT[n:0] or CTLOUT is driven, the CLKOUT signals should be driven one-half bit time later ($T_{CADV\_typical} = 312.5ps$). The margin of skew allowed between CLKOUT and CADOUT[n:0] or CTLOUT is a function of the uncertainties related to the transmitter PHY, transmitter package skew, and transmitter clock. The minimum and maximum values are defined below:

- T_{CADV_min} (CAD valid time) is the smallest time allowed between the CADOUT[n:0] or CTLOUT differential crossing point and CLKOUT differential crossing point measured at the driver. It is also the smallest time allowed between CLKOUT crossing point and CADOUT[n:0] or CTLOUT differential crossing point measured at the driver.
- T_{CADV_max} is the longest time allowed between CADOUT[n:0] or CTLOUT differential cross-over point and CLKOUT differential crossing point measured at the driver. It is also the longest time allowed between CLKOUT crossing point and CADOUT[n:0] or CTLOUT differential crossing point measured at the driver.

Table 14-12 on page 382 lists the permitted values for T_{CADV}.

Figure 14-11: T_{CADV} Minimum and Maximum Measurements

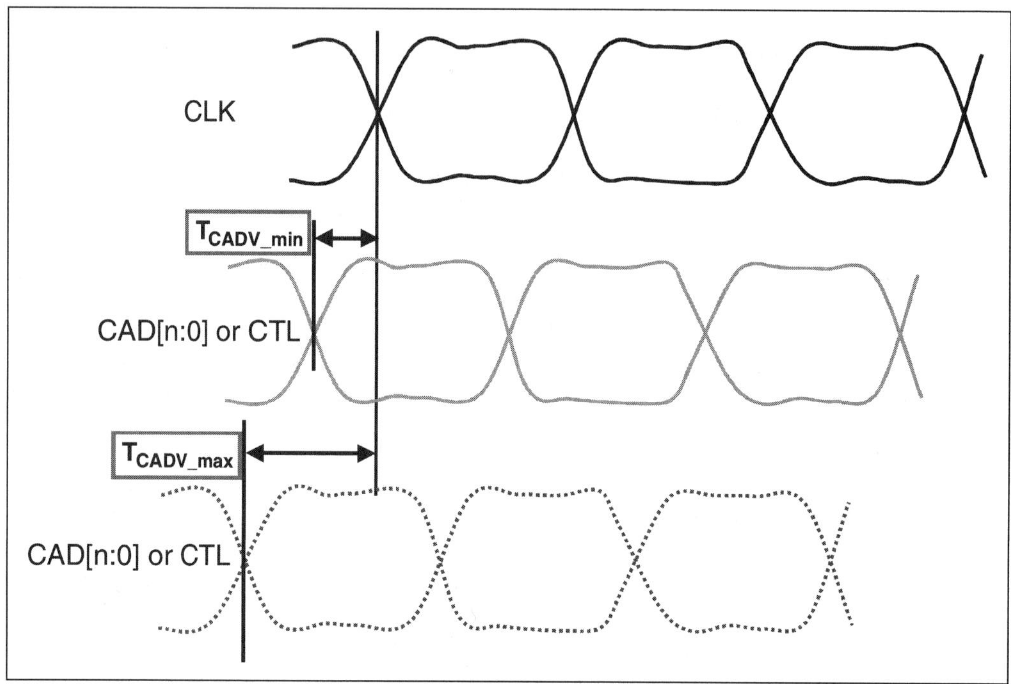

Table 14-12: Source Synchronous CLK Output Skew Values

Parameter	Description	Link Speed (Mega Transfers/s)	Minimum Skew (ps)	Maximum Skew (ps)
T_{CADV}	CADOUT valid time to/ from CLKOUT	400 MT/s	695	1805
		600 MT/s	467	1200
		800 MT/s	345	905
		1000 MT/s	280	720
		1200 MT/s	234	600
		1600 MT/s	166	459

Source Synchronous Clock Skew at the Receiver

At the receiver end, source synchronous clock skew is affected by motherboard PCB skew. The PCB CLK trace length and median of the associated CAD/CLK length must be matched so that setup and hold time for CAD/CLK signals are not violated at the receiver. Realistically, the specification allows for some motherboard skew between CAD/CTL and its associated CLK signal. Two factors contribute to this skew:

1. Route length miss-match
2. Transmission line effects which are time-variant

The source synchronous clock skew parameter at the receiver is termed CAD valid time at receiver (T_{CADVRS}).

- $\mathbf{T_{CADVRS}}$ is the CADIN/CTLIN valid time **to** CLKIN measured at the receiver inputs and includes PCB skew.

$$(T_{CADVRS} = T_{CADV\_min} - PCB\ Skew/2)$$

$$(T_{CADVRS} = T_{CADV\_max} + PCB\ Skew/2)$$

- $\mathbf{T_{CADVRH}}$ is the CADIN/CTLIN valid time **from** CLKIN measured at the receiver inputs and includes PCB skew. This is the time left over at the receiver after the CLK crossing point till the CAD/CTL crossing point. This time must be greater than the hold time.

T_{CADVRS} and T_{CADVRH} are measured at the receiver over a large number of samples and conditions which maximizes PCB skew. Table 14-13 on page 383 lists the minimum and maximum skew values.

Table 14-13: Source Synchronous CLK Input Skew Values

Parameter	Description	Link Speed (MegaTransfers/s)	Minimum Skew (ps)	Maximum Skew (ps)
T_{CADVRS}	CADIN valid time to/ from CLKIN	400 MT/s	460	90
		600 MT/s	312	90
		800 MT/s	225	90
		1000 MT/s	194	65
		1200 MT/s	166	65
		1600 MT/s	116	65

Setup and Hold Timing

Figure 14-12 on page 384 illustrates the setup and hold timing parameters and Table 14-14 on page 385 specifies the values at different transmission rates. The parameters are defined as follows:

- T_{SU} setup time is the required amount of time that the CAD[n:0] or CTL signals must be valid for prior to CLK transition crossing point at the receiver in order for the signal to be internally sampled correctly. T_{SU} is measured from the crossing point of the last CADIN transition to the CLKIN transition crossing point. T_{SU} maximum is specified to limit the amount of setup time that a device can require.
- T_{HD} hold time is the required amount of time that the CAD[n:0] or CTL signals must be valid for after CLK transition crossing point at the receiver in order for the signal to be internally sampled correctly. T_{HD} is measured from the crossing point of the earliest CADIN transition to the CLKIN transition crossing point. T_{HD} maximum is specified to limit the amount of hold time that a device can require.

Figure 14-12: Setup and Hold Time for CAD and Control

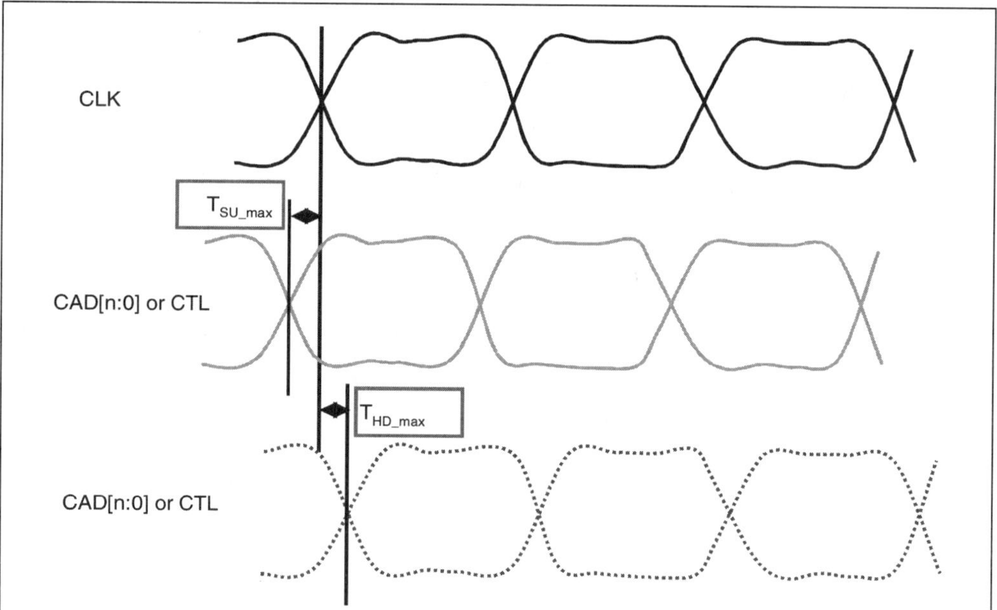

Table 14-14: Setup and Hold Times for CAD and CTL

Parameter	Description	Link Speed (Mega Transfers/s)	Minimum (ps)	Maximum (ps)
T_{SU}	Setup Time	400 MT/s	0	250
		600 MT/s	0	215
		800 MT/s	0	175
		1000 MT/s	0	153
		1200 MT/s	0	138
		1600 MT/s	0	110
T_{HD}	Hold time	400 MT/s	0	250
		600 MT/s	0	215
		800 MT/s	0	175
		1000 MT/s	0	153
		1200 MT/s	0	138
		1600 MT/s	0	110

Testing

Electrical testing of a device is grouped into three categories.

1. Voltage testing under both DC and AC conditions,
2. Output and termination impedance testing,
3. Timing testing.

Teradyne is a supplier of test equipment. DC voltage testing is done when a steady state logic '1' and logic '0' is driven. AC voltage testing is made when the transmitter is transmitting a vector pattern at a specified frequency. Impedance testing is performed by measuring the amount of current sourced by the transmitter or the amount of current sunk by the transmitter when driving a logic '1' or logic '0'. $R = V/I$ resistance can be calculated given a signal voltage of 1.2 volts.

FuturePlus designs a probe and Agilant provides a logic analyzer. Both are used for logical testing of HyperTransport bus. Packets are captured and observed using the logic analyzer. The board designer must route signals to match up with probe pin layout.

15 *Clocking*

The Previous Chapter

The high speed signaling performed by HT devices is based on point-to-point differential signaling and source synchronous clocking. Details associated with link power requirements and the driver and receiver characteristics are discussed in this chapter. Also, the characteristics of the system-related signals, including RESET#, PWROK, LDTSTOP#, and LDTREQ# are discussed.

This Chapter

This chapter focuses on the source synchronous clocking environment within HT. This involves the use of the source synchronous transmit clock to load data into a receive FIFO and the transfer of data into the receiver time domain with a receive clock that unloads data from the FIFO. Additionally, the specification defines three clocking modes that require different levels of support for passing packets between these two clock domains.

The Next Chapter

The next chapter describes the configuration of devices which use the Hyper-Transport technology *type 1* configuration header for bridges. Such devices include HyperTransport-to-HyperTransport bridges and bridges to other PCI compatible protocols (e.g. HyperTransport-to-PCI or PCI-X). In this chapter, the basic architecture of a HyperTransport-to-HyperTransport bridge is reviewed and the configuration header fields are described. Differences in usage of bit fields by HyperTransport bridge interfaces vs. PCI bridge interfaces are emphasized. The format of PCI compatible bridge headers is formally defined in the *PCI-to-PCI Bridge Architecture Specification, Revision 1.1.*

Introduction

The point-to-point high-speed transmission of data from transmitter to receiver across a HT link relies on source synchronous clocking. The specification identifies three modes of clocking:

- Synchronous
- Pseudo-Synchronous
- Asynchronous

The 1.04 version of the specification states that: "Only the synchronous clocking mode is fully specified in this revision of this specification." The specification further states that the other modes will be completely specified in later versions of the specification. All HT compliant devices are required to support the synchronous clocking mode. This introduction defines all three modes, and later sections in this chapter detail each mode. As you may have guessed, the section on the synchronous mode contains considerably more detail than the other modes.

Clock Initialization

The receive FIFO in each device must be able to absorb timing differences between the transmit and receive clocks. Data is written into the FIFO in the transmit clock domain and read in the receive clock domain.

The design and operation of this FIFO must account for the dynamic variations in phase between the transmit clock domain (Tx Clock Out) and the receive clock domain (Rx Clock). The FIFO depth must be large enough to store all transmitted data until it has been safely read into the receive clock domain. The separation from the write pointer to which the FIFO data is written and the read pointer from which the FIFO location is read (write-to-read separation) must be large enough to ensure the FIFO location can be read into the receive clock domain.

The deassertion of the incoming CTL/CAD signals across a rising CLK edge is used in the transmit clock domain within each receiver to initialize the write (load) pointer. The same deassertion CTL and CAD signals is read from the FIFO synchronous to the receive clock domain and used to initialize the read (unload) pointer. The separation between the write and read pointers is calculated based on worst-case variation between the transmit and receive clocks.

Note also that CTL cannot be used to initialize the pointers for byte lanes other than 0 in a multi-byte link, because CTL only exists within the byte 0 transmit clock domain.

Synchronous Clock Mode

The specification requires that all HT devices support the synchronous clock mode. This mode is the least complicated method of transferring data from transmitter to receiver. Synchronous clock mode requires that the transmit clock and receive clock have the same source, and operate at the same frequency. If we were to assume that the transmit clock and the receive clock always remained synchronized, then a simple clocking interface could be used as described in the following example.

A Conceptual Example

In this synchronous example, the transmit clock (Tx Clock) and receive clock (Rx Clock) are presumed to be in synchronization. Note, however, that source synchronous clocking requires that Transmit Clock Out (Tx Clk Out) be 90° phase shifted from Tx Clock. In this example all other sources of transmit to receive clock variation are ignored, including the expected clock drift associated with PLLs.

Refer to Figure 15-1 on page 390 during the following discussion. (Note that only one link direction is illustrated.) The transmitter delivers data synchronously across the link using the transmit clock. Tx Clock Out is sourced later and lags the data by 90° (or one-half bit time), thereby centering the clock edge in the middle of the valid data interval. When the data arrives at the receiver it is clocked into the FIFO using Tx Clock Out. Note that the clocked FIFO has two entries, which provides a separation of 1 between Tx Clock Out and Rx Clock. Data written into the FIFO during clock 1 would not be read from the FIFO using Rx Clock until clock 2. This one entry separation (called write-to-read separation) permits time for the sample to be stored prior to being read (i.e. the FIFO entry is not being written to and read from in the same clock cycle). In short, two FIFO entries are sufficient to provide the separation needed to ensure that data is safely stored and transferred into the receive clock domain.

Figure 15-1: Simple Synchronous Clocking Interface

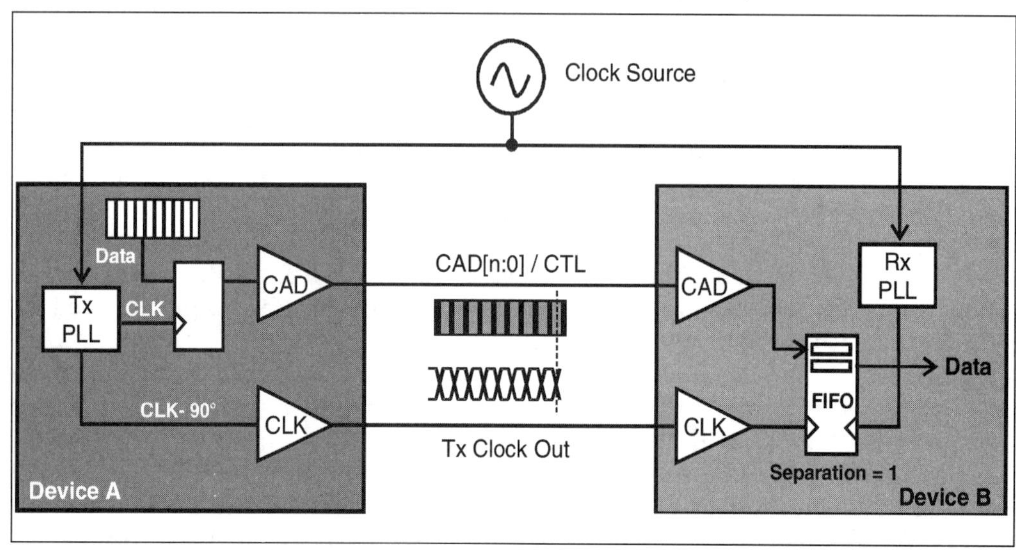

However, in the real world many factors contribute to timing differences between the transmit and receive clock that are potentially significant, even though the clocks originate from the same source. These real world perturbations result in somewhat more complicated implementations that must account for and manage the worst case variation between the transmit and receive clocks. Specifically, the specification describes the receive FIFO implementation for handling the variation between the transmit and receive clocks.

Sources of Transmit and Receive Clock Variance

The specification defines and details the sources of transmit and receive clock variation that can exist. These clock differences can create FIFO overflow or underflow if not identified and taken into account. The clock differences can be attributed to two different categories or sources:

- Invariant sources — components that represent a constant phase shift between the transmit and receive clock domain.
- Variant sources — dynamic variations in the transmit and receive time domain (these phase variations can occur even though both transmit and receive clock are running at the same frequency).

The sources of clock variation in some cases can accumulate over time, causing clock variation to increase over time. However, all of the sources of clock variation are naturally limited in terms of the maximum amount of change that can occur. For example, a PLL is designed to produce an output clock that is synchronized with the input source clock, but with certain limitations. That is, variation of output frequency is specified not to change beyond a certain phase shift. The time over which the clock phase may change can be relatively short or perhaps much longer depending upon conditions. The consideration and assessment of the sources of clock variance is done to determine a FIFO size that can absorb the worst-case clock variation. This would occur if all sources of clock variation simultaneously reach their extremes, a very unlikely circumstance.

This chapter discusses the variant and invariant sources of transmit clock to receive clock variance. It also provides an example timing budget for each source.

Invariant Sources

The time-invariant factors contribute a small proportion of the overall clock variance. The invariant factors include:

1. Cross-byte skew in multi-byte link implementations
2. Sampling Error

Cross-byte skew in multi-byte link implementations. Differences in the arrival of Tx Clock Out at the receiver (CLKIN) between each byte lane is caused by path length mismatch. This constant skew is termed $T_{bytelaneconst}$ in the specification. The specification allows up to 1000ps for this skew. Consequently, when multiple bytes are clocked into the FIFO the maximum skew could result in one of the bytes being clocked into the FIFO 1000ps later than the associated bytes. Thus, when the associated bytes are clocked out of the FIFO by Rx Clock, one byte having arrived late may be left behind. This problem is solved by adding additional entries in the FIFOs to handle the maximum lane-to-lane skew, ensuring that all associated bytes are clocked out at the same time. Note that lane-to-lane skew may change due to the effects of temperature, voltage change, etc. This parameter called $T_{bytelanevar}$ is included in the variant source list.

Sampling Error. Uncertainty in read pointer due to CTL sampling error in the receive clock domain (1 device specific Rx Clock bit time). The specification does not specifically define the source of this sampling error, but is likely caused by phase variations between the Tx Clock Out and Rx Clock that could cause a sample to be missed. Adding an additional bit time solves this problem.

Variant Sources

The phase difference between the transmit and receive clock may change significantly due to dynamic factors such as:

1. Reference Clock Distribution Skew.
2. PLL Variation in Transmitter and Receiver.
3. Transmitter and Link Transfer Variation
4. Receiver Transfer Variation
5. Dynamic Cross Byte Lane Variation

All time variant parameters must be considered in terms of their worst-case variance. The total dynamic phase variation due to these factors is called $T_{variant}$. Additionally, the transmit clock could either LEAD the receive clock by $T_{variant}$ or it could LAG the receive clock by $T_{variant}$. Consequently, the receive FIFO must be sized to accommodate both phase variations.

Reference Clock Distribution Skew. Synchronous clock mode requires that the input reference clocks to the transmitter and receiver be derived from the same time base. The distribution of the reference clock to the transmitter and the receiver results in skew between the two reference clocks. This is due to:

- differences in the output skew of the clock source, including phase error associated with Spread Spectrum Clocking in the reference clock generator, and the skew associated with the mismatch in the distribution path.
- differences in the distribution of the clocks to their PLLs due primarily to temperature and voltage changes.

This skew results in phase difference between the Transmit and Receive Clocks and must be included in the $T_{variant}$ calculation.

PLL Variation in Transmitter and Receiver. The largest contribution to the overall Tx Clock to Rx Clock variance comes from the PLLs. The PLL is constantly making adjustments to the output frequency as a result of a feedback loop. In addition, voltage and temperature changes also add to the possible output clock variation. The sample timing budget included within the specification allows a maximum PLL output phase variation of 3500ps. This represents >1 bit time at the 400 MT/s rate and approximately 5.6 bit times at the 1600MT/s rate.

Transmitter and Link Transfer Variation. The transmitter clock error (accumulated over a single bit time), the transmitter PHY, and the interconnect contribute small amounts of phase error into the link transfer clock domain through all of the parameters included in the link transfer timing. This includes noise on the PCB that affects both the clock and data in the same way causing a minor shift in frequency or phase of clock and data. (Note that if the noise affected the clock and data differently, this would affect the maximum bit transfer rate due to potential violations of T_{SU} and T_{HD}).

Receiver Transfer Variation. The receiver contributes small amounts of phase error in the received CLKIN due to distribution effects.

Dynamic Cross Byte-Lane Variation. The specification also defines the dynamic components of the byte-land variation due primarily to temperature and voltage changes ($T_{bytelanevar}$). The static elements of byte-lane variation are discussed in "Cross-byte skew in multi-byte link implementations." on page 391.

An Example Timing Budget

The specification includes an example timing budget for the identified sources of clock variation. Table 15-1 on page 393 is duplicated from the specification and lists the timing values for transfer rates ranging from 200 to 1600MT/s.

Table 15-1: Timing Variance Budget from Specification for Source of Clock Variation

Phase Recovery Timing Uncertainties	400 Mb/s	600 Mb/s	800 Mb/s	1000 Mb/s	1200 Mb/s	1600 Mb/s	Unit
Trefclk	733	733	733	733	733	733	ps
TxmtPLL	3500	3500	3500	3500	3500	3500	ps
Txmttransfer	918	592	469	358	294	227	ps
Tbytelanevar (Variant)	250	250	250	250	250	250	ps
Tbytelaneconst (Invariant)	1000	1000	1000	1000	1000	1000	ps
TrcvPLL	3500	3500	3500	3500	3500	3500	ps
Trcvtransfer	425	250	188	130	108	81	ps
Tsampling (Invariant)	2500	1667	1250	1000	833	625	ps
1 RCLK bit time (Invariant)	2500	1667	1250	1000	833	625	ps

Table 15-1: Timing Variance Budget from Specification for Source of Clock Variation

Phase Recovery Timing Uncertainties	400 Mb/s	600 Mb/s	800 Mb/s	1000 Mb/s	1200 Mb/s	1600 Mb/s	Unit
Total (2x variant + 1x invariant)	24651	21983	20799	19942	19437	18832	ps
Minimum FIFO Depth	10	14	18	20	24	32	entries
Safe Write to Read Pointer Separation	5	7	9	10	12	16	entries

Clock Variance, FIFO Size, and the Read Pointer

This section discusses the relationships between the worst-case clock variance calculation, minimum FIFO size, and unload pointer initialization for synchronous clock mode. The following example is provided to help explain these relationships. (Also, see Figure 15-2 on page 394.)

The following assumptions are made for this synchronous clocking mode example:

- 8-bit link
- 800 MegaTransfers/second (bit time = 1250ps)
- $T_{invariant}$ = 2,500ps
- $T_{variant}$ = 8,390ps

Figure 15-2: Synchronous Clock Example, Single Direction

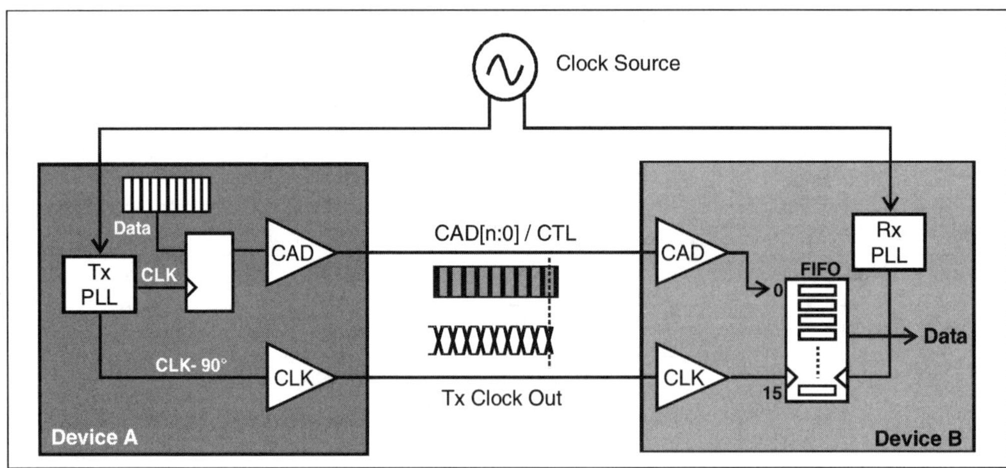

Minimum FIFO Size

Recall that the FIFO depth must be large enough to store all transmitted data until it has been safely read into the receive clock domain. The minimum FIFO size must account for the total possible variation between Tx Out Clock and Rx Clock. Note the $T_{variant}$ parameter must be doubled because Tx Clock Out may either lead or lag Rx Clock by the time variant values. Therefore, the maximum phase shift is calculated as:

```
Tvariant * 2 = Ttotal variant
```

The variance numbers in this example yield the following $T_{total\ variant}$ value:

```
8,390ps * 2 = 16,780ps
```

The minimum FIFO size can be calculated by dividing the total clock variation time by the bit time duration.

```
((Tvariant * 2) + Tinvariant) ÷ Bit time = FIFO Entries
```

For this example, the number of FIFO entries is:

```
((8,390ps * 2) + 2,500ps) ÷ 1250ps = 15.4 FIFO Entries
```

The number of entries is rounded up to the next integer value, or 16 in this example. Note also that this computation of minimum FIFO entries is different from the results shown in Table 15-1 on page 393 from the specification. The reason for the smaller FIFO size is that this example implementation does not have multiple byte lanes, therefore the $T_{bytelanevar}$ and $T_{bytelaneconst}$ parameters are not included in the worst-case clock variation.

Write-to-Read and Read-to-Write Separation

Recall that the FIFO depth must be large enough to store all transmitted data until it has been safely read into the receive clock domain. The separation from the write pointer location where data is written and the read pointer location from which data is read must be large enough to ensure the FIFO location can be read safely into the receive clock domain.

To accommodate this clock variance in this example, the read pointer within the FIFO would need to be separated from the write pointer by 8 entries (or, bit times). The following three scenarios are provided to explain the operation of the FIFO and its pointers.

Scenario 1: Tx Out Clock and Rx Clock are in Sync. In this example, the clock variation happens to be zero, with the specified separation between the write and read pointers set to 8 entries as calculated above. Figure 15-3 on page 396 illustrates the position of the pointers as a progression (labeled Stages A, B, and C). Note that the write pointer is labeled as Tx Clock Out to remind us that data is written using the transmit clock. For the same reason, the read pointer is labeled as Rx Clock.

Stage A — the write pointer has progressed from entry 0 to entry 8. Because the separation between the write and read pointer is 8, Rx Clock is prevented from clocking data from the FIFO until the separation reaches 8. At this stage, the separation has just been reached, so Rx Clock clocks data from entry 0, while the Tx Clock Out clocks data into entry 8.

Stage B — the write pointer has progressed to entry 15 and because there is still no phase difference between Tx Clock Out and Rx Clock the separation between the pointers remains at 8. Rx Clock is clocking data from entry 7 as Tx Clock Out is clocking data into entry 15.

Stage C — the write pointer has rolled from entry 15 back to entry 0 while the read pointer has advanced to entry 8. This simply illustrates that the separation is still maintained when the write pointer reaches the end of the FIFO and wraps back to entry 0.

Figure 15-3: FIFO Operation When Tx Clock Out and Rx Clock are in Sync

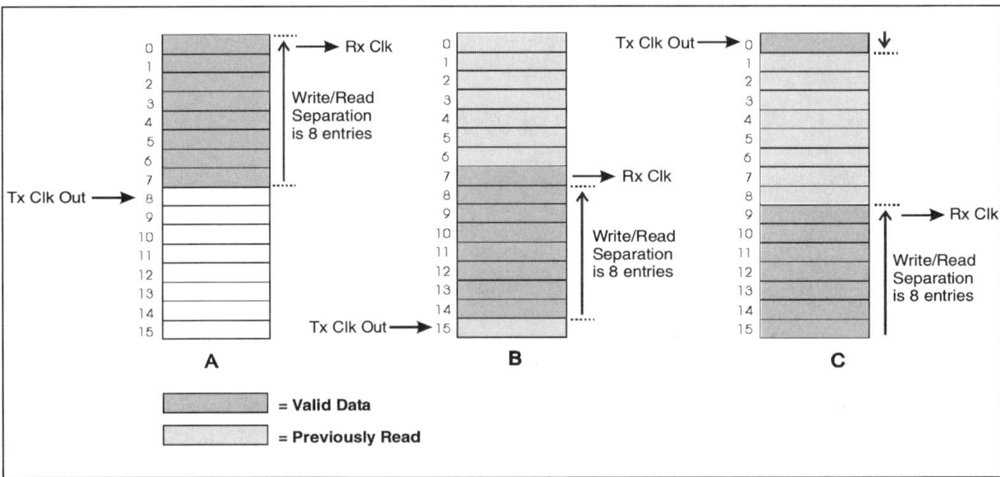

Scenario 2: Tx Clock Out Lags Rx Clock. This scenario shows the effects of the Tx Clock Out lagging the Rx Clock. The amount change in phase shift between the clocks as illustrated in Figure 15-4 on page 397 is dramatic. The amount of change illustrated would not likely have accumulated over such a small number of clocks; however, this amount of change could easily accumulate over a long interval.

Stage A — Stage A illustrates an initial write-to-read separation of 8, with the write pointer having progressed from entry 0 to entry 8. The Rx Clock clocks data from entry 0, while the Tx Clock Out clocks data into entry 8.

Stage B — Due to accumulated phase shift between Tx Clock Out and Rx Clock, Tx Clock Out now lags Rx Clock. This phase shift could be caused by phase changes in Tx Clock Out, phase changes in Rx Clock or a combination of both (if the event that the phase shift occurred in opposite directions). In this example, the write pointer has progressed to entry 13 but the read pointer has advanced more quickly to entry 7. Due to the accumulated phase difference between the transmit and receive clocks, the write-to-read separation has diminished to 6 entries.

Figure 15-4: Effects of Tx Out Clock Lagging Rx Clock

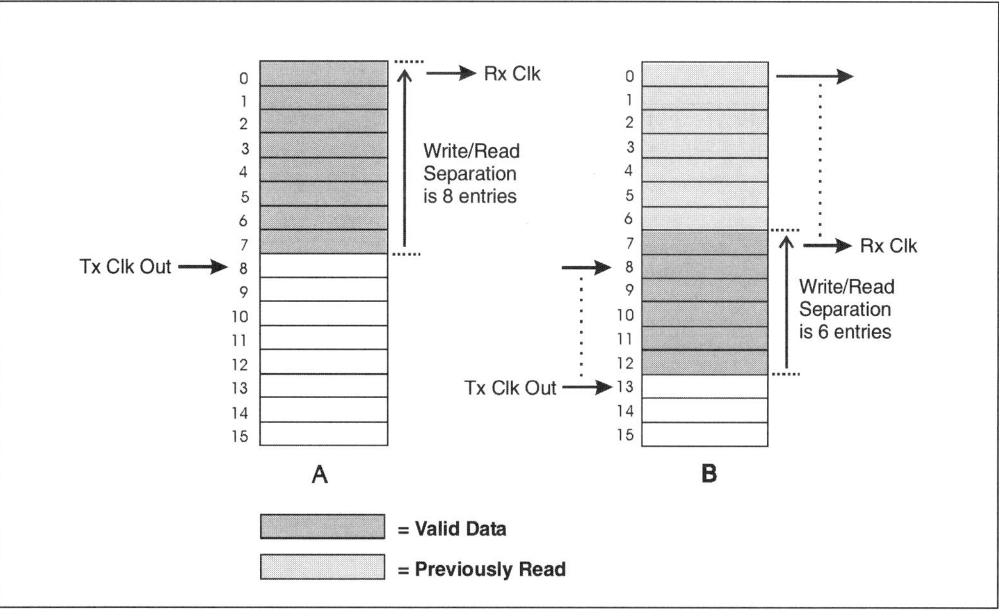

The FIFO was sized to 16 to absorb the maximum clock variation in the direction of Tx Clock Out lagging Rx clock. The maximum write-to-read separation of 8 in this example ensures the read pointer will not overtake the write pointer, which would result in FIFO underflow.

Scenario 3: Rx Clock Lags Tx Clock Out. This scenario presents the opposite condition that was illustrated in scenario 2. In this example, the receive clock lags the transmit clock. As in the previous example, the phase difference between the clocks would not likely accumulate so quickly.

Stage A — the write pointer has previously traversed all of the entries and is back at entry 0 again, while the read pointer is at entry 8 This scenario focuses on the possibility that the Rx Clock lags the Tx Clock Out clock. In this case, the read-to-write separation becomes critical. In stage A this separation is 8.

Stage B — the write pointer has advanced to entry 13, while the read pointer has only advanced to entry 15. The write pointer had moved ahead by 13 entries and the read pointer has moved only 7 entries, leaving a read-to-write separation of only 2.

Once again, the large change in clock variance over such a short period of time as illustrated in stage B would not occur. But the example does serve to illustration that over time the clock variance can accumulate and that an appropriately sized FIFO will be able to absorb the clock variance without overflow.

Figure 15-5: Effects of Rx Clock Lagging Tx Clock Out

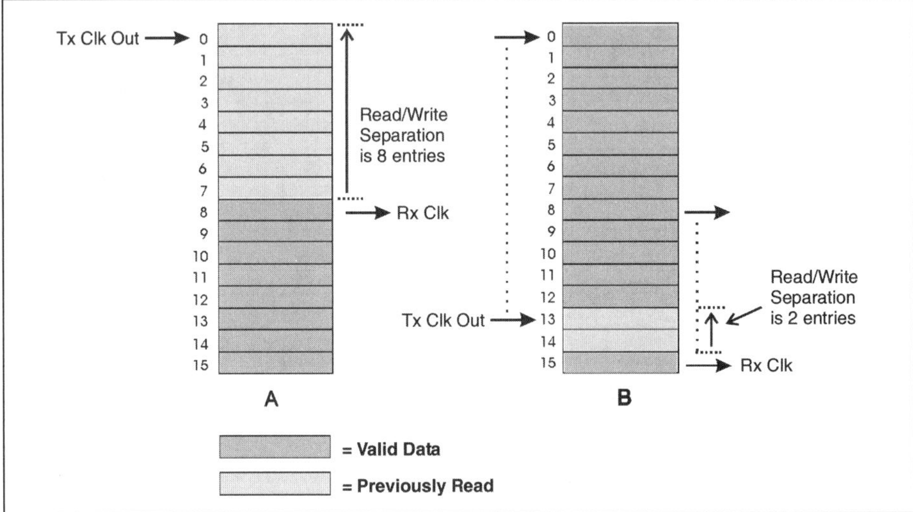

Buffering Width and Speed Differences

FIFOs may also provide buffering between a narrow high-speed link and a wider slower data path inside a receiver.

CAD/CTL synchronization time: As discussed in "Clock Synchronization (CTL=0 & CAD=0)" on page 292, the read (unload) pointer of the FIFO for byte lane zero is established during initialization when the CAD and CTL signals are sampled deasserted in the receive clock domain. Since sampling the initial CTL and CAD signals in the receive clock domain will have some synchronization delay, this device-specific synchronization delay should be removed from the initial read pointer.

Pseudo-Synchronous Clock Mode

In pseudo-synchronous mode, both Rx Clk in the receiver device and Tx Clk in the transmitter device are generated from the same time base clock just as in the synchronous mode case. During initialization, software configures each link to the maximum common frequency based on the values reported in each device's frequency capability register. The highest frequency supported by both devices is loaded into the Link Frequency register of each device. This value defines the highest frequency that both devices can use when sending packets over the link. In synchronous implementations this would be the exact frequency used by both devices. However, a device implementing pseudo-synchronous mode may arbitrarily lower the transmit clock frequency (Tx Clk or Tx Clock Out) below that specified by the Link Frequency register. Note that the receiver clock (Rx Clk) still runs at the frequency specified by the Link Frequency register.

Figure 15-6 on page 400 illustrates an example implementation in which Device A lowers its Tx Clock frequency below the value specified in its Link Frequency register. Consequently, Device B stores data in its FIFO at a lower frequency than it removes data from the FIFO.

Figure 15-6: Example Pseudo-Synchronous Mode Implementation

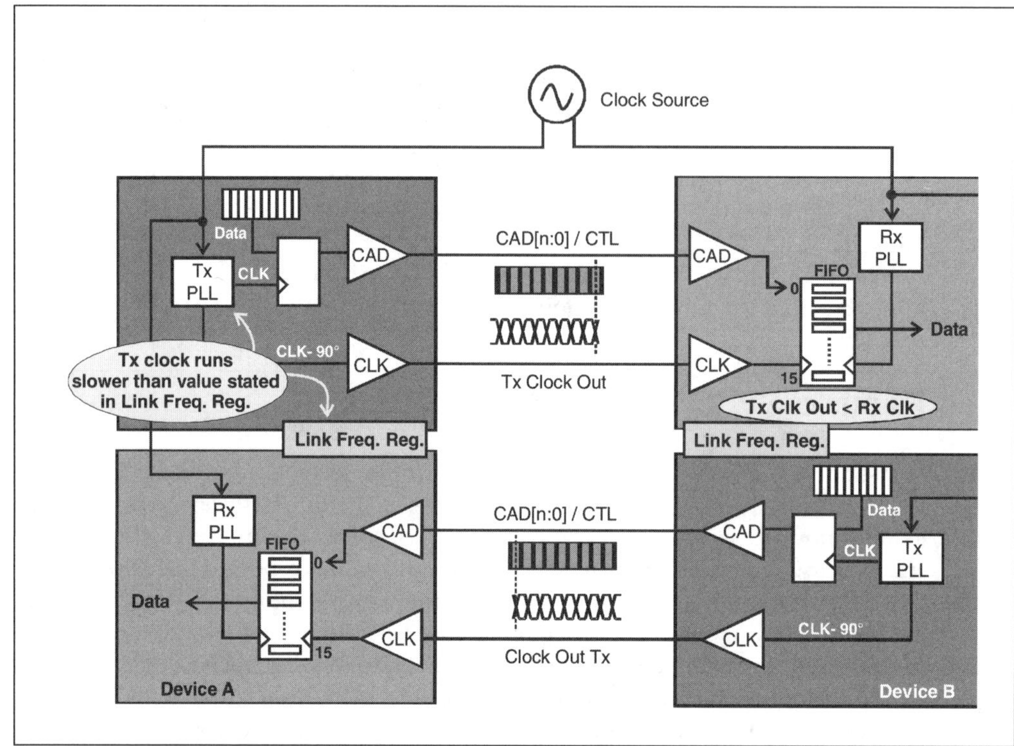

Why Use Pseudo-Synchronous Clock Mode?

The specification does not address any specific application for Pseudo-Synchronous clock mode. It appears that the main advantage is that a link is given the ability to transfer data in one direction at a higher rate than the other. But this begs the question, "Why not transfer in both directions at the highest speed possible, thereby keeping bus efficiency as high as possible?" It further raises the question of a possible advantage associated with clocking one direction at a slower rate; however, there would be power savings, reduced EMI, and reduced transmit PHY complexity.

Implementation Issues

Pseudo-synchronous clocking mode must take into account the same clock variance issued as synchronous mode. Additionally, several other key issues must be considered for pseudo-synchronous clocking mode. These issues include:

- Methods and procedures required to implement pseudo-sync mode.
- Managing the FIFOs and pointers given the different transmit and receive clock frequencies.
- Is support mandatory?

Methods and Procedures. The specification does not define a mechanism to lower the transmit clock frequency, nor does it provide a method for determining which clock modes are supported by a given HT device. The specification states that:

> "The means by which the operating mode is selected for a
> device that can support multiple modes is outside the scope
> of this specification."

Further, no definition exists regarding the level of software that would be involved in transitioning a device to the pseudo-sync mode.

FIFO Management. Pseudo-sync mode must consider the same sources of clock variation as in synchronous mode and the receive FIFOs must be sized appropriately and the separation between the write and read pointers must be established.

Because Tx Clock Out may run slower than Rx Clk in pseudo-synchronous mode, incoming packets may be clocked into the receive FIFO more slowly than they are clocked out. This situation results in a buffer underrun condition. To prevent this from happening the unload pointer occasionally must be stopped and then restarted when sufficient data is present in the receive FIFO. One approach to solving the potential underrun problem is to implement the FIFO to set a flag when the read pointer reaches the write pointer. The unload pointer could be stopped to keep additional reads from occurring until the situation is corrected. When sufficient separation between the load and unload pointers have accumulated, the flag can be cleared and reads can continue.

Is Support for Pseudo-Sync Mode Required? The HT specification clearly requires support for **synchronous** clocking mode for all devices. It further states that:

> "Devices may also implement Pseudo-sync and Async modes based on their unique requirements."

This statement suggests that Pseudo-sync mode is conditionally required; that is, it's optional unless a device has some special conditions that require the support. Further, the specification does not mention any requirement for standard synchronous devices to operate correctly when attached to devices that operate in pseudo-sync mode. It may be that it is expected that all synchronous clocking mode devices will be able to inter-operate with pseudo-sync devices. As discussed in the previous section, support for pseudo-sync mode at the receiving end simply requires that the FIFO read pointer not be allowed to advance to the same entry as the write pointer.

Asynchronous Clock Mode

The asynchronous clock mode permits the transmit and receive clocks to be derived from different sources. The specification limits the maximum difference permitted between the transmit and receive clock frequency. In this case, either the transmit clock or the receive clock may run faster than the other. So, both situations must be taken into account.

Transmit Clock Slower Than Receive Clock

In this case, a potential underrun condition can develop. The solution for preventing underrun is the same as that discussed for the pseudo-synchronous clock mode as discussed in "FIFO Management." on page 401. In summary, the FIFO read pointer is prevented from reaching the write pointer by stopping the read clock until the transmit clock has had a chance to catch up.

Transmit Clock Faster Than Receive Clock

Tx Clock Out can run slightly faster than Rx Clk in asynchronous mode (but by no more than 2000 ppm), thus incoming packets may be clocked into the receive FIFO faster than they are clocked out. This situation will result in a buffer overrun condition, and the receiver has no way of stopping or slowing the incoming packets. The following discussion describes how to prevent the buffer overrun condition from occurring.

CRC bits appear on the link for 4 bit-times (on 8-,16-, and 32-bit links) after every 512 bit-times. These CRC bits are detected by the receiver, but NOT clocked into the receive FIFO. Instead the CRC bits are routed into the CRC error checking logic. Consequently, the FIFO write pointer does not increment during the CRC bit times, but the read pointer continues to increment and data continues to be read from the FIFO. As a result, the unload pointer has sufficient time to catch-up by clock data in the receive FIFO out before the buffer over-runs.

Part Three

HyperTransport Optional Topics

Part Three details optional features of the HT bus. The chapters included in Part Three are:

- Chapter 16: HyperTransport Bridges
- Chapter 17: Double-Hosted Chains
- Chapter 18: HT Power Management
- Chapter 19: Networking Extensions Overview

16

HyperTransport Bridges

The Previous Chapter

The previous chapter focused on the source synchronous clocking environment within HT. This involves the use of the source synchronous transmit clock to load data into a receive FIFO and the transfer of data into the receiver time domain with a receive clock that unloads data from the FIFO. Additionally, the specification defines three clocking modes that require different levels of support for passing packets between these two clock domains.

This Chapter

This chapter describes the configuration of devices which use the HyperTransport technology *type 1* configuration header for bridges. Such devices include HyperTransport-to-HyperTransport bridges and bridges to other PCI compatible protocols (e.g. HyperTransport-to-PCI or PCI-X). In this chapter, the basic architecture of a HyperTransport-to-HyperTransport bridge is reviewed and the configuration header fields are described. Differences in usage of bit fields by HyperTransport bridge interfaces vs. PCI bridge interfaces are emphasized. The format of PCI compatible bridge headers is formally defined in the *PCI-to-PCI Bridge Architecture Specification, Revision 1.1*.

The Next Chapter

The next chapter describes the features of the optional HyperTransport double-hosted chain topologies. Topics include the reasons behind sharing and non-sharing chains, PCI configuration space registers used to initialize the fabric for multiple hosts, and tunnel support for upstream and downstream packets moving in both directions.

HyperTransport Bridges Uses PCI Configuration

HyperTransport bridges use the PCI configuration method and the 256 byte configuration space to set up and manage one primary bus and one or more secondary buses. The primary and secondary interfaces of a HyperTransport bridge may both be HyperTransport, or either one could bridge to another PCI-compatible protocol such as PCI or PCI-X. Because the configuration header contains bits to manage two different interfaces, there are two things to keep in mind when describing bridges between HyperTransport and another protocol:

1. For the HyperTransport interface, the meaning of some bits in the configuration header depends upon the particular interface (primary or secondary) that is implemented as HyperTransport.
2. For the interface that is not HyperTransport, bit field definitions revert back to the bus protocol being supported (e.g. PCI or PCI-X).

Basic Jobs Of A HyperTransport Bridge

As in the case of PCI bridges, a HyperTransport bridge has a number of responsibilities:

1. It extends the topology through the addition of one or more secondary buses. Each HyperTransport chain (bus) can support up to 32 UnitIDs. Because a device is permitted to consume multiple UnitID's, implementing a bridge is a reasonable way to add a new chain that can support 32 additional UnitIDs (the bridge secondary interface consumes at least one of the new UnitIDs).
2. It acts as host for each of its secondary chains. There are many aspects to this, including ordering responsibilities, error handling, maintaining a queue for outstanding transactions routed to other buses, reflecting peer-to-peer transactions originating below it, decoding memory addresses so it may claim and forward transactions moving between the primary and secondary bus, forwarding/converting configuration cycles based on target bus number, etc.
3. In cases where it bridges between HyperTransport and PCI/PCI-X, the bridge also must translate protocols for transactions going in either direction. It may also have to remap address ranges between the 40-bit HyperTransport address range and the 32/64-bit PCI or PCI-X range.

How Does The Bridge Manage It All?

HyperTransport bridges make use of the same HyperTransport Host/Primary and Slave/Secondary advanced capability blocks already defined for non-bridge devices. Non-bridge HyperTransport device configuration is described in Chapter 13, entitled "Device Configuration," on page 305. Second, they implement the *type 1* configuration space header common to all PCI-compatible bridges (with the redefinition of certain bits described shortly). This is fortunate because:

1. HyperTransport bridges maintain software compatibility with PCI bridges.
2. Bridge headers already contain two sets of control/status registers for managing two independent interfaces, making support for mixed HyperTransport and PCI/PCI-X bridges a reasonable extension. Each set of primary/secondary bus registers is programmed (and interpreted) according to whether the specific interface is HyperTransport, PCI, or PCI-X.

Same Slave/Primary And Host/Secondary Blocks

The Slave/Primary and Host/Secondary capability blocks used in non-bridge devices such as tunnels and end (cave) devices are also used for the HyperTransport interfaces of bridges. Figure 16-1 on page 410 depicts a simple HyperTransport-to-HyperTransport bridge with a single secondary chain. Note that a bridge must implement a separate Host/Secondary interface block for each secondary bus it supports.

HyperTransport Bridge Header Fields

In this section, the configuration space type 1 header format for HyperTransport bridge devices is described. For the most part, HyperTransport bridges use these fields in the same way as PCI bridges; the differences are described here. Header fields not mentioned are used in the same way as in PCI bridges. Refer to "The Type 0 Header Format" on page 322 for a description of HyperTransport non-bridge type 0 headers.

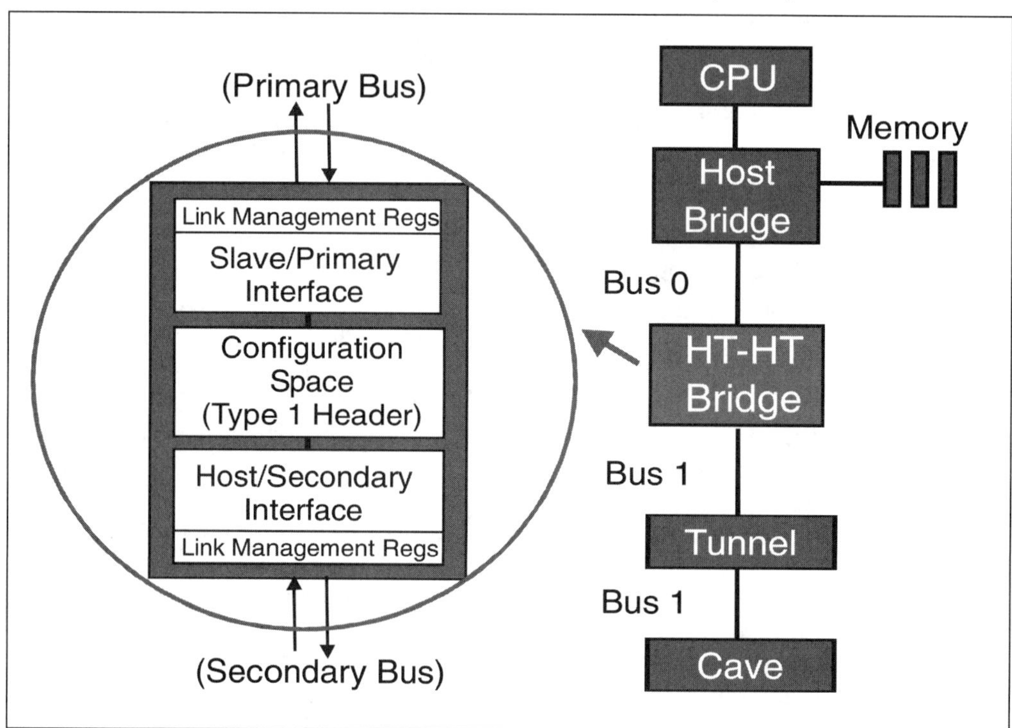

Figure 16-1: HyperTransport-HyperTransport Bridge Interfaces

Bridge Header Command Register

Lower 16 bits at dword 01. The bridge header *Command* register is used by software to enable basic capabilities of the bridge <u>on its primary bus</u>, including bus mastering, target address decoding, error responses, etc. Bits marked "0" in Figure 16-2 are not used (hardwired = 0); refer to Table 16-1 for bit definitions.

Figure 16-2: HyperTransport Bridge Header Command Register

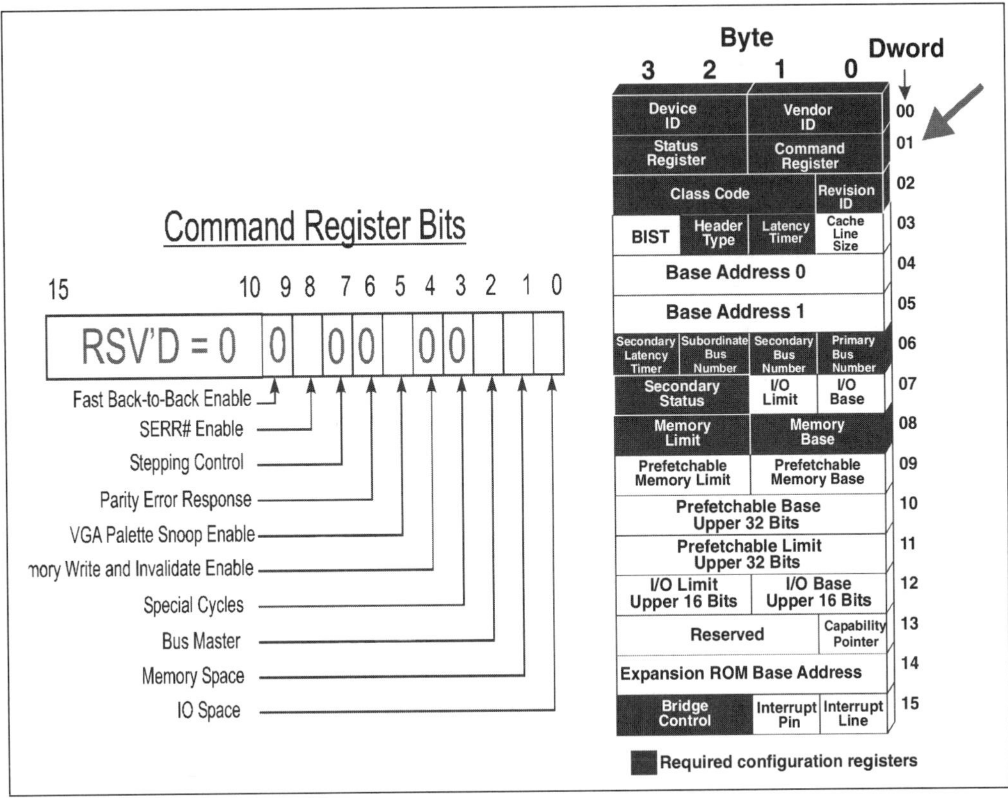

Table 16-1: HyperTransport Bridge Header Command Register Bit Fields

Bit	Function
0	**I/O Space**. When this bit is set to a one, the primary bus interface of the bridge may act as a target of requests in the **I/O** portion of the HyperTransport memory map. If this bridge is also a subtractive decoder, the setting of this bit does not affect the device's ability to claim requests with *Compat* bit set. Warm Reset to 0.

Table 16-1: HyperTransport Bridge Header Command Register Bit Fields

Bit	Function
1	**Memory Space**. When this bit is set to a one, the primary bus interface of the bridge may act as a target of requests in the **Memory** portion of the HyperTransport memory map. If this is a subtractive decoder, the setting of this bit does not affect bridge's ability to claim requests with *Compat* bit set. Warm Reset to 0.
2	**Bus Master**. When set to a one, enables the bridge to forward transactions from the secondary interface to the primary interface. If cleared, the transaction will not be forwarded. If the secondary interface is PCI and targets the primary bus while this bit is clear, it will not be claimed by the bridge (result is master abort). If the secondary interface is HyperTransport, the bridge will set EOC error and issue an error response with NXA set (for non-posted requests).Warm Reset to 0.
5	**VGA Palette Snoop Enable**. (Optional) if not implemented, tie bit = 0. If implemented and set, WrSized (byte) transactions originating on the primary bus targeting the lower 64KB of I/O address range (See System Memory Map) will have address bits 9:0 compared to 3C6h, 3C8h, 3C9h (VGA I/O registers). If there is a match the transaction will be forwarded to the secondary bus. If this bit is set, transactions originating on the secondary bus targeting these addresses will not be forwarded upstream. *Note: if address remapping is in use, the address decoding of the bridge may be undefined when this bit is set = 1.*
8	**SERR# Enable**. When set to a one, the device will flood all outgoing links with sync packets in the event of an error which has been programmed to cause a sync flood. If this bit is clear, device only generates sync packets as a part of initial link synchronization. Bit does not affect device's ability to propagate sync flood packets from one link to another in a chain. *Note: sync flood is similiar to SERR# assertion in PCI.* Warm Reset to 0.
Other Bits	All other bridge header Command Register bits are unused in HyperTransport bridges, and should be tied to a low level.

Bridge Header Status Register

Upper 16 bits at dword 01. The HyperTransport bridge header *Status* register is implemented and used in the same manner as for non-bridges to report status <u>on the primary bus</u>. As in the case of the Command register, a number of these bits are used differently than in PCI, or not at all. Bits marked "0" in Figure 16-3 are not used (hardwired = 0); refer to Table 16-2 on page 414 for bit definitions.

Figure 16-3: HyperTransport Bridge Header Status Register

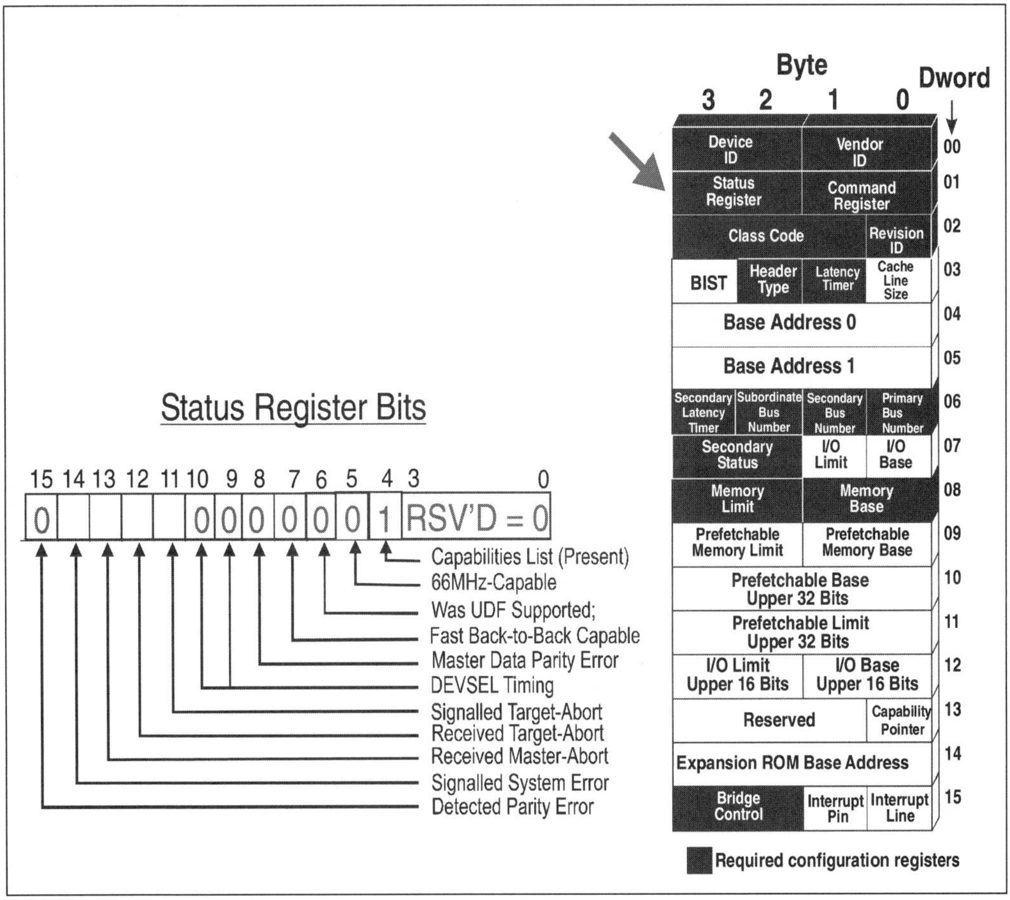

Table 16-2: HyperTransport Bridge Header Status Register Bit Fields

Bit	Function
4	**Capabilities List**. (Introduced in the PCI 2.2 Specification) When this bit is hardcoded to a one, the device indicates there are one or more advanced capability block register sets to be discovered. All HyperTransport devices are required to have at least one advanced capability block (to support HyperTransport), so this bit is always set = 1. When software reads this bit high, it indicates that the Capabilities Pointer (at dword 13) is valid, and points to the first capability register set.
11	**Signalled Target Abort**. This status bit indicates that a HyperTransport bridge, acting as a target on its primary interface, has aborted and generated a *non-NXA error response* to the requestor. Cold Reset = 0.
12	**Received Target Abort**. This status bit indicates that a HyperTransport bridge, acting as a requester on its primary interface, has received a *non-NXA error response* to a request it issued. This is equivalent to the Target Abort event on PCI. Cold Reset = 0.
13	**Received Master Abort**. This status bit indicates that a HyperTransport bridge, acting as a requester on its primary interface, has received an *NXA (non-existent address) error response* to a request it has issued. The NXA error response is returned by the end-of-chain device acting on behalf of the intended target. This is equivalent to the Master Abort event on PCI. Cold Reset = 0.
14	**Signalled System Error**. This status bit indicates that a HyperTransport device has flooded the link with Sync packets (indicating a serious error). A device merely forwarding Sync packets from another device should not set this bit; this allows identification of the offending device(s). A link reset is required before accessing the device. Cold Reset = 0.
Other Bits	All other bridge header Status Register bits are unused in HyperTransport, and should be tied to a low level.

Secondary Status Register

Upper 16 bits at dword 07. If the secondary bus in the bridge is HyperTransport, many of the *Secondary Status* register bits used by PCI-PCI bridges are not used and tied to "0" as shown in Figure 16-4 on page 415; refer to Table 16-3 on page 416 for definitions of the active bits.

Figure 16-4: HyperTransport Bridge Header Secondary Status Register

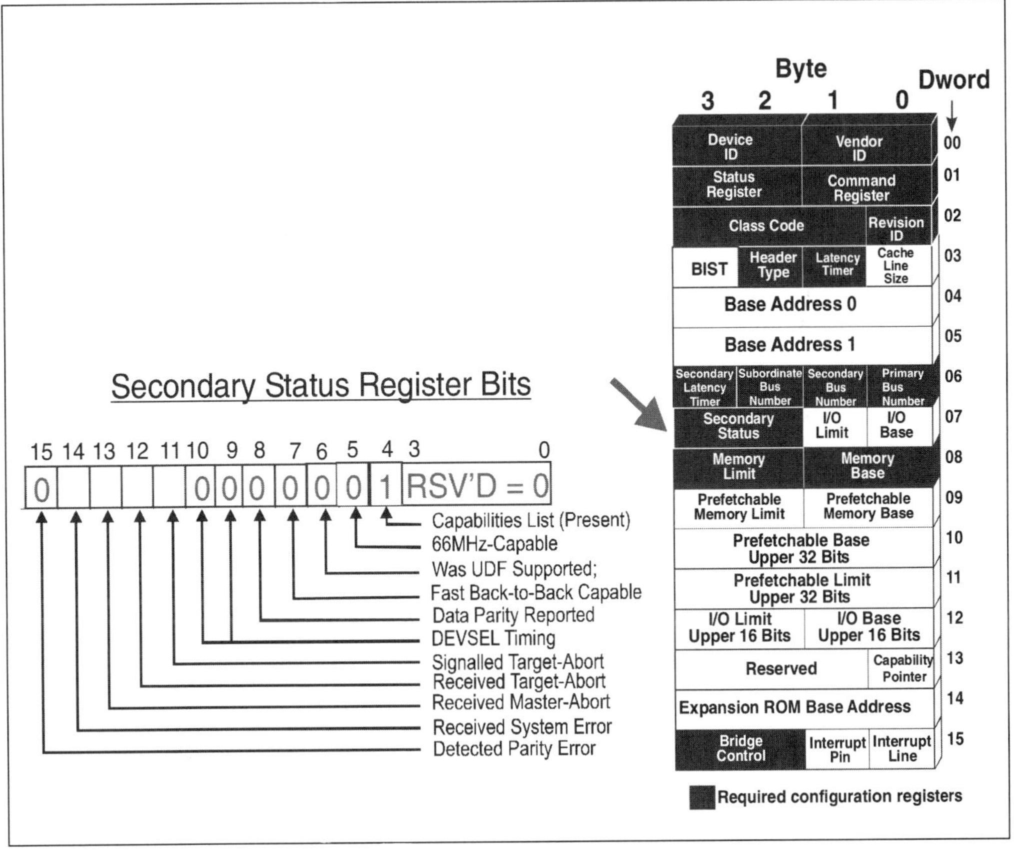

Table 16-3: HyperTransport Bridge Secondary Status Register Bit Fields

Bit	Function
11	**Signalled Target Abort**. When set =1, this bit indicates that the bridge has issued a target abort on the secondary bus. If the secondary bus is HyperTransport, an error response with NXA set was sent because a request packet was received by the bridge in error. Cold Reset = 0.
12	**Received Target Abort**. When set = 1, this bit indicates bridge received a target abort on the secondary bus. If the secondary bus is Hyper-Transport, the bridge received an error response with NXA set to one of its non-posted requests. Cold Reset = 0.
13	**Received Master Abort**. When set = 1, this bit indicates the bridge had to abort a transaction request on the secondary bus. In HyperTrans-port, this means it sent a non-posted directed request which did not find the intended target, and the device at the end of the chain returned an error response with NXA set. Cold Reset = 0.
14	**Detected System Error**. When set = 1, bridge detected a catastrophic error (SERR# in PCI) on the secondary bus. In HyperTransport, this bit indicates that a Sync flood was detected on the secondary bus. Cold Reset = 0.
Other Bits	If the secondary bus is HyperTransport, all other Secondary Status register bits are unused and should be tied to a low level.

Memory And Prefetchable Base And Limit Registers

Dwords 08-11. As is the case of PCI bridges, HyperTransport bridges support two sets of registers for decoding primary bus memory transactions targeting the secondary bus. One set defines the size and range of all prefetchable memory below the bridge and the other defines the size and range of all non-prefetchable memory below it. Figure 16-5 on page 417 illustrates the register sets, and Table 16-4 on page 418 defines the bit usage. Note that the prefetchable memory range has an extra pair of registers (Upper Base/Limit) so the address range may be extended to 64 bits. If the secondary bus is HyperTransport, only the lower 40 bits of the prefetchable memory address are valid (upper 24 bits of the 64 bit address range must be = 0).

Figure 16-5: HyperTransport Bridge Header Memory And Prefetchable Base/Limit Register

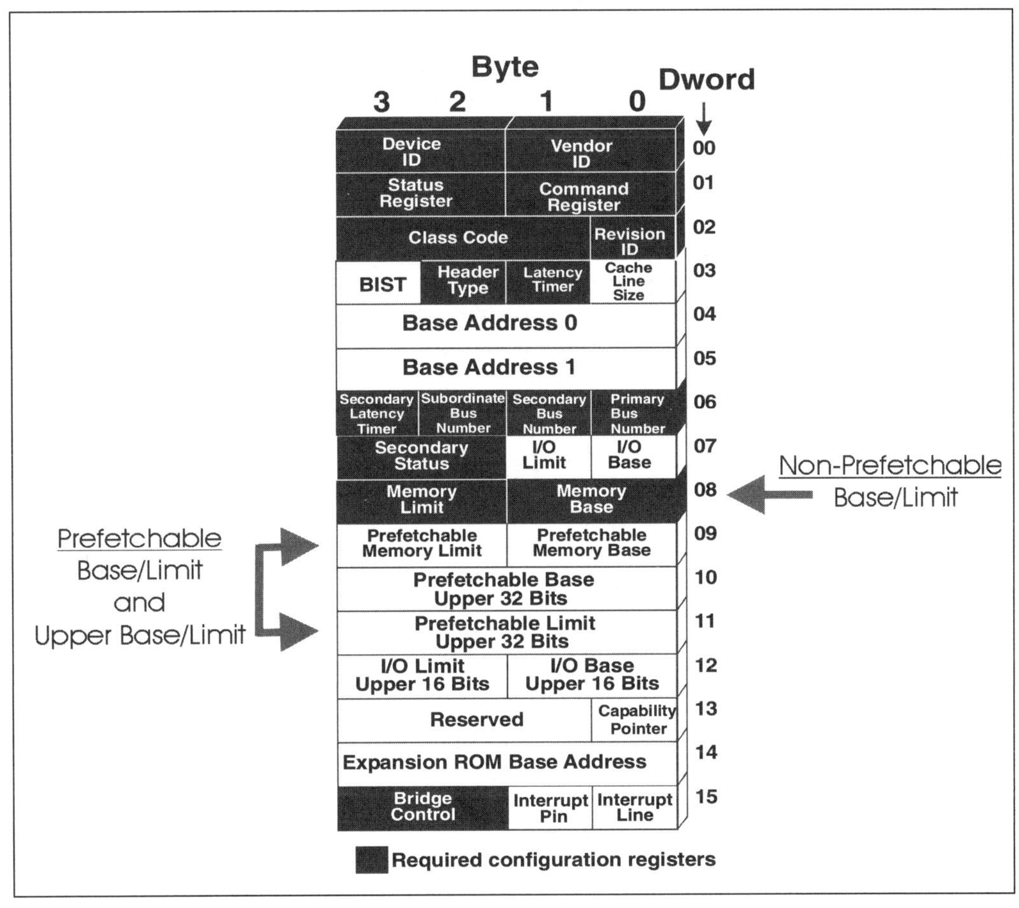

Table 16-4: Bridge Memory And Prefetchable Base And Limit Register Bit Fields

Register	Bits	Function
Memory Base	15:4	Upper 12 bits of the 32 bit start address of non-prefetchable memory. Lower 20 bits are assumed to be 00000h. (Base address is always divisible by 1MB)
	3:0	Bits 3:0 are reserved.
Memory Limit	15:4	Upper 12 bits of the 32 bit ending address for non-prefetchable memory. Lower 20 bits are assumed to be FFFFFh. (Range is always in multiples of 1MB)
	3:0	Bits 3:0 are reserved.
Prefetchable Memory Base	15:4	Upper 12 bits of the 32 bit start address of prefetchable memory. Lower 20 bits are assumed to be 00000h. (Base address is always divisible by 1MB)
	3:1	Bits 3:1 are reserved.
	0	Bit 0 = Size bit. If 1, Prefetchable Base *Upper* register provides an additional 32 bits of address (63:32). If Size = 0, *Upper* register not used.
Prefetchable Memory Limit	15:4	Upper 12 bits of the 32 bit end address of prefetchable memory. Lower 20 bits are assumed to be FFFFFh. (Range is always divisible by 1MB)
	3:1	Bits 3:1 are reserved.
	0	Bit 0 = Size bit. If 1, Prefetchable Limit *Upper* register provides an additional 32 bits of address (63:32). If Size = 0, *Upper* register not used.
Prefetchable Base Upper	31:0	If *Size* bit in the Prefetchable Memory Base register is set = 1, this register provides the upper 32 bits of the 64 bit Prefetchable Base Address. If *Size* bit = 0, this register is not used.

Table 16-4: Bridge Memory And Prefetchable Base And Limit Register Bit Fields

Register	Bits	Function
Prefetchable Limit Upper	31:0	If *Size* bit in the Prefetchable Memory Limit register is set = 1, this register provides the upper 32 bits of the 64 bit Prefetchable Limit Address. If *Size* bit = 0, this register is not used.

Memory And Prefetchable Memory Base/Limit Notes

Bus-To-Bus Forwarding Rules. The Memory and Prefetchable Base and Limit registers are used to decode memory transactions for the purposes of forwarding packets between the primary and secondary bus:

1. Matching addresses on the primary bus in the HyperTransport memory mapped I/O range (Below FD_0000_0000h) are forwarded to the secondary bus.
2. Non-matching addresses on the primary bus (including any above the memory mapped I/O range) are not forwarded downstream.
3. Matching addresses on the secondary bus in the HyperTransport memory mapped I/O range (Below FD_0000_0000h) are not forwarded to the primary bus. If the bridge supports peer-to-peer transfers, it will reissue these requests onto the proper secondary bus in the event of an address match.
4. Non-Matching addresses on the secondary bus (including any above the memory mapped I/O range) are forwarded to the primary bus.

64 Bit Addressing And The 40 Bit HyperTransport Space. If memory addressing supports 64 bit addresses, then incoming HyperTransport requests must have the 40-bit address field 0-extended to 64 bits before comparison.

To Disable Memory or Prefetchable Memory Decoding. If either range is not required, software may program the appropriate base register with a value greater than its corresponding limit register.

The Optional *Address Remapping* Registers. In addition to the Memory and Prefetchable Memory Base and Limit registers just described, Hyper-Transport also supports an optional *Address Remapping Capability Block* which can be used to define additional downstream and upstream windows for positive address decoding. If implemented, these registers augment the memory Base/Limit registers in the bridge header. Refer to Chapter 21, entitled "Address Remapping," on page 477 for a discussion of address remapping.

I/O Base And Limit Registers

Dwords 07 and 12. HyperTransport memory maps I/O transactions into a special address range (FD_FC00_0000h to FD_FD_FFFFh). Once a device performs the 40-bit address decode and determines the request targets I/O, only the lowest 25 bits of the address field are considered valid. All bits above bit 24 are treated as "0", then a comparison is made using the I/O Base/Limit registers described here. Note that the I/O Base and Limit registers at Dword 7 only cover I/O requests in the lower 64KB range; an extra pair of registers (Upper I/O Base and Limit) allow extending the bridge I/O decode range to 32 bits for better PCI compatibility. The 25 bit valid I/O address limit still applies.

Figure 16-6 on page 420 depicts the I/O Base/Limit and Upper Base/Limit registers; Table 16-5 on page 421 provides definitions for each of the register fields.

Figure 16-6: HyperTransport Bridge I/O Base And Limit Register

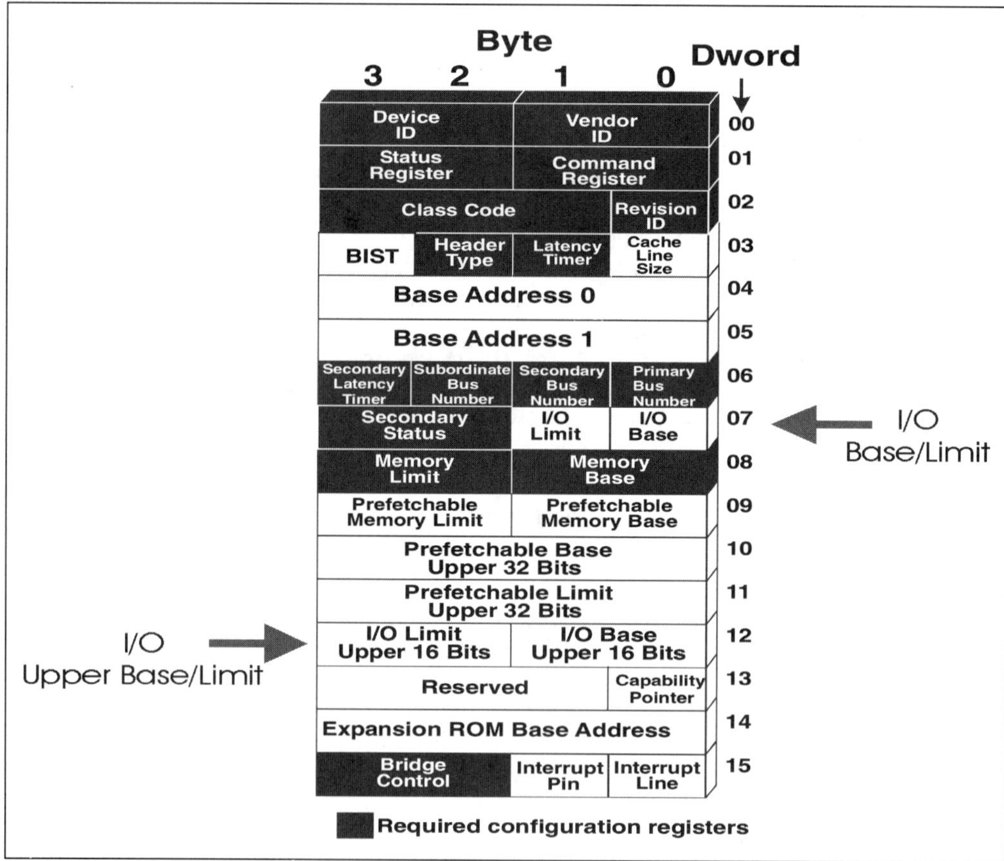

Table 16-5: Bridge I/O Base And Limit Register Bit Fields

Register	Bits	Function
I/O Base	7:4	Upper 4 bits of the 16 bit start address of I/O. Lower 12 bits are assumed to be 000h. (Base address is always divisible by 4KB)
	3:1	Bits 3:1 are reserved.
	0	Size bit. If set = 1, the I/O Base Upper Register will supply the upper 16 bits of I/O Base address (32 bits total= 4GB)
I/O Limit	7:4	Upper 4 bits of the 16 bit ending address for I/O. Lower 12 bits are assumed to be FFFh. (Range is always in multiples of 4KB)
	3:1	Bits 3:1 are reserved.
	0	Size bit. If set = 1, the I/O Limit Upper Register will supply the upper 16 bits of I/O Limit address (32 bits total= 4GB)
I/O Base Upper	15:0	If *Size* bit in the I/O Base register is set = 1, this register provides the upper 16 bits of the 32 bit I/O Base Address. If *Size* bit = 0, this register is not used.
I/O Limit Upper	15:0	If *Size* bit in the I/O Limit register is set = 1, this register provides the upper 16 bits of the 32 bit I/O Limit Address. If *Size* bit = 0, this register is not used.

Base/Limit Notes

Bus-To-Bus Forwarding Rules.

1. Matching I/O addresses on the primary bus are forwarded to the secondary bus; they are ignored if generated on the secondary bus.
2. Non-Matching I/O addresses on the primary bus are not forwarded to the secondary bus.
3. Matching I/O addresses on the secondary bus are not forwarded to the primary bus. If the bridge supports peer-to-peer transactions, it will reissue these requests downstream on the proper chain.
4. Non-Matching addresses on the secondary bus are forwarded to the primary bus.

To Disable I/O Decoding. Software may program the I/O Base Register with a value greater than the I/O Limit Register to disable I/O decoding.

Bridge Control Register

Dword 15. The Bridge Control register is used to manage the bridge secondary bus. If the secondary bus in the bridge is HyperTransport, many of the *Bridge Control* register bits used by PCI-PCI bridges are not used and tied to "0" as shown in Figure 16-7 on page 422; refer to Table 16-6 on page 423 for definitions of the active bits.

Figure 16-7: HyperTransport Bridge Control Register

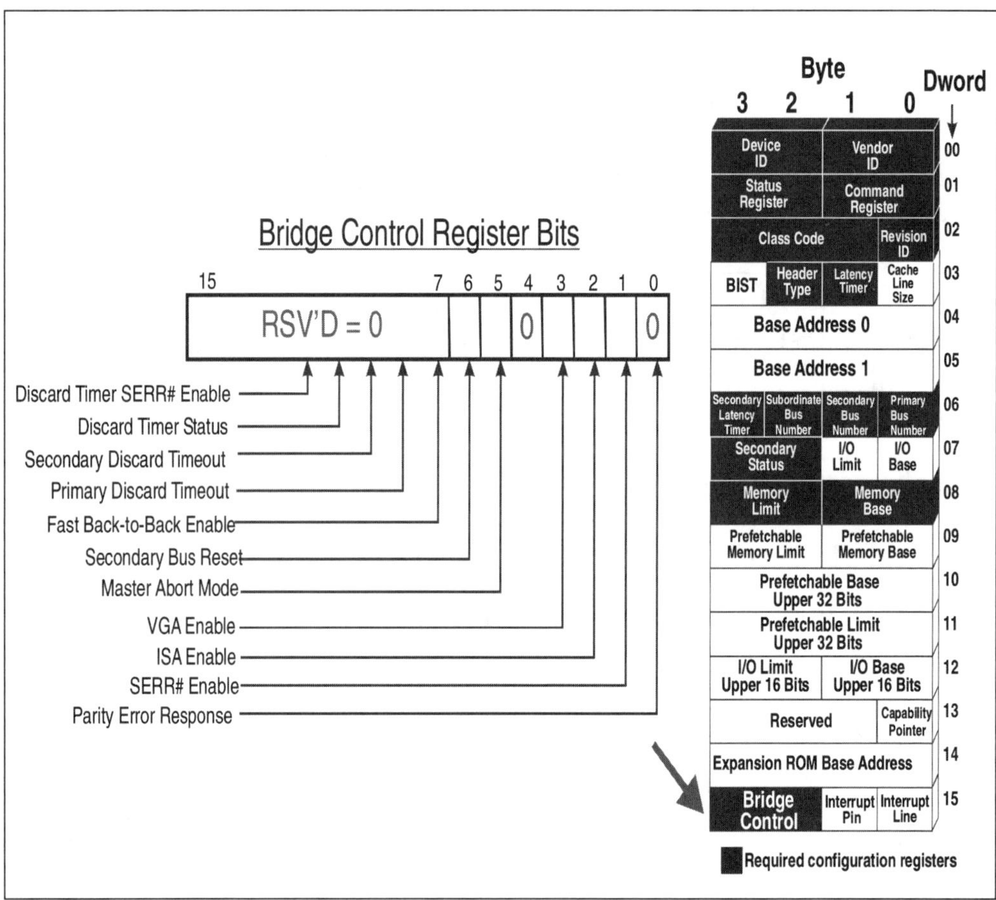

Table 16-6: HyperTransport Bridge Control Register Bit Fields

Bit	Function
1	**SERR# Enable**. When set =1, a HyperTransport Sync flood event detected by the bridge on its secondary bus will be propagated onto the primary bus. If the secondary bus is PCI or PCI-X, an SERR# detected on the secondary bus will be propagated on the HyperTransport primary bus as a Sync flood. Warm Reset = 0.
2	**ISA Enable**. (ISA alias control). When set = 1, this bit causes transactions targeting bytes 256d - 1023d of each 1KB range within the first 64KB of I/O space to be blocked from passing from primary to secondary bus (regardless of I/O Base/Limit register setting). Addresses in these ranges which originate on the secondary bus <u>will be</u> forwarded to the primary bus. This bit is required for PCI compatibility. Warm Reset = 0. *Note: if address remapping is in use, the address decoding of the bridge may be undefined when this bit is set = 1.*
3	**VGA Enable**. When set = 1, RdSized and WrSized requests in the address range 0_000A0000h - 0_000BFFFFh (video buffer) will be forwarded from primary to secondary bus; they will be ignored by the bridge if generated on the secondary bus. I/O accesses in the first 64KB which alias to the ranges 3B0-3BBh or 3C0-3DFh (bits 9:0) will similarly be forwarded from primary to secondary buses. They, too, will be ignored by the bridge if generated on the secondary bus. Warm Reset = 0. *Note: if address remapping is in use, the address decoding of the bridge may be undefined when this bit is set = 1.*
5	**Master Abort Mode**. This bit sets the policy to be used on the originating bus when a transaction forwarded to the destination bus ends in a *master abort* (error response with NXA set). When this bit set = 1, a master abort on the target bus will result in a target abort (error response with NXA set) on the originating bus. When this bit is = 0, a master abort error on the target bus will not be reported on the originating bus (normal response). If the request was a read, the bridge will drive all bytes in the data packet = FFh in either case. Warm Reset = 0.

Table 16-6: HyperTransport Bridge Control Register Bit Fields

Bit	Function
6	**Secondary Bus Reset**. This bit, required if the primary bus is Hyper-Transport, allows the generation of a secondary bus reset under software control. Writing the bit = 1 begins the reset sequence; clearing the bit removes the secondary bus reset. If the secondary bus is HyperTransport, this bit may be used in conjunction with the *Warm Reset* bit in the Host/Secondary Interface Command register to generate either a warm or cold reset event. Warm Reset = 0.
Other Bits	All other Bridge Control Register bits are unused and should be tied to a low level.

Other Fields In The Header

Primary Latency Timer Register

This register is not implemented by HyperTransport devices. Should return 0's if read by software. If primary bus is PCI or PCI-X, use of this register follows that protocol.

Base Address Registers

The two Base Address Registers (BARs) are used by bridges in much the same way as for PCI bridge devices, with the following limits if the primary interface is HyperTransport:

I/O BAR. For an I/O request, a single BAR is implemented. Only the lower 25 bits of the value programmed into the BAR is used for address comparison by the target, and the upper bits of the BAR should be written to zeros by system software. Any I/O request packet sent out on a link should have the start address bits 39-25 programmed for the I/O range in the HyperTransport memory map.

Memory BAR. A request for memory using 32-bit addressing can be accomplished using a single BAR, just as in PCI. This would limit the assigned target start address for the device to the lower 4GB of the 1 TB (40 bit) HyperTransport address map.

Optionally, a HyperTransport device may support 64 bit address decoding, and use a pair of BARs to support it. If this is done, only the lower 40 bits of the 64 bit BAR memory address will be valid, and the upper bits are assumed to be zeros.

Memory windows for HyperTransport devices are always assigned in BARs on 64-byte boundaries; this assures that even the largest transfer (16 dwords/64 bytes) will never cross a device address boundary. This is important because HyperTransport does not support a disconnect mechanism (such as PCI uses) to force early transaction termination.

Capabilities Pointer

This field contains a pointer to the first advanced capability block. Because all HyperTransport bridge devices have at least one advanced capability, this register is always implemented. The pointer is an absolute byte offset from the beginning of configuration space to the first byte of the first advanced capability register block.

Interrupt Line Register

The HyperTransport specification indicates that this register should be read-writable and may be used as a software scratch pad. The information routing information programmed into this register in PCI devices isn't required in HyperTransport because interrupt messages are sent over the links and sideband interrupts are not defined. If the primary bridge interface is PCI or PCI-X, this register is used by software to program the system interrupt mapped to this device.

Interrupt Pin Register

This register is reserved in the HyperTransport Specification. It may optionally be implemented for compatibility with software which may expect to gather interrupt pin information from all PCI-compatible devices. If the primary bus interface is PCI or PCI-X, this register is hard-coded with the interrupt pin driven by this device (if any).

Cache Line Size Register

This register is not implemented by HyperTransport devices. If both interfaces are HyperTransport, bit should be tied low and read back as 0's if read by software. If either interface is PCI, this register is read-write.

Double-Hosted Chains

The Previous Chapter

The previous chapter described the configuration of devices that use the Hyper-Transport technology *type 1* configuration header for bridges. Such devices include HyperTransport-to-HyperTransport bridges and bridges to other PCI compatible protocols (e.g. HyperTransport-to-PCI or PCI-X). The basic architecture of a HyperTransport-to-HyperTransport bridge is reviewed and the configuration header fields are described. Differences in usage of bit fields by HyperTransport bridge interfaces vs. PCI bridge interfaces are emphasized. The format of PCI compatible bridge headers is formally defined in the *PCI-to-PCI Bridge Architecture Specification, Revision 1.1.*

This Chapter

This chapter describes the features of the optional HyperTransport double-hosted chain topologies. Topics include the reasons behind sharing and non-sharing chains, PCI configuration space registers used to initialize the fabric for multiple hosts, and tunnel support for upstream and downstream packets moving in both directions.

The Next Chapter

HT provides a variety of mechanisms to support power management. These mechanisms include LDTSTOP#, LDTREQ#, STPCLK messages, and STOP_GRANT messages. While these mechanism are optional for HT devices, the specification requires this support for x86-based platforms. Note also that functions other than power management may make use of these signals and messages. This chapter discusses the strategy employed by HT for implementing power management and how a given platform can use these mechanisms to support power management.

Introduction

A HyperTransport chain consists of a host bridge at one end and some collection of devices connected to it in a daisy-chain arrangement. At the end of the chain, there is a device with a single-link connection. This could either be an end (I/O hub) device, or a multi-link device (e.g. tunnel) which has its downstream link disabled.

By contrast, a double-hosted chain has a host bridge at either end and some collection of multi-link devices between them. Figure 17-1 on page 428 illustrates a double-hosted chain. Note that there are no end (I/O hub) devices in a double-hosted chain.

Figure 17-1: HyperTransport Double-Hosted Chain Configuration

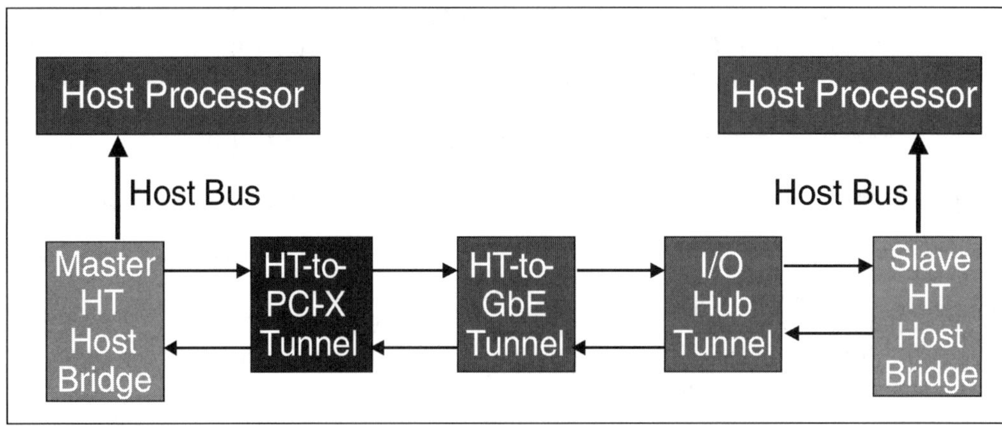

Reasons For Implementing A Double-Hosted Chain

A double hosted-chain can be useful in fault-tolerant applications where a backup host interface takes over in the event of a failure of the primary interface. It also permits the sharing of a single set of resources and inter-processor communications by two CPUs in a clustering arrangement. *Note:* the Hyper-Transport I/O Link Specification Network Extensions allow extending the multiple-host concept to broader topologies using *switch* and *router* components. Refer to Chapter 19, entitled "Networking Extensions Overview," on page 443.

PCI Configuration Plays Key Role In Chain Setup

PCI configuration cycles are used to program bridges, tunnels, and end devices in each HyperTransport all chain. Two key registers used in setting up double-hosted chain (DHC) parameters are the HyperTransport *Host Command CSR* for host bridges and the HyperTransport *Slave Command CSR* for interior devices such as tunnels. These two registers and the key fields pertaining to double-hosted chains are described below. Refer to Chapter 13, entitled "Device Configuration," on page 305 for a more complete description of PCI device configuration.

Slave Command CSR. Figure 17-2 and Table 17-1 on page 429 show the format of the Slave Command Register used by all non-host interfaces; key fields used in double-hosted chain configuration are highlighted.

Figure 17-2: Slave Command CSR: Key Fields In DHC Configuration

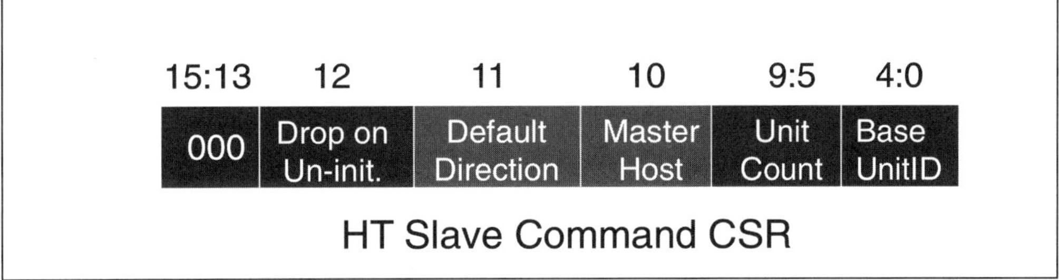

Table 17-1: Slave Command CSR: Definitions Of Key Fields In DHC Configuration

Bit	Function
10	**Master Host.** This read-write bit is set by hardware automatically to indicate which link is attached to the host. Any write to the Command register will cause this bit to be set to indicate which link the write arrived on. Warm Reset = 0.
11	**Default Direction.** This read-write bit determines the default direction requests should be sent when originating with this device. A "0" in this register indicates requests should be sent in the direction of the master host (see previous bit). A "1" indicates requests should be sent in the other direction. Bit has no meaning for a single-link device. Warm Reset = 0.

Host Command CSR. Figure 17-3 and Table 17-2 show the format of the Host Command Register used by all host interfaces; key fields used in double-hosted chain configuration are highlighted.

Figure 17-3: Host Command CSR: Key Fields In DHC Configuration

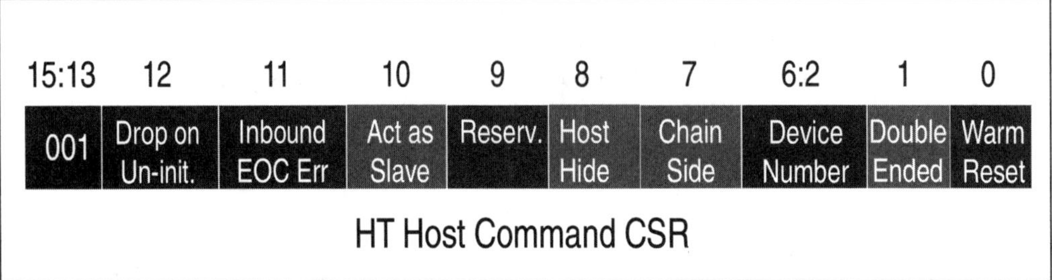

Table 17-2: Host Interface Block Host Command CSR Bit Assignment

Bit	Function
1	**Double Ended**. If this read-write bit is set = 1, there is another bridge at the other end of the chain (double-hosted chain). This bit does not affect hardware and can be used by software during initialization. If not implemented, hardwire this bit = 0. Cold reset = 0.
7	**Chain Side**. This bit is set to indicate which side of the host bridge is being accessed. A "0" in this field indicates the read is coming from within; a "1" indicates the read is coming from the chain attached to the host interface. If double-hosted chains are not supported, this bit is hardwired = 0.
8	**Host Hide**. This read-write bit is used to hide a bridge's configuration space from accesses coming from the chain side. If the bit is set = 1, the host will behave as an end-of-chain device during configuration cycles; if it is clear, configuration accesses from the chain are allowed by the device. For hosts which support DHCs, this bit is cleared on warm reset. If a host does not support double-hosted chains, this bit is hardwired = 1.
10	**Act As Slave.** This bit, when set, causes host to act as a slave, using the device number programmed in bits 6:2 as the base UnitID for requests and responses it sources. In this mode, the host also won't set the Bridge bit in its responses. If this bit is clear, interface behaves as a host — using UnitID 0, setting the Bridge bit on responses it sends, etc. If host doesn't support double-hosted chains, this bit is hardwired = 0. Cold reset = 0.

Two Types Of Double-Hosted Chains

There are two basic arrangements for double-hosted chains: *sharing* and *non-sharing*.

Sharing Double-Hosted Chain

In a sharing double-hosted chain, traffic is allowed to flow from end to end. Either host may target any of the devices in the chain, including the other host. In this arrangement, one host is the master host bridge and the other is the slave host bridge. The determination about which host is master or slave is not defined in the specification, but must be defined before reset occurs. Most likely, the system board layout will determine master/slave host bridges — possibly through a strapping option on the motherboard. Figure 17-4 on page 431 depicts a sharing double-hosted chain with master and slave host bridges.

Figure 17-4: Sharing Double-Hosted Chain With Master/Slave Host Bridges

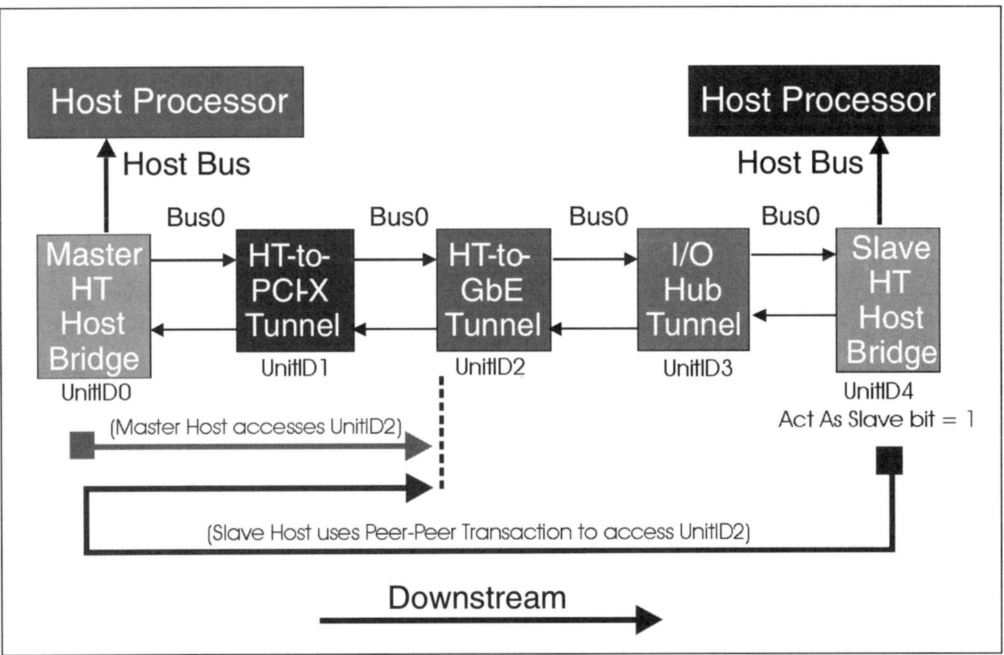

If Possible, Assign All Devices To Master Host Bridge

The HyperTransport specification recommends that all resources in a sharing double-hosted chain be assigned to the master host bridge if possible; this eliminates a potential deadlock condition in peer-to-peer transactions. The Slave Command Register *Master Host* and *Default Direction* bits in PCI configuration space are used to program tunnel devices with the information needed to recognize the "upstream vs. downstream" directions. This is important because interior devices always issue requests and responses in the upstream direction. They only accept responses in the downstream direction.

If Slave Must Access Devices, It Uses Peer-to-Peer Transfers

As illustrated in Figure 17-4 on page 431, the slave host in a sharing double-hosted chain may be required to access the devices on the link. To do so, it may have its Command Register *Act as Slave* bit set = 1. When this is done, all packets it issues travel first to the master host bridge where they are reissued back to the target devices as peer-to-peer transactions.

Non-Sharing Double-Hosted Chain

A non-sharing double-hosted chain appears logically as two distinct chains with a host bridge at each end.

Software May Break The Chain

Software chooses a point to break the chain in two parts and then:

1. While the link is idle, the link between the two tunnel devices is broken by programing the *End Of Chain* (EOC) bits in the appropriate tunnel Link Control registers on each side. The *Transmit Off* bit in each of the Link Control registers can also be set.
2. The slave host bridge writes to the Slave Command register for each device now under its control to force the *Master Host* and *Default Direction* bits in each to point at the slave host bridge.
3. Unique bus numbers are assigned to each segment in a non-sharing double-hosted chain. The bus number is used so that chains may be uniquely identified and so type 1 configuration cycles may be forwarded and/or converted to type 0 cycles by bridges.
4. If peer-to-peer transactions are not required, software link partitioning can also be used for load balancing.

Figure 17-5 on page 433 depicts a non-sharing double-hosted chain.

Figure 17-5: Non-Sharing Double-Hosted Chain

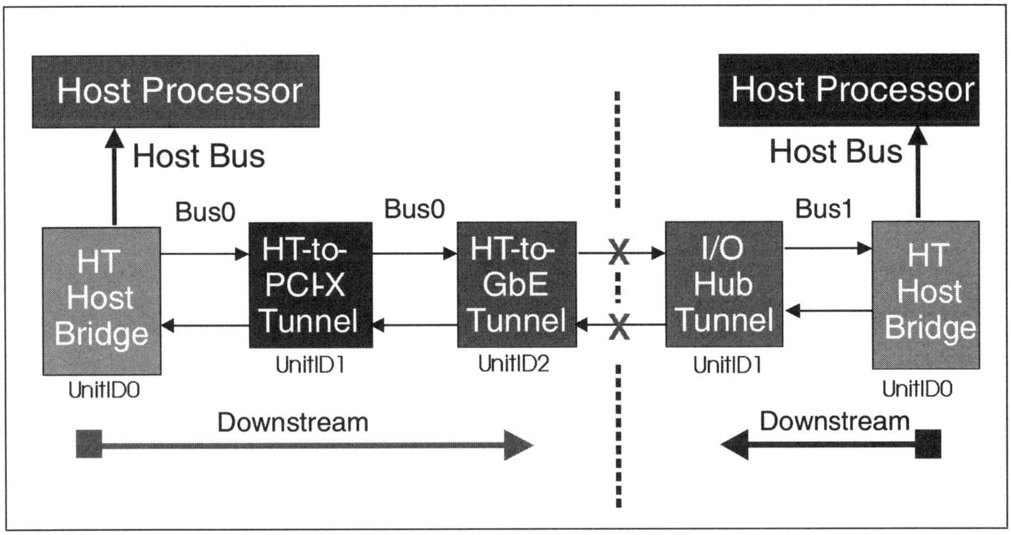

Additional Notes About Double-Hosted Chains

Initialization In A Double-Hosted Chain

One of the responsibilities of a master host bridge in a double-hosted chain is to help with initialization after reset. Following low-level link initialization, the slave host bridge "sleeps" pending set up by the master. The basic steps in master initialization include:

1. The master host bridge sets the Slave Command CSR *master host* bit to point towards the master host bridge in all slave devices it finds. This bit is set automatically whenever the Slave Command CSR is written.
2. When the master host bridge discovers the slave host bridge, it sets the Host Command CSR *Double Ended* bit in the both its own and the slave's Host Command register. This informs the slave (when it wakes up) that it is in a double-hosted chain and that it is not required to configure devices below it.
3. If the *Double Ended* bit is not set in the slave, it will initialize its end of the double ended chain when it awakens.

Type 0 Configuration Cycles In A Double-Hosted Chain

Because all host bridges tend to own UnitID 0, a configuration cycle carrying a device number field of "0" in a double-hosted chain might be misinterpreted. The direction a type 0 configuration cycle request is traveling determines which host bridge is the target. If configuration software wishes to prevent a host bridge (e.g. the slave host) in a double-hosted chain from accessing another host's configuration space, the Host Command Register *host hide* bit may be set = 1.

HT Power Management

The Previous Chapter

The previous chapter described the features of the optional HyperTransport double-hosted chain topologies. Topics include the reasons behind sharing and non-sharing chains, PCI configuration space registers used to initialize the fabric for multiple hosts, and tunnel support for upstream and downstream packets moving in both directions.

This Chapter

HT provides a variety of mechanisms to support power management. These mechanisms include LDTSTOP#, LDTREQ#, STPCLK messages, and STOP_GRANT messages. While these mechanisms are optional for HT devices, the specification requires this support for x86-based platforms. Note also that functions other than power management may make use of these signals and messages. This chapter discusses the strategy employed by HT for implementing power management and how a given platform can use these mechanisms to support power management.

The Next Chapter

The next chapter summarizes some of the major additions to the HyperTransport protocol which will be forthcoming in Release 1.05 and Release 1.1 of the specification. Collectively, these additions are referred the HyperTransport *Networking Extensions*, and target some of the special requirements of communications processing. Key features include a message passing protocol for larger packets, a formal definition for *switch* devices, a link-level error recovery method, sixteen optional additional posted write virtual channels with defined arbitration and bandwidth allocation, direct peer-peer transfers, an increase in the number of outstanding transactions for host bridges, and a 64-bit addressing option.

Background

Over the years, power management control has migrated to the operating system in many platforms. System and I/O devices designers provide registers for the OS to control power at the function, device, bus, and system levels.

- Hardware registers for power management reside in chipsets and IO devices (e.g. PCI).
- Transitions in power state typically occur under software control as chipset hardware detects inactivity time-outs and interrupts the processor, which in turn executes the power management routines.

Some buses, such as PCI, define power management registers for each function that can be programmed to cause changes in power states and to enable wakeup, if supported (e.g. modem wake-up). Other buses like the ISA bus appeared before power management was widely implemented, making it difficult to implement power management. That is, lacking a standard set of configuration registers, an ISA device doesn't play well in either power management or plug-and-play schemes. The HT specification defines the signals LDTSTOP# and LDTREQ# (LDT Request) to support power management, but does not define a standard set of registers for meeting low power requirements. Instead, to meet platform power management requirements, devices can gate clocks, stop PLLs, and power down portions of the device after the LDTSTOP# signal is asserted. Remote wakeup can be implemented using LDTREQ#.

In addition to LDTSTOP# and LDTREQ#, the STPCLK and STOP_GRANT messages can also be used to support power management activities.

Reporting Power Management Events to the Host Bridge

HT provides two System Management messages that can be used to report power management events to the host.

- STPCLK message — uses the SMAF field to define the type of event being reported and the action to be taken. The SMAF field encoding is not defined by the HT specification.
- SMI message — used to notify (interrupt) the processor when a system management event occurs. The SMI message can support power management as will as other features. (Note: SMI is an x86 legacy signal and is discussed in more detail in "X86 CPU Compatibility" on page 491.)

Message transmission may be stimulated by either hardware or software. A system that supports power management will very likely include a set of registers within the chipset (e.g., the South bridge or ICH) to support power management. For example, a chipset may give system software the ability to select which hardware events will cause the chipset to send a power management notification message to the Host Bridge. Similarly, software (e.g. ACPI software) may access a register, causing a power management to be sent. The actual entity responsible for sending these message to the Host Bridge is the System Management Controller (SMC).

Reporting Host Power Management Events to SMC

The Host Bridge may also need to report power management messages to the System Management Controller residing on the HT bus (e.g. within the South bridge). The mechanism for reporting such events is the STOP_GRANT message. Like the STPCLK message, the SMAF field defines the type of event being reported and the action required.

Processor VID/FID

The specification defines a STOP_GRANT message for Voltage ID and Frequency ID changes associated with the processor. The Athlon processors support changing the frequency and voltage used by the processor as a means of power conservation. That is, if the processor clock frequency is slowed, the voltage may also be lowered for greater power conservation. When the host initiates a VID/FID change, this indicates that the system is entering a low-power state. Consequently, the north bridge must generate a STOP_GRANT message with an SMAF code specifying the VID/FID change. This message results in the assertion of LDTSTOP#.

Reporting Power Management Events to HT Devices

In response to a STOP_GRANT message, the SMC controller may assert LDT-STOP to all HT device interfaces as illustrated in Figure 18-1 on page 438. For example, when the Host Bridge detects that the Processor is entering a VID/FID change, it will send a STOP_GRANT message indicating the change and causing the SMC to assert LDTSTOP#. Details regarding the timing relationships between the reception of STOP_GRANT and the assertion of LDTSTOP# is discussed in "The Link Initialization Disconnect Sequence" on page 227.

As described earlier, devices can gate clocks, stop PLLs, and power down portions of the device after the LDTSTOP# signal is asserted. However, LDTSTOP# can be asserted for a variety of reasons besides power management. In such cases, it may be inappropriate for the device to enter a low power state. If a device needs to differentiate between the causes of LDTSTOP# assertion, it must monitor STOP_GRANT cycles and decode the SMAF code so that it knows the reason for LDTSTOP# being driven by the SMC.

Figure 18-1: LDTSTOP# is an Input to All HT Devices Except the SMC.

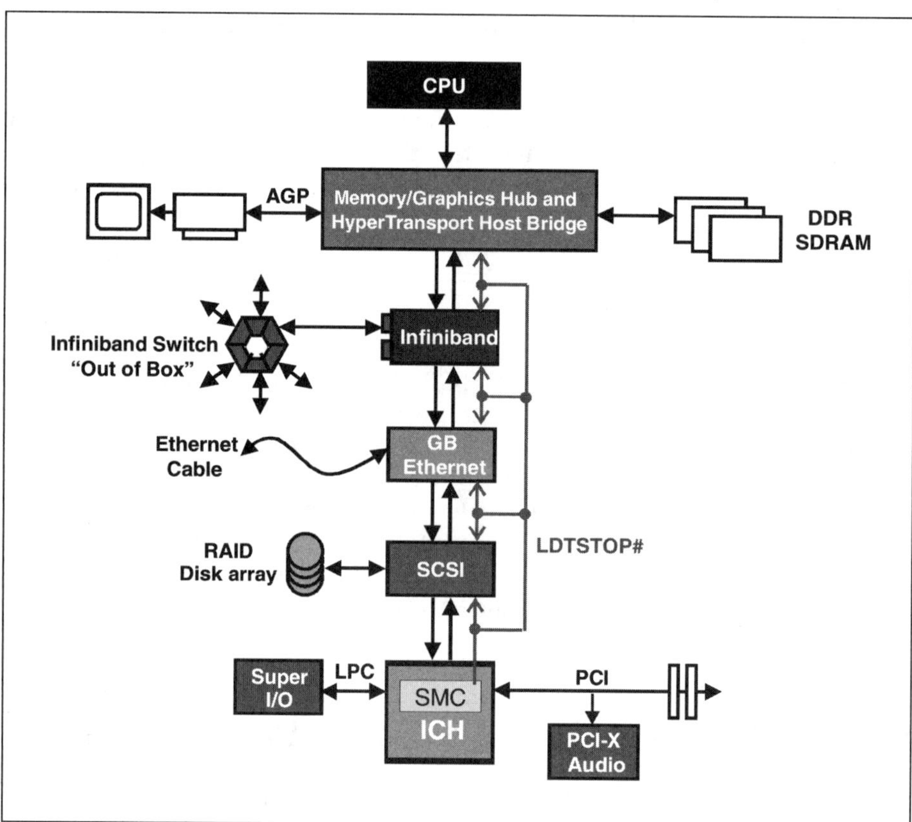

Signaling Wakeup

Most power management schemes provide a mechanism that allows a device to initiate a sequence that returns the system from a power conservation state to a fully operational state. HT provides this ability via the LDTREQ# signal. Figure 18-2 shows that the LDTREQ# signal is an output from each HT device interface and an input to the SMC.

Figure 18-2: LDTREQ# is an Output from All HT Devices and an Input to the SMC.

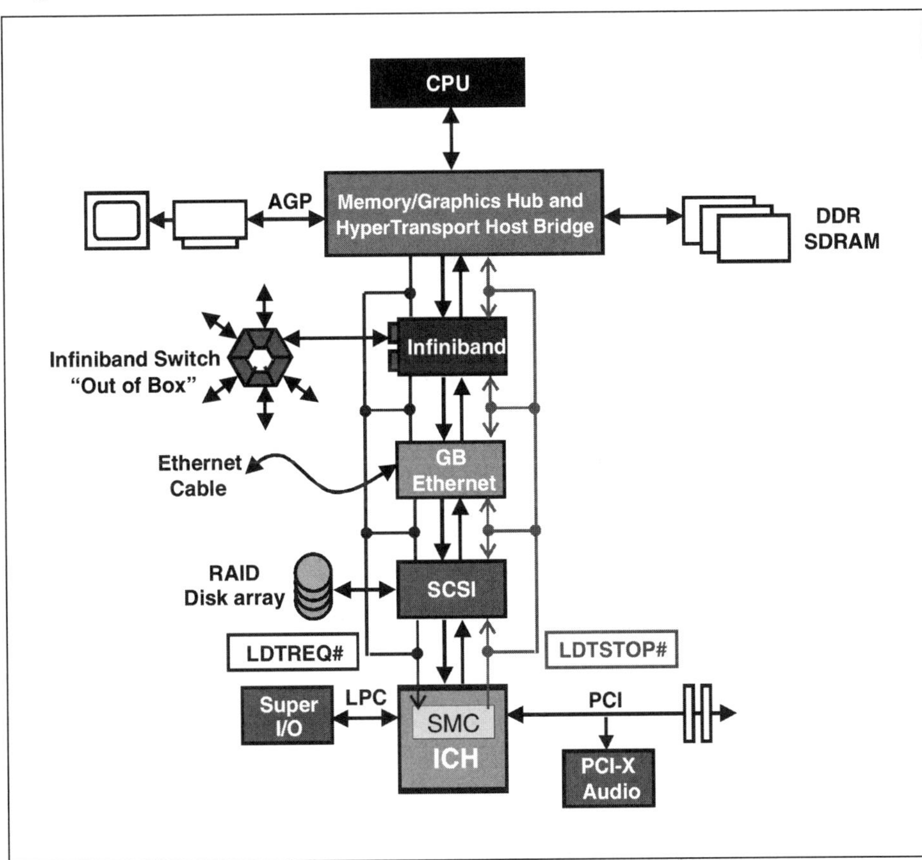

An HT device currently in a low-power state (LDTSTOP# asserted) may assert LDTREQ# to initiate a system wakeup. The actions taken by the SMC to transition the system back to normal operation includes:

- deasserting LDTSTOP#
- Request the host to return to normal operation (via a STPCLK message)

A HT device that might be required to implement LDTREQ# is an HT-to-PCI-X bridge. See Figure 18-3. PCI-X requires support for power management, including the possibility of a PCI-X device signaling a wakeup event via the SME# (System Management Event) signal. The assertion of SME# is intended to awaken the system, and in an HT platform this is accomplished via LDTREQ#. Note that if the HT SMC is located within the PCI-X bridge, then the SMC would simply deassert LDTSTOP# to initiate the system wakeup.

Figure 18-3: Example Wakeup Signaled by HT-to-PCI-X Bridge.

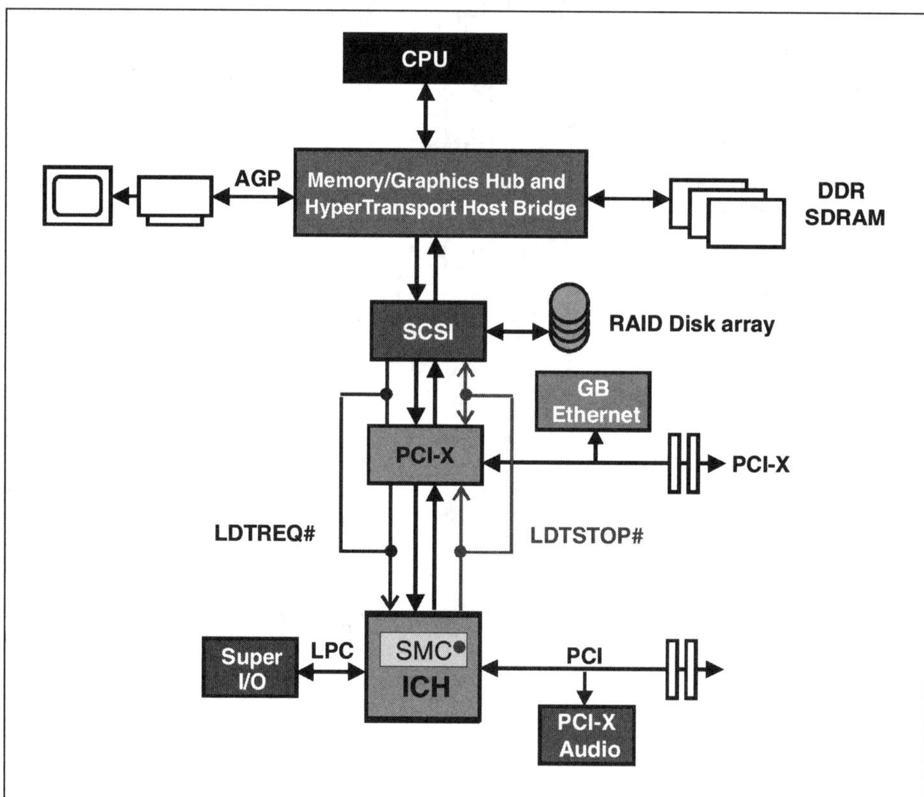

X86 Power Management Support

X86 power management is based on the ACPI specification for the Windows operation environment. The specification defines specific timing requirements associated with STPCLK and SMI message cycles related to power management events. The specification also describes ACPI-defined system state transitions that relate to wakeup event signaling via LDTREQ#. See the specification for reference information related to these events.

Stop Clock Signal

The STPCLK# is one of the basic x86 power management signals. When power management logic asserts this signal, it places the CPU into its Stop Grant State, which has the following effects (Intel PIII example). The processor:

- issues a Stop Grant Acknowledge transaction
- stops driving the AGTL FSB signals, allowing them to return to the minimum power state (pulled up by termination resistors to VTT)
- turns off clocks to internal architecture regions, except external bus (FSB) and interrupt sections (e.g. IOAPIC).
- latches incoming interrupts, but does not service them until the CPU returns to the Normal State.
- handles requests for Snoop transactions on the FSB; to do this the CPU transitions to the HALT/Grant Snoop State to perform the snoop, then returns to the Stop Grant State upon completion.

When STPCLK# is deasserted, the CPU returns to the Normal State. Many newer CPU's have an additional signal which may be used to expand the number of low power states. For example, the Intel Pentium III has a SLP# (Sleep) signal used in conjunction with STPCLK# to drive the CPU into a very deep low power state (e.g., clocks are stopped, no interrupts are recognized, and no snoops are performed). This is the next best thing to being powered down completely, and the time to recover to normal operation is much faster.

HT Method of STPCLK# Signaling

As described in Chapter 9, STPCLK is signaled by the System Management Controller. The SM Request packet content that defines STPCLK is illustrated in Figure 18-4 on page 442. Note that the lower nibble of the *SysMgtCmd* defines the SMAF (System Management Action Field). Bit 0 of this field (labeled "S" for state) defines the new state of STPCLK# signal. Bits [3:1] define the power management action to be taken in association with STPCLK.

Figure 18-4: SM Request Packet Contents for Delivering STPCLK

Bytes \\ Bits	7	6	5	4	3	2	1	0
0	SeqID[3:2]		Cmd[5:0] = 101000 (posted, Wr(sized), byte)					
1		SeqID[1:0]		UnitID[4:0]				
2	Count[1:0]		Reserved					
3	Reserved						Count[3:2]	
4	SysMgtCmd[7:0] = STPCLK 0011 xxxSb							
5	Addr[23:20] = 1h				Reserved			
6	Addr[31:24] = F9h							
7	Addr[39:32] = FDh							

19 Networking Extensions Overview

The Previous Chapter

HT provides a variety of mechanisms to support power management. These mechanisms include LDTSTOP#, LDTREQ#, STPCLK messages, and STOP_GRANT messages. While these mechanism are optional for HT devices, the specification requires this support for x86-based platforms. Note also that functions other than power management may make use of these signals and messages. The previous chapter discusses the strategy employed by HT for implementing power management and how a given platform can use these mechanisms to support power management.

This Chapter

This chapter summarizes some of the major additions to the HyperTransport protocol which will be forthcoming in Release 1.05 and Release 1.1 of the speci-fication. Collectively, these additions are referred the HyperTransport *Network-ing Extensions,* and target some of the special requirements of communications processing. Key features include a message passing protocol for larger packets, a formal definition for *switch* devices, a link-level error recovery method, six-teen optional additional posted write virtual channels with defined arbitration and bandwidth allocation, direct peer-peer transfers, an increase in the number of outstanding transactions for host bridges, and a 64-bit addressing option.

The Next Chapter

HT is designed to support a variety of I/O and processor buses via bridges. The specification defines specific requirements for supporting PCI, PIC-X, AGP, and processor buses. The next chapter discusses these support requirements.

HyperTransport System Architecture

An Important Note

At the time of this writing, the HyperTransport I/O Link Specification Revision 1.04 is released. The 1.04 revision of the specification does not deal with the networking extensions but much of the work on the HyperTransport 1.05 and 1.1 revisions has been done by the HyperTransport Technical Working Group, and quite a bit of preliminary information on this important addition to the protocol has been released.

The networking extensions are backward compatible with the 1.04 and earlier revisions. Compatibility extends to any mix of devices which may or may not support the extensions.

This chapter presents material based on information currently available. Check http://www.hypertransport.org for updated information on all revisions to the specification.

Server And Desktop Topologies Are *Host-Centric*

As illustrated in Figure 19-1 on page 445, a typical desktop or server platform is somewhat vertical. It has one or more processors at the top of the topology, the I/O subsystem at the bottom, and main system DRAM memory in the middle acting as a holding area for processor code and data as well as the source and destination for I/O DMA transactions performed on behalf of the host processor(s). The host processor plays the central role in both device control and in processing data; this is sometimes referred to as managing both the *control plane* and the *data plane*.

HyperTransport works well in this dual role because of its bandwidth and the fact that the protocol permits control information including configuration cycles, error handling events, interrupt messages, flow control, etc. to travel over the same bus as data — eliminating the need for a separate control bus or additional sideband signals.

Figure 19-1: Host-Centric HyperTransport System

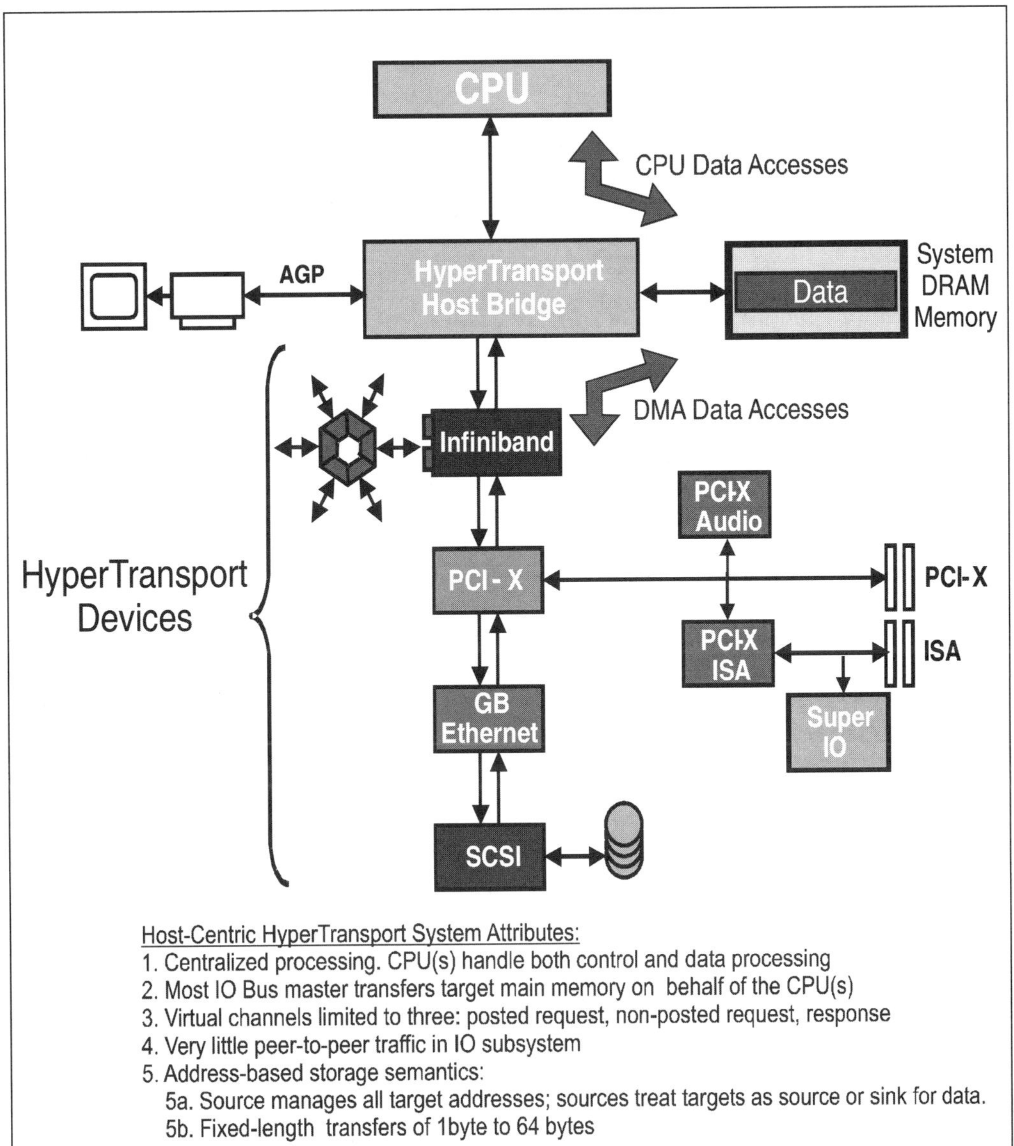

Host-Centric HyperTransport System Attributes:
1. Centralized processing. CPU(s) handle both control and data processing
2. Most IO Bus master transfers target main memory on behalf of the CPU(s)
3. Virtual channels limited to three: posted request, non-posted request, response
4. Very little peer-to-peer traffic in IO subsystem
5. Address-based storage semantics:
    5a. Source manages all target addresses; sources treat targets as source or sink for data.
    5b. Fixed-length transfers of 1byte to 64 bytes

Upstream And Downstream Traffic

There is a strong sense of *upstream* and *downstream* data flow in server and desktop systems because very little occurs in the system that is not under the direct control of the processor, acting through the host bridge. Nearly all I/O initiated requests move upstream and target main memory; peer-peer transactions between I/O devices are the infrequent exception.

Storage Semantics In Servers And Desktops

Without the addition of networking extensions, HyperTransport protocol follows the conventional model used in desktop and server busses (CPU host bus, PCI, PCI-X, etc.) in which all data transfers are associated with memory addresses. A write transaction is used to store a data value at an address location, and a read transaction is used to later retrieve it. This is referred to as associating *storage semantics* with memory addresses. The basic features of the storage semantics model include:

Targets Are Assigned An Address Range In Memory Map

At boot time, the amount of DRAM in the system is determined and a region at the beginning of the system address map is reserved for it. In addition, each I/O device conveys its resource requirements to configuration software, including the amount of prefetchable or non-prefetchable memory-mapped I/O address space it needs in the system address map. Once the requirements of all target devices are known, configuration software assigns the appropriate starting address to each device; the target device then "owns" the address range between the start address and the start address plus the request size.

Each Byte Transferred Has A Unique Target Address

In storage semantics, each data packet byte is associated with a unique target address. The first byte in the data packet payload maps to the start address and successive data packet bytes are assumed to be in sequential addresses following the start address.

The Requester Manages Target Addresses

An important aspect of storage semantics is the fact that the requester is completely responsible for managing transaction addresses within the intended target device. The target has no influence over where the data is placed during write operations or retrieved in read operations.

In HyperTransport, the requester generates request packets containing the target start address, then exchanges packets with the target device. The maximum packet data payload is 64 bytes (16 dwords). Transfers larger than 64 bytes are comprised of multiple discrete transactions, each to an adjusted start address. Using HyperTransport's storage semantics, an ordered sequence of transactions may be initiated using posted writes or including a non-zero *SeqID* field in the non-posted requests, but there is no concept of streaming data, per se.

Storage Semantics Work Fine In Servers And Desktops

As long as each requester is programmed to know the addresses it must target, managing address locations from the initiator side works well for general purpose data PIO, DMA, and peer-peer exchanges involving CPU(s), memory and I/O devices. When the target is prefetchable memory, storage semantics also help support performance enhancements such as write-posting, read pre-fetching, and caching — all of which depend on a requester having full control of target addresses.

1.04 Protocol Optimized For Host-Centric Systems

Because the HyperTransport I/O Link Protocol was initially developed as an alternative to earlier server and desktop bus protocols that use storage semantics (e.g. PCI), the 1.04 revision of the protocol is optimized to improve performance while maintaining backwards compatibility in host-centric systems:

1. The strongly ordered producer-consumer model used in PCI transactions which guarantees flag and data coherence regardless of the location of the producer, consumer, flag location, or data storage location is available in the HyperTransport protocol.
2. Virtual channel ordering may optionally be relaxed in transfers where the full producer-consumer model is not required.
3. The strong sense of upstream and downstream traffic on busses such as PCI is also preserved in HyperTransport. Programmed I/O (PIO) transactions move downstream from CPU to I/O device via the host bridge. I/O bus master transactions move upstream towards main memory.
4. Direct peer-peer transfers are not supported in the 1.04 revision of the HyperTransport I/O Link Specification; requests targeting interior devices must travel up to the host bridge, then be reissued (reflected) back downstream towards the target.

All of the above features work well for what they are intended to do: support a host-centric system in which control and data processing functions are both handled by the host processor(s), and I/O devices perform DMA data transfers using main system memory as a source and sink for data.

Some Systems Are Not Host-Centric

Unlike server and desktop computers, some processing applications do not lend themselves well to a host-centric topology. This includes cases where there are multiple levels of processing, complex look-up functions, protocol translation, etc. In these cases, a single processor (or even multiple CPUs on a host bus) can quickly become a bottleneck. Often what works more effectively is to assign control functions to a host processor and distribute data processing functions across multiple co-processors under its control. In some cases, pipeline (cascaded) co-processing is used to reduce latency.

The Need For Networking Extensions

While HyperTransport was initially developed to address bandwidth and scalability problems associated with moving data through the I/O subsystems of desktops and servers, the networking extensions bring a number of enhancements which permit the advantages of HyperTransport technology to be extended to communications processing applications. There are some major differences in the requirements of host-centric systems such as desktops and servers and communications processing systems.

Communications Processing Is Often Less Vertical

In communications applications, there may be a number of processors or coprocessors located in various corners of the topology. The host processor may assume responsibility for configuration and control of coprocessors and interface devices, while the coprocessors perform specialized data processing tasks. Because of the distributed responsibility for control and data handling tasks, these systems tend to be much less host processor-centric.

As a result of decentralizing data processing in communications systems, information flow may be omni-directional as coprocessors initiate transactions targeting devices under their control. When switch components are added to the topology, elaborate multi-port configurations are possible.

Communications Processing Example

Figure 19-2 on page 449 illustrates an example of decentralized data processing in a communications processing system. This HyperTransport-based network switch system translates and routes data between an Ethernet Local Area Network (LAN) and a Wide Area Network (WAN).

Figure 19-2: A HyperTransport-Based Communications Processing System

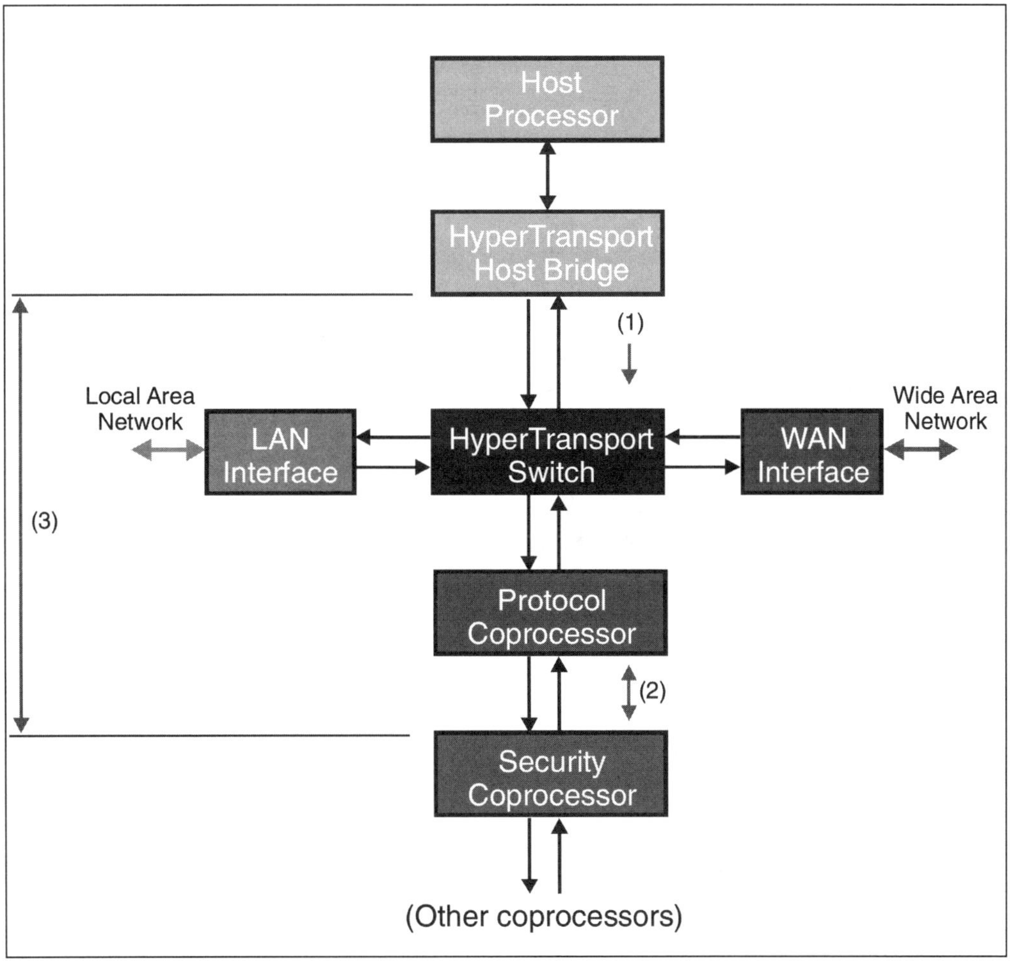

Some things to note about Figure 19-2 on page 449:

- Unlike a desktop or server system, the host processor in this system is not involved in most of the data processing. Traffic moving between the local area network (LAN) and wide area network (WAN) pass through the corresponding interface and move into the HyperTransport fabric. At the switch, packets are passed downstream to the coprocessors for protocol look up and translation, as well as a security screening. Once translated, message packets are passed back up to the switch for routing to the other interface.
- In this application, the host processor might be used to configure devices at boot time, program the coprocessors as needed, handle errors, collect statistics, etc. Host control events are represented by the tag (1) in Figure 19-2.
- Note how useful direct peer-to-peer transfers are in this type of topology. The coprocessors can pass data and messages to each other without involving other links. This greatly enhances system concurrency and efficiency.

Summary Of Anticipated Networking Extension Features

Although much of the HyperTransport task force work on the *Networking Extensions* is done, at the time of this writing, the1.05 and 1.1 revisions to the specification have not been released. The actual format of new packets and control mechanisms to be used with the network extensions are still being finalized, but the major features have been described in materials available on the Hyper-Transport web site (www.hypertransport.org), FAQ documents, and white papers on the subject. These are summarized here.

Network Extensions Adds Message Semantics

In handling the special problems of communications processing, the Hyper-Transport networking extensions add *message semantics* to the *storage semantics* used in the 1.04 revision of the HyperTransport I/O Link Specification. Storage semantics were described in the last section. Message semantics are more efficient in handling variable length transfers, broadcasting messages, etc. The 64-byte HyperTransport packets are concatenated to form longer messages, and additions to request packet fields identify the start of a message, end of a message, or may even be used to signal the abort of a scheduled transaction. Unlike storage semantics, in which the payload is data targeting an address, messages can also be sent which convey interrupts and other housekeeping events.

Another difference between message semantics and storage semantics is the concept of addressing. In storage semantics, addresses are managed by the source device, and each byte of data transferred is associated with a particular address in the system memory map. This makes sense because the locations are within (and owned by) the device being targeted. In message semantics, the message is tagged as to which stream it belongs, and the destination determines where it goes. The ultimate destination is often external to the system, where the system memory map has no meaning.

16 New Posted Write Virtual Channels

Release 1.1 adds 16 new optional Posted Write Virtual Channels to the hardware of each node (above the three already required). Each of these new virtual channels may be given a dedicated bandwidth allocation, and an arbitration mechanism is defined for managing them.

An End-To-End flow control mechanism has also been added to allow devices to put millions of *user streams* into these 16 additional virtual channels. In this way, very large numbers of independent real-time streams (e.g. audio or video) make be handled.

Direct Peer-to-Peer Transfers Added

HyperTransport supports the full producer-consumer ordering model of PCI. In cases where this strict global ordering is needed, transactions from one HyperTransport I/O device to another (called peer-to-peer transfers) must first move upstream to the host bridge where they are then reissued downstream to the target device (a process HyperTransport calls *reflection*). Release 1.1 adds the option of sending send some traffic directly from peer-to-peer when the application does not require strict global ordering (it often isn't a concern in communications processing). Tag (2) in Figure 19-2 on page 449 depicts a direct peer-to-peer transfer.

Link-Level Error Detection And Handling

With the addition of direct peer-to-peer transfers, Release 1.1 permits coprocessors and other devices to communicate directly without involvement of the host bridge. Along with this capability, network extensions provide for error detection and correction on the individual link level. In the event of an error, the receiver sends information back to the transmitter which causes a re-transmission of the packet. Obviously, the packet can't be consumed or forwarded until its validity is checked.

64 Bit Addressing Option

In keeping with the very large address space of many newer systems, Release 1.05 allows the optional extension of the normal 40-bit HyperTransport request address field to 64 bits.

Increased Number Of Host Transactions

Release 1.05 increases the number of outstanding transactions that a host bridge may have in progress from 32 to 128.

End-To-End Flow Control

In communication systems, there are occasions when devices are transferring packets to distant targets (not immediate neighbors) which may go "not ready" (or to another state which makes them unable to accept traffic) for extended periods. Prior to Release 1.1, HyperTransport devices only have flow control information for their immediate neighbors. Release 1.1 adds new end-to-end flow control packets which distant devices may send to each other to indicate their ability to participate in transfers. If a device is not ready, the source device does not start sending (or continue sending) packets; this helps eliminate bottlenecks which otherwise occur when the flow control buffers of devices in the path between source and target become full of packets which cannot be forwarded. An end-to-end flow control path is represented by tag (3) in Figure 19-2 on page 449.

Switch Devices Formally Defined

Finally, Release 1.05 formally defines the *switch* device type which may be used to help implement the complex topologies required in communications systems. A switch behaves much like a two-level HyperTransport-HyperTransport bridge with multiple secondary interfaces. The basic characteristics of a switch include:

1. A switch consumes one or more UnitIDs on its host interface. The port attached to the host is the default upstream port.
2. The switch acts as host bridge for each of its other interfaces. Each interface has its own bus number.
3. Switches, like bridges, are allowed to reassign UnitID, Sequence ID, and SrcTag for transactions passed to other busses. The switch maintains a table of outstanding (non-posted) requests in order to handle returning responses.
4. Switches may be programmed to perform address translation.
5. Switches must maintain full producer-consumer ordering for all combinations of transaction paths.
6. Switches must provide a method for configuration of downstream devices on all ports.

Part Four

HyperTransport Legacy Support

Part Four discusses the features added to HT for implementing bridges to legacy buses including: ISA/LPC, PCI, and AGP. This section also discusses the features added for supporting legacy x86 processors. The chapters included in Part Four are:

- Chapter 20: I/O Compatibility
- Chapter 21: Address Re-Mapping
- Chapter 22: X86 CPU Compatibility

20 *I/O Compatibility*

The Previous Chapter

The previous chapter summarized some of the major additions to the Hyper-Transport protocol which will be forthcoming in Release 1.05 and Release 1.1 of the specification. Collectively, these additions are referred the HyperTransport *Networking Extensions*, and target some of the special requirements of communications processing. Key features include a message passing protocol for larger packets, a formal definition for *switch* devices, a link-level error recovery method, sixteen optional additional posted write virtual channels with defined arbitration and bandwidth allocation, direct peer-peer transfers, an increase in the number of outstanding transactions for host bridges, and a 64-bit addressing option.

This Chapter

HT is designed to support a variety of I/O and processor buses via bridges. The specification defines specific requirements for supporting PCI, PIC-X, AGP, and processor buses. This chapter discusses these support requirements.

The Next Chapter

The large 1 Terabyte HT address space may be outside the limits of a given processor or expansion bus. When address locations are mapped into the HT space that exceed the processor or expansion bus address space then the addresses must be remapped to/from HT space. The next chapter discusses the HT solution for remapping prefetchable memory, MMIO, and I/O addresses.

Introduction

PC compatibility remains important in many operating environments and may include the use of several legacy buses. HT is intended to support connections to a variety of I/O buses, and the specification gives special attention to the PC legacy buses to ensure compatible support.

Several areas must be considered to ensure compatibility with I/O buses, including:

- Protocol differences
- Ordering requirements
- Command translation
- Address space and range differences
- Potential deadlocks

The buses currently supported and discussed by the HT specification include PCI, PCI-X, AGP, and host processor buses, with special attention given to issues associated with PC legacy support via the ISA/LPC buses.

PCI Bus Issues

Several features of the PCI bus must be handled in the correct fashion when interfacing with the HT bus. For background information and details regarding PCI ordering, refer to MindShare's PCI System Architecture book, 4th edition.

PCI Ordering Requirements

Transaction ordering on the PCI bus is based on the Producer/Consumer programming model. This model involves 5 elements:

1. Producer — PCI master that sources data to a memory target
2. Target — main memory or any PCI device containing memory
3. Consumer — PCI master that reads and processes the Producer data from the target
4. Flag element — a memory or I/O location updated by the producer to indicate that all data has been delivered to the target, and checked by the Consumer to determine when it can begin to read and process the data.
5. Status element — a memory or I/O location updated by the Consumer to indicate that it has processed all of the Producer data, and checked by the Producer to determine when the next batch of data can be sent.

This model works flawlessly in PCI when all elements reside on the same shared PCI bus. When these elements reside on different PCI buses (i.e. across PCI to PCI bridges, the model can fail without adherence to the PCI ordering rules.

The PCI specification, versions 2.2 and 2.3, defines the required transaction ordering rules. These ordering rules are included in this section as review and to identify rules that have may have no purpose in some HT designs. Table 20-1 on page 459 defines the ordering rules or PCI bridges. When reading the table, please note the following:

- PMW stands for posted memory write.
- DRR and DRC stand for Delayed Read Request and Delayed Read Completion, respectively.
- DWR and DWC stand for Delayed Write Request and Delayed Write Completion, respectively.
- "Yes" specifies that the transaction just latched must be ordered ahead of the previously latched transaction indicated in the column heading.
- "No" specifies that the transaction just latched must never be ordered ahead of the previously latched transaction indicated in the column heading.
- "Yes/No" entries means that the transaction just latched is allowed to be ordered ahead of the previously-latched operation indicated in the column heading, but such reordering is not required. The Producer/Consumer Model works correctly either way.

Table 20-1: PCI Ordering Rules

Transaction just latched	Posted Memory Write	Delayed Request		Delayed Completion	
	PMW Column 1	DRR Column 2	DWR Column 3	DRC Column 4	DWC Column 5
PMW (row 1)	No	Yes	Yes	Yes	Yes
DRR (row 2)	No	Yes/No			
DWR (row 3)	No				
DRC (row 4)	No	Yes		Yes/No	
DWC (row 5)	Yes/No				

Note that all of the transaction types listed under the heading, "transaction just latched" (except Delayed Write Completions, because the write has already completed) must never be reordered ahead of a previously posted memory write transaction (column 1). These rules are present to enforce proper operation of the producer/consumer model. HT support these rules providing that transactions originated from or targeting PCI devices do not use the PassPW feature in HT.

Avoiding Deadlocks

PCI ordering rules require that Posted Memory Writes (PMWs) in Row 1, be ordered ahead of the delayed requests and delayed completions listed in columns 2-5. This requirement is based on avoiding potential deadlocks. Each of the deadlocks involve scenarios arising from the use PCI bridges based on earlier versions of the specification. If all PCI bridge designs used in HT platforms are based on 2.1 and later versions of the PCI specification, the PCI ordering rules with "Yes" entries in row 1 can be treated as "Yes/No."

Table 20-1 also specifies that Delayed Read Completions and Delayed Write Completions in rows 4 and 5, must be ordered ahead of the Delayed Requests in Columns 2 and 3. These ordering rules arise from potential deadlocks that can occur when two hierarchical bridges are implemented as illustrated in Figure 20-1 on page 460. Refer to MindShare's PCI System Architecture book for a detailed explanation of this deadlock. If a given platform avoids this topology, then the "Yes" entries in rows 4 and 5 can be treated as "Yes/No."

Figure 20-1: Topology Causing Deadlock Scenario for Rows 4 and 5

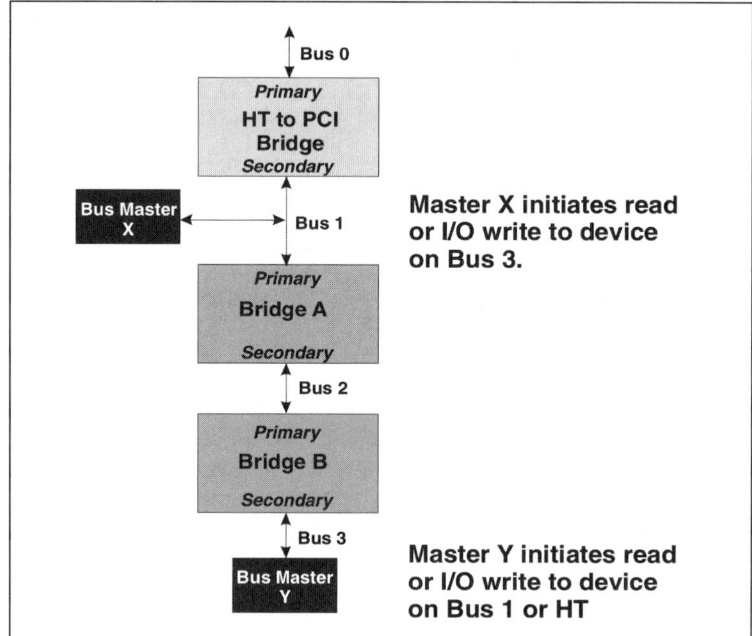

Subtractive Decode

PCI employs a technique referred to as subtractive decode to handle devices that are mapped into memory or I/O address space by user selection of switches and jumpers (e.g. ISA devices). Consequently, configuration software has no knowledge of the resources assigned to these devices. Fortunately, these PC legacy devices are mapped into relatively small ranges of address space that can be reserved by platform configuration software.

Subtractive Decode: The PCI Method

Subtractive decode is a process of elimination. Since configuration software allocates and assigns address space for PCI, HT, AGP and other devices, any access to address locations not assigned can be presumed to target a legacy device, or may be an errant address.

All PCI devices must perform a positive decode to determine if they are being targeted by the current request. This decode must be performed as a fast, medium, or slow decode. The device targeted must indicate that it will respond to the request by signaling device select (DEVSEL#) across the shared bus. When device driver software issues a request with an address that has not been assigned by configuration software, no PCI device is targeted (i.e. no DEVSEL# is asserted within the time allowed) By process of elimination, the subtractive decode agent recognizes that no PCI device has responded and therefore it asserts DEVSEL# and forwards the transaction to the ISA bus, where the request is completed.

Subtractive Decode: The Simple HT Method

An HT system with a single chain can possibly implement subtractive decode without extra host support required of more complex HT systems. Figure 20-2 on page 462 illustrates a simple system with a single-hosted chain. Note that the subtractive decode agent is at the end of the chain. If a request initiated at the host reaches the South Bridge, then the bridge knows that no other HT agents have claimed the transaction based on positive decode; therefore, a subtractive decode is safe.

Figure 20-2: Subtractive Decode in a Simple HT System

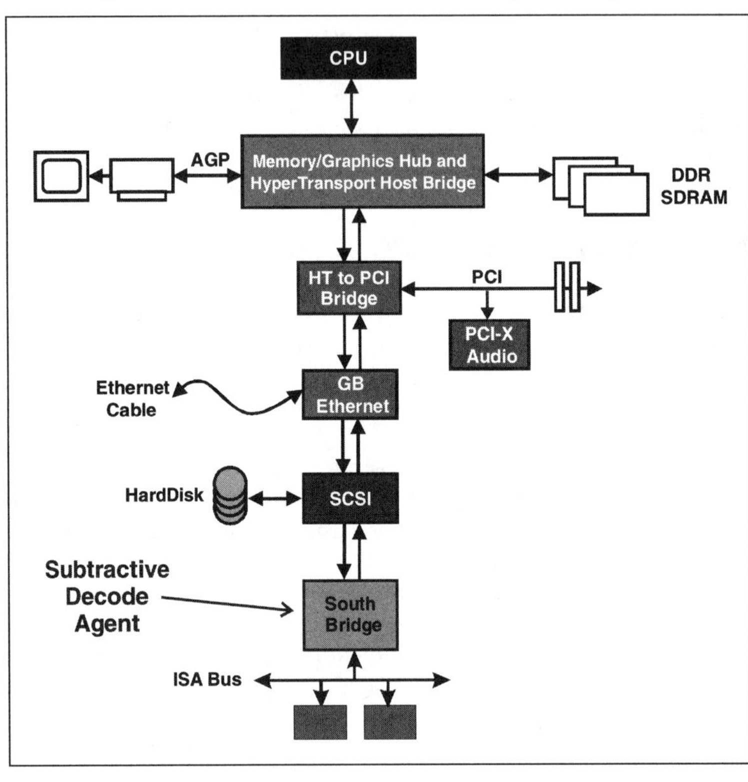

Subtractive Decode: HT Systems Requiring Extra Support

When the subtractive decode agent is not at the end of a single-hosted chain, or when more than one HT I/O chain is implemented in a system, subtractive decode becomes more difficult.

The Problem. HyperTransport devices in a chain do not share the same bus as in PCI, so a subtractive decode agent cannot detect if a request has not been claimed by other devices on the chain.

The Solution. As described previously, configuration software assigns addresses to all HT, PCI, and AGP devices. Therefore, the host knows when a request will result in a positive decode and when it will not. The specification requires that all hosts connecting to HyperTransport I/O chains implement registers that identify the positive decode ranges for all HyperTransport technology I/O devices and bridges (except as noted in the simple method). One of these I/O chains may also include a subtractive bridge (typically leading to an ISA, or LPC bus). Requests that do not match any of the positive ranges must be issued with the *compat bit* set, and must be routed to the chain containing the subtractive decode bridge. This chain is referred to as the compatibility chain.

The Compat bit indicates to the subtractive decode bridge that it should claim the request, regardless of address. Requests that fall within the positive decode ranges must not have the Compat bit set, and are passed to the I/O chain upon which the target device resides. The target chain may be the compatibility or any other I/O chain.

Subtractive Decode: Behind PCI Bridge

Figure 20-3 on page 463 illustrates the subtractive decode agent residing on a PCI bus behind a HT-to-PCI Bridge. When the host initiates a request that falls outside the assigned positive address ranges, it will set the *compat* bit and deliver the request to the HT-to-PCI Bridge. The bridge will detect the Compat bit set and forward the transaction on to the PCI bus where the South Bridge will perform the subtractive decode.

Figure 20-3: Subtractive Decode Agent Behind PCI Bridge

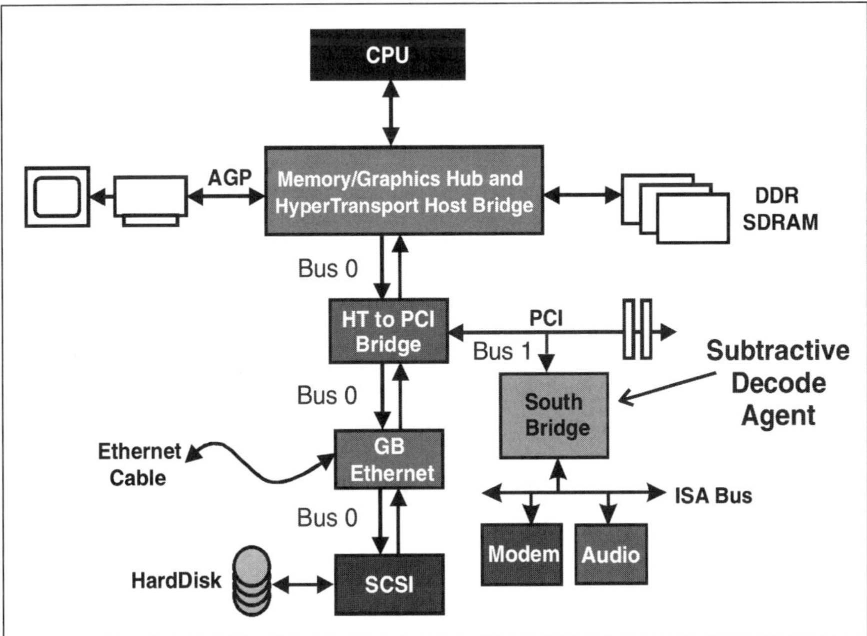

Subtractive Decode: Legacy System Considerations

Some legacy software requires that the subtractive decoder reside on Bus 0. Figure 20-4 on page 464 illustrates a system where the subtractive decode agent resides on the PCI bus. Normally, this topology would assign Bus 0 to the HT chain and the PCI bus would be numbered as Bus 1. This would violate the software requirement that the compatibility bus be Bus 0. To solve this problem, the HT-to-PCI bridge could be implemented with a regular device configuration header rather than a bridge configuration header. In this way, configuration software will view the devices on the PCI bus as residing on bus zero along with the HT devices on chain 0. The bridge will simply forward all transactions to the PCI bus as well as downstream to other devices residing on the chain. Any transaction with the compat bit set would be sent to the PCI bus.

Figure 20-4: Subtractive Decode Agent on PCI Bus 0

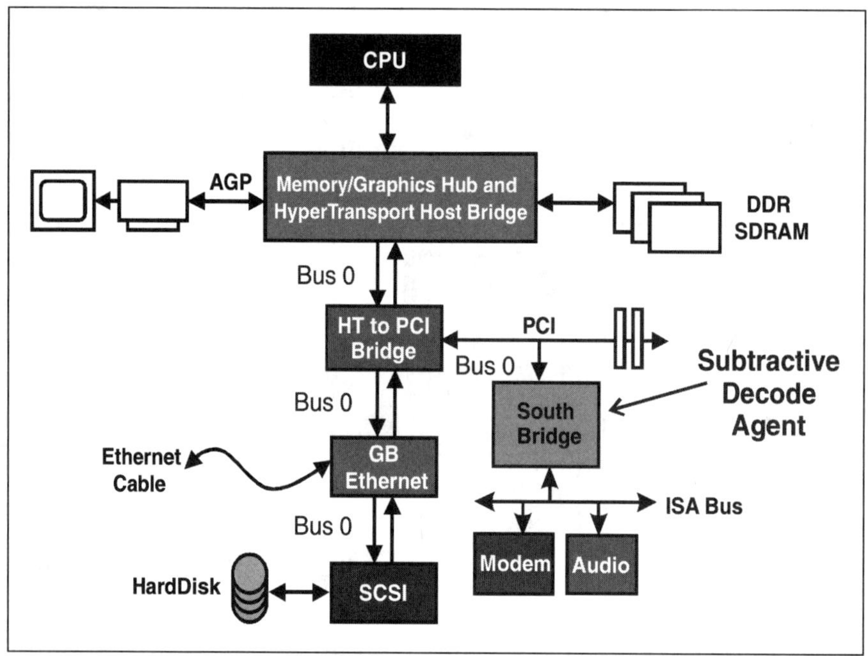

Subtractive Decode: Without Software Initialization

Systems may optionally set up the subtractive decode path in hardware so that software initialization is not required to enable subtractive decode. This remains true even though devices on the HyperTransport chain do not yet have their UnitIDs programmed. By setting the Compat bit accesses will be routed to the subtractive decode device. This approach may be useful when accessing the boot ROM following powerup.

HT-to-PCI Address Remapping

Address remapping between HT address space and PCI space is discussed in Chapter 21, entitled "Address Remapping," on page 477.

Transaction Translation

When transactions are passed between the HT and PCI buses, the command type must be translated for the other protocol. Table 20-2 lists the PCI to HT command conversion that must take place when bridging from PCI to HT.

Table 20-2: PCI to HT Command Conversion

PCI Transaction Type	HyperTransport Packet Type
Posted Memory Write	WrSized, Posted, PassPW = 0, Data Error=PERR#[1]
Delayed Read Request	RdSized, PassPW = 0, RespPassPW = 0
Delayed Write Request[2]	WrSized, Nonposted, PassPW = 0
Delayed Read Completion	RdResponse, PassPW = 0 (from request packet), Data Error=PERR# [3]
Delayed Write Completion	TgtDone, PassPW = 1[4]

1. *DataError is set if the bridge detected a data parity error.*
2. *DataError is set if a data parity error is detected in a non-posted write, and the write is discarded (only if Parity Error Response Enable is set).*
3. *DataError is set when PERR# is detected during a Delayed Write Completion (only if the Parity Error Response Enable is set for the PCI interface).*
4. *To ensure correct ordering of some message sequences (e.g., Interrupt and STPCLK virutal signaling) the PassPW bit must be cleared in the TgtDone.*

Table 20-3 lists the command conversion required when bridging from HT-to-PCI.

Table 20-3: HT to PCI Command Conversion

HyperTransport Packet Type	PCI Transaction Type
WrSized to Memory Space[1]	Posted Memory Write[3]
WrSized to Configuration or I/O Space	Delayed Write Request[2]
RdSized from Memory, Configuration, or I/O Space	Delayed Read Request[2]
RdSized due to X86 Interrupt Acknowledge Cycle	Interrupt Acknowledge
RdResponse	Delayed Read Completion[3]
TgtDone	Delayed Write Completion

1. *PCI requires that all memory write be posted, thus HT non-posted writes will be treated as posted operations on PCI.*
2. *Delayed Read Requests are never posted operations and will always result in a Read Response or Target Done.*
3. *If the DataError bit is set, the bridge should send incorrect parity to alert the receiver that the data is corrupt.*
4. *Assertion of PERR for a data error is gated by the Data Error Response Enable for the Hyper-Transport interface.*

PCI Burst Transactions

PCI permits long burst transactions with either contiguous or discontiguous byte masks (byte enables) that may not be supported by HT. These long bursts must be broken into multiple requests to support the HT protocol as follows:

- PCI read requests with discontiguous byte masks that cross aligned 4-byte boundaries must be broken into multiple 4-byte HT RdSized (byte) requests.
- PCI write requests with discontiguous byte masks that cross 32-byte boundaries must be broken into multiple 32-byte HT WrSized (byte) requests. Note that the resulting sequence of write requests must be strongly ordered in ascending address order.
- PCI write requests with contiguous byte masks that cross 64-byte boundaries must be broken into multiple 64-byte HT WrSized (dword) requests.

PCI-X Bus Issues

PCI-X is a superset of the PCI bus that provides a relaxed ordering bit, which changes the PCI-X ordering rules from PCI. PCI-X also uses different transaction types and protocols, requiring a very different bridge design than PCI; however, the bridge must be compatible with PCI protocols.

PCI-X Ordering Requirements

PCI-X ordering is based on the same principles as PCI; thus, the ordering rules are very similar. In fact, the only differences are due to the Relaxed Ordering supported by PCI-X. Table 20-4 on page 468 lists the PCI-X ordering rules. Note that Posted Memory Writes (PMW) and Split Request Completions (SRC) support Relaxed Ordering, yielding to possible ordering scenarios for each case in Column 1.

PCI-X also supports split transactions rather than delayed transactions. The terms and acronyms below reflect the split transaction types:

- PMW stands for posted memory write.
- SRR and SRC stand for Split Read Request and Split Read Completion, respectively.
- SWR and SWC stand for Split Write Request and Split Write Completion, respectively.
- "Yes" specifies that the transaction just latched must be ordered ahead of the previously latched transaction indicated in the column heading.
- "No" specifies that the transaction just latched must never be ordered ahead of the previously latched transaction indicated in the column heading.
- "Yes/No" entries means that the transaction just latched is allowed to be ordered ahead of the previously-latched operation indicated in the column heading, but such reordering is not required.

Table 20-4: PCI-X Ordering Rules

Transaction just latched	Posted Memory Write	Split Requests		Split Completions	
	PMW Column 1	SRR Column 2	SWR Column 3	SRC Column 4	SWC Column 5
PMW (row 1)	RO=0: No RO=1: Yes/No	Yes	Yes	Yes	Yes
SRR (row 2)	No	Yes/No			
SWR (row 3)	No				
SRC (row 4)	RO=0: No RO=1: Yes/No	Yes		Yes/No	
SWC (row 5)	Yes/No				

Transaction Translation

PCI-X replaces delayed transactions with split transactions, and supports a Relaxed Ordering bit and the No Snoop bit to enhance performance, causing command conversion to vary slightly from PCI. Table 20-5 on page 468 lists the command conversion from PCI-X to HT and Table 20-6 on page 469 lists the command conversion from HT-to-PCI-X. Please note that later versions of the specification were pending completion at the time of this writing and may add new information to the transaction tables.

Table 20-5: PCI-X to HT Command Conversion

PCI-X Transaction Type	HyperTransport Packet Type
Posted Memory Write	WrSized, Posted, PassPW = RO, Data Error=PERR#[1]
Split Read Request	RdSized, PassPW = 0, RespPassPW = RO, Coherent = Snoop

Table 20-5: PCI-X to HT Command Conversion

PCI-X Transaction Type	HyperTransport Packet Type
Split Write Request	WrSized, Nonposted, PassPW = 0^2
Split Read Completion	RdResponse, PassPW = RO, Data Error=PERR#[1]
Split Write Completion	TgtDone, PassPW = 1^4, Data Error=PERR#[3]

1. *DataError is set if the bridge detected a parity error or (in mode 2) an unrecoverable ECC error.*
2. *Split Write requests with data errors are discarded if the Parity Error Response Enable is set.*
3. *DataError is set if a data parity error is detected during a Split Write Completion, and the write is discarded (only if Parity Error Response Enable is set).*
4. *To ensure correct ordering of some message sequences (e.g., Interrupt and STPCLK virutal signaling) the PassPW bit must be cleared in the TgtDone.*

Table 20-6: HT-to-PCI-X Command Conversion

HyperTransport Packet Type	PCI-X Transaction Type
WrSized to Memory Space[2]	Posted Memory Write, RO=PassPW[1]
WrSized to Configuration or I/O Space	Split Write Request[2]
RdSized	Split Read Request[3],RO = RespPassPW[1]
RdResponse	Split Read Completion, RO = PassPW
TgtDone	Split Write Completion

1. *RO is the Relaxed Ordering bit in PCI-X. If RO=0, PassPW and RespPassPW must be 0. RO must never be set in requests, regardless of the value of PassPW or RespPassPW*
2. *All memory writes on PCI are posted regardless of whether they are posted on HT or not.*
3. *Split Requests are never posted operations and will always result in a Read Response or Target Done.*
4. *The No Snoop bit in PCI-X requests from HT bridges is always 0.*

AGP Bus Issues

AGP Configuration Space Requirements

Some legacy operating systems require that the AGP capability registers be mapped at Bus 0, Device 0, and Function 0. Also, The AGP aperture base address configuration register must be at Bus 0, Device 0, Function 0, Offset 10h. In a legacy system, these registers are located within the Host to PCI bridge configuration space (Host to HT bridge in our example). See Figure 20-5.

*Figure 20-5: Legacy Configuration Mapping for Host to HT Bridge
with AGP and DRAM Controller*

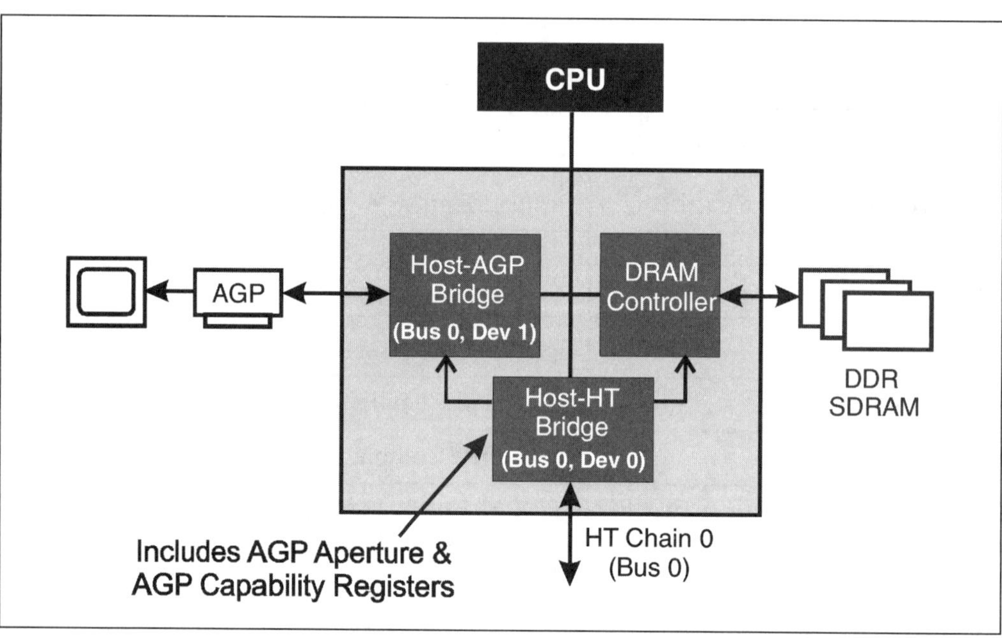

For complete legacy software support, the specification recommends that the AGP subsystem be designed as follows:

- AGP bridges are placed logically on HyperTransport chain 0 (Bus 0).
- The AGP interface uses multiple UnitIDs due to AGP configuration being split between the Host to HT bridge and the Host to AGP bridge (i.e., vir-

tual PCI to PCI bridge).

- During initialization the base UnitID of an AGP device must be assigned a non-zero value to support configuration of chain 0. Following HT initialization the base UnitID should be changed to zero.
- Device number zero, derived from the base UnitID register value, should contain the capabilities header and the AGP aperture base address register (at Offset 10h), as pictured in Figure 20-5 on page 470.
- Device number 1, derived from the base UnitID+1, should be used for the Host to AGP bridge.
- The UnitID that matches the base (0) is not used for any AGP-initiated I/O streams or responses so that there is no conflict with host-initiated I/O streams or responses. Only UnitIDs greater than the base may be used for I/O streams.
- Legacy implementations place the AGP graphics address remapping table (GART) in the host. Thus, the AGP aperture base address register and any other registers that are located in the AGP device but required by the host are copied by software into implementation-specific host registers. These implementation-specific registers should be placed somewhere other than Device 0, to avoid conflicts with other predefined AGP registers. In a sharing double-hosted chain, this requires the hosts to implement the Device Number field so that the hosts may address each other after the AGP bridge has assumed Device 0. See "Case 3: Initializing A Double Hosted Chain" on page 320 for more information.

Note that if legacy OS support is not required, the AGP device's base UnitID register may be programmed to any permissible value.

AGP Ordering Requirements

Three categories of AGP transaction types lead to three separate sets of ordering rules. These categories can be thought of as three separate transaction channels. These three channels are completely independent of each other with respect to ordering, and should have their own UnitIDs. The transaction types are:

- PCI-based
- Low Priority
- High Priority

The specification makes the following observation that leads to HT-based AGP ordering requirements being slightly less complex that PCI-based requirements:

> The ordering rules presented here for reads are somewhat different from what appears in the AGP specification. That document defines ordering between reads in terms of the order that data is returned to the requesting device. We are concerned here with the order in which the reads are seen at the target (generally, main memory). The I/O bridges can reorder returning read data if necessary. This leads to a slightly relaxed set of rules.

See MindShare's AGP System Architecture book for details regarding the AGP ordering rules.

PCI-Based Ordering

AGP transactions based on the PCI protocol follow the same rules as PCI. Therefore, the ordering rules discussed in Figure 20-1 on page 459 also apply to these types of AGP transactions.

Low Priority Ordering

Ordering rules for the low priority AGP transactions are:

- Reads (including flushes) must not pass writes.
- Writes must not pass writes.
- Fences must not pass other transactions or be passed by other transactions.

High Priority Ordering

High priority transactions only carry graphics data using split transactions. Consequently, the Producer/Consumer model has no relevance and ordering requirements can be reduced to the following single rule:

- Writes must not pass writes.

Transaction Translation

AGP is also allowed to generate requests with discontiguous byte masks, and require the same transaction handling as PCI. Refer to "PCI Burst Transactions" on page 466.

Command conversion from AGP-to-HT is listed in Table 20-7 on page 473. Note that the AGP graphics device is always the initiator of AGP transactions; therefore, HT-to-AGP conversion is not defined.

Table 20-7: AGP-to-HT Command Conversion

PCI-X Transaction Type	HyperTransport Packet Type
High Priority Write	WrSized, Posted, PassPW = 1 or 0 (alternate)
High Priority Read	RdSized, PassPW = 1 or 0 (alternate), RespPassPW = 1
Low Priority Write	WrSized, Nonposted, PassPW = 1
Low Priority Read	RdSized, PassPW = 1, RespPassPW = 1
Low Priority Flush	None (wait for all outstanding writes to complete) Flush, PassPW = 0 (alternate)
Low Priority Fence	None Wait for all outstanding read responses (alternate)

ISA/LPC Buses

The ISA and LPC buses reside typically on the PCI bus. These buses support bus mastering and legacy DMA transfers. These devices differ from HT and PCI devices in that they do not support either split transactions or retries.

Deadlocks

The specification defines two possible deadlock conditions that can occur because the ISA and LPC (Low Pin-Count) buses do not support transaction retry. For example, if an ISA (LPC) Master initiates a transaction that requires a response, the bus cannot handle a new request prior to the current transaction having completed. This type of protocol is extremely simple from an ordering perspective because all transactions must complete before the next one begins; thus, no ordering rules are required. Of course the downside to this approach is that all other devices are stalled while they wait for the current transaction to complete. Delayed transactions supported by the PCI bus and split transactions supported by PCI-X and HyperTransport can handle new transactions while a response to a previous transaction is pending. The price — complex ordering rules to ensure that transactions complete in the intended order.

Deadlock Scenario 1

Consider the following sequence of events as they relate to the limitations of the ISA/LPC bus as discussed above and to the PCI-based Producer/Consumer transaction ordering model. Figure 20-6 on page 475 illustrates the topology.

1. An ISA/LPC Master initiates a transaction that requires a response from the Host-to-HT Bridge (e.g., a memory read from main memory).
2. The CPU initiates a write operation targeting a device on the ISA/LPC bus, and the Host Bridge issues this write as a posted operation.
3. The posted write reaches HT-to-PCI bridge where it is sent across the PCI bus to the south bridge.
4. The south bridge cannot accept the write targeting the ISA bus because the ISA/LPC bus is waiting for the outstanding response. So, the south bridge issues a retry.
5. The read response reaches the HT/PCI bridge. However, the Producer/Consumer model requires that all previously-posted write headed to the PCI bus be completed before sending a read response. The read response is now stuck behind a posted write that cannot complete prior to the read response. Result: Deadlock!

The recommended solution to this problem is to require that all requests targeting the ISA/LPC bus be non-posted operations. This eliminates the problem because non-posted operations can be forwarded to the PCI bus in any order.

Figure 20-6: Deadlock Scenario 1 Topology

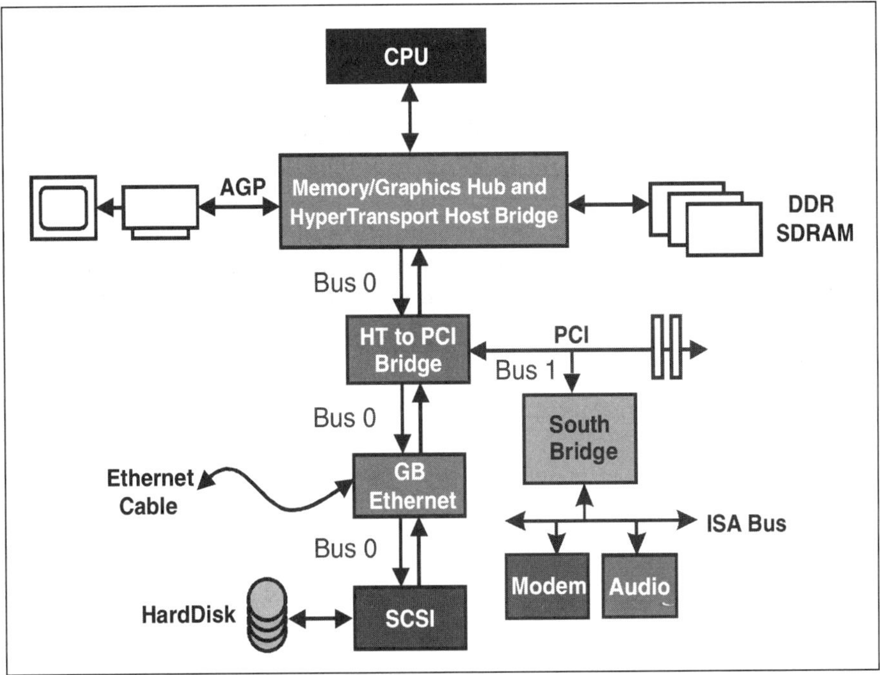

Deadlock Scenario 2

Once again because the ISA or LPC bus is unable to accept any requests while it waits for a response to its own requests a possible deadlock can occur. This deadlock can occur when the downstream non-posted request channel fills up while awaiting a response to an ISA DMA request. The sequence of events is as follows:

1. A DMA request is issued by an ISA/LPC device to main memory.
2. Downstream requests targeting the ISA bus are initiated but stack up because they are not being accepted by the south bridge, because its's waiting on a response from the previously issued DMA request. Consequently, it is possible for the downstream nonposted request channel to fill.
3. A peer-to-peer operation is initiated to a device on the same chain that is in the non-posted request queue ahead of the ISA/LPC request (in step 1) This peer-to-peer transaction is sent to the Host, which attempts to reflect the transaction downstream to the target device. However, because the down-

stream request channel is full; the upstream nonposted peer request stalls as does the request from the ISA bus. This prevents the ISA/LPC bridge from making forward progress.

The solution to this deadlock is for the host to limit the number of requests it makes to the ISA/LPC bus to a known number of requests (typically one) that the bridge can accept. Because the host cannot limit peer requests without eventually blocking the upstream nonposted channel (and causing another deadlock), no peer requests to the ISA/LPC bus are allowed. Peer requests to devices below the ISA/LPC bridge on the chain (including other devices in the same node as the ISA/LPC bridge) cannot be performed without deadlock unless the ISA/LPC bridge sinks the above mentioned known number of requests without blocking requests forwarded down the chain. This can be implemented with a buffer (or set of buffers) for requests targeting the bridge, but separate from the buffering for other requests.

21 *Address Remapping*

The Previous Chapter

HT is designed to support a variety of I/O and processor buses via bridges. The specification defines specific requirements for supporting PCI, PIC-X, AGP, and processor buses. The previous chapter discussed these support requirements.

This Chapter

The large 1 Terabyte HT address space may be outside the limits of a given processor or expansion bus. When address locations are mapped into the HT space that exceed the processor or expansion bus address space, then the addresses must be remapped to and from HT space. This chapter discusses the HT solution for remapping memory, MMIO, and I/O addresses.

The Next Chapter

Many HT platforms may be based on x86 processors. Compatibility support for these processors is defined by the HT specification. This chapter discusses the x86 features that require specific support by HT technology and details the HT methods of signaling.

Introduction

An HT-based system may include processors or expansion buses that have smaller address ranges than the large 1 TeraByte space used by HT. In such cases address translation may be required. For example, the x86 CPU might have a maximum memory address range of 64GB and a PCI bus implementation may be limited to 4GB of memory address space. If addresses are allocated in HT address space beyond the range of the CPU and Expansion bus, then address translation is required.

To illustrate the possible remapping requirement, Figure 21-1 on page 479 depicts an implementation similar to that described in the previous paragraph. This example does not depict a typical implementation, but rather illustrates an extreme case where address remapping is required by both the CPU to HT Bridge and the HT-to-PCI Bridge. The system allocates a 4GB range of processor memory address space for PCI devices that is near the top of the 64GB range. This processor address space is mapped into the HT space above the 64GB range of the processor and beyond the address range of the PCI bus. When software executing on the CPU accesses a memory location within a PCI device two address translations must occur:

1.   The CPU to HT Bridge must translate the CPU address within the 60 - 64GB address range to the 1,008 to 1,012GB range.
2.   The HT-to-PCI Bridge must translate the HT address down to the 0-4GB range of the PCI bus.

A similar address translation must take place to support I/O address space. HT maps I/O space very high in its address space and outside the range of both the CPU and PCI bus, thereby creating the need for address remapping.

HT provides an Address Remapping Capability Block that support remapping HT addresses to expansion buses. However, the specification does not define a mechanism for remapping CPU to HT address space.

The Address Remapping Capability Block

This capability block allows software to control the HT to expansion bus bridge address for:

*   remapping HT I/O address space down to a lower range within an expansion bus.
*   remapping HT MMIO space to expansion bus address space.
*   remapping HT Memory space to expansion bus address space.
*   remapping Bus Master memory transactions from expansion bus address space to HT address space.

Figure 21-2 on page 479 illustrates the format of the Address Remapping capability block. The following sections describe each register and its remapping function.

Figure 21-1: HT Address Space May Exceed that of the Processor and Expansion Bus

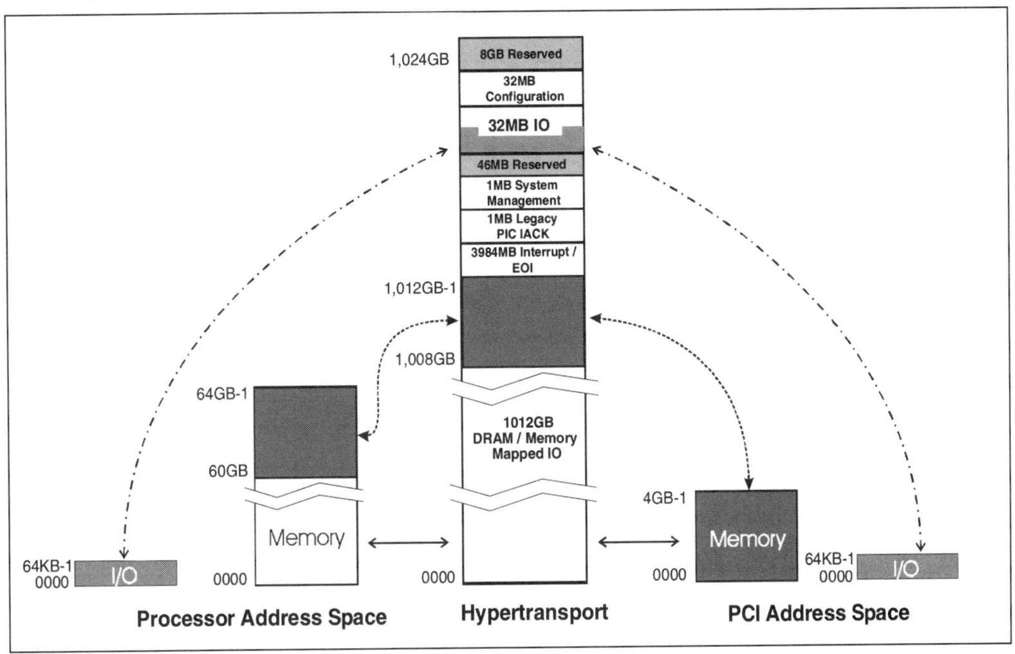

Figure 21-2: Format of the Address Remapping Capability Block

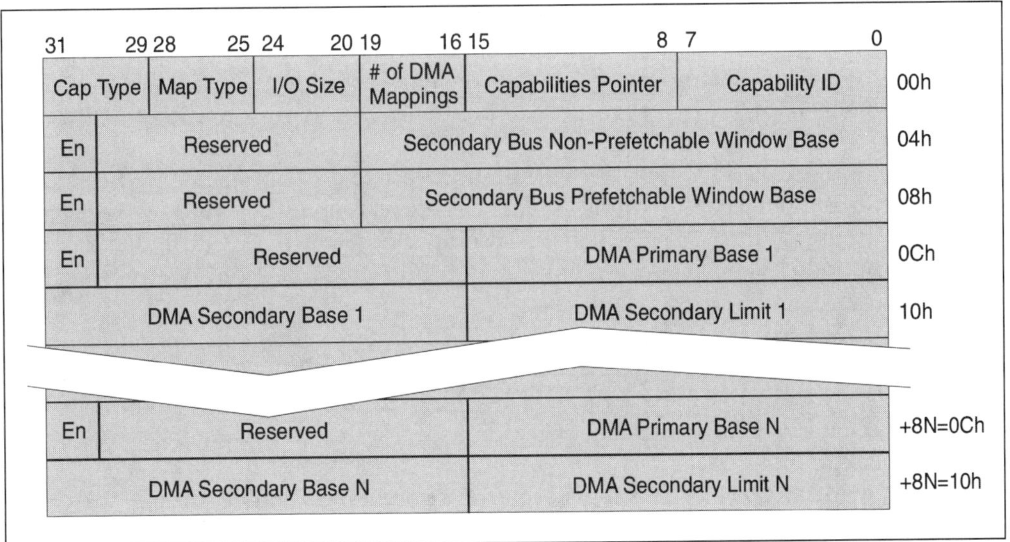

31	29 28	25 24	20 19	16 15	8 7	0	
Cap Type	Map Type	I/O Size	# of DMA Mappings	Capabilities Pointer		Capability ID	00h
En	Reserved		Secondary Bus Non-Prefetchable Window Base				04h
En	Reserved		Secondary Bus Prefetchable Window Base				08h
En	Reserved			DMA Primary Base 1			0Ch
DMA Secondary Base 1				DMA Secondary Limit 1			10h
En	Reserved			DMA Primary Base N			+8N=0Ch
DMA Secondary Base N				DMA Secondary Limit N			+8N=10h

The registers within the remapping capability block are briefly described below. Detailed discussion of their implementation is discussed in the following sections.

- **Number of DMA Mappings** — This read-only field indicates how many DMA Primary/Secondary register sets (if any) are defined by the remapping capability block.
- **I/O Size** — This register is intended to permit configuration software to limit the I/O address range supported by the expansion bus to a range smaller than the 32MBs supported by HT.
- **Mapping Type** — The HT specification does not currently define a mapping type. This read-only register must be zero (0) and is reserved for future extensions.
- **Capability Type** — A read-only value of 01000b indicates an address remapping capability block.
- **Secondary Bus Non-Prefetchable Window Base** — This register is written by configuration software to define the HT address range that is mapped to MMIO space on the secondary (expansion) bus. Contents of this register are address bits 39:20, providing a minimum HT range of 1MB for expansion bus MMIO addresses.
- **SBNPCtl** — This register permits software to enable/disable the Secondary Bus Non-Prefetchable Window Base register. When the window register is enabled, the *SBNPCtl* register also specifies which HT attributes are supported by the expansion bus. (i.e. HT requests passing through the window may have these attributes active). The attributes include NonCoherent, Isochronous, and Compatibility.
- **Secondary Bus Prefetchable Window Base** — This register is written by configuration software to define the HT address range that is mapped to prefetchable memory address space on the secondary (expansion) bus. Contents of this register are address bits 39:20, providing a minimum HT range of 1MB for expansion bus prefetchable memory addresses.
- **SBPreCtl** — This register permits software to enable/disable the Secondary Bus Prefetchable Window Base register. When the window register is enabled, the *SBPreCtl* register also specifies which HT attributes are supported by the expansion bus. The attributes include NonCoherent, Isochronous, and Compatibility.
- **DMA Primary Base 1-N** — The *DMA Primary Base* register permits configuration software to define the HT base address (bits 39:24) for each DMA mapping. These mappings are used by bus masters residing on the expansion bus when accessing HT address space that is outside the range of the expansion bus.
- **DMACtl 1-N** — This register permits software to enable/disable the corresponding DMA mapping. When the mapping is enabled, the *DMACtl* regis-

ter also specifies whether NonCoherent and Isochronous transactions are supported on the expansion bus.

- **DMA Secondary Base 1-N** and **DMA Secondary Limit 1-N** — These registers create a DMA memory window within the secondary (expansion) bus address range for DMA transfers to HT address space. The starting address of the HT range is specified by the DMA Primary Base register.

I/O Address ReMapping

Because HT does not directly support I/O address space, it reserves a 32MB of memory address space for transporting I/O addresses across the bus. When the processor initiates and I/O transaction, the address must be translated by the CPU to HT bridge to an address within the reserved HT I/O address range (FD_FC00_0000h- FD_FDFF_FFFFh). HT transports the transaction across the HT chain to the HT-to-PCI bridge which must translate the HT I/O address so that is falls within the expansion bus I/O address range.

The *IO Size* register within the Address Remapping Capability block supports the I/O address space remapping. This 5-bit register defines the size of the I/O address space on the expansion bus, up to 32MBs. Note that the upper 15 bits of the HT I/O address range (FD_FC00_0000h- FD_FDFF_FFFFh) identifies the address as I/O, while the lower 25 address bits define any location within the 32MB I/O address range. The *I/O Size* register specifies the number of upper address bits (Addr 24:0) that will not be used when generating an I/O address on the expansion bus. For example, an I/O size value of 01101b (13d) results in an expansion bus I/O range of 4KB (Addr 11:0). The default value of zero causes all 25 bits of a HyperTransport I/O request to be passed to the expansion bus, thereby supporting the maximum 32MB I/O range. The next two sections give example implementation of I/O address remapping.

X86 Processor and PCI I/O Remapping Example

This example assumes that the x86 processor and the PCI bus support a maximum I/O address space of 64KB as illustrated in Figure 21-3 on page 483. The Host to HT bridge translates each processor initiated I/O address up to the reserved HT I/O address range and the HT-to-PCI bridge translates the address back down to the PCI I/O address range. HT-to-PCI bridges use the PCI to PCI bridge's I/O Base and Limit configuration registers that are defined by the PCI 2.3 and earlier specifications, along with the I/O size field of the Address Remapping capability block.

In this example, the processor performs a read from I/O address 058Ah that targets a location within a PCI device. The following actions are taken:

1. The x86 processor initiates an I/O read transaction from address 058Ah.
2. The Host to HT bridge forwards the processor read to the HT bus but translates the I/O address to memory location FD_FC00_058Ah.
3. The HT-to-PCI bridge recognizes that the address falls within the 32MB address range reserved by HT for I/O addresses starting at location FD_FC00_0000h. Once the address is recognized, only the lower 25 bits of the address are relevant (Addr[24:0]).
4. Configuration software has previously established a value in the I/O size field that represents the maximum address space supported by the PCI bus (64KB in this example). The I/O size value (01001b) specifies that the upper 9 bits of the remaining 25-bit address must be ignored and delivered as all zeros to the PCI bus. This restricts the I/O space to 64KB.
5. The HT-to-PCI Bridge's I/O Base and Limit registers have also been programmed by configuration software to encompass all locations assigned to PCI devices. In this example addresses within the lower 8KB of I/O address space have been assigned (0000h - 1FFFh). The bridge verifies that the I/O address (058Ah) falls within the address window specified by the I/O Base and Limit registers and forwards the address to the PCI bus.

Figure 21-3: X86 Processor I/O Mapping Example

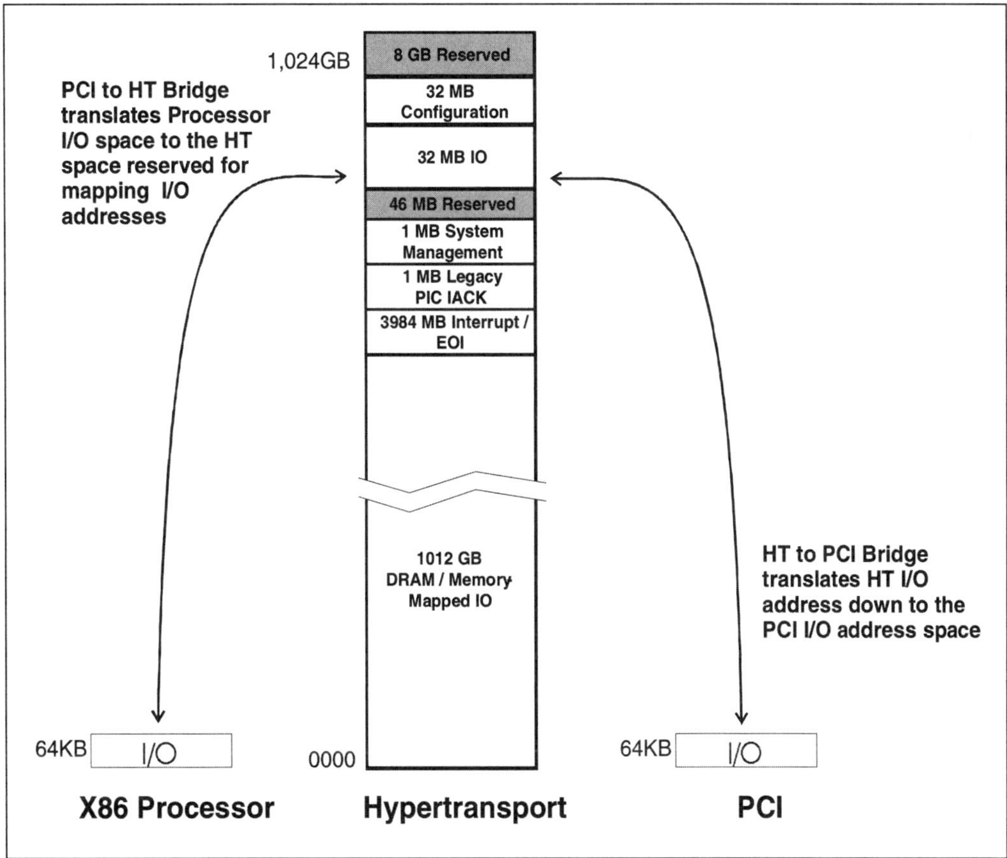

PowerPC and PCI I/O Remapping Example

Some processors such as the PowerPC may not support I/O address space. In this case, the CPU to HT bridge must translate a range of processor memory address space to the HT I/O address range. The translation from HT-to-PCI I/O address space is accomplished using the I/O Size field of the Address Remapping capability block.

In this example, a 16MB block of processor memory space is dedicated for I/O address mapping. Consequently, the PCI can support up to 16MB of I/O space that can be accessed by software executing on the CPU. Figure 21-4 on page 484 illustrates this address remapping. The I/O Size value in this example would be 00001b to indicate that only the upper address bit of the 24-bit I/O address will be ignored, yielding a valid PCI I/O address of Addr[23:0].

Figure 21-4: PowerPC I/O Mapping Example

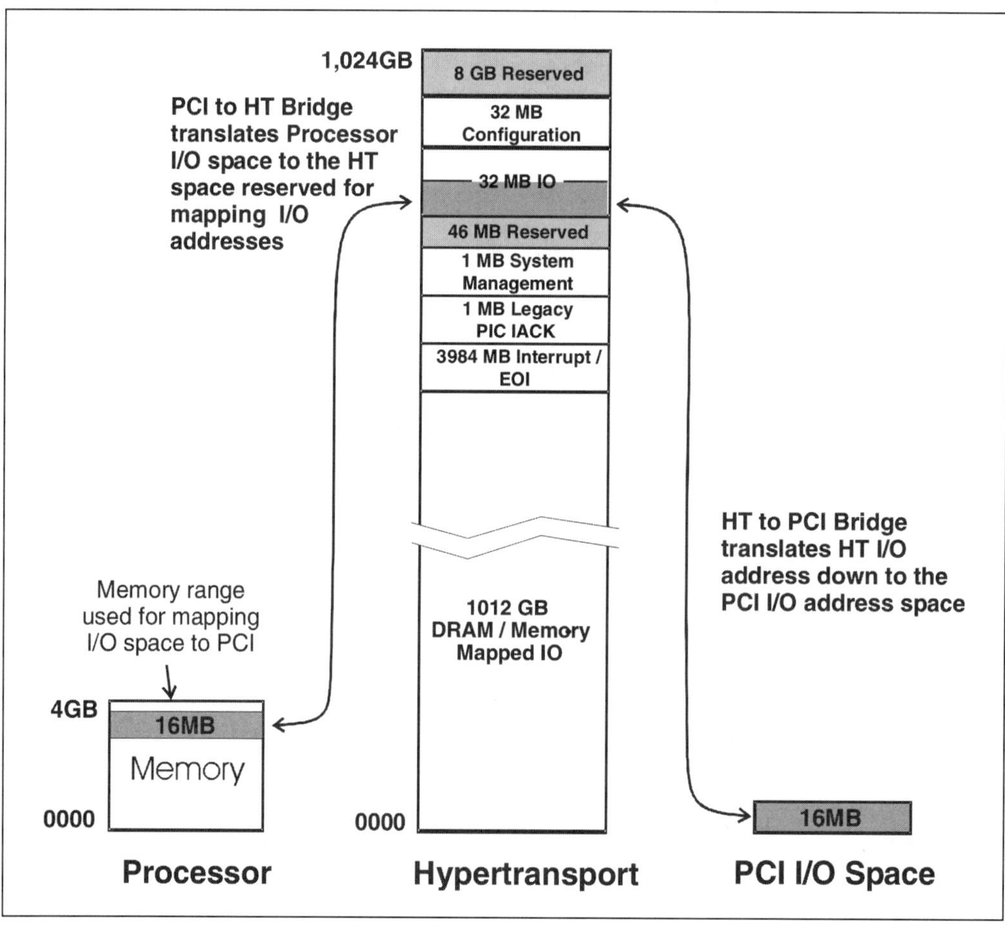

DownStream HT to Expansion Bus Memory Mapping

Downstream memory transactions may be accomplished solely on the basis of mapping registers defined by the bus architecture (e.g. PCI), or may be aided by the Address Remapping Capability block registers when the HT addresses are outside the range of the expansion bus memory address range. Because HT defines a large range of address space for mapping memory and MMIO locations (00_0000_0000h-FC_FFFF_FFFFh), its address space may extend beyond the address range of the secondary bus.

Downstream Memory Access Without Remapping

The first example represents the simple case where no remapping of memory addresses is required. This example assumes that the processor memory space is directly mapped to HT address space and to the PCI bus memory address space. This mapping is illustrated in Figure 21-5 on page 486. Note that no remapping is required in this example because the memory addresses assigned to the PCI devices are mapped to the same memory locations of both the processor and HT bus.

In this implementation, the Address Remapping Capability block registers defined by HT for remapping memory can be disabled using the associated control registers. The bridge can simply use the PCI Memory Base and Limit registers for both prefetchable and non-prefetchable memory address space.

Figure 21-5: Example of Direct Memory Mapping between CPU, HT, and PCI.

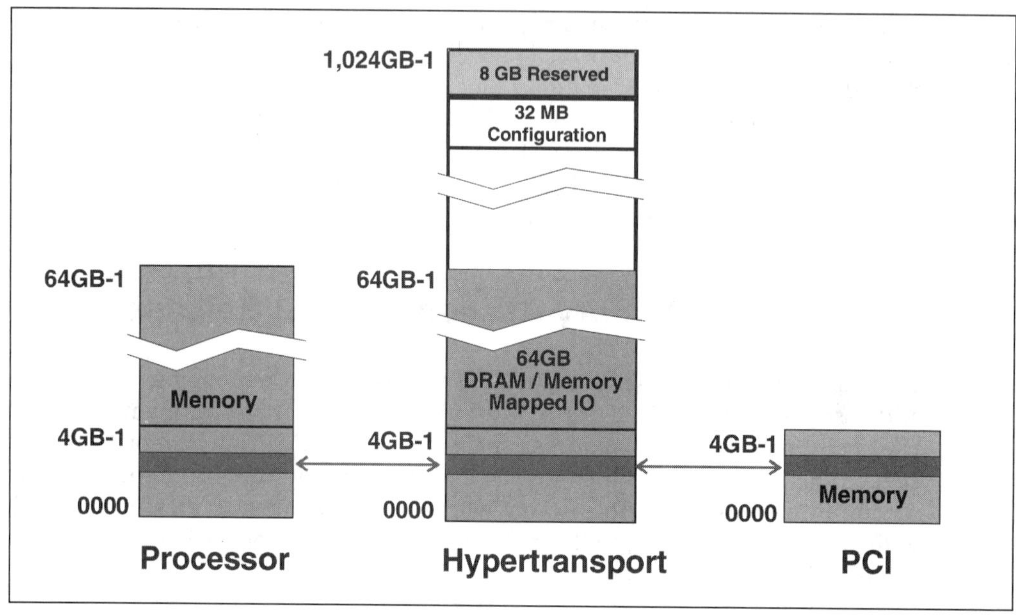

Memory Accesses with Remapping

The mechanism used for remapping memory addresses includes support for prefetchable and non-prefetchable address space. Two mapping registers are defined for each type of address space:

- Secondary Bus Non-Prefetchable Window Base
- Secondary Bus Prefetchable Window Base

Each register has the same format and function except for the different memory address types that they support. Both registers permit the HT to expansion bus bridge to remap HT memory address space to the expansion bus address space. This capability is needed only in the event that HT space is beyond the range of the expansion bus.

The 20-bit value written to these registers by configuration software defines HT address bits 39:20, which is the base address of the HT memory range that is assigned to expansion bus memory. The range of addresses associated with the base address is determined by other registers within the bridge. For example, the PCI Prefetchable Memory Base and Prefetchable Memory Limit registers define the range of prefetchable address space allocated for the PCI bus.

Figure 21-6 on page 487 illustrates the registers that would be used when remapping prefetchable memory from HT-to-PCI. In this example, an HT memory address range has been allocated for PCI prefetchable memory. The base address of the address range is specified in the Secondary Bus Prefetchable Window Base register. The 4GB PCI address space allocates a 1GB range of addresses for prefetchable memory. This range is defined by the Prefetchable Base and Prefetchable Limit registers. The bridge uses these registers to recognize an HT address that falls within the prefetchable range and remaps the address to PCI based on the address offset. Remapping of the non-prefetchable address space is based on the same principles, but uses the PCI Memory Base and Memory Limit registers.

Figure 21-6: Prefetchable Memory Address Remapping Registers

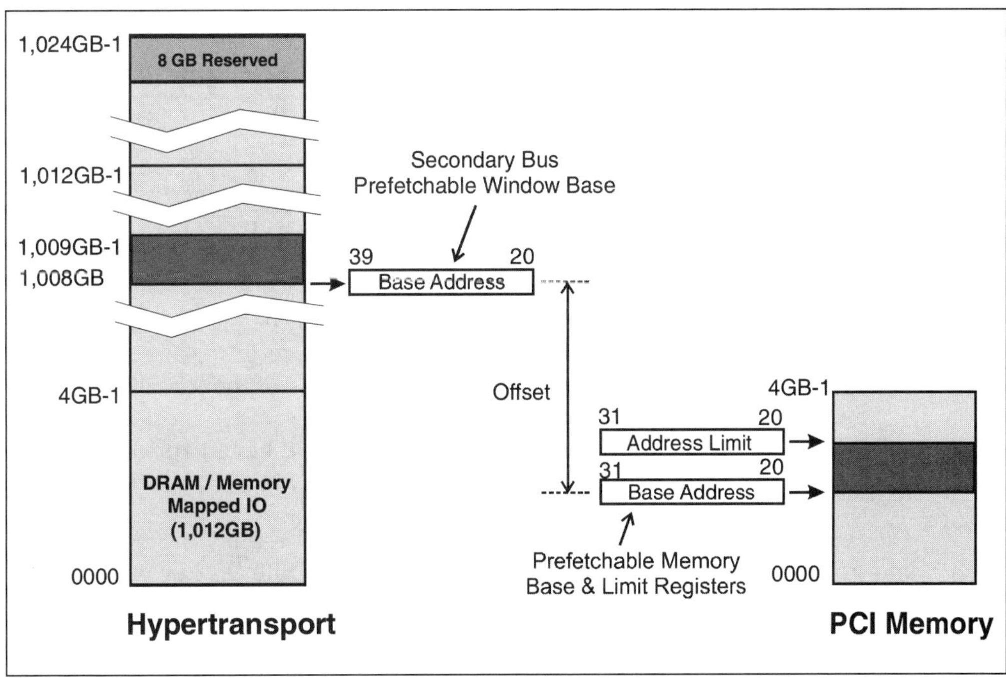

SBNPCtl and SBPreCtl

The *Secondary Base Non-Prefetchable Control (SBNPCtl)* and *Secondary Base Prefetchable Control (SBPreCtl)* fields define attributes associated with the memory range specified and permits software to enable and disable the downstream remapping windows. Figure 21-7 on page 488 illustrates the format and definition of each field within the control register.

Figure 21-7: SBNPCtl and SBPreCtl Register Format

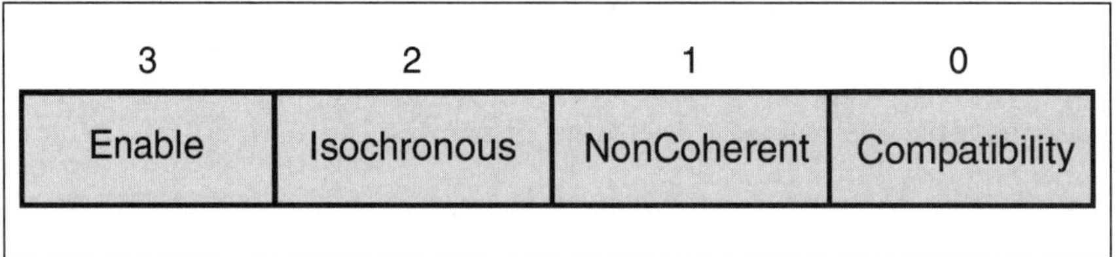

The following list defines and specifies the use of each field:

- The Compatibility field indicates whether the downstream request passing through this window may have the *compat* bit set. This bit supports the implementation of the subtractive decode residing on the secondary bus. When the subtractive decode agent resides on the secondary bus, transactions with the compat bit set may pass through this window.
- NonCoherent indicates whether downstream requests that pass through this memory window will have the coherent bit cleared. This allows hosts to relax memory ordering. Relaxed ordering may be supported on the expansion bus, which is the case with PCI-X.
- Isochronous indicates if downstream requests that pass through this memory window will have the *Isoc* bit set. If this device has Isochronous Flow Control enabled, the requests will be issued in one of the Isoc virtual channels.
- Enable controls whether downstream requests will be remapped using the Secondary Bus Window Base registers.

The following sections provide examples of memory remapping between HT and PCI. Specific examples have been chosen to illustrate the key implementations.

DMA Mapping

Expansion bus master accesses to memory may need to target HT and main memory locations that are beyond the range of the expansion bus. Five registers within the address remapping block combine to define one or more DMA mappings within HT address space. Each DMA mapping is intended to provide a separate range of memory space for each bus master on the expansion bus. Note also that the corresponding *DMA Control* field enables or disables each DMA mapping range. The definition and use of each DMA register is defined below.

Number of DMA Mappings

This read-only register defines the number of DMA mappings implemented by the bridge. The specification recommends that one set of DMA fields be allocated for each device (bus master) on the expansion bus that can access system memory. The HT specification suggests that HyperTransport-to-PCI bridges, support at least one mapping per REQ/GNT pair.

DMA Secondary Base N and DMA Secondary Limit N

These fields, when enabled, define a range of expansion bus memory space used by a bus master when accessing memory locations upstream. These registers specify a range of addresses in 16MB blocks. Each DMA secondary base and DMA secondary limit field defines address bit 39:24. The lower 24 address bits of the secondary base are specified to be all zeros and the lower address bits of the limit register are specified to be all ones, thereby creating a minimum address range of 16MB when the base and limit values are the same. Note, also, that the upper bits these fields will not be used if the expansion bus address is smaller than HT address space.

DMA Primary Base N

The *DMA primary base* field defines the base HT address where the address within the expansion bus DMA address range is to be mapped. The format of this field is the same as the DMA base field. The HT address is formed by adding the difference between the primary base and secondary base values to the actual address.

DMA Control Field

In addition to the DMA enable/disable function, the DMA Control field also establishes HT attributes associated with DMA accesses. Expansion bus masters may not define attributes that are supported by HT. The DMA control register permits software to define characteristics and attributes associated with each DMA address range. Figure 21-8 on page 490 illustrates the format and defines each bit of the DMA Control field.

- Enable — permits software to enable or disable a range of DMA address space
- Isochronous — specifies that this DMA range is associated with Isochronous transfers
- NonCoherent — indicates that this range of address space is not cacheable

Figure 21-8: DMA Control Field

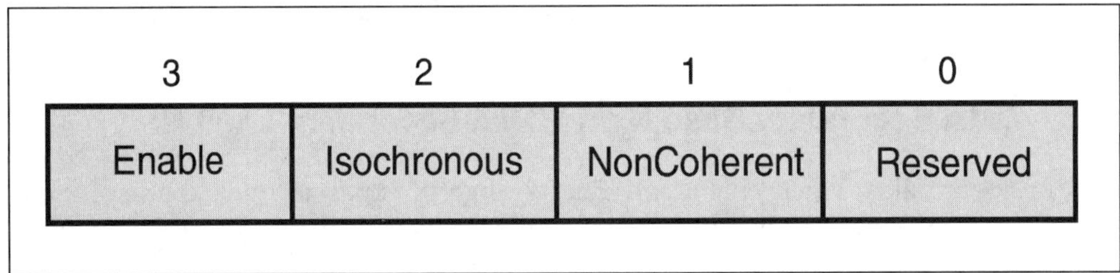

22 *X86 CPU Compatibility*

The Previous Chapter

The large 1 Terabyte HT address space may be outside the limits of a given processor or expansion bus. When address locations are mapped into the HT space that exceed the processor or expansion bus address space, then the addresses must be remapped to and from HT space. The previous chapter discussed the HT solution for remapping memory, MMIO, and I/O addresses.

This Chapter

Many HT platforms may be based on x86 processors, and the HT specification defines the necessary compatibility support. This chapter discusses the x86 features that require specific support by HT and details how HT provides the capability.

Background

Several x86 processor-specific features require support for both hardware and software compatibility, including:

- A method for supporting x86 CPU legacy signals
- A method for supporting x86 Special Cycles
- x86 legacy interrupt support — PCI (8259) and APIC
- Power Management

Other legacy issues related to Industry Standard Architecture (ISA) platforms deal with I/O bus compatibility, rather than processor-related issues. These topics are discussed in the chapter entitled, "I/O Compatibility."

Legacy Signals

Support for the features mentioned above involves a number of X86 signals. These signals include those preserved throughout the x86 processor evolution to maintain compatibility:

- INTR (Interrupt Request from the 8259 Interrupt Controllers)
- APIC bus (Advanced Programmable Interrupt Controller bus)
- A20M (Processor Address Line A20 Mask)
- SMI (System Management Interrupt)
- SMIACT (System Management Interrupt Active)
- STPCLK (Stop Clock)
- FERR (Floating-Point Error)
- IGNNE (Ignore Numeric Error)

Figure 22-1 on page 492 illustrates the typical implementation of these signals in a non-HT system. Note that these signals are typically routed directly between the CPU and South bridge in a legacy platform with PCI.

Figure 22-1: CPU Signals Routed Between South Bridge and CPU

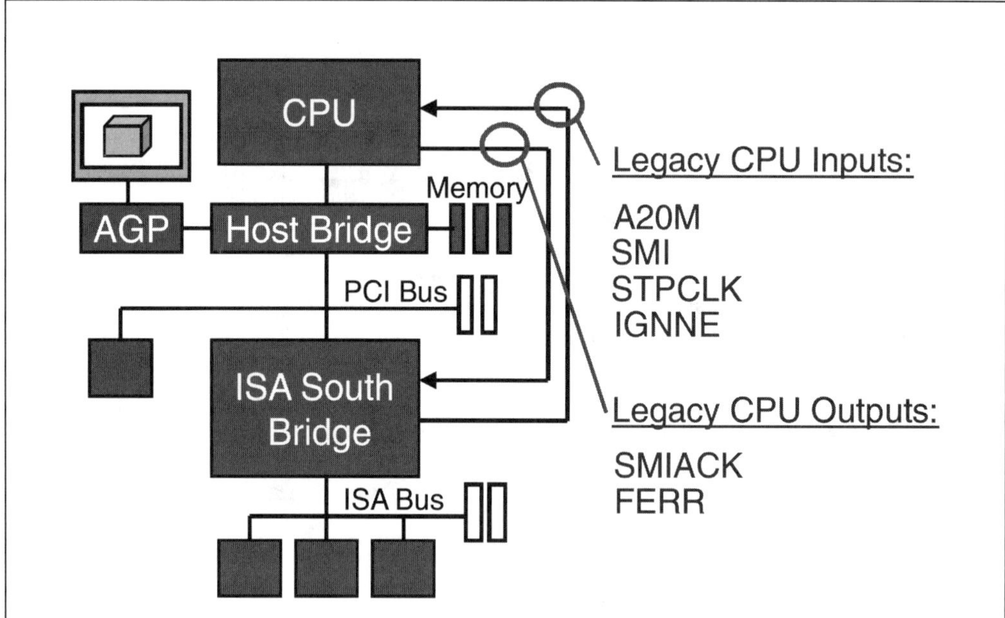

Legacy Special Cycles

Some x86 CPU events are signaled via special cycles. The CPU uses its system interface (e.g. front-side bus) to signal special cycles to the Host Bridge. The events signaled by an x86 CPU include:

- SHUTDOWN
- HALT
- INTA (Interrupt Acknowledge)
- STOP_GRANT
- Voltage ID/Frequency ID Change (Athlon processor)
- WBINVD (write-back and invalidate)
- INVD (invalidate)

Special cycles have various jobs related to x86 functions such as Interrupts, System Management Mode (SMM), power management, and cache coherency.

Note that each of the x86 signals and special cycles are discussed within the section that describes the x86 function to which the signal or special cycle relates.

System Management Messages

HyperTransport eliminates the direct signal routing between the x86 CPU and the compatibility bridge (South Bridge, ICH, etc.) used in legacy platforms. Instead, HT defines System Management (SM) requests that serve to convey information that otherwise would be conveyed via signals. These messages act as virtual wires that signal INTR, FERR#, IGNNE#, A20M#, STPCLK#, SMI# and SMIACT#. Note that the default state of virtual wires is deasserted. Figure 22-2 on page 494 illustrates that messages may move in either direction.

Delivery of special cycle messages is also done via SM messages. When the Host Bridge receives a special cycle from the CPU, it sends an SM message that delivers the special cycle message to interested parties residing on the HT bus. Refer to Chapter 22, entitled "X86 CPU Compatibility," on page 491 for a detailed explanation of the SM messages.

Figure 22-2: SM Request Sources

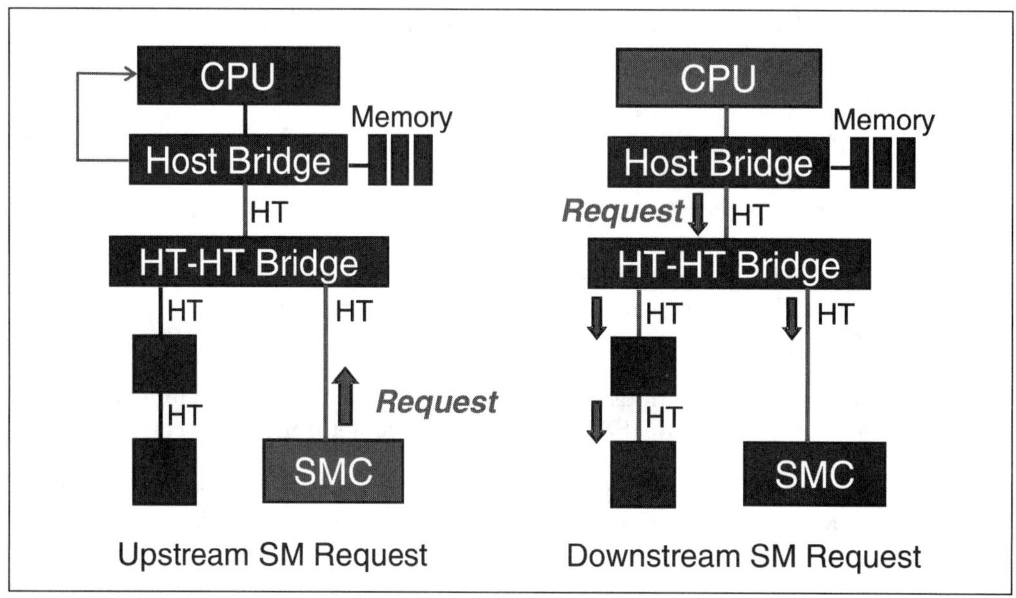

X86 Interrupt Support

The HT specification defines the mechanism necessary to support x86 compatible interrupt handling. This mechanism supports both single and multiprocessor interrupt handling:

- Cascaded PIC (8259) Interrupt controllers for legacy support with single processor systems.
- APIC support for x86 multiprocessor systems.

The APIC solution includes support for delivering 8259 interrupt requests to a single processor, and HT takes advantage of this support for delivering interrupt requests in single processor platforms. Therefore, the HT method of supporting APIC interrupt controllers is discussed first, followed by a discussion of 8259 support.

APIC Interrupt Support

The APIC subsystem was designed to support multi-processing implementations. The subsystem permits an interrupt to be directed to a particular processor for handling or may deliver an interrupt to a group of processors and allow them to arbitrate to determine which one will accept and service the interrupt. In addition to standard interrupt delivery, the APIC bus supports delivery of other interrupts including 8259 interrupts, NMI (Non Maskable Interrupts), and SMI# (x86 System Management Interrupts).

Legacy Method of Handling APIC Interrupts

Figure 22-3 on page 495 illustrates a standard APIC subsystem implemented in a legacy-based system with a PCI-X expansion bus. The APIC subsystem is split between the IO APIC located in the ICH and local APIC modules associated with each CPU. A synchronous APIC bus connects the IO APIC with the local APIC modules of all processors.

Figure 22-3: Legacy APIC Implementation

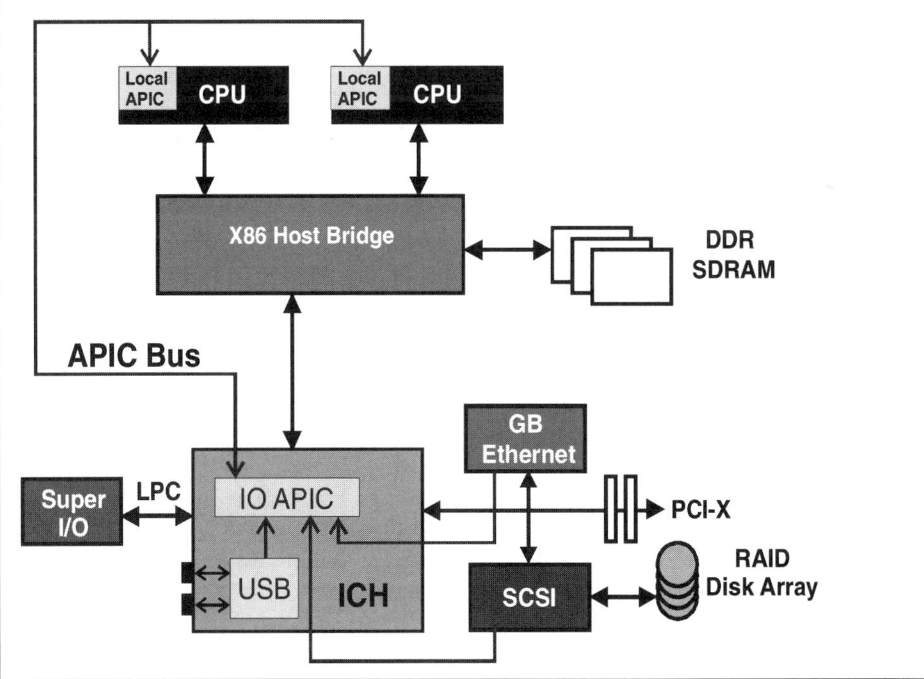

Interrupts are send via the APIC bus to one or more processors. The destination of the interrupt and the type of interrupt being sent (Delivery Mode) is a function of how the Interrupt Redirection Register within the IO APIC is programmed by software. Each input pin to the IO APIC has an associated Interrupt Redirection Register that specifies how interrupts received on that pin will be handled. The name for this register comes from its primary job of directing each interrupt to a specific entry in the interrupt table. The table entry contains the memory address of the interrupt service routine handler associated with the interrupt input. Figure 22-4 on page 497 illustrates the redirection function along with the format of each register. A look at each field clarifies how the IO APIC knows the type of interrupt to send, its destination, and the entry point of the interrupt handler.

- Vector — points to an entry in the x86 interrupt vector table containing the entry point of the interrupt service routine.
- Delivery Mode — Defines the type of interrupt being signalled.
- DM (Destination Mode) — Specifies the number of bits in the destination field that identifies the target processor(s).
- Interrupt Status — Specifies that an interrupt has been sent but delivery is being delayed due to the APIC bus busy condition, or because the local APIC being targeted is temporarily unable to accept the interrupt.
- Polarity — Specifies whether an interrupt input is active high or low.
- Remote IRR (Interrupt Request Register) — Supports level triggered sharable interrupts and has no meaning for edge triggered interrupts. This bit and the EOI message together tell the I/O APIC when an interrupt for this pin has been serviced; thus, if the input pin is still active (low) another interrupt is pending and can be signaled. This bit is reset (0) when an EOI message is received from a local APIC and set (1) when Local APIC(s) accept the level interrupt sent by the I/O APIC.
- Trigger Mode — Determines whether the interrupt input is edge or level triggered.
- Mask — Inhibits the transfer of interrupts associated with this pin.
- Destination — Specifies the target processor(s) with an APIC ID value. The number of bits used depends on the destination mode selected. The lower nibble of this field is used for static destination mode, while the entire 8-bit field is used for logical destination mode.

For a review of APIC operation, refer to MindShare's Pentium Processor System Architecture and Pentium Pro and Pentium 3 System Architecture (www.mindshare.com).

Figure 22-4: Format of the Interrupt Redirection Register

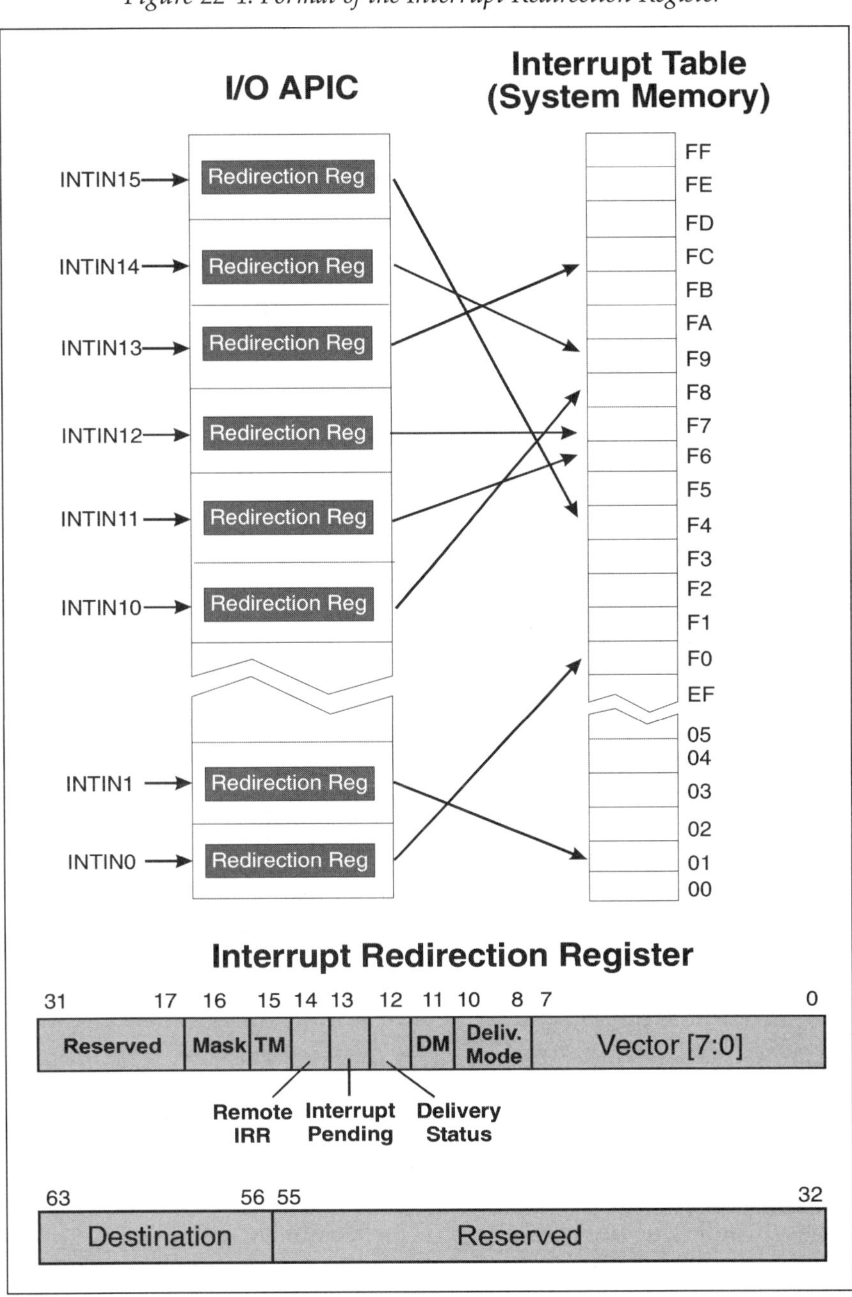

HT Method of Handling APIC Interrupts

HT replaces the APIC bus by sending interrupts via the SM interrupt message. An actual APIC is typically implemented in an x86-based multiprocessor platform as illustrated in Figure 22-5. In this example, the HT bus provides the interconnect between the Host Bridge and I/O Controller Hub (ICH). The IO APIC receives interrupt requests from PCI-X devices and from imbedded I/O devices within the ICH. The IO APIC in turn delivers interrupt requests to the SMC for transfer to the target processor(s) via the Host Bridge and APIC bus.

Figure 22-5: HT-Based Platform with IO APIC

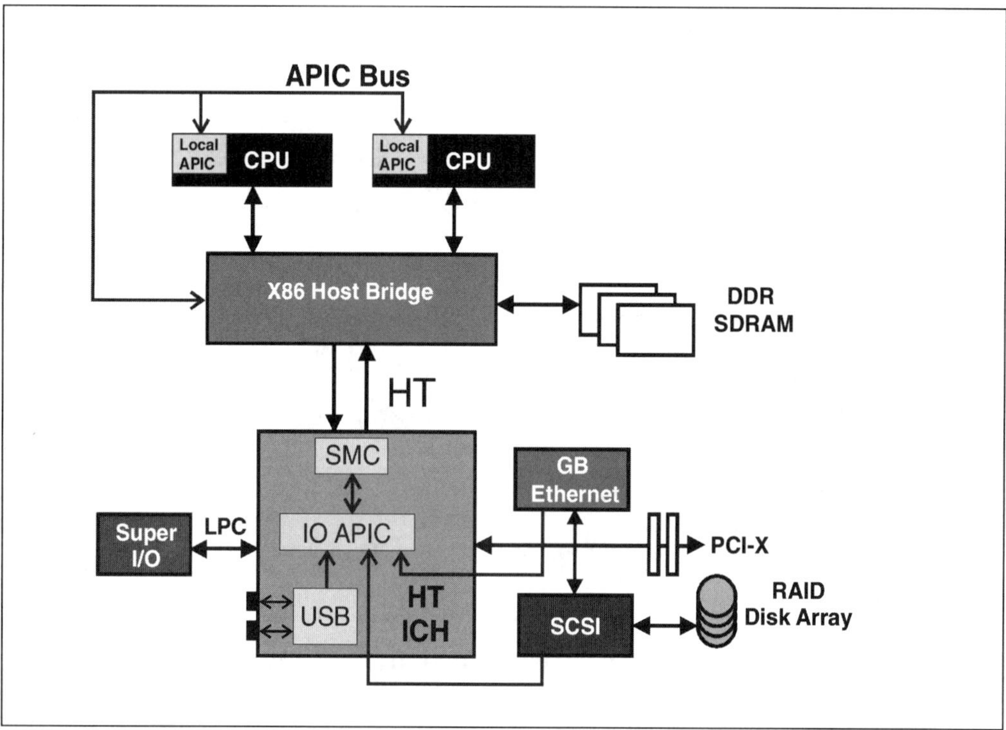

HT I/O Device Delivery of X86 Interrupts

HT I/O devices may also need to signal interrupts, and must use the message signaled interrupt method of delivery. To signal an interrupt request, HT I/O devices must issue an SM interrupt message directly. The interrupt message sent by an HT device contains the APIC-compliant information. Note that HT I/O installed in x86 platforms are required to use the APIC definition when sending interrupt request messages whether an IO APIC is implemented within the system or not. Figure 22-6 illustrates a system in which SM interrupt requests will be issued HT I/O devices and by the SMC (in response to IO APIC interrupt requests).

Figure 22-6: Example X86 Platform Containing HT I/O Devices and an IO APIC

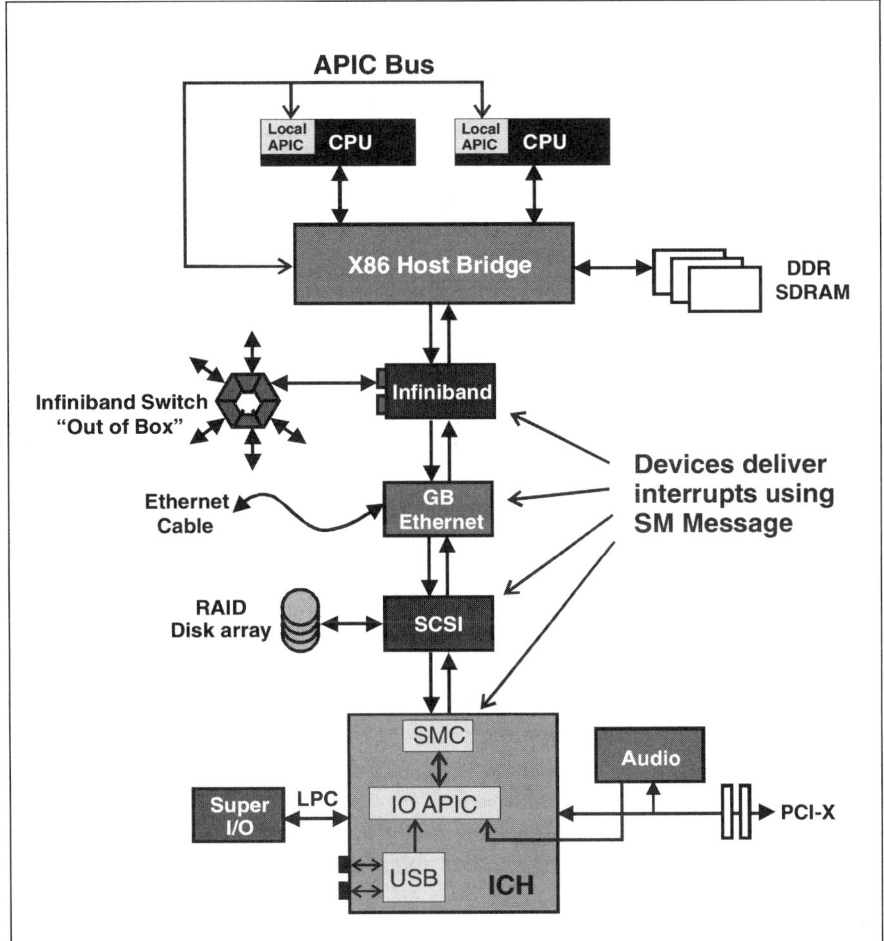

Figure 22-7 on page 502 illustrates the format of the x86 SM interrupt message that contains APIC-compliant data. Software must set up the redirection registers associated with each APIC input pin to define the characteristics of the interrupt. Similarly, each x86-compatible HT device must emulate a redirection register for each interrupt it supports. This is accomplished through the interrupt definition registers. (See "Interrupt Definition Registers" on page 211.) The content of the x86-compatible SM interrupt message comes from the interrupt definition register.

The portion of the message that is APIC-specific include:

- MT[3:0] — Message type indicates the type of interrupt being delivered. This field represents the Delivery Mode specified by the APIC definition. The HT specified values include:
  - o Arbitrated - Used to deliver an interrupt to two or more processors that will arbitrate between themselves to determine which is currently executing the lowest priority code. That processor will accept and service the interrupt. These interrupts can be either edge or level triggered.
  - o Fixed - This interrupt targets the processor(s) specified in the destination field. It can target a single processor, a group of processors, or can be broadcast to all processors. These interrupts can be either edge or level triggered.
  - o SMI - Specifies that a System Management Interrupt is being delivered by an HT device. No interrupt vector is specified for SMIs. These interrupts must use edge triggering.
  - o NMI - Non-Maskable Interrupts are typically sent by devices that have incurred fatal (non-recoverable) errors. These interrupts must also use edge triggering.
  - o INIT - The Initialize delivery type goes to all processor cores listed in the destination. All addressed local APICs will transition to their INIT state. INIT is always treated as an edge triggered interrupt.
  - o ExtInt - External Interrupt delivery indicates that an interrupt request has been received by a legacy 8259 interrupt controller. Consequently, no vector is associated with the interrupt. The processor must perform and Interrupt Acknowledge special cycle upon receiving this interrupt.
- DM — Destination Mode, same as defined by APIC, physical or logical. The mode defines the number of bits within the destination field that will be used.
- RQEOI — Request EOI indicates whether an EOI message is required to clear the interrupt and is comparable to the Remote IRR field defined by the APIC.
- Interrupt Destination — Specifies the processor(s) targeted by the interrupt. The number of bits used depends on the Destination Mode.
  - o Physical destination uses bits [7:0]. A value of FFh selects all processors. Note that some x86 platforms use only bit [3:0] as the physical ID.
  - o Logical destination uses all 32 bits of the destination field. However, some x86 implementations are restricted to [7:0] for the logical ID.

Table 22-1 taken from the specification summarizes each interrupt message type and lists the permissible values for the other APIC-related fields.

Table 22-1: Summary of Interrupt Request Bit Field Encoding Values

MT[3:0]	Name	RQEOI	Vector	DM	Destination
0_000	Fixed	0 or 1	0-FFh	0	0-FF
				1	0-FFFF_FFFFh
0_001	Arbitrated			0	0-FF
				1	0-FFFF_FFFFh
0_010	SMI	0	0	0	FFh
0_011	NMI				
0_100	INIT				
0_101	Startup (Host Only)				
0_110	ExtInt	0	0	0	FFh
1_011	Legacy NMI (LINT1)*				
1_110	Legacy PIC ExtInt (LINT0)*				
x_111	Reserved for EOI				

* The Legacy NMI and PIC delivery is typically related to a single processor implementation as discussed in the respective sections entitled, "Legacy NMI Signaling" on page 508 and "Legacy Interrupts (8259 Interrupt Controllers)" on page 503.

Figure 22-7: Format of X86-Based Interrupt Request Message

Bits Bytes	7	6	5	4	3	2	1	0
0	SeqID[3:2]		Cmd[5:0] = 101000 (posted, Wr(sized), Byte					
1	PassPW	SeqID[1:0]		UnitID[4:0]				
2	Count[1:0]		Reserved					
3	MT[3]	DM	RQEOI	MT[2:0]			Count[3:2]	
4	Interrupt Destination[7:0]							
5	Interrupt Vector [7:0]							
6	Addr[31:24] = F8h							
7	Addr[39:32] = FDh							
8	Interrupt Destination[15:8]							
9	Interrupt Destination[23:16]							
10	Interrupt Destination[31:24]							
11	Reserved							

EOI (End of Interrupt) Message

When an HT device (I/O device or ICH /South bridge) sends an interrupt request that requires confirmation of interrupt service completion, it must set the RQEOI bit in the interrupt request message. Subsequently, the interrupt service routine clears the EOI bit within the local APIC and the local APIC returns an EOI message back to the Host Bridge. The bridge then initiates an SM EOI message that is broadcast to all HT devices downstream. Figure 22-8 on page 503 illustrated the format of the EOI message.

Figure 22-8: EOI Request Message Format

Bytes / Bits	7	6	5	4	3	2	1	0
0	SeqID[3:2]		Cmd[5:0] = 111010 (posted, broadcast)					
1	PassPW	SeqID[1:0]		UnitID[4:0]				
2	Reserved							
3	Reserved			MT[2:0] = 111b			Reserved	
4	Interrupt Information[15:8]							
5	Interrupt Information[23:16]							
6	IntrInfo[31:26] = Addr[31:26]*						IntrInfo[25:24]	
7	Addr[39:32] = FDh							

* Note that the specification defines byte 6 as Interrupt Information[31:24]

Legacy Interrupts (8259 Interrupt Controllers)

Many x86 systems employ a single processor and use a cascaded pair of 8259 interrupt controllers for managing interrupts. This section briefly describes the legacy method of handling these interrupts for background information, and details the HT method for providing compatibility.

The Legacy Method of Handling 8259 Interrupts

Figure 22-9 on page 504 illustrates a single x86 processor system with a cascaded pair of 8259 interrupt controllers. Note that the Interrupt Request (INTR) signal connects directly between the ICH/South bridge and the processor. The sequence of events associated with signaling and servicing an interrupt delivered by the 8259 interrupt controller is enumerated below:

1. The 8259 Interrupt controllers signal an interrupt request (INTR) to the processor upon receiving an interrupt request from an I/O device.
2. The processor detects the INTR, but has no knowledge of which interrupt routine to execute. To obtain this information the processor performs an

interrupt acknowledge special cycle to read the interrupt vector from the interrupt controllers.

3. The Host Bridge detects the interrupt acknowledge special cycle and forwards it to the HT device that contains the 8259 interrupt controllers (typically the ICH/South bridge).
4. The interrupt controller (via the ICH/South bridge) returns the vector number associated with the highest priority interrupt pending.
5. The Host Bridge receives the vector and forwards it on to the processor.
6. Having retrieved the interrupt vector, the processor fetches and executes the interrupt service routine corresponding to the vector and interrupting device.

Figure 22-9: Legacy x86 Platform — Single Processor & 8259 Interrupt Controllers

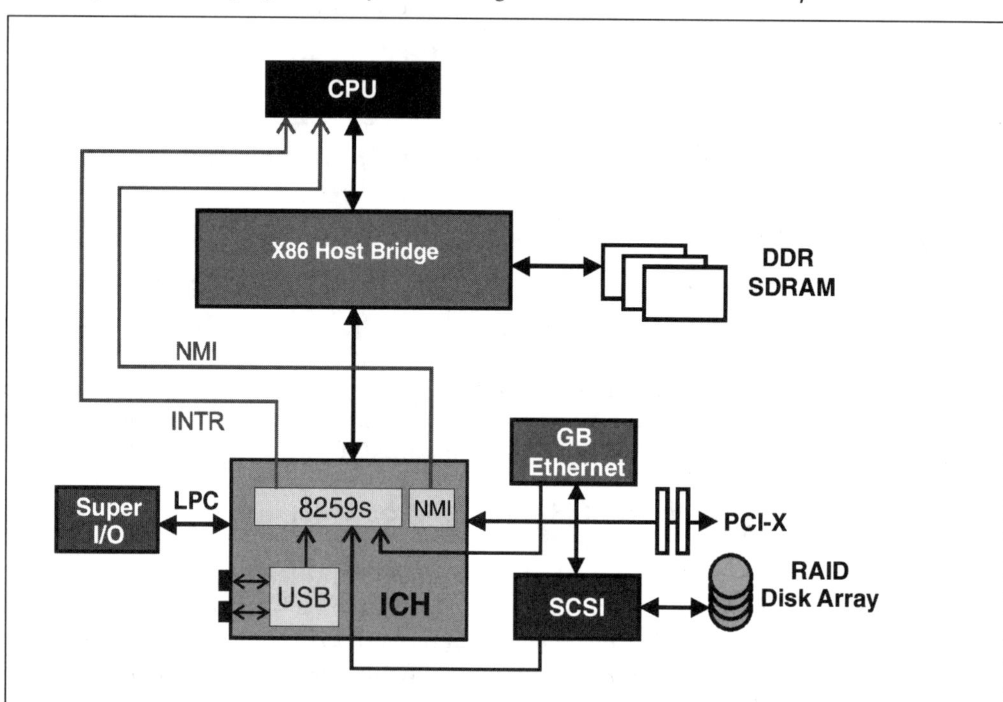

HT Method of Handling 8259 Interrupts

Figure 22-10 on page 505 illustrates a single processor HT system that employs a cascaded pair of 8259 interrupt controllers. The sequence of events described in the previous section are the same for the HT system with two differences:

- The ICH delivers INTR as a virtual wire via an SM Interrupt Message.
- The Host Bridge delivers the interrupt acknowledge special cycle by performing a memory read from the address space reserved for interrupt acknowledge transactions.

Figure 22-10: HT-based System with 8259s

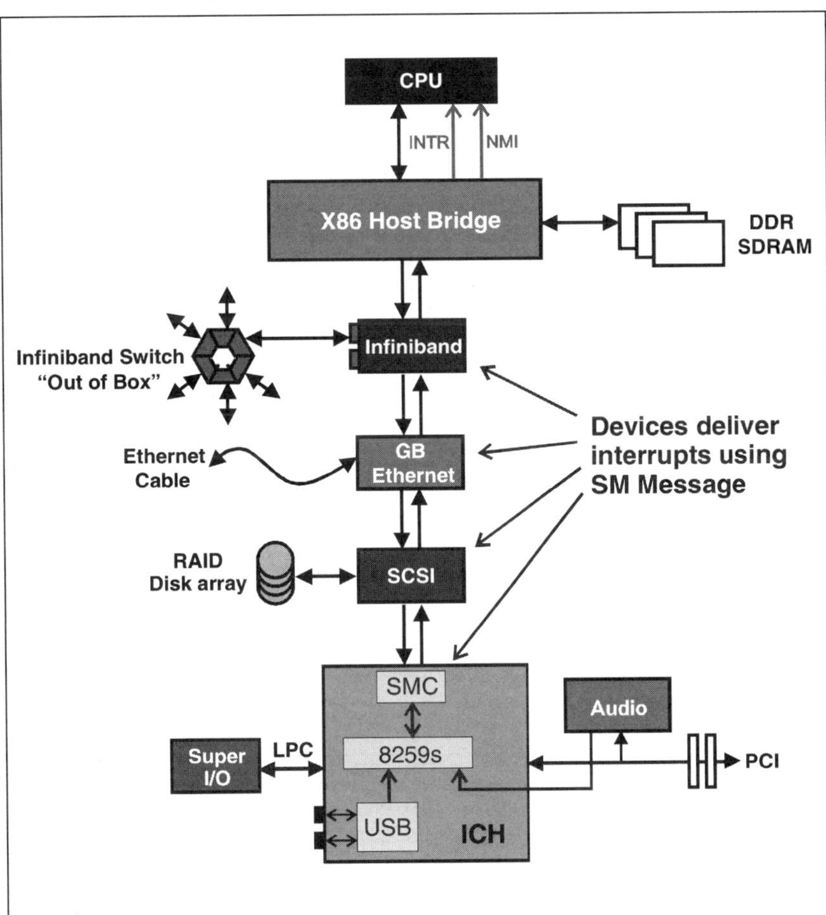

Signaling INTR. The SMC routes interrupt request from the 8259s to the Host Bridge via an SM Interrupt Request message. This message acts as a virtual wire for signaling INTR from the ICH to the Host Bridge, which signals INTR to the CPU. Figure 22-11 on page 506 illustrates the interrupt request packet. The SM Interrupt Request message defines the ExtInt (external interrupt) type for delivering interrupts that originate at the 8259 interrupt controller. Note that two methods of delivering the ExtInt type are defined:

- MT[3:0] = 0_110 — Notifies the processor of the interrupt request via the APIC bus. This method requires that the APIC bus be enabled to accept interrupts. This approach is normally supported only by multiprocessor-based platforms and operating systems.
- MT[3:0] = 1_110 — Notifies the processor of the interrupt request via its INTR input. This is the legacy method of interrupt delivery to the processor that is used by single processor-based platforms and operating systems.

In both cases, processors must obtain the interrupt vector from the 8259 interrupt controllers by performing interrupt acknowledge transactions.

Figure 22-11: Legacy INTR — Interrupt Request and Data Packet Format

Bytes \ Bits	7	6	5	4	3	2	1	0
0	SeqID[3:2]		Cmd[5:0] = 101000 (posted, Wr(sized), Byte					
1	PassPW	SeqID[1:0]		UnitID[4:0]				
2	Count[1:0]		Reserved					
3	MT[3]=1	DM	RQEOI	MT[2:0] = 110			Count[3:2]	
4	Interrupt Destination[7:0]							
5	Interrupt Information[23:16]							
6	Addr[31:24] = F8h							
7	Addr[39:32] = FDh							
8	Interrupt Destination[15:8]							
9	Interrupt Destination[23:16]							
10	Interrupt Destination[31:24]							
11	Reserved							

Signaling Interrupt Acknowledge. Once an x86 CPU recognizes an INTR, it performs an Interrupt Acknowledge cycle to obtain the interrupt vector from the interrupt controller. When the Host Bridge receives the interrupt acknowledge special cycle, it forwards the special cycle to the HT bus in the form of a Sized byte read with a target address that falls within the reserved Interrupt Acknowledge address range (a 1MB range from FD_F900_0000 to FD_F90F_FFFF). The device containing the interrupt controller (typically the ICH/South bridge) returns a read response packet containing an 8-bit vector value. The Host bridge recognized the Interrupt Acknowledge response and forwards the vector on to the processor. Figure 22-12 illustrates the HT Interrupt Acknowledge Request packet.

Figure 22-12: Interrupt Acknowledge Request Packet Format

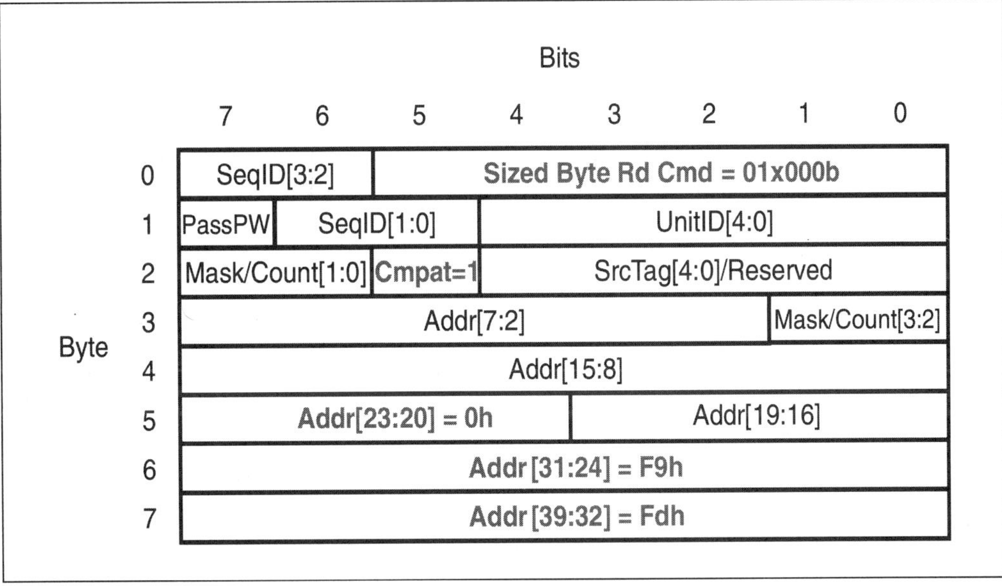

8259 EOI Command

End of Interrupt for the 8259 interrupt controllers is performed via I/O writes to the legacy address locations:

- I/O write to location 20h (Master Interrupt Controller)
- I/O write to location A0h (Slave Interrupt Controller)

These I/O writes are encoded into the HT address space that is reserved for I/O transactions. Translation of the HT address to I/O addresses is discussed in Chapter 21, entitled "Address Remapping," on page 477.

The HT-defined EOI message is reserved for the APIC EOI and not used for the legacy 8259 interrupt controllers.

Legacy NMI Signaling

Figure 22-9 on page 504 illustrates the legacy Non-Maskable Interrupt (NMI) signaling between the ICH/South bridge and the CPU. When the processor recognizes the NMI, it knows that the NMI handler is stored within entry 2 of the interrupt table.

HT delivers NMOS to the processor via the SM interrupt request message as expected. The message types are defined for delivering an NMI to the processor:

- MT[3:0] = 0_011 — Notifies the processor of the NMI via the APIC bus. This method requires that the APIC bus be enabled to accept interrupts. This approach is normally supported only by multiprocessor-based platforms and operating systems.
- MT[3:0] = 1_011 — Notifies the processor of the NMI via its NMI signal input. This is the legacy method of NMI delivery to the processor, and is used by single processor-based platforms and operating systems.

Figure 22-10 on page 505 illustrates the delivery of the SM NMI message from the ICH to the Host Bridge, and the Host Bridge signaling NMI directly to the CPU.

The A20 Mask

The A20M# input forces the processor to emulate the address wrap-around at the 1MB boundary that occurs on the 8086/8088 processors. This pin should only be asserted by external logic (under software control) when the microprocessor is in Real Mode. The x86 microprocessors mask physical address bit 20 (forces it to a zero) before performing a lookup to the internal cache or driving a memory bus cycle onto the buses. For additional information, see MindShare's *ISA System Architecture* book, published by Addison-Wesley.

The Legacy Method of Signaling A20M#

Figure 22-13 illustrates the legacy method of signaling A20 Mask to the processor, along with methods that software has to change the A20M# signal state:

- Output pin from the keyboard microcontroller. (original source) — The A20M# signal is connected to one of the keyboard microcontroller's local output pins. Software issues a legacy command sequence to micro controller, causing this output pin to change state.
- I/O Port 92h (introduced by IBM in PS/2 products as an alternate source) — A bit position within a register mapped at I/O location 92h also determines the state of the A20M# signal. Software can set or clear the bit, causing the A20M# state change.

Figure 22-13: Legacy A20 Mask Signal

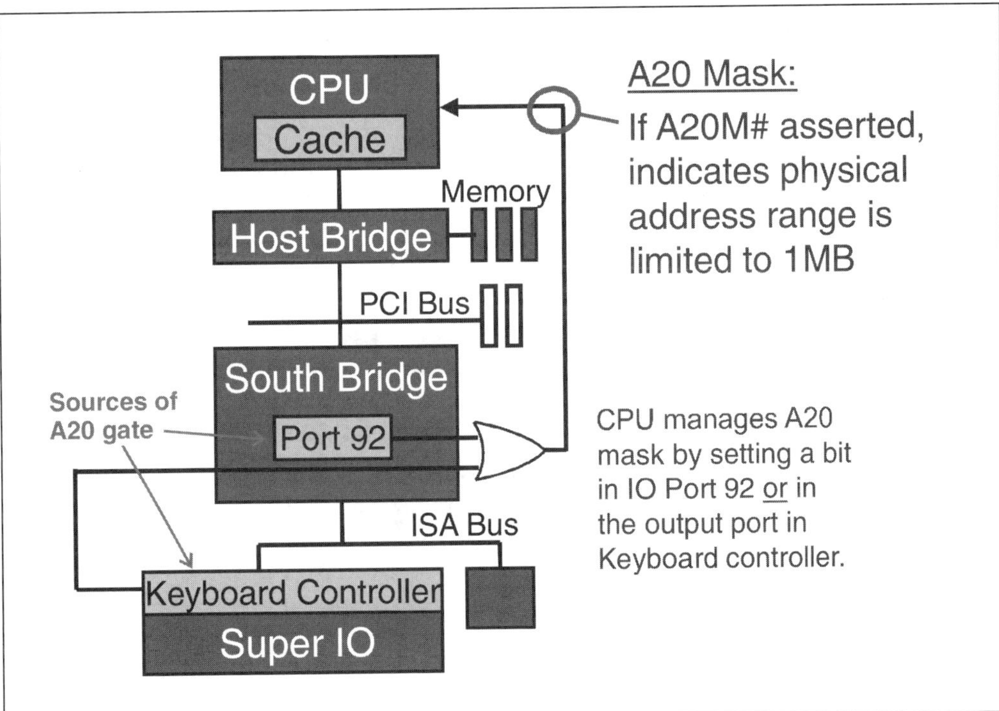

The HT Method of Signaling A20M#

The System Management Controller delivers state changes associated with the A20M# signal via SM Requests to the Host Bridge, which in turn delivers the change to the processor via the physical A20M# signal. Figure 22-14 illustrates this mechanism and Figure 22-15 on page 511 illustrates the contents of the SM request packet, including the value used in the SysMgtCmd field.

Figure 22-14: HT A20M# Delivery

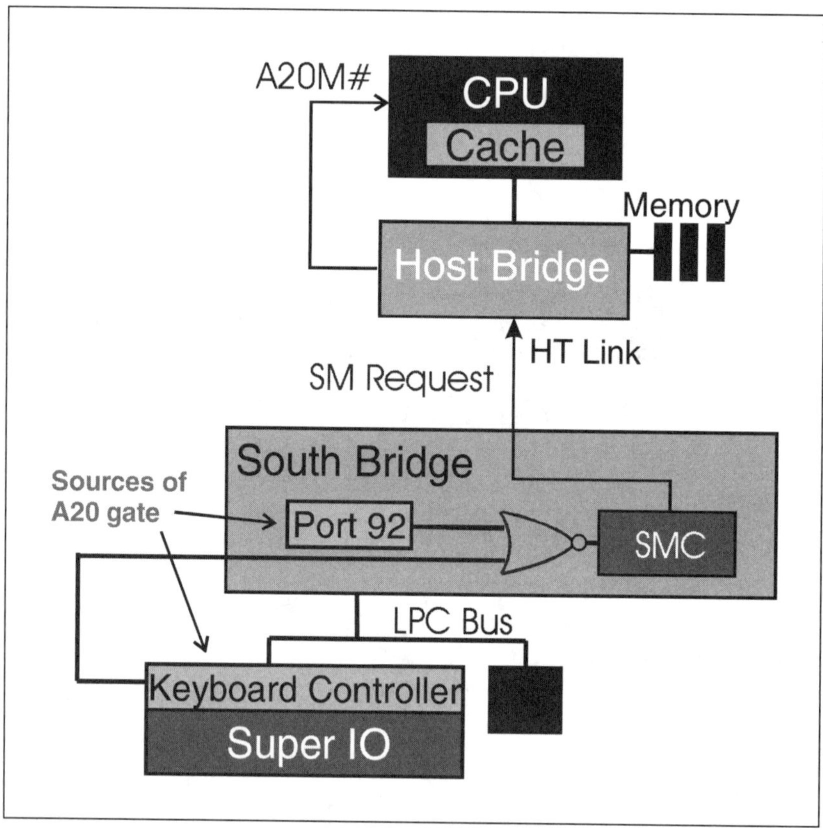

Figure 22-15: Format and Contents of the A20M SM Request Packet

Bytes \ Bits	7	6	5	4	3	2	1	0
0	SeqID[3:2]		Cmd[5:0] = 101000 (posted, Wr(sized), byte)					
1		SeqID[1:0]		UnitID[4:0]				
2	Count[1:0]		Reserved					
3	Reserved						Count[3:2]	
4	SysMgtCmd[7:0] = 0001 xxSx S= new state of A20M							
5	Addr[23:20] = 1h				Reserved			
6	Addr[31:24] = F9h							
7	Addr[39:32] = FDh							

System Management Mode (SMI# & SMIACT#)

System Management Mode (SMM) is a special mode of operation supported by x86 microprocessors in addition to their real and protected modes of operation. SMM was introduced originally to support power management in mobile PCs. Today power management is largely handled by the operating system (O.S.), but SMM is still used in a variety of other applications.

SMM Applications

System management mode (SMM) allows the processor to execute code from system management memory that is separate from the regular memory used by the operating system and applications. This permits the system designer to implement features that are transparent to the operating system, making SMM extremely valuable for applications not supported by the O.S., or where O.S. transparency is important. Some example applications are:

- Devices that have no native O.S. support — for example, USB device support exists only in later editions of operating systems. Support for booting USB devices and running them under older operating system such as DOS can be handled by running support code in SMM.

- System security — when managed within SMM, security code is not accessible via the operating systems and applications making it less accessible to hackers.
- Fault-tolerant solutions — applications running in SMM can monitor the health of the operating system and applications. When problems are detected SMM, code can be used to recover from many types of failures.

The HT specification supports SMM by permitting HT devices to generate a system management interrupt (SMI), which puts the processor into SMM. This type of action is common to a south bridge or I/O Controller Hub in an x86-based system (e.g. USB support). The HT specification expressly prohibits any HT device other than the ICH/South bridge from sending an SMI Request message.

Legacy SMM Signals

The x86 processors implement two signals to support SMM.

- **SMI#** (system management interrupt) input pin informs the processor that a system management handler, residing in System Management (SM) address space, needs to be executed.
- **SMIACT#** (system management interrupt active) informs external logic that the processor is in system management mode. The SMIACT# signal specifies access to system management RAM (SMRAM) so that SMRAM can be accessed. (Note that later x86 processors signal SMIACT# to the Host Bridge via a special cycle operation, while earlier processors implemented an SMIACT# pin.)

Figure 22-16 on page 513 illustrates SMM signaling in a legacy system. The system management interrupt (SMI#) causes the processor to enter system management mode. When SMI# is recognized asserted on an instruction boundary, it completes all buffered write transactions, and signals SMI acknowledge (SMIACT#). The processor then saves the contents of its internal registers to system management RAM (SMRAM) and begins to fetch and execute SMM code.

Figure 22-16: SMM Signaling in Legacy Systems

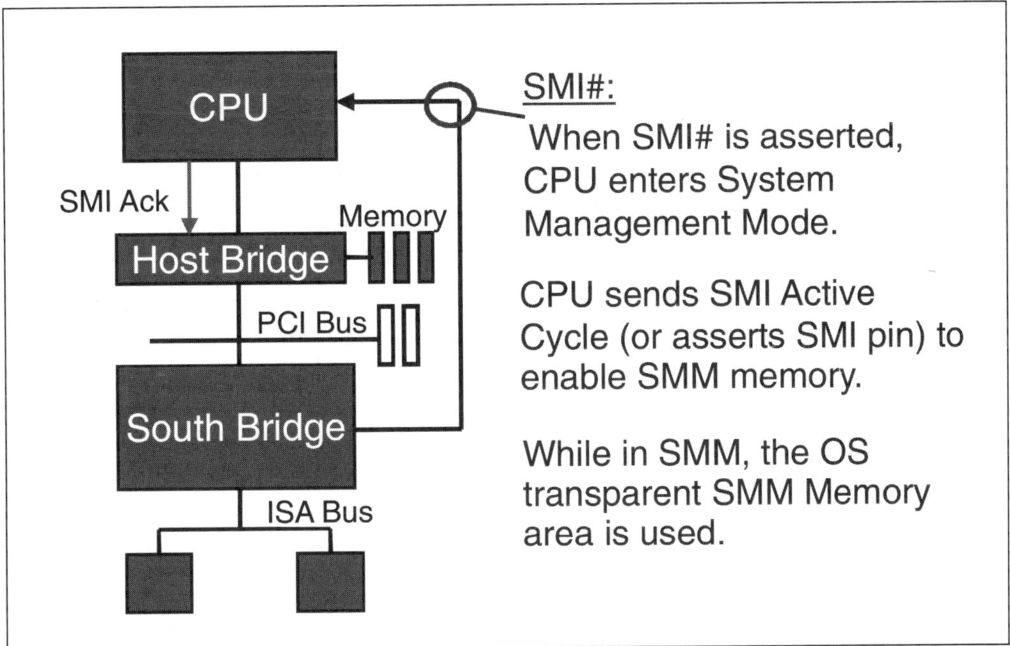

The HT Method of Signaling SMI# & SMIACT#

HT conveys SMI# to the Host Bridge via an SM interrupt message cycle. The bridge in turn asserts the SMI# signal to the processor, or optionally can send the SMI# request via the APIC bus in a multiprocessor environment. An x86 processor responds with an SMI Active indication using one of two possible methods depending on the model:

- asserts the SMIACT# signal that connects to the Host Bridge
- delivers an SMIACT special cycle message to the Host Bridge

In either case, the Host Bridge broadcasts the SMIACT indication to the SMC via an HT special cycle. Figure 22-17 on page 514 illustrates the contents of the Interrupt Request packet that the SMC uses when signaling SMI, and Figure 22-18 on page 515 illustrates the SM Request packet contents that the Host Bridge broadcasts when signaling SMIACT.

Figure 22-17: SM Request Content for SMI Message

Bytes \ Bits	7	6	5	4	3	2	1	0
0	SeqID[3:2]		Cmd[5:0] = 101000 (posted, Wr(sized), Byte					
1	PassPW	SeqID[1:0]		UnitID[4:0]				
2	Count[1:0]		Reserved					
3	MT[3]=0	DM=0	RQEOI=0	MT[2:0] = 010			Count[3:2]	
4	Interrupt Destination[7:0]							
5	Interrupt Vector = 00h							
6	Addr[31:24] = F8h							
7	Addr[39:32] = FDh							
8	Interrupt Destination[15:8] = 00h							
9	Interrupt Destination[23:16] = 00h							
10	Interrupt Destination[31:24] = 00h							
11	Reserved							

Figure 22-18: SM Message Content for SMIACT

Bytes \ Bits	7	6	5	4	3	2	1	0
0	SeqID[3:2]		Cmd[5:0] = 111010 (posted, Broadcast)					
1		SeqID[1:0]		UnitID[4:0]				
2	Reserved							
3	Reserved							
4	SysMgtCmd[7:0] = **1010** xxxx x= implementation specific							
5	Addr[23:20] = 1h				Reserved			
6	Addr[31:24] = F9h							
7	Addr[39:32] = FDh							

Numeric Error Handling (FERR# and IGNNE#)

The x86 processors that have integrated floating-point units provide support for DOS compatible error-handling. These processors have a numeric exception (NE) control bit in Control Register 0 that allows the programmer to select the error-handling scenario to be used by the processor when a floating-point error is detected. The error handling options are:

- Native error handling — this method of numeric error handling involves the processor reporting the error via processor exception (16d), which calls the error handling code.
- DOS-compatible error handling — this method ensures that numeric error code designed for IBM PC-AT compatible systems (ISA machines) running DOS will work correctly with later processors. This code was written to handle floating-point errors associated with the external numeric co-processors. Later x86 processors (starting with the i486™) incorporate the floating-point unit within the processor, thereby creating a legacy problem.

Prior to discussing the DOS-compatible method of error handling by the i486 and later processors, a brief description of floating-point operation and error handling is provided as background.

Numeric Error Handling (The Original Method)

The x86 floating-point solution prior to the i486 microprocessor included a separate numeric co-processor as illustrated in Figure 22-19 on page 517. Note that interface pins between the microprocessor and the numeric co-processor permit some communication, while other communication is initiated by the processor via I/O writes. For example, instructions are fetched only by the microprocessor, so when a floating-point instruction is detected, the processor forwards the instruction to the co-processor via a series of I/O write operations. During the execution of a floating-point instruction, the co-processor asserts the BUSY# signal to notify the processor that it has not yet completed the execution of the instruction. If an error occurs while executing the instruction, BUSY# remains asserted and the floating-point error is reported via the numeric co-processor error# signal. When the floating-point error is signaled, the following primary actions are taken:

1. The ERROR# signal goes to the interrupt controller's IRQ13 input, causing a maskable interrupt to occur (i.e., INTR is signaled to the processor).
2. In response, the processor interrogates the interrupt controller (via interrupt acknowledge bus cycles) to determine which of the 15 interrupts to service.
3. The interrupt controller returns a vector number (75h), which corresponds to IRQ13 (assuming it is the highest interrupt currently pending).
4. The processor reads entry 75h in the interrupt table where the floating-point error handler entry point resides (this was previously set up during initialization).
5. The processor fetches and executes the error handler code in an attempt to determine the nature of the error and correct the problem.
6. A final step performed by the error handler is to clear the BUSY# signal, thus permitting the processor to continue forwarding floating-point code to the co-processor. The error handler clears BUSY# by writing to a register within the co-processor (i.e., data value of 00h is written to I/O location F0h).

The error handler then executes an Interrupt Return instruction (IRET) allowing the processor to continue executing code from the point of the error.

Figure 22-19: Legacy Numeric Coprocessor Error Reporting

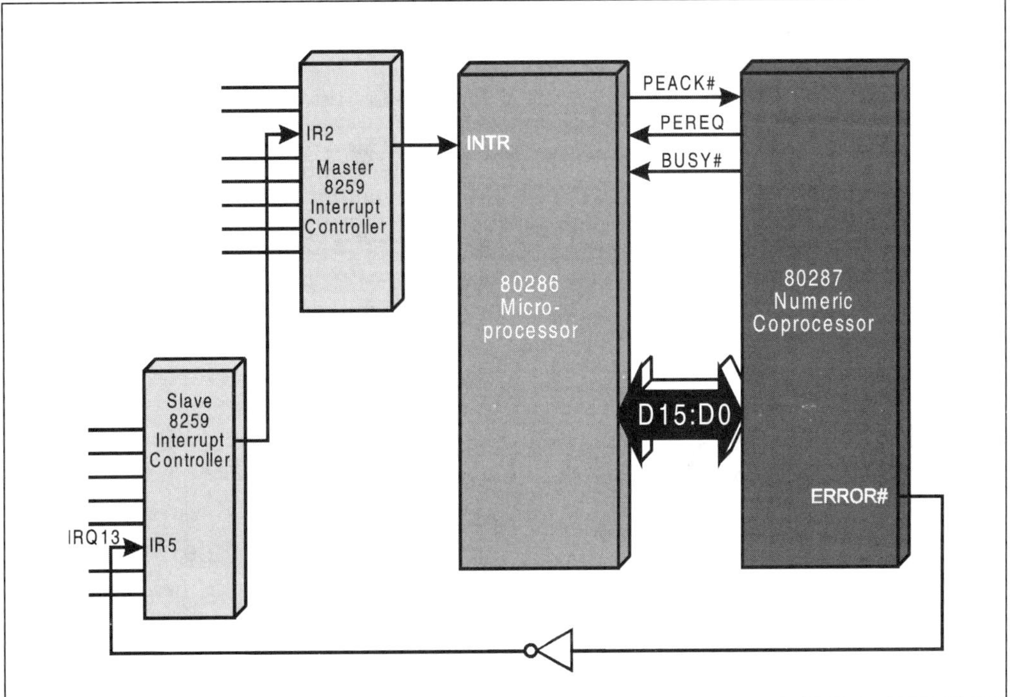

DOS-Compatible Error Handling

X86 processors that have an integrated floating-point unit implement a feature that allows PC-DOS-compatible floating-point error reporting and handling. The NE (Numeric Error) bit in Control Register 0 (CR0) must be cleared (0) to enable DOS-compatible error handling. When compatible error-handling is selected, two signals come into play as illustrated in Figure 22-20 on page 518:

- FERR# (Floating-point Error) — comparable to the co-processor's ERROR# output, this signal reports a floating-point error. FERR# is connected to the IRQ13 input of the interrupt controller.
- IGNNE# (Ignore numeric error) — this input is normally deasserted and controls the behavior of the internal Floating Point Unit (FPU) when an error is detected. External logic controls the state of this signal as described in the following discussion.

When a floating-point error occurs, the processor asserts FERR# and freezes so that no further floating-point instructions are executed. This behavior is similar to that of the numeric co-processor in the previous legacy example. Note that steps 1-6 outlined in the legacy example are the same as those performed in this example. However, writing a data value of 00h to I/O location 0Fh as described in step 6 has a different effect using this solution. In the legacy solution, the I/O write targets the co-processor, but in this case the I/O write has no effect on the integrated FPU. Instead, external logic recognizes the I/O write and asserts IGNNE# in response. Recall that the processor freezes when an error is detected, but upon seeing IGNNE# asserted, the FPU restarts and continues to execute floating-point instructions.

Figure 22-20: DOS Compatible Floating-Point Error Signaling

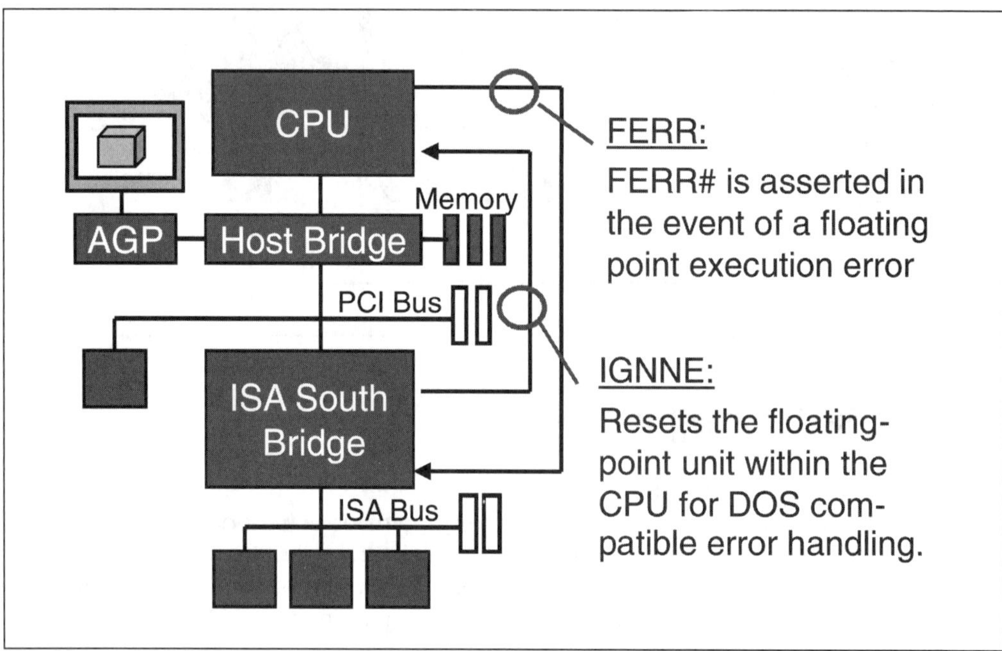

HT Method of Signaling FERR# and IGNNE#

Figure 22-21 on page 519 illustrates an example HT system's implementation of FERR# and IGNNE#. The processor FERR# and IGNNE# signals connect directly to the Host Bridge.

Figure 22-21: Example X86 System with Required DOS-Compatible FPU Error Handling

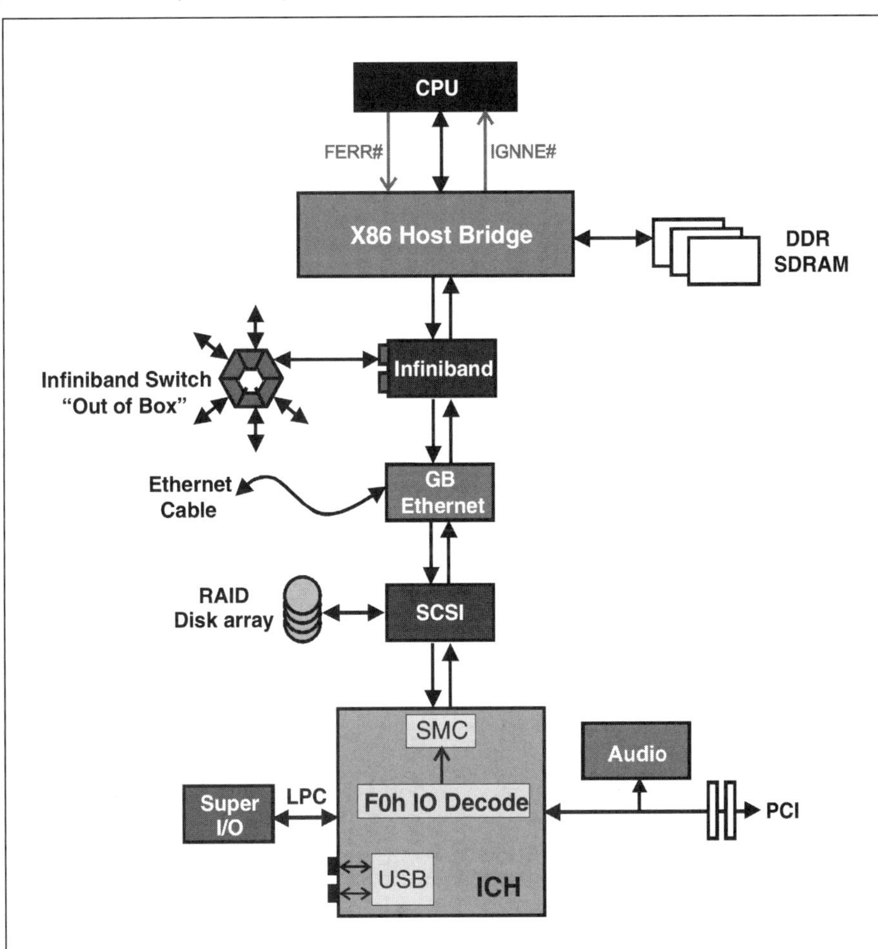

HT Method of FERR# Signaling

The CPU asserts the FERR# signal that is monitored by the Host Bridge. The bridge responds by initiating a broadcast transaction to all downstream HT devices. Figure 22-22 illustrates the FERR message delivered by the Host Bridge.

Figure 22-22: Format and Content of the SM FERR Message

Bytes \ Bits	7	6	5	4	3	2	1	0
0	SeqID[3:2]		Cmd[5:0] = 111010 (posted, Broadcast)					
1		SeqID[1:0]		UnitID[4:0]				
2	Reserved							
3	Reserved							
4	SysMgtCmd[7:0] = **0010** _xxx**S** (S= state of FERR#)							
5	Addr[23:20] = 1h				Reserved			
6	Addr[31:24] = F9h							
7	Addr[39:32] = FDh							

HT Method of IGNNE# Signaling

DOS-compatible floating point error handlers perform writes to I/O location F0h to clear the floating point error (based on the legacy numeric coprocessor design). The ICH/South bridge includes an I/O decoder that detects the attempt to clear the FPU error and responds by issuing a IGNNE message to the Host bridge. The bridge then asserts the IGNNE# signal to notify the internal FPU that it can resume execution of floating point instructions.

X86 Instructions and Special Cycles

Another area of x86 compatibly are CPU special cycles. Most IO devices have no interest in special cycle messages. HT supports the delivery of x86 special cycles via SM broadcast messages. These messages are propagated throughout the HT topology, but are mainly of interest to chipset components, memory/cache controllers, etc. The special cycles supported by HT include:

- SHUTDOWN — x86 processors beginning with the 80286 generate a shutdown cycle when it experiences a triple fault conditions, some models of x86 processor may also generate shutdown cycles as a result of other failures such as internal processor errors.
- HALT — All X86 processors generate a halt special cycle when a HLT (Halt) instruction is executed.
- STOP_GRANT — The stop grant special cycle is performed when the processor recognizes the STPCLK# (stop clock) interrupt.
- SMAF Message — System Management Action Fields are programmed by BIOS software to define specific actions to be taken with a given message. For example, the SMAF field associated with the STPCLK message defines the power management action to be taken (e.g., a transition to an ACPI-defined power-management state).
- Voltage ID/Frequency ID Change — Issued by the CPU if software changes settings related to Voltage (VID) or Frequency (FID)
- WBINVD — The WBINVD (write-back and invalidate) instruction causes all modified data within the processor's internal cache memory to be written back to mail memory prior to invalidating the cache entries. As each modified line is written back, the processor invalidates that entry. Once all lines are written back and invalidated, the processor runs a WBINVD special cycle to notify external logic that all lines in the internal caches have been written back and flushed and that the external L2 cache (if present) should also perform these actions.
- INVD — When the INVD (invalidate) instruction executes, the processor sets all cache entries to the invalid state and runs the INVD special cycle. This special cycle notifies external logic that all internal cache lines have been invalidated, and tells the external L2 cache (if present) that it should also invalidate its cache entries.

Appendix

Glossary
of
Terms

Address Map. The 1.04 revision of the HT specification defines a 40-bit memory map. All resources which can be targeted with directed packets are mapped into this space. In addition, reserved portions of the memory map are used for special purposes, including IO accesses, broadcast messages, interrupts, configuration cycles, etc.

AGP (Accelerated Graphics Port). A high-performance, point-to-point interface which connects a graphics adapter to the main memory controller in the Memory Controller Hub (MCH) or North Bridge (Host/PCI Bridge).

Atomic Read-Modify-Write Transaction. A hybrid read and write operation issued from one source, targeting main memory. An Atomic RMW is guaranteed to complete without intervening accesses of the same location by any other device. This command is useful when a memory semaphore is being updated by one of the devices sharing it. HT supports two Atomic RMW variants: *Fetch and Add* and *Compare and Swap*.

Base Address Registers (BARs). Device configuration registers that define the start address, length, and type of address space required and owned by a device. The type of space implemented will be either memory or I/O. The value written to this register during device configuration will program its address decoders to detect and accept accesses within the indicated range. Because HT memory maps IO accesses, an IO request in the Base Address Register will result in the assignment of a starting address in the memory map range reserved for IO.

Bit Time. One half of a link clock period. As a double data rate (DDR) interface, HT CAD and CTL information is sent during each bit time, resulting in two bits transferred, per signal, per clock.

Bridge. A device that provides a logical and electrical interface between two independent buses. Examples would be the bridge between the host processor bus and HyperTransport, a bridge between PCI/PCI-X and HT, or a bridge between two HT buses. Each secondary interface hosts a new bus (chain).

Broadcast Message. A special case of a posted write request which is used to deliver a message to all devices that see it. There is no specific target; each device accepts it (based on the command type) and forwards it downstream. The end-of-chain device accepts the message, and drops it.

Broadcast Request. See Broadcast Message.

Bus Concurrency. Separate transfers occurring simultaneously on two or more buses. Because HT is implemented with independent point-to-point connections instead of a shared bus, HT concurrency can occur within the same chain.

Byte. 8 bits of digital information.

Byte Mask. In HT, there are no separate byte enable signals; data bus usage is implied in the command type and the mask/count field which accompanies request and response packets. For WrSized (byte) requests, a 32 bit "byte mask" precedes the 1-16 dword data packet and indicates valid bytes being sent — much like byte enables in PCI.

Byte Read. Sized read requests (RdSized) in HT carry a bit indicating whether the data to be transferred is in bytes or dwords. For byte reads, the maximum transfer size is one dword (four bytes); the mask field in the request indicates the valid bytes being transferred. Any byte combination is valid.

Byte Write. Sized write requests (WrSized) in HT also carry a bit indicating whether the data to be transferred is in bytes or dwords. For byte writes, any combination of bytes within a 32-byte, address aligned group may be transferred. The count field in the request indicates the total dwords sent; valid bytes within those dwords are indicated in the 32-bit byte mask which precedes the data.

Cache. A relatively small amount of high-speed Static RAM (SRAM) that is used by CPUs to keep copies of code/data information recently read from system DRAM memory. Data from internal CPU caches may be accessed at full internal clock speed, avoiding a bus cycle to main memory.

Cache Line. When data is moved into or out of a cache, the transfer occurs in fixed amounts called *cache lines*. The size of a cache line is cache design dependent, but typically is 32, 64, or 128 bytes.

CAD Bus. The CAD (*Command, Address* and *Data*) bus carries all information, control, and data packets between two devices on a link. There is one CAD bus in each link direction, and bus width ranges from 2 bits to 32 bits wide.

Capability Registers. HT devices implement one or more sets of advanced capability registers to support features beyond basic PCI compliance. Capability register defined include: Host/Secondary interfaces, Slave/Primary interfaces, Interrupt Discovery and Configuration, Address Remapping, Revision ID, etc.

Cave Device. A single-link HT device. These always reside at the end of a chain and are also referred to as *end devices*.

Chain. In HT, a logical bus may be comprised of multiple devices daisy-chained together. At the top of the chain is a host bridge, in the middle there may be dual-link tunnel devices. At the end of a chain is a device with a single link connection to the chain. This may be a cave device, a tunnel device using only one interface, or the primary interface of a bridge to a new chain (bus).

Chain Down Error. In HT, each device is required to track outstanding requests until a response is returned. A chain down error is said to have occurred if a link goes down between the time a non-posted request is issued and its response returns. A reset will flush any pending requests after a chain down error.

Chain Fail. A chain fail occurs when a Sync flood or an event that can cause one is detected. Each device which detects the event sets the *chain fail* bit in its Error Handling register. The bit is cleared on reset.

Clocking Modes. HT supports three clocking modes of a link receiver interface with respect to the corresponding transmitter: synchronous clocking, pseudo-synchronous clocking, and asynchronous clocking.

Coherency. If the information resident in a cache accurately reflects the original information in DRAM memory, the cache is said to be coherent or consistent. In HyperTransport, transactions targeting DRAM main memory may either require action to guarantee coherency, or not. The *coherent* bit in WrSized and RdSized request packets indicates whether or not the host bridge must take coherency actions (cause a snoop of the CPU caches, etc.).

Coherent Bit. See Coherency.

Command Code. Each HyperTransport control packet contains a 6-bit command code in the first byte. This information informs other devices of the intended operation and the format of the remainder of the packet. Some of the 6-bit command codes include options bits which may be used to indicate whether the packet is to travel in the isochronous channel, is posted or not, etc.

Compatibility Bit. The compatibility bit (*Compat*) in an HT request is set by the host bridge to indicate the packet must be forwarded down the compatibility chain in the direction of the compatibility bridge where it will be accepted, regardless of the address it carries. The use of this bit provides compatibility with the South Bridge subtractive decoder in PCI systems where transactions which are otherwise unmapped in the system may be claimed by the subtractive decoder and forwarded to the compatibility bus (e.g. ISA).

Configuration Cycle. A link transaction to read or write the contents of a device's configuration registers is called a configuration cycle. In HT, configuration accesses are performed using RdSized and WrSized requests targeting addresses reserved for configuration type 0 and type 1 cycles.

Configuration Space. Each HT device is required to implement the 256 byte configuration space required of all PCI-compliant devices. Because PCI permits 256 busses in a system, 32 logical devices per bus, and 8 functions per device, the total configuration address space to reach all possible devices and functions is 16MB (256 busses x 32 devices x 8 functions x 256 bytes = 16MB). HT memory maps the entire configuration address space in a 32 MB reserved address range; the lower 16MB of the reserved range is for type 0 configuration cycles and the upper 16MB of the reserved range indicates type 1 configuration cycles.

Configuration Header Region. The first one-fourth of the configuration space (64 bytes) has a well-defined format and is referred to as the configuration header region. The two key header formats are type 0 (non-bridge) and type 1 (HT bridges or bridges between HT and other compatible protocols (PCI, PCI-X, or AGP).

Consistency. See Coherency.

Control Packet. Control packets include information, request, and response types. Information packets are used for local communication between the transmitter-receiver pairs on each link. Request packets may be 4 or 8 bytes, and are used to initiate transactions. Response packets are 4 bytes and are returned by the target of each non-posted request.

CRC. In HT, a Cycle Redundancy Code (CRC) is used to assure the integrity of transmitted data. Starting after reset, each transmitter on each link interface calculates a 32-bit CRC value and periodically sends it to the corresponding receiver where it is checked against the value calculated as CAD packets arrive. CRC is calculated independently for each 8 bits of CAD width. The error handling strategy for handling CRC errors is programmable.

CRC Testing Mode. HT provides a method for stress-testing CRC checking at the receiver. If both devices on a link support the CRC test mode, a transmitter can enter the CRC test mode under software control and generate dummy packets which are checked for CRC validity by the receiver, then dropped.

CTL Signal. The link control (CTL) signal is driven by each transmitter to indicate to the receiver that control packet information is in transit over the CAD bus; when CTL is deasserted by the transmitter, a data packet is in transit. The receiver uses the CTL signal to demultiplex control and data packets.

Cycle Redundancy Code. See CRC.

Data Packet. In HT, the data transfer payload is sent in data packets. All HT packets are multiples of four bytes and a data packet may contain from 1 to 64 valid bytes depending on the request type and transfer count that caused it.

Device. A device is a physical IC package. In HT, a device may contain from one to eight independent logical functions, each of which may consume multiple Unit IDs. PCI compatible device types include host bus bridges, bridges from HT-to-HT (or HT-to-PCI/PCI-X), and tunnel and single-link peripherals.

Direct Memory Access (DMA). A class of transactions in which the source is an IO device accessing main memory directly. Devices commonly using DMA transfers include disk adapters, graphics adapters, USB controllers, etc. DMA transfers relieve processors from much of the burden of large data transfers.

Disconnect. A bus event in which a target forces a source to break off the transfer of data early because it cannot sink or source the next data in the time allowed by bus latency rules. HT uses flow control to assure that no packet is sent across a link which cannot be accepted in its entirety by the corresponding receiver; **HT does not support a disconnect mechanism**.

DHC. See Double-Hosted Chain.

DMA. See Direct Memory Access.

Double-Hosted Chain. A HT topology which includes a host bridge on both ends of a chain. One host bridge is designated the master host, the other is the slave host bridge. These topologies include two variants: sharing double-hosted chains and non-sharing double-hosted chains. Under software control, chain resources may be allocated to either bridge.

Double Word (Dword) Read. Sized read requests (RdSized) in HT carry a bit indicating whether the data to be transferred is in bytes or dwords. For dword reads, any sequence of consecutive dwords in a 16 dword address-aligned group may be transferred. The count field in the request indicates the number of dwords being transferred.

Double Word (Dword) Write. Sized write requests (WrSized) in HT carry a bit indicating whether the data to be transferred is in bytes or dwords. For dword writes, any sequence of consecutive dwords in a 16 dword address-aligned group may be transferred. The count field in the request indicates the number of dwords being transferred.

Doubleword (Dword). Four bytes (32 bits) of digital information.

End Device. A single-link HT device. These always reside at the end of a chain and are also referred to as *cave devices* or *IO hubs*.

End Of Chain (EOC). Devices at the end of each HT chain have special responsibilities for handling mis-directed packets, including the return of responses for non-posted requests received in error. An end-of-chain device may be a single-link peripheral or a tunnel device which is either unconnected on one interface or has been programmed to act as an end-of-chain device.

End Of Interrupt (EOI) Message. Certain interrupts (e.g. level sensitive, shared interrupts) may require an acknowledgement of servicing by the CPU. In HT, the EOI broadcast message is sent downstream to provide this confirmation. Each device that decodes the EOI message clears outstanding interrupts indicated in the *IntrInfo* field, and passes the message on to the next downstream device.

Error Handling. HT defines a number of error types, including CRC, protocol, receive buffer overflow, end-of-chain, response, and chain down errors. Only the generation and checking of CRC errors is required. One or more logging bits in configuration space is defined for each supported error type, and the reporting method for each is programmable in software. Reporting methods include response with error bit set, fatal and non-fatal interrupts, and Sync flood.

Fairness Algorithm. HT specifies a fairness algorithm that tunnel devices are required to observe when inserting their own packets into a packet stream that they are forwarding on behalf of devices below them.

Fabric. The collection of devices in a HyperTransport topology, including a bridge to the host system.

Fatal Interrupt. A fatal interrupt message is sent by a device which detects an error condition it may not be able to recover from. It is roughly equivalent to the non-maskable interrupt (NMI) seen in PCI.

Fence Request. A request which is sent to a bridge and acts as a barrier to subsequent posted writes. Posted writes received before the Fence must be sent on to memory before later posted writes are processed by the bridge. Fence applies to posted writes from all transaction streams.

Flow Control. A credit scheme in which each link receiver uses NOP packets to inform the transmitter of released space in virtual channel buffers. Flow control guarantees that no packet is sent by a transmitter that cannot be accepted in its entirety by the receiver.

Flush Request. A non-posted request sent to a bridge which causes all previous posted writes from that source to be pushed to memory. A target done response is returned to the source when the flush operation is complete.

Functional Devices. PCI-compatible devices, including HT, may support from one to eight logical entities called functions. Each function has its own configuration space and behaves as an independent logical device.

Header. HT device functions implement the 256 byte PCI configuration space. The first one-fourth of configuration space is called the header, and two formats are used: the HT technology type 0 *device header* is for non-bridges; the HT technology type 1 *bridge header* is for HT-HT, HT-PCI, and HT-PCIX bridges.

Hierarchy. In a topology based on point-to-point connections and bridges which add additional secondary busses (chains), the relative positions of devices upstream and downstream from each other is referred to the hierarchy; hierarchy must be taken into account when determining latencies, applying ordering rules, etc.

Hit. Refers to a hit on the cache. This may result from a processor initiating a read or write to a cache, or from a host bridge submitting an address associated with an IO access of memory to the cache controller(s) for look-up. Actions taken by the system depend on the state of the cache line. In HT, transactions targeting main memory may or may not be submitted to caches for look-up, depending on the state of the *coherent* bit in the request packet.

Host Bridge. A device that provides the bridge between the host processor's bus and the first HT chain. The bridge buffers information and provides protocol translation in both directions; it also acts as host to the HT chain below it. Note: In HT, the secondary bus interfaces of each HT-to-HT bridge or PCI(X)-to-HT bridge act as host bridge for their respective chains.

Host/Secondary Interface Block. The advanced capability register block implemented in configuration space by bridges to manage a secondary bus. This set of registers would be implemented by a HT host bridge, HT-HT bridge, or PCI(X)-HT bridge for each secondary interface.

Idle State. A transaction is not currently in progress on a link. In HT, the idle condition is indicated when NOP packets are transferred by each link transmitter. The NOP packets also contain flow control update information.

Information Packet. The NOP and Sync/Error information packets are used for nearest-neighbor communication on each link. They are not forwarded to other links and are not flow-controlled. When sent by a transmitter, they must be accepted by the corresponding receiver.

Interrupt. HT devices requiring service generate interrupts using messages. A WrSized request packet is sent upstream using an address range reserved for interrupt traffic. The method the host bridge uses for delivering them to the CPU(s) is implementation-specific.

Interrupt Acknowledge. In x86-compatible systems, the host processor may respond to an interrupt request on its INTR input by generating an *interrupt acknowledge* transaction. If the system interrupt controller resides in (or below) the HT topology, the interrupt acknowledge cycle travels downstream through HT to the interrupt controller as a RdSized (byte) read request targeting the reserved IACK address range. The South Bridge or other component containing the interrupt controller later returns the vector of the highest priority interrupting device with the HT read response.

Interrupt Controller. In many systems, interrupts from IO devices are gathered and prioritized by a motherboard *interrupt controller* (e.g. IOAPIC). While some systems would always implement an interrupt controller, it is not required in HyperTransport. HT defines an Interrupt Discovery and Configuration capability block for functions. Using this capability block, software can program individual devices with information required to manage each interrupt.

I/O Chain Initialization. The sequence used to bring up the devices on an HT chain. Initialization starts at reset, and includes low-level link negotiation and synchronization by devices, sending of initial buffer release packets to establish flow control buffer size, followed by software configuration of device header and advanced capability registers.

I/O Hub. A type of HT device used to perform one or more functions generally associated with a legacy I/O controller (e.g. South Bridge or ICH). An I/O hub may or may not reside at the end of a chain.

I/O Stream. A collection of transactions in HT sourced by a single node and terminating at the same destination. HT ordering treats separate I/O streams independently.

Isochronous Traffic. Traffic with special latency or bandwidth requirements. Devices which support isochronous packets give them priority over the standard posted request, non-posted request, and response virtual channels.

LDTREQ# Signal. This open drain, wire or'd power management signal is bussed to devices supporting power management state changes. When asserted (low), the link is either active or being requested by device(s). If it is de-asserted, the link is inactive.

LDSTOP# Signal. This signal is driven by the system management controller (e.g. South Bridge) and is used to enable and disable HT links during system state transitions. It is an input to HT devices which support the signal; when asserted, device transmitters finish sending any packets in transit and then start disconnecting from the link. After sending disconnect NOP packets, the transmitter shuts down its drivers. Receivers detecting the disconnect NOPs also may be shut down to save power. Upon deassertion of LDTSTOP#, devices commence reconnection to their links.

Link. The connection between two HT devices. A link is comprised of two sets of high speed, unidirectional signals (CAD, CTL, CLK) and a single set of bussed, low speed signals: PWROK, RESET#, and optional LDTSTOP# and LDTREQ# signals.

Master Abort. If an HT source attempts to perform a non-posted transaction with a target, and the response packet (with error and NXA bits set) is returned instead by an end-of-chain device, the source considers the event a master abort. This is equivalent to unclaimed PCI or PCI-X transactions which are characterized by no assertion of the DEVSEL# signal. The source sets the *received master abort* bit in its device configuration header Status register. It also may inform its driver of the event by generating an interrupt.

Miss. Refers to a miss during a cache look-up. In HT, transactions targeting main memory may or may not be submitted to caches for look-up, depending on the state of the *coherent bit* in the request packet.

Multi-Function Devices. A PCI-compatible physical device may have one to eight independent functions integrated within the package. A component that incorporates more than one function is referred to as a multi-function device.

Node. A physical attachment to one end of an HT link.

Non-Sharing Double-Hosted Chain. See Double-Hosted Chain.

Ordering Rules. HyperTransport defines upstream, downstream, and host ordering rules for packets moving through the topology. While packets in different I/O streams generally have no ordering relationships with each other, virtual channels within the same I/O stream follow either the full producer-consumer ordering model used in PCI, or may use optional relaxed ordering if strict ordering is not required.

Packet. All information transmitted across each HT link is sent in multiples of 4-byte blocks called packets. The two classes of packets are Control and Data. Control packets include information, request, and response variants; data packets carry a payload of 1-64 bytes.

PCI. See Peripheral Component Interconnect.

Peer-to-Peer. HT defines transaction types: programmed IO (PIO), direct memory access (DMA), and peer-to-peer. A peer-to-peer transfer is one in which one IO devices target one another directly, without involving main memory.

Peripheral Component Interconnect (PCI). A widely accepted interconnect and bus transfer protocol which provides a relatively fast, plug-and-play, parallel bus used to connect high speed peripherals to each other and to main memory. The PCI specification also defines connectors which may used to support add-in cards. The PCI configuration and control methods are also used in newer PCI-compatible protocols including HT, PCI-X, and AGP.

Point-To-Point Signals. Point-to-point signals provide a direct interconnect between two devices. These connections are simpler electrically than shared bus signals such as those found on PCI; if speed is required, they can be clocked faster. With only two devices, the arbitration latency which associated with multiple masters on a shared bus is also eliminated in point-to-point connections. HT is comprised of a series of point-to-point links.

Posted-Writes. The immediate completion of write transactions through the use of dedicated write buffers at the receiver. Posted writes are fast because they do not require a response; they are considered complete by a sender as soon as they are accepted by the corresponding receiver. Because HT is comprised of a series of point-to-point connections, posted writes are critical to eliminating latencies in reaching the ultimate target. HT write request packets include a "posted" bit which indicates whether the request should be posted or not. Non-posted writes always result in the return of a *target done* response by the target.

PWROK Signal. Used in conjunction with RESET# to indicate a cold or warm reset event on a chain. If PWROK is asserted with RESET#, warm reset is signalled; if PWROK is deasserted and RESET# asserted, cold reset is underway.

Primary Bus (Chain). For bridge devices, the primary bus (chain) is the one in the direction of the host CPU.

Quadword (Qword). Eight bytes (64 bits) of digital information.

RdSized Request. See Double Word Read and Byte Read.

Read Request. See Double Word Read and Byte Read.

Read Response (RdResponse) A control packet returned by targets of RdSized or Atomic RMW requests; the *read response* precedes the data packet and acts as a tag to assure delivery to the original source. In the event of an error, a read response carries *NXA* and *error* bits which are used to indicate the nature of the error. All requested data is sent after the read response, regardless of any error condition.

RESET# Signal. As in PCI, HT hardware or software reset events may be either cold or warm. If PWROK is asserted with RESET#, a warm reset is indicated; if PWROK is deasserted and RESET# asserted, a cold reset is underway. Cold and warm reset have different implications on persistent areas of configuration space, including errors which may have been logged there prior to the reset.

Retry. In some bus protocols (e.g. PCI), a target may signal "retry" to the initiator to indicate it cannot sink or source the first data in the time allowed by bus latency rules. HT uses flow control to assure that no packet is sent across a link which cannot be accepted in its entirety by the corresponding receiver; **HT does not support a retry mechanism**.

Request. An HT control packet type used to initiate a transaction. Request packet size is four bytes or eight bytes, and carries attribute bits which may include I/O stream, target address, virtual channel, whether it should be posted or not, byte or dword data transfer, transfer size, coherency requirements, etc.

Routing. HT devices accept, forward, or reject packets based on the type of packet, its transaction stream, and the direction it is moving. Basically, directed requests are routed on the basis of the 40-bit address field, responses are routed based on UnitID and the bridge bit state, and broadcast messages are accepted and forwarded by all devices which decode them. Flush and Fence requests are routed upstream to the host bridge based on the command type.

Scalability. Depending on bandwidth needs, HyperTransport allows scaling of the CAD bus signal group width from 2-32 bits in each link direction. Asymmetrical CAD bus widths on a link is also permitted. The link clock frequency is also scalable from 200MHz-800MHz.

Secondary Bus (Chain). For bridge devices, the secondary bus (chain) is the one away from the direction of the host CPU. In HT, bridge devices are permitted to support multiple secondary busses; each secondary bus has its own bus number and acts as host bridge to the devices on the chain below it.

SeqID [4:0]. See Sequence ID.

Sequence ID. Used to override general ordering rules. All request packets carrying the same 5-bit, non-zero sequence ID will be handled as being part of the same strongly ordered sequence by tunnels and bridges in the path to the target.

Sharing Double-Hosted Chain. See Double-Hosted Chain.

Sized Read. See Double Word Read and Byte Read.

Sized Write. See Double Word Write and Byte Write.

Slave. A tunnel or cave device which implements the Slave/Primary Interface block advanced capability register set.

Slave Host Bridge. See Double-Hosted Chain.

Slave/Primary Interface Block. The advanced capability register block implemented in configuration space by non-bridges to manage HT interfaces. The register set is used to configure each link, log errors, set up error reporting, etc.

Snooping. The process of presenting transaction addresses to cache controllers so that a check can be made to see if the information being accessed is also resident in the cache(s). If a cache contains information from the target address, a "snoop hit" occurs. Depending on the cache line state, action may have to be taken to assure cache coherency. Cache snoops take time and affect performance; in HT, request packets targeting main memory include a *coherent bit* indicating whether coherency actions are required.

Source. The HT node initiating a transaction.

Source Tag. A 5-bit field in request packets used to uniquely identify one of the 32 outstanding non-posted transactions allowed for one I/O stream.

SrcTag [4:0)] See Source Tag.

Stale Cache Line. When caches are in use, there always is a hazard that a local cached copy may become stale if the memory location that sourced it is modified by another CPU or I/O device, or that memory may become stale if a cache line is modified and not written back. It is a system responsibility to use snooping or other mechanisms to guarantee that both of these cases are avoided.

Subordinate Bus Number. The highest numbered bus (chain) below a bridge. A bridge contains three bus number registers which are programmed at boot time: primary bus number, secondary bus number, and subordinate bus number. These bus number registers are used to help the bridge claim/pass configuration transactions.

Subtractive Decode. In x86 compatible systems, the PCI-to-expansion bus bridge (South Bridge) may claim transactions not accepted by other devices on the PCI bus. This is permitted because a non plug-and-play expansion bus (e.g. ISA) may host devices which own addresses not otherwise mapped in the system. For compatibility, HT also supports a subtractive decoder. Packets sent downstream by a host bridge with the *Compat* bit set are routed to the subtractive decoder, where they are accepted and forwarded to the compatibility bus.

Sync Flood. A drastic method of reporting errors in which a device transmits Sync packets continuously until reset occurs. Each device on the chain that detects the Sync flood repeats the pattern on all links. A reset is required to recover from a Sync flood. Sync flood is analogous to SERR# on a PCI bus.

Sync Packet. A control packet generated on each link interface during initialization to synchronize devices. It also is issued during Sync flood error reporting.

System Management. HT supports a number of x86-compatible system management (SM) features. System management functions make use of system management messages sent to and from the *system management controller* (typically in the South Bridge), and the two SM signals: LDTREQ# and LDTSTOP#. SM functions include generating special cycles and disconnecting and reconnecting links.

Target. A generic term used to indicate the ultimate destination in a transaction. Tunnels and bridges in the target path forward packets along. In HT, directed request packets are accepted by the target device when it decodes the address field. Some packets are accepted by a host bridge target based on the command type rather than a specific address (Flush, Fence). Finally, broadcast messages target all devices that see them.

Target Abort. If an HT source attempts to perform a non-posted transaction with a target and the response packet returns with the error bit set and NXA bit cleared, the source considers the event a target abort. This is equivalent to the target abort in PCI which is characterized by the target asserting STOP# and de-asserting DEVSEL#. The source sets the *received target abort* bit in its device configuration header Status register. It also may inform its driver of the event by generating an interrupt.

Target Done (TgtDone) Response. A control packet returned by the target of a non-posted write request or a Flush request. The *TgtDone response* acts as a tag to assure delivery to the original source. The response may indicate normal completion or that an error occurred at the target or at an end-of-chain device.

Transaction. A sequence of packets accomplishing a transfer of information (e.g. read request packet followed by a read response and data packet)

Transaction Stream. See I/O Stream.

Tree. A HT chain with at least one bridge device. This results in more than one system chain (bus).

Tunnel Device. A dual-link, non-bridge HT device which implements an internal peripheral function and is capable of forwarding traffic on behalf of other devices in the chain. Tunnels are the basic building blocks in HT chains.

Type One Configuration Access. PCI compatible devices recognize two types of configuration cycles. The type one configuration access is used to indicate a cycle which has not yet reached the bus hosting the target device. Only bridges accept type 1 configuration cycles, and then only to pass them either upstream or downstream based on bus number. HT memory maps type 1 configuration cycles into the upper half of the address range reserved for configuration cycles.

Type Zero Configuration Access. The type zero configuration access is used to indicate a cycle which has reached the bus hosting the target device. Upon decoding a type 0 configuration cycle, a device compares the device number field with its own. If the device number matches, it accepts the type 0 configuration cycle and uses the *function* and *dword offset* fields to access the proper internal configuration space location. HT memory maps type 0 configuration cycles into the lower half of the address range reserved for configuration cycles.

Unit ID. A logical entity within a node. In HT, a functional device is permitted to consume multiple UnitIDs; packets from each UnitID are considered as independent I/O streams.

Virtual Channel (VChan). For ordering purposes, HT defines three required *virtual channels*: posted requests, non-posted requests, and responses. Each receiver maintains a pair of flow control buffers for each virtual channel, one for command information and one for data associated with it. If isochronous traffic is supported, devices implement another set of six flow control buffers.

Word. Two bytes (16 bits) of digital information.

Write Request. See Double Word Write and Byte Write.

WrSized Request. See Double Word Write and Byte Write.

Numerics

Count[3:0] 76, 91
Mask[3:0] 76, 78
Addr[39:2] 22
Command 69, 74, 146, 151, 154, 156, 162, 164, 168, 170, 175, 176, 180, 183, 185, 190, 192, 194, 195
8259 EOI Command 507
8259 Interrupt Compatibility 505

A

A20 Mask Support 508
Accepting Packets 264
Address Map 23, 202, 312, 446
Address Remapping 477
Address Remapping Capability Block 419, 478, 479
Address Remapping Capability Block Format 479
Address-Based Semantics 21, 22
Advanced Capability Registers 323
AGP Compatibility 470
AGP Ordering 471
APIC Interrupt Support 495
Asynchronous Clock Mode 402
Atomic Operations 46
Atomic Read-Modify-Write 46, 66, 75, 89, 91, 182, 250, 261

B

Bandwidth per Pin 17
Base Address Register 329, 424
Bit-Time 294, 381
Broadcast 263
Broadcast Message 46, 70, 75, 82, 83, 207, 261, 266, 521
Broadcast Request 220, 263, 268
Bus Numbering 318

Byte Mask 141, 159, 160, 252, 466
Byte Read 37, 78, 79, 96, 167
Byte Write 42, 78, 79, 156

C

Cache Line Size Register 329, 425
CAD Signal Group 55
CAD Signals 28, 30, 35, 55, 291, 292, 294, 365, 388
Capabilities Pointer 209, 328, 330, 334, 355, 414, 425
Capability Command Register. 266
Capability ID 209, 324, 331, 334, 355
Capability Registers 50, 225, 245, 276, 299, 300, 308, 323, 470
Capability Type 209, 332, 336
Cave 20, 33
Cave Device 243, 245, 259, 321, 333
Chain 20, 33, 82
Chain Down Error 230, 247
Chain Fail 247, 350, 351
Chain Initialization 316, 318
CLK Signal 28, 35, 55, 383
Clock Distribution Skew 392
Clock Frequency Tuning 299
Clock Initialization 388
Clock Signals 55
Clock Variance Timing Budget 393
Clocking Mode 388, 394, 401
Coherent Bit 488
Cold Reset 276
Command Register 325, 410
Compare and Swap 70, 90, 184, 188
Compatibility Bit 82, 152, 260, 264, 269, 326, 463, 488
Configuration 434
Configuration Cycle Types 309

Index

Configuration Cycles 75, 82, 267, 306, 308, 309, 310, 311, 313, 315, 316, 408, 429, 434, 444

Configuration Space 117, 200, 230, 306, 308, 310, 311, 408, 434, 470

Configuration Space Mapping 312

Consortium URL 16, 17

Control Packet 28, 35, 55, 65, 140, 238

Control Packet Format 69

Control Packet Types 36

CRC 50, 58, 231, 232, 234, 235, 242, 254, 282

CRC Test Mode 237, 337

CRC Window 223, 282, 294

Cross-Byte Skew 391

CTL Signal 35, 55, 67, 68, 104, 231, 237, 294, 349, 365

D

Data Coherency 120

Data Packet 28, 35, 66, 67, 140, 263, 506

Data Packet Types 37

Direct Memory Access 25, 121

Directed Requests 263, 266, 267, 268

DMA Primary Base Register 480

DMA Remapping 480

DMA Secondary Base and Limit Registers 481

DMA Transactions 121

DMA Transfers 25

DMA Window Control Register 480

Double-Host Chain Initialization 320

Double-Hosted Chain 137, 320, 428, 431, 432, 433, 434

Double-Hosted Chain Ordering 137

Downstream Ordering Rules 123, 136

Dword Read 37, 79, 161, 164

Dword Write 67

E

End of Chain 230, 243, 265, 432

End of Interrupt 200, 207, 502, 507

Enumeration Scratchpad Register 360

EOI Packet Format 208, 209

EOI Request Message Format 503

Error Bit 42, 49, 241, 243, 250, 251

Error Conditions 230

Error Handling Register 230, 234, 247, 350, 360

Error Interrupts 230

Error Logging 234, 239, 241, 243, 245, 249, 307

Error Reporting 60, 234, 239, 242, 246, 250, 259, 517

Error Response 230, 250, 251

Error Signaling 58

F

Fairness Algorithm 48, 258, 272

Fairness and Forward Progress 271, 272

Fatal Interrupt 231, 234, 252

Feature Capability Register 347, 348, 359

Fence Request 45, 87, 88, 179, 180

Fence Transactions 45

Fetch and Add 90, 184

FIFO Size Calculation 394
flow 101
Flow Control 36, 42, 47, 58, 65, 68, 71, 99, 101, 103
Flow Control Buffers 71, 75, 104, 105, 106, 107, 145, 241, 452
Flow Control Counters 107
Flush Request 39, 44, 66, 84, 86, 97, 173, 175, 176
Flush Transactions 44
Forwarding Packets 259, 265, 271
Frequency Capability Register 299

G

Globally Ordered 134, 136
Globally Visible 84, 124, 134, 135

H

High-Speed Signals 28, 55
Host Bridge 24, 34, 39, 82, 121, 132, 137, 200, 216, 220, 226, 258, 267, 270, 307, 316, 320, 428, 432, 436, 446, 493
Host Ordering 132
Host/Secondary Interface 354, 355, 409
HyperTransport Consortium 16
HyperTransport Protocol Characteristics 16
HyperTransport-to-HyperTransport Bridge 318, 408, 409

I

I/O Address ReMapping 481
I/O Base and Limit Registers 420, 482
I/O Stream 37, 39, 137
Information Packet 36, 65, 71, 115, 145, 259

Initialization Complete 266, 295, 317
Interleaving Control and Data Packets 67
Interrupt Capability Block Format 209
Interrupt Definition Register 209, 210, 211, 500
Interrupt Discovery and Configuration Capability Block 200, 209, 210
Interrupt Information Fields 205
Interrupt Line Register 330, 425
Interrupt Pin Register 330, 425
Interrupt Request Data Packet 206
Interrupt Request Packet 204
Interrupt Request Packet Format 205
Interrupt Requests 204
Interrupt Signaling 58
Interrupt Type Field 205
Isoc Bit 116, 360
Isochronous Channel 37, 48, 73
Isochronous Flow Control 75, 116, 348
Isochronous Transactions 48

L

Latency Timer Register 329, 424
LDTREQ# 29, 30, 57, 366, 436, 439, 440, 441
LDTSTOP# 29, 30, 57, 72, 216, 223, 225, 226, 282, 300, 304, 365, 436, 437
LDTSTOP# Signaling 438
Legacy Numeric Error Handling 516
Legacy Signal Support 492

Index

Link Configuration Register 282, 288, 295, 337, 339, 358

Link Control Register 73, 225, 227, 234, 244, 295, 317, 336, 358, 432

Link Error Register 238, 239, 241, 245, 345, 359

Link Frequency Capability Register 282, 299, 346

Link Frequency Register 300, 301, 344, 399

Link Initialization 282, 288

Link Initialization Disconnect 225

Link Management Overview 47

Link Support Signals 29

Link Width Initialization 282, 283

Link Width Tuning 295

Low-Level Link Initialization 282, 291, 433

Low-Speed Signals 56

M

Master Host Bridge 320, 431, 432

Max_Latency Register 330

Memory Base and Limit Register 419, 485

Memory Base Upper Register 352

Memory Map 23, 312, 446

Message Semantics 21, 450

Min_Grant Register 330

Minimum FIFO Size 395

N

Networking Support 17

Non-Fatal Interrupt 234, 249

Non-NXA Error 247

Non-Posted Request 66, 94, 105, 108, 125, 132, 135, 250

Non-Posted Sized Writes 42

Non-Sharing Double-Hosted Chain 432

NOP Packet 67, 71, 104, 109, 116, 145, 146, 148, 223, 259

NXA Error 251

O

Ordering Rules, HT 122

P

Packet Framing 294

Packet Rejection 265, 266

Packetized Transfers 35

PassPW Bit 45, 124, 126, 127, 128

PCI Ordering 458, 459, 460

PCI-X Ordering 467

Pin Count 31

Point-to-Point Interconnect 14

Posted Request 37, 44, 45, 58, 70, 77, 80, 81, 105, 108, 111, 125, 127, 130, 131, 136, 241

Posted Sized Writes 43

Posted Write 37, 43, 44, 45, 70, 84, 125

Prefetchable Memory Base and Limit Registers 419, 486, 487

Primary Latency Timer Register 424

Processor VID/FID 437

Producer/Consumer Model 458, 459, 472

Programmed I/O 24, 41

Programmed I/O Transactions 121

Protocol Errors 230, 237, 238, 239, 345, 350

Pseudo-Synchronous Clock Mode 399

Pseudo-Synchronous Clocking Mode 401
PWROK 29, 49, 56, 223, 280, 281, 302, 356, 365, 378

R

Read Request 66
Read Response 36, 37, 66, 70, 94, 262
Receive Buffer Overflow 50, 230, 241
Receive FIFO 110, 282, 290, 388, 401
Receiver Flow Control Counters 108
Request EOI Field 206
Request packet 65
Request Packet Types 36
Reset 27, 29, 49, 56, 74, 110, 223, 276, 278
RESET# 29, 49, 56, 223, 247, 277, 278, 302, 304, 365, 378
Response Errors 50, 230
Response packet 66
Response Packet Types 36
Revision ID Capability Block 361
Revision ID Register 343
Routing Packets 258

S

Scalability 15, 20, 30
Scalable Bandwidth 20
Scalable Clock Speed 32
Scalable Data Width 31
Secondary Bus Non-Prefetchable Window Base 480
Secondary Bus Non-Prefetchable Window Control Register 480
Secondary Bus Prefetchable Window Control Register 480
Secondary Bus Window Base register 488
Secondary Status Register 254, 415
Sequence ID 39, 75, 124, 125
Signal Groups 54
Signaling Interrupt Acknowledge 507
Signaling INTR 506
Sized Byte Read 37, 79, 507
Sized Byte Write 37, 79
Sized Dword Read 37, 79
Sized Read 70, 75
Sized Read Transactions 41
Sized Write 75
Sized Write Transactions 42
Slave Host Bridge 320, 431
Slave/Primary Interface 332, 333
SM Request Packet Format, Downstream 220
SM Request Packet Format, Upstream 218
SM Request, Sources of 216
Source Tag 39, 94
Special Cycles 493, 521
Specification 1.04 20
Status Register 230, 307, 326, 413, 415
Stop Clock Signal 441
STOP_GRANT Cycles 438
Storage Semantics 446, 447
STPCLK# 441
STPCLK# Signaling 442
Strongly Ordered Sequence 126

Index

Subtractive Decode 77, 82, 264, 412, 461
Sync Flood 230, 235, 253
Sync Pattern 74
Sync/Error Packet 74
Synchronous Clock Mode 389
Synchronous Clocking Interface 390
System Block Diagram 21
System Management 216, 436, 493
System Management Address Range 217
System Management Commands, Upstream 221
System Management Mode 493, 511

T

Target Done 36, 37, 39, 66, 94, 97
Transaction Requests 40
Transaction Responses 40
Transaction Stream 39, 125, 137
Transaction Types 41
Transmit and Receive FIFOs 110
Transmitter Flow Control Counters 109
Tunnel 20, 33, 105, 125, 244, 271, 300, 321, 333
Type 0 Configuration Cycle 267, 310, 314, 315, 434
Type 1 Configuration Cycle 309, 314, 432

U

Unit ID 77, 94, 125, 154, 179, 243, 268, 316
Upstream Ordering Rules 122, 123, 125

V

Virtual Channel 37, 38, 69, 105, 125, 447, 451
Virtual Channel Buffers 38

W

Wakeup Signaling 439, 440
Warm Reset 302
Write Request 42, 43, 66, 75

X

X86 Power Management Support 441
X86-Based Interrupt Request Message Format 502

The PC System Architecture Series

The PC System Architecture Series is a crisply written and comprehensive set of guides to the most important PC hardware standards. Each title illustrates the relationship between the software and hardware, and thoroughly explains the architecture, features, and operation of systems built using one particular type of chip or hardware specification.

MindShare, Inc. is one of the leading technical training companies in the computer industry, providing innovative courses for dozens of companies, including Intel, IBM, and Compaq.

> *"There is only one way to describe the series of PC hardware and architecture books written by Tom Shanley and Don Anderson: INVALUABLE."*
> —*PC Magazine*'s "Read Only" column

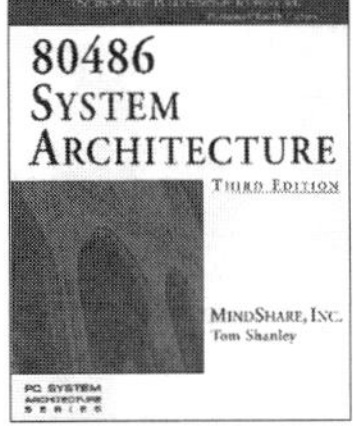

ISBN 0-201-40994-1

ISBN 0-201-70069-7

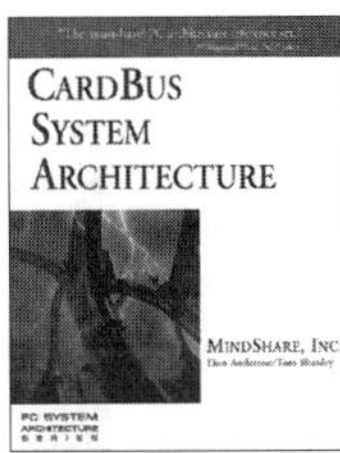

ISBN 0-201-40997-6

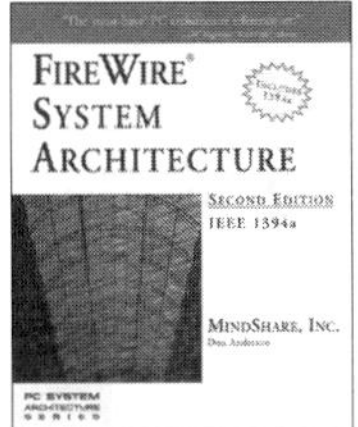

ISBN 0-201-48535-4

ISBN 0-321-11765-4

ISBN 0-201-40996-8

ISBN 0-201-30974-2

ISBN 0-201-72682-3

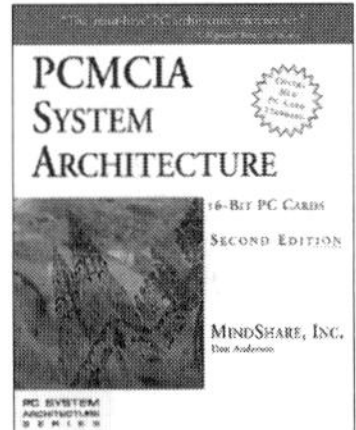

ISBN 0-201-40991-7

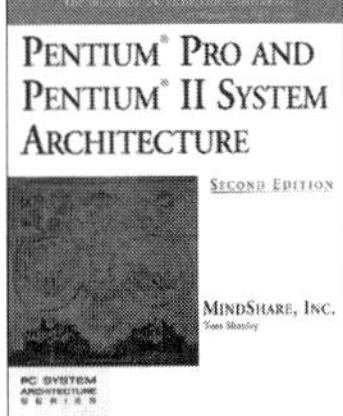

ISBN 0-201-30973-4

ISBN 0-201-40992-5

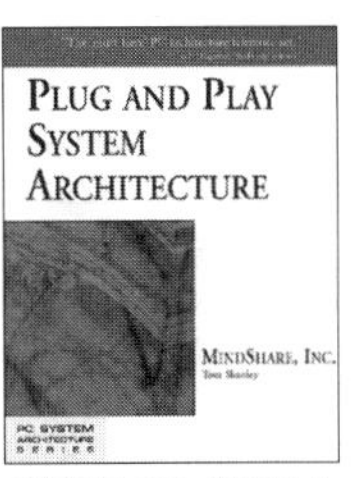

ISBN 0-201-41013-3

ISBN 0-201-55447-X

ISBN 0-201-30975-0

♦ Addison-Wesley